32nd Air Infiltration and Ventilation Centre Conference (AIVC 2011) and 1st TightVent Conference 2011

Towards Optimal Airtightness Performance

Brussels, Belgium
12 – 13 October 2011

ISBN: 978-1-5108-3760-7

Printed from e-media with permission by:

Curran Associates, Inc.
57 Morehouse Lane
Red Hook, NY 12571

Printed by Curran Associates, Inc. (2017)

International Network for Information on Ventilation and Energy Performance (INIVE)
Lozenberg 7
B-1932 Sint-Stevens-Woluwe
Belgium

Fax: +32 (0)2 529 81 10

info@inive.org

Additional copies of this publication are available from:

Curran Associates, Inc.
57 Morehouse Lane
Red Hook, NY 12571 USA
Phone: 845-758-0400
Fax: 845-758-2633
Email: curran@proceedings.com
Web: www.proceedings.com

TABLE OF CONTENTS

PARALLEL SESSION 2B WITH SHORT ORAL PRESENTATIONS AND POSTERS - AIRTIGHTNESS OF BUILDINGS AND DUCTWORK

PARALLEL SESSION 3A - VENTILATION AND COOLING

PARALLEL SESSION 3B - DEVELOPMENT OF AIR LEAKAGE DATABASES

PARALLEL SESSION 4A WITH SHORT ORAL PRESENTATIONS AND POSTERS - VENTILATION IN HIGH PERFORMANCE BUILDINGS - SPECIAL APPLICATIONS

PARALLEL SESSION 4B - QUALITY FRAMEWORKS FOR AIRTIGHTNESS ASSESSMENT

PARALLEL SESSION 5A WITH SHORT ORAL PRESENTATIONS AND POSTERS – IAQ ANALYSIS AND SIMULATION OF AIRFLOW AND POLLUTANT TRANSPORT

PARALLEL SESSION 5 B - PHILOSOPHY FOR DEFINING AIRTIGHTNESS REQUIREMENTS

TightVent Europe

BUILDING AND DUCTWORK AIRTIGHTNESS PLATFORM

AIVC

Brussels, Belgium
Hotel Crowne Plaza Brussels
12–13 October 2011

Joint Conference
32nd AIVC Conference and 1st TightVent Conference

Towards Optimal Airtightness Performance

PROCEEDINGS

In cooperation with :

International Energy Agency
Energy Conservation in Buildings and Community Systems Programme

International BUILDAIR Symposium

REHVA
Federation of European Heating, Ventilation and Air-conditioning Associations

Joint Conference

32nd AIVC Conference and 1st TightVent Conference

Towards Optimal Airtightness Performance

Programme

Wednesday 12 October 2011

08.00 Registration and welcome coffee

09.15 – 10.45 Opening session

- The new AIVC – The TightVent Europe initiative (Peter Wouters, Manager INIVE, Belgium)
- REHVA Actions for Ventilation and NZEB (Francis Allard, Former President REHVA, France)
- High-impact R&D Initiatives for Near-zero Energy Consumption in the Built Environment: The IEA Perspective (Morad Atif, IEA ECBCS Chairman, Canada)
- Presentation by Tudor Constantinescu, Principal adviser, DG Energy, European Commission

10.45 – 11.15 Coffee break

11.15 – 12.45 Combined session – Ventilation and infiltration challenges for NZEB

- Lessons learned from the concerted action (Eduardo Maldonado, Chairman Concerted Action EPBD, Portugal)
- Why we ventilate? (Max Sherman, LBNL, USA)
- Trends in national Nearly Zero-Energy Buildings approaches (Hans Erhorn / Heike Erhorn-Kluttig, Fraunhofer-IBP, Germany)

12.45 – 13.45 Lunch break

13.45 – 15.15 Parallel Session 1A - Ventilation system performance: energy and indoor air quality

- Formaldehyde and Relative Humidity in High-Performance Homes with Outdoor Air Intakes and exhaust ventilation (Jonathan Coulter, USA)
- Decentralized mechanical ventilation with heat recovery (Jean Lebrun, Belgium)
- Filters for balanced ventilation systems: design, long-term performances and energy considerations (Alain Ginestet, France)
- Analysis and implications of the revision of the Spanish regulation regarding ventilation and infiltration (José Manuel Salmerón Lissén, Spain)
- Ventilation rates and IAQ in European standards and national regulations (Nejc Brelih, Belgium)
- Window opening in high performance buildings (Piet Jacobs, Netherlands)

13.45 – 15.15 Parallel Session 1B - How tight and insulated ducts should be?

For energy and indoor climate reasons, it is important that ductwork have a good airtightness an insulation but the levels that can or should be achieved are often ignored. This session, which is part of an AIVC-TightVent project, will include several presentations to discuss these issues.

- Ductwork airtightness requirements in Portugal (Eduardo Maldonado, Portugal)
- Hands-on training courses for ventilation systems installers within the PRAXIBAT initiative (Anne-Marie Bernard, France)
- Feasibility study of ventilation system air-tightness (Jeroen Soenens, Belgium)
- Case Study: Effect of Excessive Duct Leakage in a Large Pharmaceutical Plant (David Dyer, USA)

15.15 – 15.30 Room change

15.30 – 16.30 Parallel Session 2A with short oral presentations and posters – Assessment of ventialtion system performance

- Ventilation rates and indoor air humidity depending on local climate – Simulations and measurements of 9 European countries (Rainer Pfluger, Germany)
- Whole year simulation of humidity based demand controlled hybrid ventilation in multiapartment building (Jerzy Sowa, Poland)
- Method to assess the performance of ventilation systems in dwellings considering the influence of uncertainties (Zhiming Yang, Netherlands)
- Evaluation of some DCV control strategies based on building types (Ke Xu, Norway)
- Demand-controlled Ventilation: an outline of assessment methods and simulations tools (Jean-Luc Savin, France)

15.30 – 16.30 Parallel Session 2B with short oral presentations and posters - Airtightness of buildings and ductwork

- Behavior of leakages exposed to dynamic wind loads. A numerical study using CFD on a single zone model (Dimitrios Kraniotis, Norway)
- Influence of Air Leakage on Indoor Air Quality in Low Energy Buildings: a case study (Juslin Koffi, France)
- The use of building own ventilation system in measuring airtightness (Timo Kauppinen, Finland)
- The use of a sampling method for airtightness measurement of multi-family residential buildings – an example (Jiri Novak, Czech Republic)
- Application Of Airtightness To Healthcare Buildings (William Booth, UK)
- Class C air-tightness: proven ROI in black and white (Peter Stroo, Belgium)
- The Power of Quality (Christophe Debrabander, Belgium)
- Quality Management Approach to Improve Buildings Airtightness, Requirements and Verification (Valérie Leprince, France)

16.30 – 17.00 Room Change and coffee break

17.00 – 18.30 Parallel Session 3A - Ventilation and cooling

This session is organized within the scope of a starting AIVC-TightVent project that deals with various aspects of ventilation for cooling.

- Air/Ground heat exchangers for heating and cooling: dimensioning guidelines (Pierre Hollmuller, Switzerland)
- Future climate effect on building refurbishment using ventilation for cooling: a case study (Maria Kolokotroni, UK)
- Ventilation solutions in net zero energy buildings, the Elithis Tower case study (Oscar Hernandez, France)
- Low-energy buildings with night and air-to-air heat exchangers – case studies and analysis (Jens Pfafferott, Germany)

17.00 – 18.30 Parallel Session 3B - Development of air leakage databases

There are several national initiatives to collect air leakage data from field measurements. The objective of this session is to begin structuring communication between some of these initiatives. It falls within the scope of a AIVC-TightVent project.

- Preliminary analysis of U.S. Residential Air Leakage Database Update v.2011 (Wanyu R. Chan, USA)
- U.S. Commercial Building Airtightness Requirements and Measurements (Andrew Persily, USA)
- The Web@set project: reasons behind, objectives and on-going developments (Andrés Litvak, France)
- Experience with the development of an air leakage database in Germany (Oliver Solcher, Germany)

18.30 – 19.30 Poster and industry exhibition

19.30 End of the first day

Thursday 13 October 2011

09.00 – 10.30 Parallel Session 4A with short oral presentations and posters - Ventilation in high performance buildings - Special applications

- Measured public benefits from energy-efficient homes (Jonathan Coulter, USA)
- Measurement of pollutant emissions in two similar very low energy houses with cast concrete and timber frame (Franck Alessi, France)
- Low pressure drop air transfer between rooms in buildings with balanced ventilation - A commonly ignored issue (Peter Schild, Norway)
- Impact of the filtration system on the indoor-outdoor particles concentration relationships in an air conditioned office building (Alain Ginestet, France)
- Measured performance of three types of energy program houses in two US cities (Jonathan Coulter, USA)
- Experimental evaluation of Supply-Only ventilation effectiveness (Mireille Rahmeh, France)
- The Applicability of Glazing System with Dynamic Insulation for Residential Buildings (Shinsuke Kato, Japan)
- Integrated Approach of CFD and SIR Epidemiological Model for Infectious Transmission Analysis in Hospital (Hiroaki Asanuma, Japan)
- Shelter-in-place effectiveness in the event of toxic gas releases: French and Catalan assessment approach (Montoya Maria Isabel, Spain)

09.00 – 10.30 Parallel Session 4B - Quality frameworks for airtightness assessment

Rewarding or imposing good airtightness in a regulation directly calls into question the reliability and accuracy of the measurements that are performed in practice. Several schemes will be presented to increase or assess the quality of the measurements and execution. This session is part of an AIVC-TightVent project.

- Quality system for airtightness measurement of buildings (Oliver Solcher, Germany)
- The quality framework for Air-tightness measurers in France: assessment after 3 years of operation (Valérie Leprince, France)
- Interlaboratory tests for the for the determination of repeatability and reproducibility of airtightness measurements (Christophe Delmotte, Belgium)
- Pressure distribution inside large buildings during airtightness tests (Stefanie Rolfsmeier, Germany)

10.30 – 11.00 Room Change and coffee break

11.00 – 12.00 Parallel Session 5A with short oral presentations and posters – IAQ analysis and simulation of airflow and pollutant transport

- Performances of decentralized air handling terminals connected to building airtightness and indoor hygro-thermal climate (Gabrielle Masy, Belgium)
- Basis study about prediction of air flow environment in cross-ventilated room by neural network (Tomoyuki Endo, Japan)
- Sensitivity study for architectural design strategies of office buildings in Central Chile: effectiveness of nocturnal ventilation (Felipe Encinas Pino, Belgium)
- Nano-scale Aerosol Deposition Model for CFD in Indoor Environmental Analysis (Jun Narikawa, Japan)
- Exposure Concentration Prediction by Multi-Nesting Approach Connecting Building Space-Virtual Manikin- Nasal Airway Model (Kazuhide Ito, Japan)

11.00 – 12.00 Parallel Session 5 B - Philosophy for defining airtightness requirements

Should there be specific airtightness requirements? If so, what level is to be required? Should there be a minimum level of air leakage? The objective of this session is to review critical aspects that have to be considered to tackle such questions. It falls within the scope of a AIVC-TightVent project.

- Optimal air tightness levels of buildings (Willem de Gids, Netherlands)

12.00 – 12.15 Room Change

12.15 – 13.15 Parallel Session 6A - DCV and sensor technology - CLEAR UP project

Demand Controlled Ventilation (DCV) is usually seen as an effective way for reducing the energy consumption. The aim of this workshop is to present new sensors for DCV developed in the framework of the EU project Clear-up and to discuss with ventilation experts their possible applications in buildings.

- Introduction to DCV - Willem de Gids, TNO, the Netherlands
    - What is DCV and why its application
    - The need for demand control in dwellings
    - Indoor air pollutants in general and the most dominant ones for dwellings
    - A strategy for controlling most domestic pollutants
- Introduction to DCV – Anne-Marie Bernard, Allie Air, France
    - Potential of energy saving
    - Different types of sensors
    - Success of hygro-regulated DCV in France
    - Impact of cost reduction
- Energy-efficient Demand-Controlled Ventilation using Micromachined Metal Oxide Semiconductor Gas Sensor Technology - Simone Herberger, Applied Sensor, Germany
- Panel discussion on the potential applications of these new sensors

12.15 – 13.15 Parallel Session 6B – Uncertainties in airtightness measurement - field data

- Modernizing ISO, EN and ASTM air leakage standards ... more accuracy in less time (Colin Genge, Canada)
- Evaluation of selection criteria of an air tightness measurement method for multi-family buildings (Bassam Moujalled, France)
- Improvement of air tightness of communities (Markku Hienonen, Finland)

13.15 – 14.00 Lunch break

14.00 – 15.30 Parallel Session 7A - Evaluation of ventilation strategies

- Numerical validation for natural ventilation design (François Demouge, France)
- Performance of low pressure mechanical ventilation concept with diffuse ceiling inlet for renovation of school classrooms (Søren Terkildsen, Denmark)
- Definition of occupant behaviour patterns with respect to ventilation by means of multivariate statistical techniques (Felipe Encinas Pino, Belgium)
- Control and performance of innovative Ventilation systems in Low Energy Buildings: a case study (Juslin Koffi, France)
- Shelter-in-place strategy: CONFINE, an airtightness level calculation tool to protect people against accidental toxic releases (Gaëlle Guyot, France)
- Liabilities of Vented Crawl Spaces And Their Impacts on Indoor Air Quality in Southeastern U.S. Homes (Jonathan Coulter, USA)

14.00 – 15.30 Parallel Session 7B - IAQ and energy impacts of envelope leakage

- Laboratory investigation of timber frame walls with an exterior air barrier in a temperate climate (Jelle Langmans, Belgium)
- State of Art of Non-Residential Buildings Airtightness and Impact on the Energy Consumption (Valérie Leprince, France)
- Investigations on the effects on airtight performance improvement and energy consumption of insulation retrofit in detached houses (Hiroshi Yoshino, Japan)
- The influence of air transport on the hygrothermal performance of an inclined roof (Paul Steskens, Belgium)
- Impacts of Airtightening Retrofits on Ventilation and Energy in a Manufactured Home (Andrew Persily, USA)

15.30 – 15.50 Room Change

15.50 – 17.00 Closing session

- Summing up of airtightness track (Willem de Gids, Expert, VentGuide, Netherlands)
- Summing up of ventilation track (Martin Liddament, Expert, Veetech Ltd, UK)
- Perspectives for AIVC and TightVent projects (Rémi Carrié, Senior Consultant, INIVE, France)
- Can we meet the ventilation required in international standards in an energy efficient way? (Bjarne Olesen, Professor, Technical University of Denmark

17.00 End of the conference

Wednesday 12 October 2011

08.00 Registration and welcome coffee

09.15 – 10.45 Opening session

- The new AIVC – The TightVent Europe initiative (Peter Wouters, Manager INIVE, Belgium)
- REHVA Actions for Ventilation and NZEB (Francis Allard, Former President REHVA, France)
- High-impact R&D Initiatives for Near-zero Energy Consumption in the Built Environment: The IEA Perspective (Morad Atif, IEA ECBCS Chairman, Canada)
- Presentation by Tudor Constantinescu, Principal adviser, DG Energy, European Commission

10.45 – 11.15 Coffee break

Wednesday 12 October 2011

11.15 – 12.45 Combined session – Ventilation and infiltration challenges for NZEB

- Lessons learned from the concerted action (Eduardo Maldonado, Chairman Concerted Action EPBD, Portugal)
- Why we ventilate? (Max Sherman, LBNL, USA)
- Trends in national Nearly Zero-Energy Buildings approaches (Hans Erhorn / Heike Erhorn-Kluttig, Fraunhofer-IBP, Germany)

12.45 – 13.45 Lunch break

LESSONS LEARNED FROM THE EPBD CONCERTED ACTION

Eduardo Maldonado

Universidade do Porto
Faculdade de Engenharia, R. Dr. Roberto Frias
4200-465, Porto, Portugal
ebm@fe.up.pt

ABSTRACT

The EPBD (EU Energy Performance of Buildings Directive) Concerted Action is a cooperative effort from all 27 EU Member States (MS), plus Norway and Croatia, together with the European Commission. It is running since 2004, under the Intelligent Energy programme since 2007. It aims at providing support to MS in their effort to transpose and then implement the EPBD, identifying the best solutions and practices that MS can then adopt and thus move towards harmonization throughout the EU. This report briefly describes the lessons learned, with a special highlight and focus on the issues related to ventilation and indoor air quality, as they relate to the preparation of a 2020 scenario of "nearly-zero energy buildings".

KEYWORDS

Energy efficiency; indoor climate; ventilation; building regulations.

INTRODUCTION

The transposition and implementation of the EU Directive 2002/91/EC, on the Energy Performance of Buildings (EPBD) [1], posed important challenges and difficulties to the EU Members States. It required new building regulations using a more complex common methodology for the energy performance of buildings, as well as the setting up of Certification and Inspections schemes by Qualified Experts. Therefore, the national experts in charge of preparing the technical portions of the EPBD-derived regulations and organizational schemes for compliance checking, certification and inspections decided to informally gather and help one another in finding solutions well-suited for their own countries. The Concerted Action was created with precisely this objective and, in the process, it established close cooperation with CEN as it developed a new set of supporting standards, most notably EN ISO 13790 that is the formal or informal basis for most of the current EU building thermal regulations.

Is is relevant to underline that the EPBD clearly states, as its basic objective, the improvement of energy efficiency without any compromises to comfort and indoor air quality, as well as other essential functional and safety requirements. However, still today, very few countries in Europe (and around the World) have integrated indoor air quality requirements in their building energy efficiency regulations [2].

LESSONS LEARNED 2010

The Concerted Action published a set of conclusions and recommendations in 2010 (http://www.epbd-ca.org/Medias/Pdf/CA_Summary%20report_Feb2010.pdf). They had been its contribution in 2009 to the EC and MS in the then ongoing effort to prepare the recast

EPBD. Most of the recommendations, especially those dealing with procedural aspects, were adopted and are now part of the legal requirements included in the recast EPBD [3]. The *lessons learned* are too many to be included here (the reader is advised to refer to the full report), but a few lessons that interact with ventilation rates as they relate to the quantification of the building behaviour can be highlighted next:

- *Intercomparisons of national requirement levels are almost impossible to achieve. Technical and climate conditions have to be included along traditional building styles and user behaviour. Comparisons will become even more complicated if cost aspects are added.*
- *Weather normalisation is a crucial task for both measured and calculated data. Heating and cooling needs cannot be simply corrected by degree days.*
- *The calculated energy rating is influenced by complexity in zoning and accreditation of software. Simplifications in the calculation methods have to be found whereas other methods need to be made more detailed in order to give reliable results with acceptable accuracy in specific situations.*
- *National procedures do not simply refer to CEN standards but, instead, the content is incorporated. Thus CEN standards should be further developed into general framework documents that fix general principles only and allow a maximum of flexibility for the countries' implementations.*

Checking compliance with building codes was identified as a special difficulty, both in terms of logistics and in technical aspects. If calculation methodologies are not clearly defined, it becomes very difficult to check compliance. Accounting for ventilation is a particularly hard task, especially when buildings are naturally ventilated, and this becomes especially important during summer and in the milder Southern European countries, where natural ventilation is an integral part of building tradition and where ventilation is used for controlling both indoor air quality and comfort (temperature and humidity).

CONCLUSION

Much progress has been achieved in the last decade. Building regulations have been involving more and more parameters, but some of the most difficult ones, e.g., natural ventilation, or daylighting, are still either missing or being included in very simplified ways in most countries. EU MS are now being called to transpose and implement the new recast EPBD which, among other items, requires that all new buildings designed and built in the EU must be "nearly-zero energy" buildings by 2018 (Public buildings) or 2020 (other buildings). Ventilation and air-tightness become a critical aspect in the design of such buildings, requiring innovative solutions and highly-skilled work-forces to build them and their technical systems. The Concerted Action continues its work until 2015 to help participating countries address these new challenges.

REFERENCES

[1] Directive 2002/91/EC, on the Energy performance of Buildings (EPBD), EU Official Journal, 16 December 2002.

[2] Implementing the Energy Performance of Buildings Directive (EPBD), including Country reports 2010. EU Report EA-30-11-026-EN-C, ISBN 978-92-9202-090-3, Brussels, April 2011.

[3] Directive 2010/31/EU, on the Energy performance of Buildings (recast), EU Official Journal, 18 June 2010.

WHY WE VENTILATE

Max Sherman, Jennifer M. Logue, Phillip N. Price, and Brett C. Singer

Lawrence Berkeley National Laboratory
1 Cyclotron Road
Berkeley, California, USA

ABSTRACT

It is widely accepted that ventilation is critical for providing good indoor air quality (IAQ) in homes. However, the definition of "good" IAQ, and the most effective, energy efficient methods for delivering it are still matters of research and debate. This paper presents the results of work done at the Lawrence Berkeley National Lab to identify the air pollutants that drive the need for ventilation as part of a larger effort to develop a health-based ventilation standard. First, we present results of a hazard analysis that identified the pollutants that most commonly reach concentrations in homes that exceed health-based standards or guidelines for chronic or acute exposures. Second, we present results of an impact assessment that identified the air pollutants that cause the most harm to the U.S. population from chronic inhalation in residences. Lastly, we describe the implications of our findings for developing effective ventilation standards.

KEYWORDS

Indoor air quality; hazard analysis; residential; DALYs; ventilation

INTRODUCTION

The primary purposes of ventilation in buildings are to provide a sufficient oxygen supply for the occupants and to remove any hazardous substances or noxious odors in the indoor air. For thousands of years societies have realized the need to set or adjust ventilation for specific indoor tasks. The initial inception of residential ventilation is unknown, but likely was from neolithic times and used to remove combustion gases from indoor heating and cooking such as introducing vents for fires. According to Kuhnl-Kinel [1], ancient Egyptians noticed that stone cutters working outdoors had fewer respiratory problems, people in the Middle Ages realized that air in building could transmit disease, and in 1600 the king of England required buildings to be a certain height with tall, slim windows to facilitate the removal of smoke from heating and cooking.

Traditionally in residences the dominant form of ventilation has been natural ventilation including infiltration. In older, leakier homes infiltration from weather driven flows through cracks in the building's exterior may provide sufficient ventilation for residents. In the 1960s and 1970s home construction shifted from natural materials to new synthetic materials and new construction products; and there was increasing interest in tightening homes to conserve energy due to the energy crisis of the 1970s. The increased tightness in homes reduced ventilation that, along with synthetic materials, led to dramatic increases in residential mold related problems and potential issues with combustion spillage. There was also increasing concern about the impact of material emissions on the health of occupants as new materials were introduced.

People spend the majority of their time in residences [2], making indoor air quality an increasing concern. It has been widely recognized that the health burden of indoor air is significant [3-4]. Current ventilation standards are ostensibly set to protect the health of residents. The American Society of Heating, Refrigerating and Air Conditioning Engineer's (ASHRAE's) Standard 62.2 is the most widely accepted residential ventilation standard in the United States. ASHRAE developed Standard 62.2 "Ventilation and Acceptable Indoor Air Quality in Low-Rise Residential Buildings" to address indoor air quality (IAQ) issues (ASHRAE 2010). ASHRAE 62.2 is now required in some building codes, such as California's Title 24, and is treated as a standard of practice in many energy efficiency programs and by organizations that train and certify home performance contractors. The standard specifies an overall, residence-level outdoor air ventilation rate as a function of floor area (a surrogate for material emissions) and the number of bedrooms (a surrogate for occupant-related emissions) and requires bathroom and cooking exhaust fans. The focus of the standard generally is considered to be the overall ventilation rate. This emphasis has been based on the idea that risks indoors are driven by continuously emitted, distributed sources such as formaldehyde from furnishings and bio-effluents (including odors) from humans. The required level of whole residence mechanical ventilation was based on the best judgment of experts in the field, but was not based on any analysis of chemical pollutant concentrations or other health-specific concerns.

While whole residence ventilation has been recognized as an effective method for reducing many indoor risks, there are significant costs associated with high ventilation rates due to moving and conditioning the air. Certain human needs likely set the minimum for ventilation, based on the requirements for providing sufficient oxygen and removing CO_2. However, energy demands and associated greenhouse gas emissions can be reduced by using source control and efficient task ventilation to remove other contaminants of concern. To effectively design residential ventilation systems to maximize health while minimizing ventilation costs, we first need to specify our objectives for ventilation.

This paper presents a summary of the ongoing work at the Lawrence Berkeley National Laboratory to develop a health-based ventilation standard. This work focuses on non-biological indoor air pollutants. Ventilation affects moisture in the indoor environment, and moisture affects mold development. However, ventilation is not an effective method of controlling whole residence moisture loads (although it is effective in bathrooms) because many locations have higher outdoor than indoor humidity. First we discuss a hazard assessment of indoor pollutants that identified the air pollutants in residences that exceed health-based standards and guidelines. Second, we present the results of a study that determined the relative importance of different pollutants to health. Lastly, we discuss the impact of these results on ventilation standards.

HAZARD ASSEMENT OF INDOOR POLLUTANTS

The initial step in this broad effort was to conduct a hazard assessment of non-biological air pollutants – e.g. including chemical gases and particles – in residences [5]. The analysis compiled data from published studies reporting measurements of air pollutants in residences. That literature review identified 86 articles that were relevant to acute and chronic exposure in residences and considered a broad collection of contaminants measured indoors regardless of pollutant source. The contaminants included some emitted purely from indoor sources, some that enter predominantly from outdoors, and some having both indoor and outdoor sources.

Summary results were compiled and used to calculate representative mid-range and upper-bound concentrations relevant to chronic exposures for over 300 pollutants and peak concentrations relevant to acute exposures for a few pollutants. For over 100 pollutants, measured concentrations were compared to available chronic and acute health-hazard standards and guidelines from the U.S. Environmental Protection Agency (USEPA), California Office of Environmental Health Hazard Assessment (OEHHA), the U.S. Occupational Safety and Health Administration (OSHA), the Agency for Toxic Substances and Disease Registry (ATSDR), and the World Health Organization. Fifteen diverse pollutants were identified as potential chronic or acute health hazards for many homes. A subset of pollutants were identified as priority chemical pollutants based on the prevalence of the pollutant in homes and the quality of available measurements in homes. Table 1 lists the identified priority hazards.

Priority Pollutants for Chronic Exposure	Potential Acute Exposure Concerns
Acetaldehyde	Acrolein
Acrolein	Chloroform
Benzene	Carbon Monoxide
Butadiene, 1,3-	Formaldehyde
Dichlorobenzene,1,4-	NO_2
Formaldehyde	
Naphthalene	
NO_2	
$PM_{2.5}$	

Table 1. Pollutants that potentially pose an adverse indoor health risks.

The hazard assessment narrowed the list of hundreds of chemicals to a much smaller group of pollutants of concern. But this approach considered only disease incidence for cancer standards and disease potential for non-cancer standards; it did not consider disease severity. Prioritizing mitigation efforts among residential indoor air pollutants, and comparing their cumulative health damage to other environmental hazards requires a consistent and comparative metric that accounts for both disease incidence and the severity or costs of the health endpoints. This need motivated development of an impact assessment methodology for indoor air pollutant inhalation.

HEALTH DAMAGE OF CHRONIC INDOOR AIR EXPOSURE

We synthesized disease incidence and health damage models to develop a methodology for quantifying indoor air quality and then applied the methodology to calculate the population average health damage due to chronic inhalation of non-biological air pollutants in U.S. residences [6]. We first analyzed published data to calculate mean exposure concentrations and then estimated age-dependent inhalation air intake over the course of a year. We used disease incidence and disease damage models to predict the pollutant-specific and total health damage in Disability Adjusted Life Years and to identify the pollutants that dominate impacts on human health.

Determining Annual Population Health Damage

To determine the annual population health damage we compared estimates of current air pollutant intake in U.S. homes (using measurement based estimates of population-averaged, residential chronic exposure concentrations) to the theoretical case of a home with no indoor pollutant sources and no pollutants infiltrating from outdoors, i.e. with homes having no pollutants in the indoor air. Population intake via other micro-environments was held constant as a baseline for which inhalation in residences adds an increment of harm.

The Disability Adjusted Life Year (DALY) metric is a powerful tool for quantifying and inter-comparing the damages from health endpoints that can result from specific pollutant intake [7]. DALYs quantify overall disease damage including both mortality and morbidity. DALYs are the equivalent years of life lost to illness or disease and include years lost to premature death (YLL) and equivalent life years lost to reduced health or disability (YLD).

$$DALY = YLL + YLD \tag{1}$$

The years of reduced health are weighted from 0 to 1, based on the severity of disease, to calculate equivalent years lost. For example, a 5 year illness that reduces quality of life to 4/5 that of a healthy year is valued at 1 DALY lost.

Several authors have determined the DALYs lost per incidence of specific diseases using the preeminent work of Murray and Lopez [7-11]. Multiplying a disease incidence rate by a "damage factor" yields a rate of lost DALYs per disease incidence.

$$DALYs = \frac{\partial Damage}{\partial Disease\ incidence} * Disease\ Incidence \tag{2}$$

Damage rates multilplied by available disease incidence statistics , integrated over all diseases of interest, are often used to determine the total burden of disease in a community. This method was used by the World Health Organization to determine the disease damage for 192 countries [11].

Our analysis used the compilation of measured concentration data to calculate total DALYs lost due to inhalation of air pollutants in residences. We approached this using three different methods. The first method was for criteria pollutants, which are more extensively studied and have a larger body of available epidemiological studies. We aggregated the available Concentration-Response (C-R) functions in the literature to determine disease incidence as a function of a change in airborne concentrations. For each health outcome for each criteria pollutant we multiplied the change in disease occurrence rate by the damage factor for that disease. This level of epidemiological data was not available for the majority of remaining pollutants. The second method that we used was primarily for air toxics or hazardous air pollutants which have limited epidemiological data, but extensive data from toxicological studies. This method used the work of Huijbregts et al. [7] to calculate the health damage associated with the intake of non-criteria pollutants. Huijbregts et al. [7] determined cancer and non-cancer mass intake-based damage factors by synthesizing disease damage factors and animal toxicology based disease incidence rates. This method is much more uncertain than using C-R functions which is reflected by significantly larger uncertainties. The third method was used for pollutants that had already had been significantly studied and had available literature studies apportioning specific disease rates to exposure. This applied to radon and secondhand tobacco smoke (SHS). The population average DALYs lost due to radon, acute carbon monoxide (CO) and SHS were determined based on estimates of disease incidence by multiplying them by the damage factors for those diseases.

Figure 1 shows the damage in DALYs per year per 100,000 people from exposure to the 15 pollutants with the highest central estimate of damage. The whiskers indicate the aggregate uncertainty (95th percentile confidence interval) in the disease incidence and disease damage factors. Figure 1 shows the clear result of our analysis: on a population average, the most harmful pollutants in residential indoor air are $PM_{2.5}$, SHS, formaldehyde, acrolein, radon and ozone. The hazards of SHS and radon are more widely recognized and focused in a smaller fraction of homes. By contrast, $PM_{2.5}$, acrolein, and formaldehyde are present at substantial levels in most homes yet there may be less widespread recognition of these hazards. Formaldehyde is primarily emitted from materials throughout the home. Acrolein is primarily emitted from materials and cooking [12]. $PM_{2.5}$ concentrations indoors, unlike acrolein and formaldehyde, are due to both indoor and outdoor sources and outdoor concentrations may exceed indoors in many locations [4].

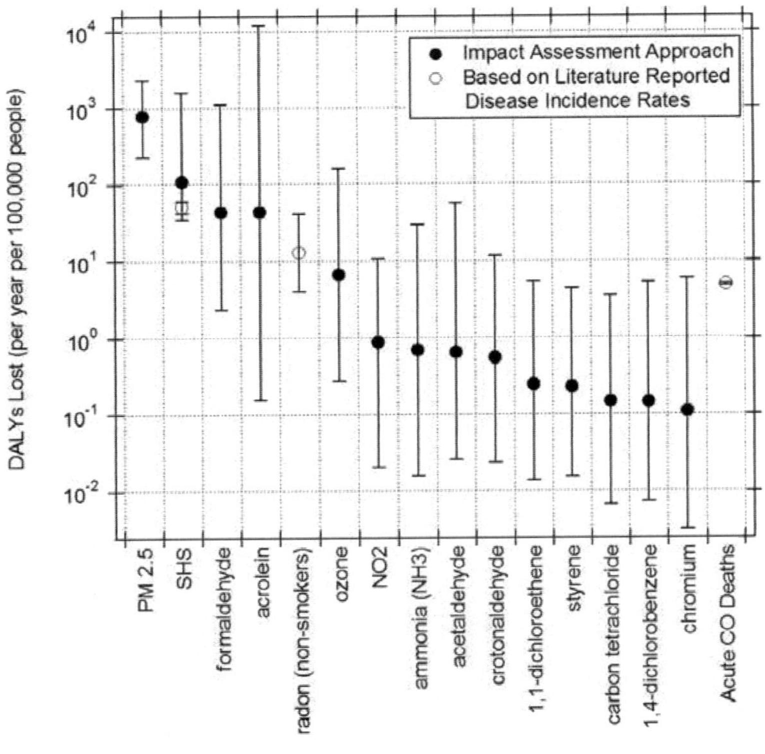

Figure 1. Estimated population averaged annual cost, in DALYs, of chronic air pollutant inhalation in U.S. residences; results for the 15 pollutants with highest mean damage estimates.

To explore possible variations in the health impact rankings of pollutants across homes, we used a Monte Carlo approach to calculate the total chronic health damage from exposure to all pollutants included in our analysis, except radon and SHS. For each model run, we sampled with replacement from the distribution of estimated damage for each pollutant and calculated an estimate of total health damage for the home. We assumed independent variability of all pollutants. This was repeated for a sufficient number of homes to yield a stable mean and standard deviation for the total health damage. We assumed that individual pollutant damages vary independently. This approach did not explicity account for any synergistic or antagonistic interactions of pollutant health effects. The resulting distribution of total health damage and the characteristics of each set of individual pollutant contributions to the total health damages were analyzed. For 80% of the sample sets (calculated damages for individual homes), $PM_{2.5}$ was the largest contributor. For 16% of the sample sets acrolein was the dominant contributor and for 4% of the sample sets it was formaldehyde. The dominant contributor was a compound other than these three in less than 0.25% of the sample sets. For 90% of the sample sets, acrolein, formaldehyde, and $PM_{2.5}$ contributed more than 80% of the total health damage. This reinforces the finding that these three pollutants account for the majority of chronic health from intake of air pollutants in non-smoking homes. We estimate that the current indoor air quality related heath damage to the U.S. population from all sources, excluding SHS and radon, is in the range of 4-11 mili-DALY/p/yr (mili-DALYs per person per year). This indicates that the damage attributable to indoor air is, comparatively, somewhere between the health effects of road traffic accidents (4 mili-DALY/p/yr) and all-cause heart disease (11 mili-DALY/p/yr) in the U.S. The compounds that dominate that total are $PM_{2.5}$, acrolein, and formaldehyde.

IMPLICATIONS FOR VENTILATION STANDARDS

Ventilation standards have the potential to significantly improve indoor air quality (IAQ) in the vast majority of homes. Identifying the pollutants that drive the risks will allow us to make suggestions for modifying the current ventilation standards and identify areas where further research is needed. This section describes how two particular elements of ventilation standards can improve IAQ: overall air exchange rate and localized exhaust ventilation.

Current ventilation standards focus primarily on providing the right amount of overall ventilation for a home based on the idea that the main drivers for pollutant concentrations are furnishings and occupants themselves. A reasonable lower bound for the overall ventilation rate would likely be the airflow needed to control for body odor [13]. Additional air flow is needed to control concentrations of pollutants that have diffuse emission sources in residences. Our analysis indicated that material emissions of acrolein and formaldehyde are the main pollutants that need to be controlled with an overall ventilation rate and the rate should be set at levels that would provide safe indoor concentrations of these pollutants.

There is insufficient material emission data currently to set a ventilation rate based on acrolein, however an appropriate ventilation rate for formaldehyde has been suggested based on California health standards of 0.3 air changes per hour for existing homes and 0.5 for new homes [14]. There are two main concerns with providing ventilation at these levels: 1) the cost of conditioning the extra airflow and 2) bringing in outdoor pollutants.

One way of reducing the needed overall ventilation for a home, and the associated energy and cost penalty, would be source control. Currently in the U.S. there is not sufficient information to estimate the benefits of source reduction by simulating the replacement of specific materials or applying specific existing standards or guidelines for material emissions [15]. Developing these databases could aid in the reduction of material loading of formaldehyde and acrolein. Implementing standards that reduced material loading in homes would reduce the required ventilation rate and save energy.

Increasing air flow through the home can increase the rate at which outdoor pollutants are brought indoors. Our study identified $PM_{2.5}$ as the most important pollutant for health in residential environments. While indoor sources such as combustion and chemistry significantly impact indoor $PM_{2.5}$ concentrations, a significant fraction of homes may have higher concentrations outdoors than indoors indicating that more ventilation may actually increase health risks [4]. Providing ventilation air via filtered supply or filtered balanced ventilation using heat/enthalpy recovery ventilators is one potential solution. Another option is to filter the indoor air independent of the ventilation system to reduce indoor $PM_{2.5}$ concentrations. Including measures to reduce indoor particle concentrations in ventilation standards could greatly improve IAQ from a health perspective.

Our analysis indicates that removing pollutants near their point of release using effective localized exhaust ventilation is key to maintaining good IAQ. The two main types of localized exhaust in ventilation standards are kitchen and bath ventilation. Effective kitchen ventilation is needed to mitigate acute pollutant events resulting from combustion based cooking appliances and food preparation activities. Task ventilation can also significantly mitigate chronic exposures by removing pollutants at their source. ASHRAE 62.2 requires a kitchen exhaust fan that is above the cooktop and provides at least 100 cubic feet per minute (roughly 50 $m^3 h^{-1}$) of airflow while producing 3 sones or less of noise. The standard doesn't specify a minimum pollutant capture efficiency or sound limits at higher flow rates. Requiring a high pollutant capture efficiency and potentially requiring automatic fan use when the range is operated could significantly improve indoor air quality. Four out of five of the identified acute contaminats of concern (except chloroform) are emitted by combustion or cooking. It is critically important to make sure that there is effective ventilation for all indoor combustion. Research is needed to determine if the health benefit of adding a commissioning requirement to ventilation standards is worth the cost.

Effective bath fans are also critical for providing good indoor IAQ. Bath fans remove bio-effluence, moisture and pollutants generated in bathroom activities such as personal care product use and showering. Showering has been shown to elevate concentration of chloroform above acute thresholds[16]. Bathroom exhaust flow rate requirements should be designed to keep chloroform levels below acute thresholds. Further research is needed to determine which episodic activities in bathrooms may lead to acute exposures.

CONCLUSION

The main air pollutants of concern for regulators setting residential ventilation standards are formaldehyde, acrolein, and $PM_{2.5}$. This implies that whole-residence ventilation rates should be based on controlling formaldehyde and acrolein. Filtration of incoming or house air to remove $PM_{2.5}$ would substantially improve indoor air quality.

Effective task ventilation is critical for controlling acute exposures in residences. All combustion in homes should be effectively vented and cooking exhaust systems should be required to meet minimum pollutant capture efficiency standards.

The identification of formaldehyde, acrolein and $PM_{2.5}$ as the highest priority pollutants for chronic exposure opens opportunities to improve energy efficiency through consideration of control measures complementary to ventilation.

ACKNOWLEDGEMENTS

Funding was provided by the U.S. Dept. of Energy Building Technologies Program, Office of Energy Efficiency and Renewable Energy under DOE Contract No. DE-AC02-05CH11231; by the U.S. Dept. of Housing and Urban Development Office of Healthy Homes and Lead Hazard Control through Interagency Agreement I-PHI-01070, by the U.S. Environmental Protection Agency Office of Air and Radiation through Interagency Agreement DW-89-92322201-0 and by the California Energy Commission through Contract 500-08-061.

REFERENCES

[1] Kuhnl-Kinel, J. *The History of Ventilation and Air Conditioning: Is CERN up to date with the latest technological developments?* in *Proceedings of the Third ST Workshop-CERN*. 200. Chamonix, France: European Organization for Nuclear Research (CERN).

[2] Klepeis, N.E., et al., *The National Human Activity Pattern Survey (NHAPS): a resource for assessing exposure to environmental pollutants.* Journal of Exposure Analysis and Environmental Epidemiology, 2001. **11**(3): p. 231-252.

[3] Edwards, R.D., et al., *VOC concentrations measured in personal samples and residential indoor, outdoor and workplace microenvironments in EXPOLIS-Helsinki Finland.* Atmospheric Environment, 2001. **35**: p. 4531-4543.

[4] Weisel, C.P., et al., *Relationships of Indoor, Outdoor, and Personal Air (RIOPA) Part I. Collection Methods.* 2005, Health Effects Institute, Mickely Leland National Urban Air Toxics Research Center and Descriptive Analysis.

[5] Logue, J.M., et al., *Hazard assessment of chemical air contaminants measured in residences.* Indoor Air, 2011. **21**(2): p. 92-109.

[6] Logue, J.M., et al., *Chronic Health Damage of Residential Indoor Air.* Submitted to Environmental Health Perspectives, 2011.

[7] Huijbregts, M.A.J., et al., *Human-toxicological effect and damage factors of carcinogenic and noncarcinogenic chemicals for life cycle impact assessment.* Integrated Environmental Assessment and Management, 2005. **1**(3): p. 181-244.

[8] Lvovsky, K., et al., *Environmental costs of fossil fuels: a rapid assessment method with application to six cities.* 2000, The World Bank Environment Department: Washington, D.C.

[9] Murray, C.L. and A.D. Lopez, *The global buden of disease: a comprehensive assessment of rmortality and disability from diseasees, injuries, and risk factors in 1990 and projected to 2020.* Global Burden of Disease and Injury Series. Vol. 1. 1996, Cambridge, MA: Harvard University. 990.

[10] Murray, C.L. and A.D. Lopez, *Global health statistics: A compendium of incidence, prevalence, and mortality estimates for over 200 conditions*, in *Global burden of disease and injury series.* 1996, Harvard University: Cambridge, MA. p. 906.

[11] WHO, *Global Health Risks: Mortality and burden of disease attributable to selected major risks.* 2009, World Health Organization: Geneva, Switzerland.

[12] Seaman, V.Y., D.H. Bennett, and T.M. Cahill, *Origin, occurrence, and source emission rate of acrolein in residential indoor air.* Environmental Science & Technology, 2007. **41**: p. 6940-6946.

[13] Cain, W.S., et al., *Ventilation requirements in buildings: I. Control of occupancy odor and tobacco smoke odor.* Atmospheric Environment, 1983. **17**(6): p. 1183-1197.

[14] Sherman, M.H. and A.T. Hodgson, *Formaldehyde as a basis for residential ventilation rates.* Indoor Air, 2004. **14**(1): p. 2-8.

[15] Willem, H. and B.C. Singer, *Chemical Emissions of Residential Materials and Products: Review of Available Information.* 2010, Lawrence Berkeley National Laboratory: Berkeley, CA. p. 48.

[16] Kerger, B.D., C.E. Schmidt, and D.J. Paustenbach, *Assessment of airborne exposure to trihalomethanes from tap water in residential showers and baths.* Risk Analysis, 2000. **20**(5): p. 637-651.

TRENDS IN NATIONAL NEARLY ZERO-ENERGY BUILDING APPROACHES

Heike Erhorn-Kluttig*[1], Hans Erhorn[1], Kirsten Engelund Thomsen[2], and Ulrich Nyffenegger[3]

1 Fraunhofer Institute for Building Physics
Nobelstr. 12
70569 Stuttgart, Germany
**Corresponding author: hk@ibp.fraunhofer.de*

2 Danish Building Research Institute, SBi, AAU
Dr. Neergaards Vej 15
2970 Hørsholm, Denmark

3 Amt für Umweltkoordination und Energie
Bau-, Verkehrs- und Energiedirektion des Kantons Bern
Reiterstrasse 11
3011 Bern, Switzerland

ABSTRACT

This paper presents some first approaches for the national application of the nearly zero-energy building definition according to the Energy Performance of Buildings Directive by summarising the current plans of Germany, Denmark, Ireland and the Netherlands. As a contribution from a 5th country, the planned national energy performance requirements of Switzerland for the phase from 2018 onwards were included. It was also analysed whether any of the countries will set specific requirements to the air-tightness of NZEBs and if there are specific requirements for ventilation techniques.

KEYWORDS

Nearly zero-energy building, Germany, Denmark, the Netherlands, Ireland, Switzerland, energy surplus houses

INTRODUCTION

The Energy Performance of Buildings Directive (EPBD, [1]) in the version of 2010, the so-called recast, demands in article 9 'Nearly zero-energy buildings' that *'Member States shall ensure that (a) by 31 December 2020, all new buildings are nearly zero-energy buildings; and (b) after 31 December 2018, new buildings occupied and owned by public authorities are nearly zero-energy buildings. Member States shall draw up national plans for increasing the number of nearly zero-energy buildings. [...] The national plans shall include [...] the following elements: the Member States detailed application in practice of the definition of nearly zero-energy buildings, reflecting their national, regional or local conditions, and including a numerical indicators of primary energy use expressed in kWh/m² per year. ...'*

In Article 2, the Directive also includes an overall definition of the nearly zero-energy building (NZEB): '*nearly zero-energy building means a building, that has a very high energy performance, as determined in accordance with Annex I. The nearly zero or very low amount of energy required should be covered to a very significant extent by energy from renewable sources, including energy from renewable sources produced on-site or nearby;*'.

Annex I of the Directive presents a general framework for the calculation of the energy performance of buildings. It lists the energy aspects that need to be included in the calculation, which also cover ventilation and air-conditioning. The air-tightness is mentioned here by *' The methodology shall be laid down taking into consideration at least the following aspects [...] (d) natural and mechanical ventilation which may include air-tightness...'*.

Currently, the Member States are working on the detailed application in praxis of the definition of nearly zero-energy buildings. The laws, regulations and administrative provisions necessary to comply with articles 2 to 18 (including the NZEB article 9) shall be adopted and published by 9 July 2012 and be applied from 9 January 2013, at the latest.

This paper presents some first approaches for the national application of the NZEB definition by summarising the current plans of Germany, Denmark, Ireland and the Netherlands. As a contribution from a 5th country, the planned national energy performance requirements of Switzerland for the phase from 2018 onwards were included. Switzerland is not an EU Member State. Its development in energy performance requirements is however in line with the leading countries of the EU in the same climate region. It was also analysed whether any of the countries will set specific requirements to the air-tightness of NZEBs and if there are specific requirements for ventilation techniques.

GERMANY

National approach for the NZEB definition

For along time, Germany's development of energy performance requirements for buildings has been accompanied by research and demonstration projects that showed further strengthenings to be technically feasible, which became due to market adaptions some year later also economically feasible. Figure 1 shows the minimum energy performance requirements (in 6 steps) as the upper line, the pilot projects (solar houses, low-energy buildings, three-liter houses, zero-heating energy houses and plus energy houses) as the lower line and the actual building practice in between. The requirements followed the pilot projects with 10 to 20 years time difference.

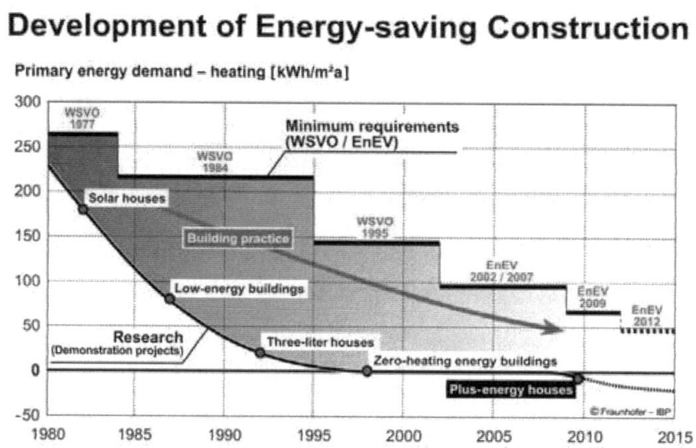

Figure 1. The historic development of pilot projects, minimum requirements and the building practice in the field of energy saving buildings in Germany.

With the last tightening of the minimum energy performance requirements by the energy decree (EnEV 2009, [2]), Germany has again become one of the countries in Europe with the strictest requirements for new buildings, see also the results of the *ASIEPI* project [3]. The national application of the definition of NZEB will be based on the results of a project called *‚Untersuchung zur Novellierung der Gebäude-Richtlinie – Identifikation und Analyse von Hemmnissen beim Neubau von hocheffizienten Gebäuden und Entwicklung eines Konzepts zur Marktdurchdringung bis 2020‘* [4] within the "Zukunft Bau" initiative of BBSR, in which the researchers analysed different possibilities for the German NZEB definition taking into account the current boundary conditions and requirements as well as the perspective of the market players such as industry, building owners, etc.

The assessment method which was proposed by the study partners Fraunhofer IBP, IB Hauser, Schiller Engineering and iTG is as follows:

- Assessment parameters (energy performance indicators): Both delivered energy and primary energy (non-renewable part).
- Balancing period: One year of operation
- Energy aspects to be included: Heating energy, ventilation and cooling energy all incl. auxiliary, lighting for non-residential buildings, energy generated from renewables (self-used and fed into the grid) if produced on-site.

There is a difference compared to the current requirements for new buildings, namely the inclusion of a maximum amount of delivered energy and the possibility to credit energy generated from renewables that is fed into the grid. The detailed quantitative requirement will be derived from an additional economical study and announced by the building ministry.

There will be no specific requirement regarding the airtightness of a NZEB in Germany. It is however clear that such a building should be designed and constructed in an airtight way in order to secure the low energy demand. The German calculation method does for a long time and will continue to credit improved airtightness. Neither will there be a specific requirement regarding the type of ventilation system. Natural or mechanical ventilation systems will be possible, all definitions will have the aim to be 'technology-open'. However it can be expected that many of the NZEBs will include a mechanical ventilation system including a heat recovery system with a high heat recovery rate. The ventilation industry (just like other manufacturers of energy efficient building components or material) will surely use the push in general awareness and will further develop their products.

Trends in high performance buildings in Germany

In September 2011 the Federal Ministry of Transport, Building and Urban Development (BMVBS) has started a subsidy programme for energy surplus houses (Effizienzhaus Plus). Energy surplus houses are defined to be buildings that produce more energy than they use. More specifically, both the annual balance of the delivered energy and the primary energy during one year shall be negative. The definition, the calculation method (which includes a default value for lighting and household electricity), and further information is available in a brochure published by the ministry [5].

A pilot project for the energy surplus houses is currently being built in Berlin, designed by the University of Stuttgart. The construction can be followed on the website of the BMVBS. The aim of this specific house is that enough of excess energy is being generated to cover the electricity necessary for driving with an electric car. In September 2011 an exhibition of several prefabricated energy surplus houses that are ready to be bought was opened in

Cologne. The buildings realise different energy concepts for achieving the energy surplus house standard.

Figure 2. Design of the energy surplus house currently under construction in Berlin.

DENMARK

In Denmark the minimum energy performance requirements for buildings are set by so-called energy frames. The energy frames for new buildings are fixed for 2010, 2015 and 2020. They divide into residential buildings (including those with similar types of use like hotels) and non-residential buildings. For buildings with a special use resulting in for example high ventilation rates there are additions to the allowed energy frames. The energy frame limits the delivered energy and includes the energy use for heating, ventilation, cooling, domestic hot water and the necessary electricity for operating the building. In the case of non-residential buildings it also includes the lighting energy. Electricity use has to be multiplied by a conversion factor and excess temperatures are punished by an addition to the calculation.

Characteristic values		Energy frame 2010	Energy frame 2015	Energy frame 2020
Maximum of total delivered energy to	Residential buildings (houses, hotels, etc.)	52.5 + 1650/A in kWh/m²yr	30 + 1000/A in kWh/m²yr	20 kWh/m²yr
	Non-residential buildings (offices, schools, institutions and other buildings)	71.3 + 1650/A in kWh/m²yr	41 + 1000/A in kWh/m²yr	25 kWh/m²yr
Conversion factors	Electricity	2.5	2.5	1.8
	District heating	1.0	0.8	0.6
Airtightness	Tested at a pressure difference of 50 Pa	1,5 l/s per m²	1.0 l/s per m²	0.5 l/s per m²
Heat recovery rate		0.70/0.80*	0.70/0.80*	0.75/0.85*
Fan power		1800 J/m³**	1800 J/m³**	1500 J/m³ / 800 J/m³*

where A is the heated gross floor area

* for ventilation systems for a single dwelling

** for installations with constant air volume, the power consumption for air transport must not exceed 1800 J/m³ fresh air

Table 1. Comparison of an extract of the energy performance requirements set in the Danish energy frames for 2010, 2015 and 2020.

Supplementary requirements exist for:

- Thermal losses in W/m^2 dependent on the number of building storeys
- U-values incl. a E_{ref} -value for windows dependent on the total solar energy transmittance and the transmission coefficient
- Minimum boiler efficiency
- Pipe insulation
- Automatic control
- Low temperature heating

As the 2020 levels will be difficult to achieve solely through improvements at the building envelope, heating, ventilation and lighting systems it is expected that the use of renewables will be increased.

IRELAND

Ireland plans to define the nearly zero energy building differently for dwellings and for other types of buildings. They want to implement a carbon neutral framework for dwellings in 2013 and to introduce a low energy standard for other buildings in 2016.

In order to reduce the net total carbon dioxide emissions to zero they propose first to optimise fabric, space heating and domestic hot water measures. As there will always be a base electrical load that will need to be compensated for they think of 3 possible solutions for this:

- Offsetting CO_2 emissions at the unit or dwelling level: use of onsite microgenerators to produce electricity from renewable sources and fed-in to the grid in order to balance the electricity that is provided by the grid.
- Offsetting CO_2 emissions at the development level: local wind turbines, biomass or gas CHPs as district/local/community schemes to reduce CO_2 emissions from the dwelling to zero.
- Offsetting CO_2 emissions at national level: paying a level to relevant local authorities based on the volume of CO_2 likely to be emitted. This revenue would be ring fenced for works on CO_2 reducing projects such as retrofitting existing building stock, renewable power generation, forestry, etc.

A voluntary standard for carbon neutral dwellings shall be introduced in 2013 and the adoption shall be encouraged through incentive schemes. A carbon neutral standard for dwellings requires a cross-departmental multi agency approach with buy-in and collaboration from industry. Critical success factors will be training and up-skilling on new technology, a standard for new products and technologies, research and development, appropriate commissioning and maintenance schemes and the learning from international experiences.

There are current requirements on the airtightness of buildings and on heat recovery for mechanical ventilation systems defined in Part F of the Irish Building Regulations. If and in which way they will be further tightened for carbon neutral dwellings is unclear.

THE NETHERLANDS

The Netherlands have strengthened their energy performance requirements for buildings by 40 % between 1995 and 2011. This corresponds to a development of the EPC (energy performance indicator for buildings) in the residential sector from 1.4 in 1995 to 0.6 in 2011.

Further tightenings are planned for 2015 (EPC 0.4), 2018 (EPC 0.2) and 2020 (EPC 0). Also the non-residential sector shall reach EPC 0 in 2020. For the governmental buildings EPC 0 is planned for 2018.

The energy performance calculation procedure was changed in order to have a higher degree of accuracy, to include new techniques such as biomass, micro power and solar and to allow for power generation by area or zone. An additional standard was integrated for determining the air flow and the ventilation. The new procedure takes the European standards into account.

There are some pilot zero-energy buildings in the Netherlands. They also have several existing financial instruments supporting energy efficient buildings. A special subsidy scheme for energy neutral pilot projects is planned.

SWITZERLAND

Though Switzerland is not part of the European Union, their energy performance requirements for buildings show a development similar to the neighboring countries such as Austria and Germany. Based on studies and experiences they found out that the cost-optimum for energy efficient buildings is dependent on many different influence parameters including:

- The development of the energy prices
- Investment costs which are based on the offers
- Capital costs which are based on the specific and the national financial situation
- The further development of the building technologies that can be triggered by requirements and subsidies
- Tax reductions based on the specific income
- The personal evaluation.

The cost-optimum for each building is different, depending on the size, the location and the building owner. Therefore the Swiss authorities have decided to offer a free choice of how to realise two goals:

1. Requirement for the efficiency of the building envelope
2. Requirement for the overall energy efficiency.

The free choice of either improved insulation or the integration of more renewables without any restrictions concerning the used technologies increases the public acceptance, and by stimulating competition among manufacturers it boosts innovations. To simplify the procedure, standard solutions with 11 different technology sets are offered.

Instead of primary energy factors they use so-called national (political) weighting factors. The maximum energy use of new buildings has been reduced from 120 kWh/m^{2}a in 1992 to 48 kWh/m^{2}a in 2008. The next tightening is planned for 2014. Additionally, Switzerland has a voluntary standard called 'Minergie' for different building types with significantly stronger minimum energy performance requirements and subsidies for the building owner. The national requirements serve as push while the Minergie label serves as pull to the development in energy saving buildings. As next steps nearly zero energy buildings and energy surplus buildings are being considered. Similar to Germany some energy surplus buildings have been built which were financially supported by the state.

For buildings with the Minergie label A and P a blower door test is obligatory and maximum air change rates of 0.6 1/h at 50 Pa overpressure shouldn't be topped. For the lowest Minergie label the airtightness measurement is voluntary and a target value of 1.0 1/h is given. Minergie also includes requirements regarding the household equipment (best practice) and in case of the Minergie-A label a default value for the embodied energy of 50 kWh/m².

CONCLUSION

The 5 summarised developments in energy performance requirements reflect a strong tightening over the last two decades. The national applications of the nearly zero-energy building definition are ambitious in all 5 countries, including steps towards the buildings being carbon neutral. Pilot buildings of net zero energy buildings or energy surplus buildings exist in at least 3 of the countries, with Germany just having started a support programme for the 'Effizienzhaus Plus' that includes a specific calculation method for energy surplus houses.

The important impact of airtightness on the energy balance of such high performance buildings has been recognized, and specific requirements exist in several countries. There is a strong tendency towards allowing for technology-open solutions which also applies to ventilation strategies.

ACKNOWLEDGEMENTS

The summary of the national approaches for the NZEB definition in all 5 countries is based on presentations given at the international workshop 'Cost-optimal ways to Nearly Zero-Energy Buildings' on September 27, 2011 (organised by the Federal Office for Building and Regional Planning, BBR) [6].

REFERENCES

[1] European Parliament and Council of the European Union 2010. *Directive 2010/31/EU of the European Parliament and the Council of 19 May 2010 on the energy performance of buildings (recast).* Official Journal of the European Union L 153/13-35 of 18.6.2010.

[2] Regierung der Bundesrepublik Deutschland. *Verordnung über energiesparenden Wärmeschutz und energiesparende Anlagentechnik bei Gebäuden. (Energieeinspar-verordnung - EnEV). Energieeinsparverordnung vom 24. Juli 2007 (BGBl. I S. 1519), die durch Artikel 1 der Verordnung vom 29. April 2009 (BGBl. I S. 954) geändert worden ist.* BGBl. I p. 954.

[3] *ASIEPI – Assessment and Improvement of the EPBD Impact.* EIE/07/169/SI2.466278. http://www.asiepi.eu/wp2-benchmarking/reports.html

[4] *Untersuchung zur Novellierung der Gebäude-Richtlinie – Identifikation und Analyse von Hemmnissen beim Neubau von hocheffizienten Gebäuden und Entwicklung eines Konzepts zur Marktdurchdringung bis 2020.* Research programme ‚Zukunft Bau'. Filenumber: 10.08.17.7-10-14.1. Report soon to be available.

[5] Bundesministerium für Verkehr, Bau und Stadtentwicklung: *Wege zum Effizienzhaus Plus.* Broschüre des BMVBS. Soon to be available at http://www.bmvbs.de.

[6] Federal Office for Building and Regional Planning (BBR): *Cost-optimal ways to Nearly Zero-Energy Buildings.* International workshop in Bonn, Germany. September 27, 2011. http://www.buildup.eu/events/15846

Wednesday 12 October 2011

13.45 – 15.15 Parallel Session 1A - Ventilation system performance: energy and indoor air quality

- Formaldehyde and Relative Humidity in High-Performance Homes with Outdoor Air Intakes and exhaust ventilation (Jonathan Coulter, USA)
- Decentralized mechanical ventilation with heat recovery (Jean Lebrun, Belgium)
- Filters for balanced ventilation systems: design, long-term performances and energy considerations (Alain Ginestet, France)
- Analysis and implications of the revision of the Spanish regulation regarding ventilation and infiltration (José Manuel Salmerón Lissén, Spain)
- Ventilation rates and IAQ in European standards and national regulations (Nejc Brelih, Belgium)
- Window opening in high performance buildings (Piet Jacobs, Netherlands)

For energy and indoor climate reasons, it is important that ductwork have a good airtightness an insulation but the levels that can or should be achieved are often ignored. This session, which is part of an AIVC-TightVent project, will include several presentations to discuss these issues.

- Ductwork airtightness requirements in Portugal (Eduardo Maldonado, Portugal)
- Hands-on training courses for ventilation systems installers within the PRAXIBAT initiative (Anne-Marie Bernard, France)
- Feasibility study of ventilation system air-tightness (Jeroen Soenens, Belgium)
- Case Study: Effect of Excessive Duct Leakage in a Large Pharmaceutical Plant (David Dyer, USA)

15.15 – 15.30 Room change

FORMALDEHYDE AND RELATIVE HUMIDITY IN HIGH-PERFORMANCE HOMES WITH OUTDOOR AIR INTAKES AND EXHAUST VENTILATION

Melissa Malkin-Weber[1], Jonathan Coulter[1*], Shaun Hassel[1], Tracy Dixon[1], Benjamin Hannas[1]

[1]*Advanced Energy*
909 Capability Drive, Suite 2100
Raleigh, North Carolina 27523, USA
**Email: jcoulter@advancedenergy.org*

ABSTRACT

Twenty new energy-efficient homes were equipped with an outdoor air intake and kitchen and bath exhaust fans. The outdoor air intake was equipped with a timer that operated the intake 20 minutes each hour. A closed crawl space was added to suppress moisture entering the home. The homes were compared to a control group of comparable building-code-compliant homes. Indoor formaldehyde levels were statistically not different between the intervention and non-intervention groups. Indoor relative humidity levels in all these houses averaged 63.8% during the humid season, indicating that any additional ventilation added to further reduce formaldehyde could push the homes into a problematic RH range.

KEYWORDS

Ventilation, healthy homes, formaldehyde, moisture, residential, bath exhaust, kitchen exhaust

INTRODUCTION

This paper describes outcomes from a ventilation strategy used in high-performance home programs to dilute pollutants in indoor air, and discusses the tradeoff between ventilation and moisture management in a humid climate.

METHODS

This study examined 20 intervention and 16 non-intervention homes, all built by the charitable group Habitat for Humanity in central North Carolina (a mixed-humid climate zone). Non-intervention homes were not altered from their original building code compliant construction. The intervention package was based on the *System**Vision***™ high-performance home program, administered by Advanced Energy in Raleigh, North Carolina. The program includes specifications, inspections and a quality assurance feedback loop. Intervention specifications include building envelope tightness, duct tightness, right-sized HVAC and other energy efficiency measures. No source control of formaldehyde was specified. A closed crawl space was added to this study for moisture suppression [1]. The ventilation components of the intervention package specify an outdoor air intake and spot exhaust in the kitchen and bathrooms. They are described below.

Ventilation Configuration

The outdoor air intake consisted of a six-inch flex duct connected to a vent at the foundation at one end, and to the return air plenum at the other end. A six-inch manual balancing damper was installed in the flex duct near the outside intake, and a fiberglass mesh filter was installed at the intake in order to exclude insects and large particles.

The outdoor air was controlled with a timer that turns on the air handler for twenty minutes out of every hour when the thermostat is not calling for heating or cooling. This brings in a consistent amount of outdoor air, even when the resident does not use heating or cooling during mild-weather times of the year.

The amount of outdoor air introduced to the house was calculated at 4.7 liters/second (L/s) per bedroom plus 4.7 L/s. Thus, a 3-bedroom house received added mechanical ventilation of 18.8 L/s (4.7 L/s x 3 + 4.7 L/s) on an intermittent basis.

Kitchen and bath exhaust fans were installed with exhaust ducts, and all connections were sealed with mastic. Each bathroom fan exhausted at least 23.5 L/s directly to the outdoors. The kitchen fan was hard-ducted to the outdoors with a backdraft damper installed that exhausts a minimum of 47 L/s. The kitchen and bath exhaust fans and the outdoor air ventilation system were performance tested with an exhaust flow pan meter to ensure the flows of these items met program standards.

Other House Performance Characteristics

The intervention and non-intervention homes were similar in size, but had different home performance characteristics. The non-intervention homes had more leaks in the building envelope and ducts. They also had less effective spot ventilation. None of the non-intervention homes had spot ventilation in the kitchen – all had recirculating fans, which forced odors and moisture from cooking back into the kitchen by means of a fan located in the vent hood. In contrast, all of the kitchen exhaust fans in the intervention homes were vented to the outside by running a duct from the vent hood to the exterior of the home.

Formaldehyde Measurement

A seven-day formaldehyde sample was taken in the living space of each home. All samples were taken during a 40-day period beginning August 2005. This period fell during the humid season. Due to construction delays, the intervention homes were built later than the non-intervention homes. At the time of sampling, the median age of the intervention homes was 10 months and all had been occupied for at least 5 months. The median age of non-intervention homes was 20 months. Methods for this study have been described elsewhere [2].

Passive formaldehyde samplers from a major manufacturer were used. Samplers were deployed in the central hall of the homes. After exposure, the badges were analyzed using high-performance liquid chromatography.

Relative humidity in the houses was measured hourly with a data logger placed in the return box of the forced air system.

RESULTS

No significant difference was found between formaldehyde levels in the intervention and non-intervention groups. Formaldehyde values averaged 69 parts per billion (ppb) inside the homes, as summarized in Table 1.

Average Formaldehyde Levels and Standard Deviation				
	Weight [μg/m³]	St Dev* [μg/m³]	Weight [ppb]	St Dev* [ppb]
Intervention	85	27	69	22
Non-Intervention	79	31	64	25
All	82	29	67	25

Table 1. Average Formaldehyde Levels Found in Homes by Group
*Standard Deviation

Formaldehyde levels in individual homes varied between 20ppb and 122ppb, as shown in Figure 1 below.

Guidelines for non-occupational levels of formaldehyde vary. Health Canada and the World Health Organization recommend a 100ppb action level [3] [4]. Health Canada also recommends a 40ppb eight-hour residential exposure limit. The state of California recommends an eight-hour and chronic Reference Exposure Level guideline of 7ppb (State of California Office of Environmental Health Hazard Assessment, or OEHHA).

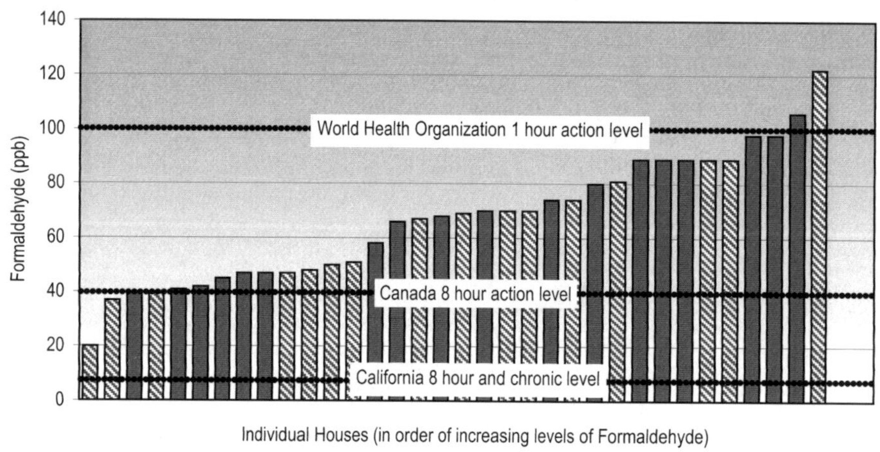

Figure 1. Formaldehyde levels in each home. Solid bars represent the intervention (I) homes and bars with diagonal lines represent the non-intervention (NI) homes.

Relative humidity in the homes averaged 63.8% during humid season, as shown in Table 2. Full moisture results have been discussed at length previously [5]

	Mean Indoor RH (%)	Standard Deviation Indoor RH (%)
Intervention	62.3	4.4
Non-Intervention	64.3	4.4
All	63.8	4.5

Table 2. Indoor Relative Humidity During Humid Season

CONCLUSION

The intervention homes, while measurable tighter compared to the control homes, did not demonstrate measurably higher indoor levels of formaldehyde. It seems that the ventilation package compensated for reducing the infiltration of outdoor air through the ductwork and building envelope. The high-performance homes program, thus, succeeded in its mandate to improve the energy efficiency performance of the homes without reducing indoor air quality. However, the program cannot claim to improve formaldehyde in indoor air compared to other leaky houses.

A recent conference on barriers to ventilation in the United States identified the need for research to identify the ventilation rate that allows adequate dilution of pollutants without unacceptable moisture and energy penalties [6]. Results from this study suggest that for humid climates, intermittent ventilation without pollutant source control does not provide low levels of indoor formaldehyde. In addition, intermittent ventilation without additional moisture control (such as an energy recovery ventilator or dehumidifier) resulted in average relative humidity exceeding 60%, a commonly used threshold for risk of mold growth.

In the United States, some reductions in indoor formaldehyde levels are expected when the U.S. Environmental Protection Agency completes its rulemaking process to regulate formaldehyde emissions from wood products. However, wood products are not the only source of formaldehyde, and formaldehyde is just one of many volatile organic compounds in the indoor air. Therefore, effective dilution strategies must be developed in tandem with this important regulatory effort.

ACKNOWLEDGEMENTS

U.S. Office of Housing and Urban Development (HUD) Office of Healthy Homes and Lead Hazard Control, The U.S. National Institutes of Environmental Health Sciences, and University of North Carolina Center for Environmental Medicine, Asthma and Lung Biology.

REFERENCES

[1] Davis B, Dastur, C. Moisture performance of closed crawl spaces and their impact on home cooling and heating energy in the southeastern United States. U.S. Department of Energy Oak Ridge National Laboratory and ASHRAE Performance of the Exterior Envelopes of Whole Buildings IX International Conference 2004. Clearwater Beach, Florida.

[2] Malkin-Weber M, Coulter J, Hannas B, Dixon T. Measured formaldehyde in high performance homes with outdoor air intakes. Indoor Air Quality Association 2009 Annual Meeting. Fort Worth, Texas.

[3] WHO - World Health Organization. Environmental health criteria for formaldehyde. 1989. Geneva, Switzerland: World Health Organization; Volume 89

[4] Health Canada. Residential Indoor Air Quality Guideline: Formaldehyde. Ottawa, Ontario, April 2006.

[5] Malkin-Weber M, Coulter J, Hannas B, Dixon T. High performance intervention houses in North Carolina – an attempt to control indoor relative humidity at 50% year round. Indoor Air 2008, Copenhagen, Denmark. http://www.indoorair2008.org/.

[6] Sherman, M. Trends in the U.S. ventilation market and driver for change. Air Infiltration and Ventilation Centre (AIVC), Brussels, Belgium. 2008. Retrieved at http://www.aivc.org/frameset/frameset.html?../Air/air_2009.html~mainFrame.

Centers for Disease Control and Prevention (CDC). Final report on formaldehyde levels in FEMA-supplied travel trailers, park models, and mobile homes July 2, 2008

Office of Environmental Health Hazard. Assessment Technical Support Document For the Derivation of Noncancer Reference Exposure Levels, Appendix D.1 Individual Acute, 8-Hour, and Chronic Reference Exposure Level Summaries. Oakland: California 2008

Offerman F, Brennan S, Roberson J, Springer D Woo T. Window Usage, Ventilation, and Formaldehyde Concentrations in New California Homes: Summer Field Sessions. ASHRAE IAQ 2007 Conference, Baltimore, Maryland.

DECENTRALIZED MECHANICAL VENTILATION WITH HEAT RECOVERY

Cleide Aparecida Silva [1], Samuel Gendebien [2], Jules Hannay [1], Nicolas Hansens [3], Jean Lebrun [1*], Marc Lengele [4], Gabrielle Masy s [3], Luc Prieels [5]

1 JCJ Energetics
Liège, Belgium
j.lebrun@ulg.ac.be

2 Thermodynamics Laboratory
University of Liège
Liège, Belgium

3 CECOTEPE
Liège, Belgium

4 WOW Technology
Nannine, Belgium

5 Greencom Development SCRL
Liège, Belgium

ABSTRACT

A new local ventilation device is designed in such a way to procure ventilation “on demand” in each room, with a maximum of effectiveness and a minimum of energy waste.
It consists in a parapipedic box to be located in one external wall (for example, just above a window) and containing two (injection and extraction) fans and a recovery heat exchanger.
The design of the heat exchanger is associated to the selection of the two fans in view of the best compromise between heat recovery effectiveness and “auxiliary” consumptions.
Great attention is paid to supply and exhaust air openings on both indoor and outdoor sides of the device, in order to get the highest ventilation effectiveness.
A fair compromise is looked for between air flow control “authority” and “auxiliary” consumption.

KEYWORDS

Ventilation, air diffusion, heat recovery, control, simulation

1. INTRODUCTION

Indoor climate control means satisfying some requirements on the following items, in order of decreasing priorities:

- Air quality
- Environmental temperature
- Air moisture

Developing more and more efficient ventilation techniques is a key issue, in the design of new “low” (and “zero”) energy buildings, as well as in the retrofit of existing ones.
Such development should include the following considerations:
1) Making ventilation “on demand” in such a way to maintain the required air quality in each occupied building zone;
2) Maximizing the ventilation effectiveness, i.e. keeping the required air quality inside the occupancy zones with as little air renovation as possible;
3) Minimizing the global energy consumption of the ventilation system;

4) Paying attention to all possible "side effects", as risk of draught effect, noise, sound transmission (and privacy loss)...

These considerations are hereafter further developed, as a continuation of a previous presentation [1].

2. STEP BY STEP DESIGN OF A VENTILATION SYSTEM: BACK TO THE BASIC PRINCIPLES

The following steps can be distinguished:

2.1. Identify the contaminants and the tolerable concentrations

2.2. Capture the contaminants as near as possible to the contamination sources, before they may be dispersed in the air of the occupancy (or breathing) zone, thanks to velocity control and local extraction. This is feasible if the contamination source is fixed, as, for example, in an industrial process or above a cooking appliance, but not if the contamination is issued from occupants, moving inside the room.

2.3. Bring fresh air as directly as possible into the breathing zone, thanks to velocity control and local supply. This is feasible if the occupant to be protected has fixed position inside the room (as, for example, in an office, in a theatre, or in a surgery room).

Both actions 2.2 and 2.3 can be (partially) realized with or without mechanical system. One or the two actions might be sufficient, but both can also be partially combined, in so-called "displacement" ventilation. But caution has to be paid to some side effects, as, for example, the risk of draught...

2.4. Having a fair idea of achievable ventilation effectiveness, identify the best combination of *air regeneration* (with contaminants capture by filtering and other separation methods) and *air renewal*.
On earth (but not in a space vehicle!) the re-introduction of O_2 (and elimination of CO_2) is usually performed by air renewal and CO_2 is the current tracer, taken as reference to determine the required fresh air flow rate.

Fresh air can be supplied to the occupants by "passive", "active", or "hybrid" methods.
The choice among these possibilities should result from a careful analysis of building tightness (including all fixed and variable openings), indoor and outdoor climates (buoyancy and wind effects) and resulting "natural" infiltrations.

2.5. Decide which part of the fresh air flow rate (if any) has to be provided by "active" means, i.e. by mechanical ventilation, with careful considerations to main advantages, inconveniences and side effects. Modify some of the building characteristics, if desirable and feasible; for example, improve its tightness and/or take control of some of the variable openings.

Free heating and free cooling are examples of advantages and inconveniences to be considered, according to the season. Noise generation, noise transmission and draughts are again among side effects to be taken in consideration.

2.6. Decide how to organize the mechanical ventilation, with central and/or local supply and exhaust systems.

2.7. Decide how to control the ventilation system, in such a way to get the correct air quality in each zone, with minimal fresh air flow rate.

Control design requires identifying the "controllability" of the whole (building and mechanical ventilation) system considered, which includes:

- The controllability of the mechanical ventilation itself, which can be defined as the ratio between ventilator head and total head (fan + wind + buoyancy);

- The mechanical ventilation "authority", i.e. the pressure range in which mechanical ventilation is able to "impose" any (minimal or maximal) air flow rate. The ventilation authority is satisfactory if this pressure range is larger than all combined wind and buoyancy perturbations.

- The controllability of the whole (building and mechanical ventilation) system is fully satisfactory if the total (mechanical and natural) fresh air flow rate can be controlled in each building zone and if all zone-to-zone air circulations are also under control. This last requirement can be very important, for example if having to deal with the risk of odour propagation from one zone to another one. It may justify some supply/exhaust disequilibrium inside a same zone (supplying more fresh air to "clean" zones and extracting more from "dirty" ones).

2.8. Minimize the (energy and/or environmental) impacts of the ventilation, thanks to heat recovery.

3. THE SYSTEM UNDER DEVELOPMENT

The new local ventilation system still under development is designed in such a way to procure ventilation "on demand" in each room, with a maximum of effectiveness and a minimum of energy waste.
It consists in a parapipedic box to be located in one external wall (for example, just above a window as shown in Figure 1) and containing two (injection and extraction) fans and a recovery heat exchanger [1], [2], [3].

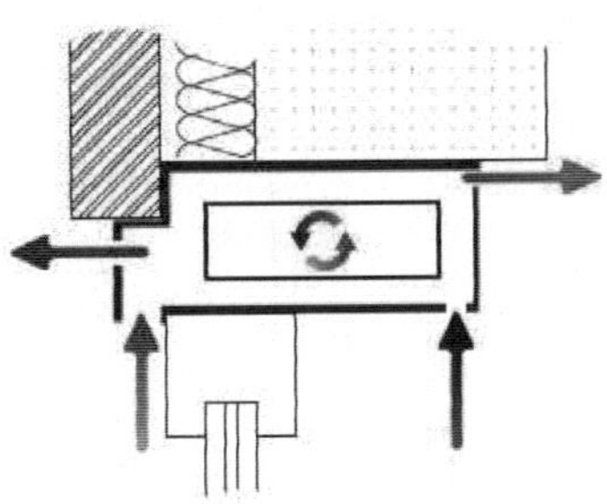

Figure 1. Typical configuration of the local ventilation system

The first applications considered are in the residential buildings, but other buildings, as offices, schools and health services are also concerned.
Above-presented design principles are here-after illustrated on this example…

3.1. Air diffusion analyses

3.1.1. Air diffusion inside the ventilated enclosure can be described by the model presented in Figure 2. Dominant air movements are here supposed to be induced by the following sources [1]:

- Ventilation and infiltration jets
- Buoyancy jet(s) issued from radiator(s)
- Buoyancy jet(s) issued from the occupant(s)
- Free convection boundary layers along cold walls (mainly the windows).

Three horizontal zones are distinguished in the model, as suggested in Figure 2, where the left and right sides are only distinguished for readability (the radiator is usually not located in opposition to the frontage).

In principle, fresh air can be supplied and exhausted to and from one, two or three different zones. But, in the case considered supply and exhaust openings are only located in zone 3.

If there is no phase change inside the room, the distribution of the air moisture can be considered as similar to the distribution of the contaminant.

The model consists in two sets of equations describing the induction effects and the balances of air, contaminant, sensible heat and moisture.

For example, the air mass balance of the first zone is described by the following equation:

$$\dot{M}_{su,1} + \dot{M}_{2,1,inductsu1} - \dot{M}_{1,2,inductsu2} - \dot{M}_{1,rad} + \dot{M}_{displ,2,1} + \dot{M}_{cbl,1} - \dot{M}_{ex,1} = 0$$

The different terms correspond to the air supply to this zone, the flow rate from zone 2 to zone 1 induced by the supply air jet of zone 1, the flow rate from zone to zone 2 induced by the supply air jet of zone 2, the flow rate "aspirated" by the radiator(s), the (positive or negative) displacement flow from zone 2 to zone 1, the flow rate supplied to zone one by free convection along the cold wall(s) and the air flow rate extracted from the zone.

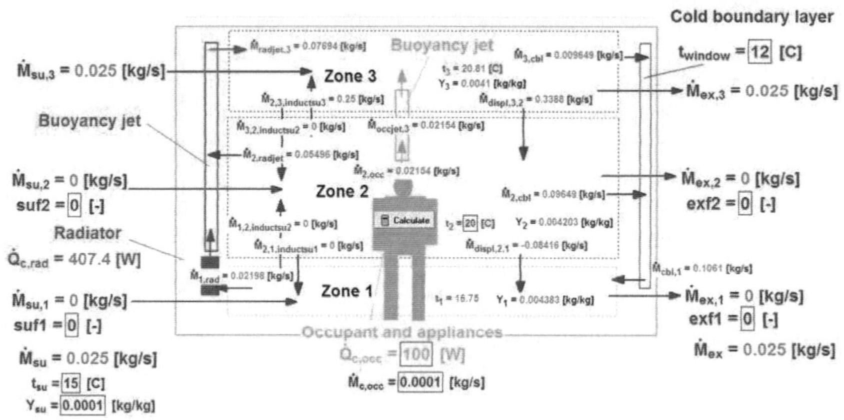

Figure 2. Air diffusion model

3.1.2. Avoiding local short-circuits on both indoor and outdoor sides requires, not only requires a careful design of the shapes of both openings, but also a sufficiently high velocity at box exhaust, in such a way to avoid that the exhaust jet would be deviated toward the aspiration opening by wind (on outdoor side) and/or buoyancy effects (on both sides). Rain water capture has also to be avoided at outdoor fresh air aspiration.

On both side of the system, there is a need of ensuring enough air mixing before getting any risk of jet deflexion. An example of verification is shown in Figure 3: it corresponds to the rejection of 60 m3/h of air, at 4 m/s of initial speed, through a circular opening. In this case, an induction factor of the order of 8 is already reached, when the jet axial velocity is reduced to 1 m/s (considered as a still sufficient "control" velocity). This means that the risk of short circuit is very reduced…

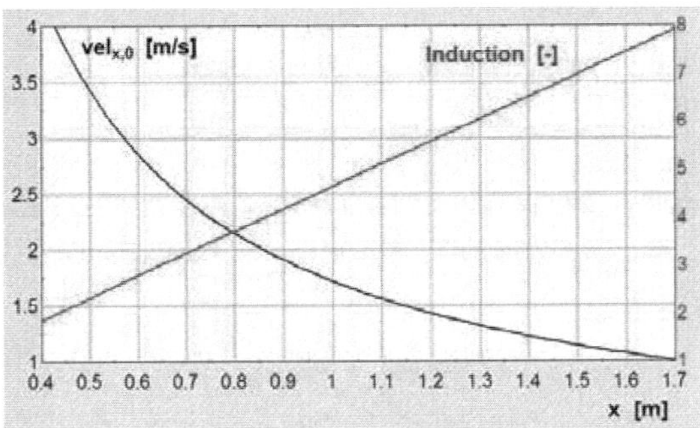

Figure 3. Diffusion of a jet issued from a circular orifice

3.1.3 Avoiding local discomfort

Inside the room, there is a risk of draught discomfort, if the jet is not diluted enough before entering the occupation zone.
The deviation due to buyancy can be verified by simulation as shown in Figure 4. In this example the air is supplied trough a rectangular opening at the rate of 35 m3/h and at a temperature of 10 C, with an idoor air temperature of 20 C. Three different initial speeds are considered: 1, 2 and 3 m/s. In this case, the initial speed must be maintained above 2 m/s, in order to avoid any risk of discomfort.

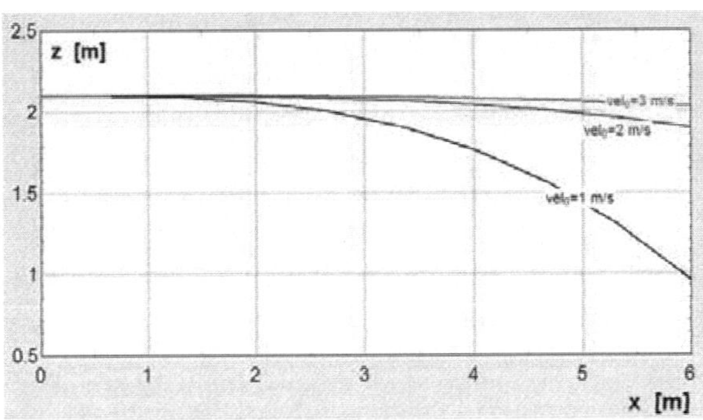

Figure 4: Deviation of the ventilation jet due to buoyancy

Of course, such simulation models have all to be tuned by experiments. A series of tests is also being performed in a climatic room…

3.2. Air diffusion effectiveness

In order of decreasing importance, air diffusion effectiveness can be defined according to its effects on air quality, environmental temperature and air moisture.
Each effectiveness results of the combination of two phenomena:
- The "*displacement*" (also called "plug flow"), whose effect is a fictitious reduction of contamination flow rate;
- The *mixing*, whose effect is to eliminate any risk of short circuit, which would itself produce a fictitious reduction of air renewal

Displacement and mixing effectiveness's can be defined by the following equations:

Global mass balance based on actual exhaust air state:

$$\dot{M}_c + \dot{M}_{su} \cdot (Y_{su} - Y_{ex}) = 0$$

Global balance based on indoor air state and fictitious flowrates, taking into account of both displacement and mixing effects:

$$[1 - \varepsilon_{d,c}] \cdot \dot{M}_c + \varepsilon_{m,c} \cdot \dot{M}_{su} \cdot (Y_{su} - Y_{in}) = 0$$

Indoor contamination flowrate:

$$\dot{M}_c = \dot{M}_{c,occ} + \dot{M}_{c,materials} + \dot{M}_{c,tu}$$

(Considering three sources of contamination: the occupants, the materials and the terminal units).

Similar equations can be used for sensible heat and moisture (with the same sources as for the contamination, plus the sun for sensible heat)…
In the case considered here, it seems that the air diffusion effectiveness is, most of the time, near to 1, but this has also has still to be experimentally verified…

3.3. Control

A high authority is provided by the system considered [2] [3], as shown in Figure 5: each variable speed fan is selected as able to keep the required air flow rate in a broad domain of pressure drop: from +110 to – 30 Pa. Above 110 Pa, the fan stays at its maximal speed; below -30 Pa, the fan is stopped.

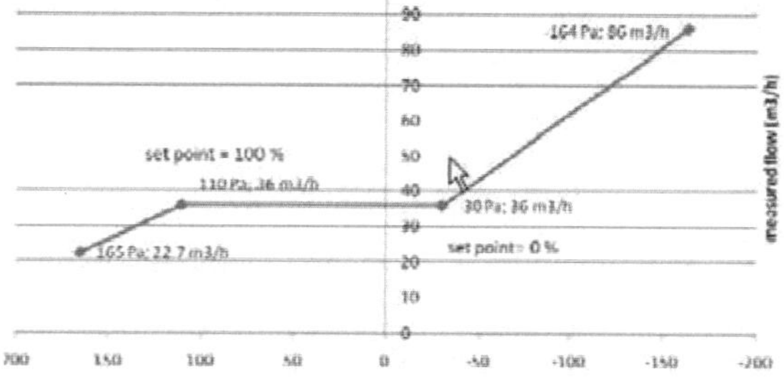

Figure 5: Fan authority

A feed back control of the air flow rate is performed on the basis of a signal given by a thermal anemometer located inside each ventilation channel. Each probe is calibrated is such a way to get a reliable correlation between air speed and corresponding flow rate, as shown in Figure 6.

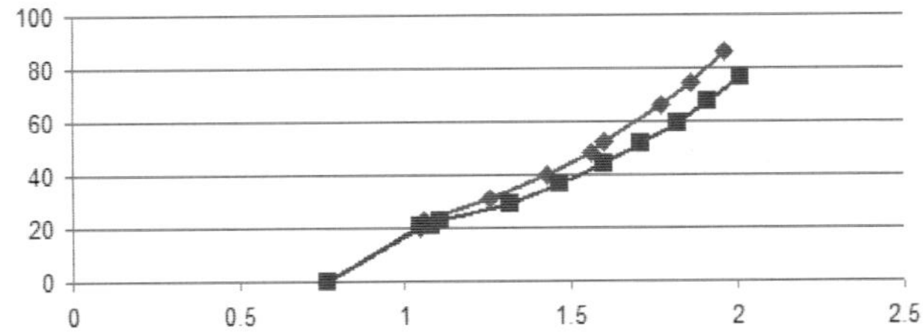

Figure 6: Example of correlation between flow rate (y) and air speed (x)

3.5 Recovery

Three « local » recovery effectiveness's can be defined at the level of the mechanical ventilation system:

$$\varepsilon_{c,rec} = \frac{Y_{su,mv} - Y_{out}}{Y_{ex,mv} - Y_{out}}$$

$$\varepsilon_{h,rec} = \frac{t_{su,mv} - t_{out}}{t_{ex,mv} - t_{out}}$$

$$\varepsilon_{w,rec} = \frac{\omega_{su,mv} - \omega_{out}}{\omega_{ex,mv} - \omega_{out}}$$

A « good » ventilation system is usually characterized by a very low value of the first effectiveness and high values of the two other ones [1] [4].

But this is not the whole of it: a global energy balances need to be establiched, by considering, not only the heat recovery, but also the "auxiliary" consumption of the system. Examples of heat exchanger tests results are presented in Figure 7.
At reference flow rate of 36 m3/h, the pressure drop of this heat exchanger is of the order of 60 Pa and its thermal effectiveness might over-pass 85 %.

But these results were obtained on the heat exchanger *alone*, i.e. not yet installed in the box. Some extra-pressure drops and losses of effectiveness have to be taken into account because of the non-uniform distribution of supply air at both sides of the heat exchanger.

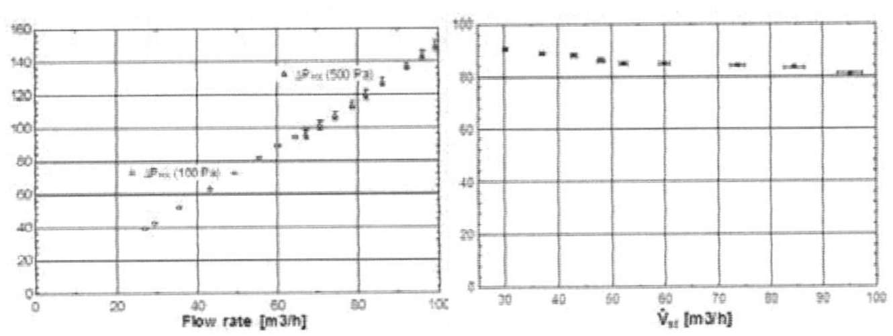

Figure 7. Example of heat exchanger test results [4]: pressure drop in Pa and thermal effectiveness in %.

The centrifugal fans selected for this system are over-sized in order to reduce their noise. Examples of fans characteristics, at nominal rotation speed, are shown in Figures 8 and 9.

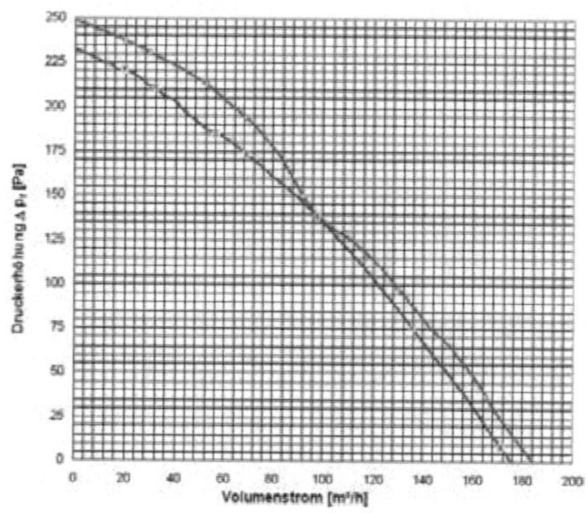

Figure 8: Pressure-flow characteristic of the fan

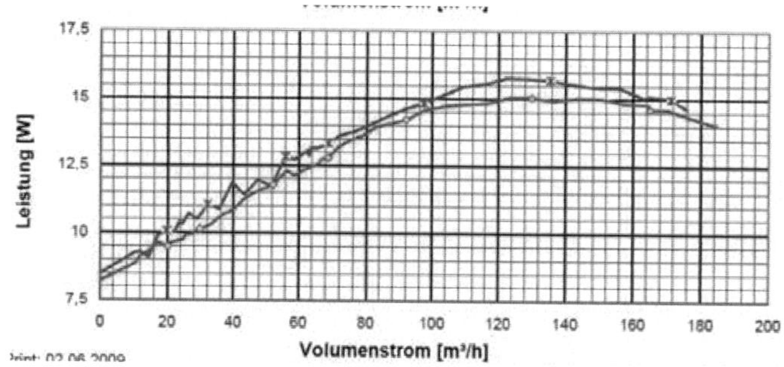

Figure 9: Electrical power consumed by the fan

Thanks to the over-sizing, the fans can be runned at low rotation speed (and low noise generation). Acoustic tests are also being made on the fans and on the whole system.

Not only the two fans, but also the control unit is consuming energy.

At nominal flow rate of 36 m3/h the electrical consumption the smallest unit considered should not over-pass 5 W.

If runned continuously on a full reference year of 2000 degree-days, such unit should be able to recover about 480 kWh of sensible heat and would not consume more than about 45 kWh. This would give an average COP higher than 10.

And this COP can be significantly increased if the unit is stopped whenever the heat recovery is too small, i.e. in mid season and summertime, when natural ventilation might be preferred.

Examples of more detailed results are presented in tables 1 and 2 and in Figure 10.

In table 1 are presented the average performances (one the whole year and on the winter season) as functions of the fans rotation speeds, with the following nomenclature:

“consigne”: ratio between actual and nominal fan speed [%]

Conso2ventilo: electrical current consumed by both fans [mA] (under 24 V DC)

Pelect: electrical power [W]

Pth recup moy: mean thermal power recovered on the whole year [W]

Pth recup hiver: mean thermal power recovered on winter time [W]

Epsilon: thermal effectiveness [%]

V: air flow rates (on both sides) [m3/h]

1.2 kg/m³

1000 J/kg°C

21°C est la moyenne des températures intérieures issue de la campagne de mesure réalisée par l'observatoire de la qualité d'air (France)

	COP moy	COP hiver
T in	21	21
Tout	9	5
ΔT	12	16

consigne	conso 2 ventilo	P elect	P th recup moy	P th recup hiver	ε	V	COP moy	COP hiver
%	mA	W	W	W	%	m³/h		
20	101.6	2.4	45.9	61.2	82	14	18.83	25.11
30	158.4	3.8	51.8	69.1	81	16	13.64	18.19
40	215.1	5.2	73.6	98.1	80	23	14.26	19.01
50	278.0	6.7	99.8	133.1	78	32	14.96	19.95
60	340.9	8.2	124.6	166.2	76	41	15.23	20.31
70	588.7	14.1	142.1	189.4	74	48	10.06	13.41
80	836.5	20.1	166.4	221.9	73	57	8.29	11.05
90	1089.2	26.1	187.2	249.6	72	65	7.16	9.55
100	1341.8	32.2	221.2	294.9	70	79	6.87	9.16

Table 1: Average performances calculated for different fans rotation speeds

The same results are presented month by month in table 2 and in Figure 10, with:

P(x)th rec= mean thermal power recovered with both fans at x % of their nominal speed.

These very encouraging results might be a bit optimistic because they are still based on separate tests of the different components and also because they don’t consider the actual building energy balance which would make useless a part of the recovery…

moyenne à Uccle 1981 à 2010

6.7 99.84

mois	*Temperature*	*P (20) th rec*	*P (50)th rec*	*P (100)th rec*	*20%*	*50%*	*100%*
janvier	3.3	67.732	147.264	326.270	27.78	22.07	10.13
fevrier	3.7	66.201	143.936	318.897	27.15	21.57	9.90
mars	6.8	54.339	118.144	261.753	22.28	17.71	8.13
avril	9.8	42.859	93.184	206.453	17.58	13.97	6.41
mai	13.6	28.317	61.568	136.407	11.61	9.23	4.24
juin	16.2	18.368	39.936	88.480	7.53	5.99	2.75
juillet	18.4	9.949	21.632	47.927	4.08	3.24	1.49
août	18	11.480	24.96	55.300	4.71	3.74	1.72
septembre	14.9	23.343	50.752	112.443	9.57	7.61	3.49
octobre	11.1	37.884	82.368	182.490	15.54	12.35	5.67
novembre	6.8	54.339	118.144	261.753	22.28	17.71	8.13
décembre	3.9	65.436	142.272	315.210	26.84	21.32	9.79

Table 2: Month by month performances for three different rotation speeds

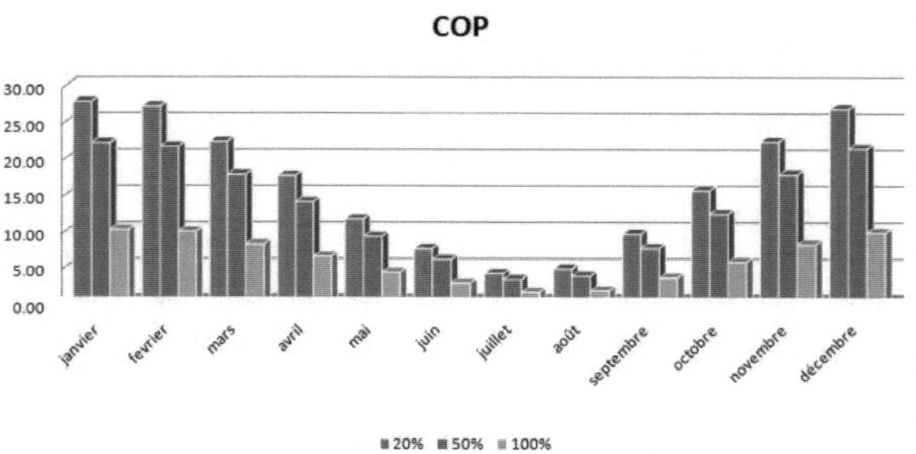

Figure 10: Monthly COP's for three different rotation speeds

"Passive" and "active" recovery techniques were also previously compared according to their respective impacts on the energy demand on a typical dwelling [1].

Examples of results are presented in Figure 10:

The dwelling is equipped with a mechanical ventilation system, without any heat recovery. The living room is supplied with 75 m3/h and two of the three sleeping rooms with 30 m3/h of fresh air. Only one sleeping room is occupied and all the surrounding dwellings are fully occupied. The space heating demand without any recovery is plotted on left side of Figure 9. It corresponds to 2667 kWh per year.
A (very) small heat pump is then supposed to be added inside each local ventilation system. This heat pump is acting downstream of the heat exchanger: it under-cools the exhaust air and over-heats the fresh air. The fresh air supply temperature is supposed to be raised until a maximum of 40°C.
The global contribution of these three small heat pumps is very significant, as shown on the right side of Figure 10: the remaining space heating demand to be covered by another heating source is marginal: it doesn't overpass 600 W and corresponds to only 65 kWh per year.
But such very small heat pumps are not yet available on the market and such solution will probably not be cost effective in near future…

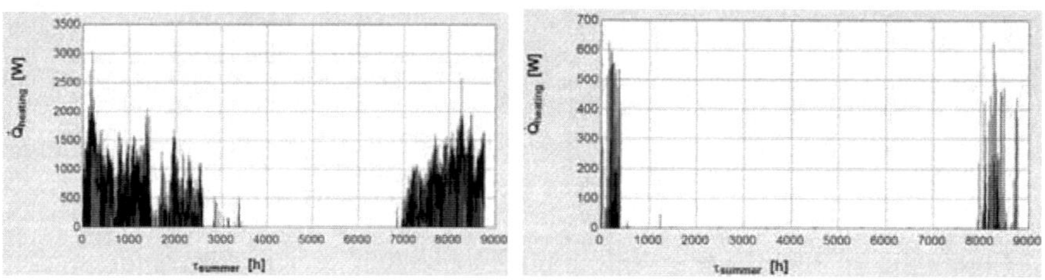

Figure 10 : Heating demands of a dwelling without and with full heat recovery

CONCLUSIONS

The local ventilation system appears as a very promising solution. It should allow an effective control the indoor air quality, with satisfactory thermal comfort and significant energy savings.

But many experimental verifications are still necessary, mainly about the correct air diffusion inside the room. This is so whenever dealing with a new terminal unit: simulation models need always to be tuned.

The system considered is of particular interest when having to deal with a very variable building occupancy…

ACKNOWLEDGEMENTS

The support of the Walloon Region for funding the Green+ project in the framework of the "Marshall Plan" to the work related in this project is gratefully acknowledged.

REFERENCES

[1] Aparecida Silva Cl., Hannay J., Lebrun J., Prieels L. *Development of an ew ventilation system with heat recovery* Roomvent 2011

[2] Lebrun J., Prieels L., Vincent B., Hansen N. *A model of decentralized air handling terminals (DAHT) interacting with building infiltration* Roomvent 2011

[3] Masy G., Lebrun J., Gendebien S., Hansen N., Lengele M., Prieels L. *Performances of DAHT conntected to building airtightness and indoor air hygrothermal climate*
AIVC TIGHTVENT Conference 2011

[4] Samuel Roomvent Gendebien S., Bertagnolio St., Georges B., Lemort V. *Investigation on an air-to-air heat recovery exchanger: modeling and experimental validation* Roomvent 2011

FILTERS FOR BALANCED VENTILATION SYSTEMS : DESIGN, LONG-TERM PERFORMANCES AND ENERGY CONSIDERATIONS

Alain Ginestet[*1], Dominique Pugnet[1] and Laure Mouradian[1]

CETIAT, Centre Technique des Industries Aérauliques et Thermiques
Domaine Scientifique de La Doua, 25 avenue des Arts
69100 Villeurbanne, France
[*] *alain.ginestet@cetiat.fr*

ABSTRACT

Due to the recent development of the balanced ventilation systems in dwellings, almost no information is available on the characteristics and performances of the filters used by these systems.
In order to control the energy consumption of the balanced ventilation systems, it is necessary to define a maximum value for the pressure drop of the filters used to clean the air for coils protection (air provided and exhausted by the systems) and indoor air quality enhancement (air provided by the systems).
The objective of our study was to determine the pressure drop increase of filters as they were continuously used for long-term (1 year) testing with outdoor air (constant airflow). It has been shown that regarding the energy consumption point of view, it is much better to use an F7 filter protected by a G4 filter installed upstream instead of the same F7 filter used alone because the increase of pressure drop is much lower (respectively 41 and 20 Pa after 1 year of use for the 2 different kind of F7 filters studied instead of respectively 325 and 611 Pa if no prefilter is used).
The energy point of view is today a major concern. Anyway, it shall not be done to the detriment of the indoor air quality. The efficiency of the filters has to be high enough to insure that the balanced ventilation systems will provide clean air to the buildings and their occupants. In our study, the filtration efficiency has shown a dramatic decrease (from 90 to 30/40 % at 0.4 µm after 4 months of use for both kind of F7 filters) which cannot be accepted. Engineers have to design filters with acceptable pressure drop increase and filtration efficiency and this compromise is still difficult to achieve. Results of the measurements have been analysed considering the quality factor (qF) of the filters.
Other types of filters are under study since May 2011 in order to provide a more representative set of data.

KEYWORDS

Balanced ventilation, air filter, filter pressure drop, electret

INTRODUCTION

Due to the energy crisis, buildings have to reduce their energy consumption. As a consequence, they become more and more tight and the characteristics of the ventilation must be well controlled in order to maintain the indoor air quality. Compared to the traditional mechanical ventilation system (mechanical exhaust and natural introduction through calibrated open devices), the balanced ventilation system (2 separate ductworks for fresh air introduction indoors and used air extraction outdoors) allows to reduce the energy consumption (heat recovery between the 2 ductworks) and to enhance the indoor air quality (filtration of the introduced air).

In order to control the energy consumption of the balanced ventilation systems, it is necessary to define a maximum value for the pressure drop of the filters used to clean the air for coils protection (air provided and exhausted by the systems) and indoor air quality enhancement (air provided by the systems).

After a survey on the typical filters available on the market (mainly G4 and F7 filters according to EN 779 filter classification), the objective of our study was to determine the pressure drop increase of filters as they were continuously used for long-term testing with outdoor air. The efficiency of the filters is also measured and the general performance of the filters is finally expressed using the quality factor (qF) parameter.

METHOD

Two different balanced ventilation systems for single dwellings were installed at CETIAT (Villeurbanne, France) for continuous running at constant air flow rate (about 120 m^3/h) in order to allow long time (1 year) filter testing according to natural loading. The 2 ductworks of each ventilation system normally used for respectively outdoor and indoor air flows were connected to a same outdoor air inlet. So the method allows to challenge 4 different filtration configurations with the same outdoor air.

For each ventilation system (named respectively system A and system B), the F7 filter under study is used alone in one of the ductworks and protected by a G4 filter installed upstream in the other ductwork. G4 and F7 filter classes have been chosen because they represent the most common filter classes available on the market for balanced ventilation systems. Also, most of the F7 filters available on the market are minipleated like those used for our study. Pictures of the filters are given on Figures 1a, 1b, 2a and 2b and the main characteristics of the F7 filters are provided in Table 1 (they use synthetic electrostatically charged fibrous medium).

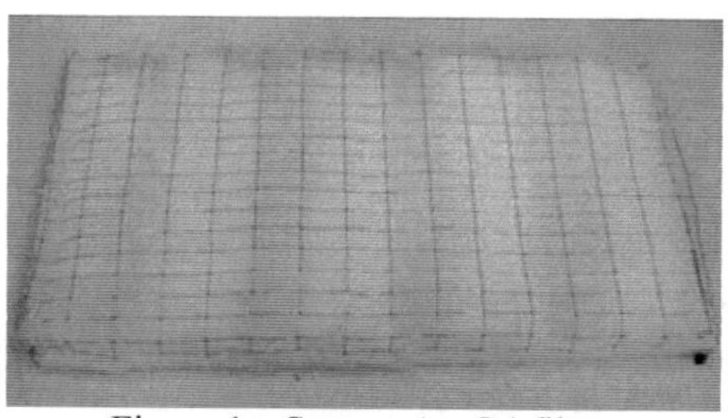
Figure 1a. System A : G4 filter.

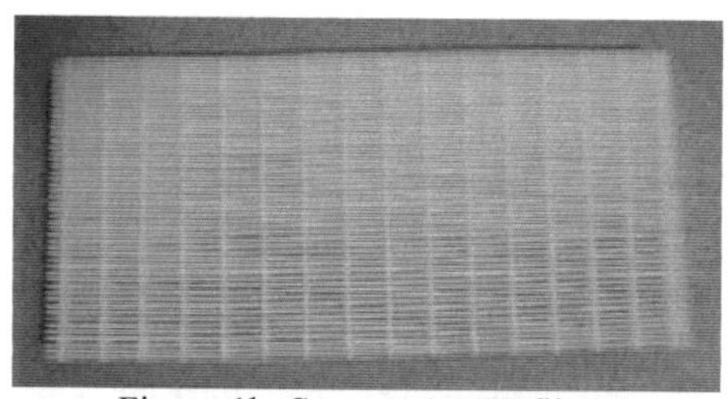
Figure 1b. System A : F7 filter.

Figure 2a. System B : G4 filter.

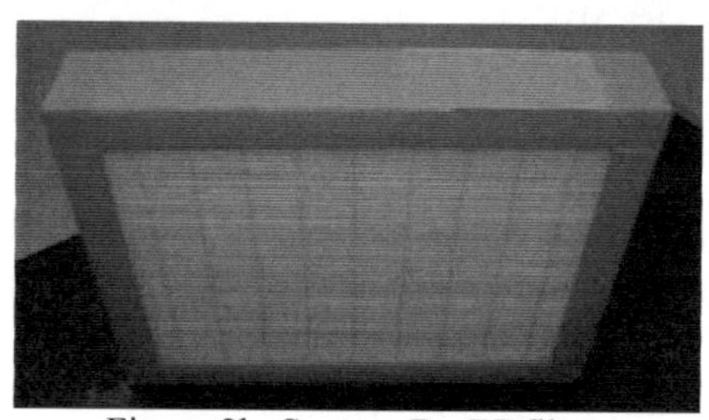
Figure 2b. System B : F7 filter.

	System A	System B
Filtration medium	Synthetic electret	Synthetic electret
Front dimensions (frame not included)	400 mm x 192 mm	233 mm x 165 mm
Thickness	20 mm	44 mm
Number of pleats	47	41
Filtration area	0.75 m²	0.84 m²

Table 1. Main characteristics of the 2 studied F7 filters.

The 2 balanced ventilation systems were regularly stopped in order to remove the filters for performance measurements (pressure drop, mass of retained dust and efficiency by particle size).

The amount of dust retained by the filters was determined by direct weighing of the filters. A test rig designed for EN 779 filter testing was used for pressure drop and fractional efficiency (by particle size with DEHS aerosol in the particle size range 0.2 to 5 µm) measurements.

RESULTS AND DISCUSSION

Filtration efficiency

The fractional efficiency (by particle size) of the different filter configurations is shown on Figure 3 for system A and Figure 4 for system B.

Not depending on the use of a G4 filter or not, the efficiency first begins to sharply decrease (from 90 to 30/40 % at 0.4 µm after several months of use). This is a well known phenomenon often described and explained in the literature ([1] and [2]) dealing with the decrease of the electret effect of the electrostatically charged fibrous medium used by the F7 filters as they become loaded by particles.

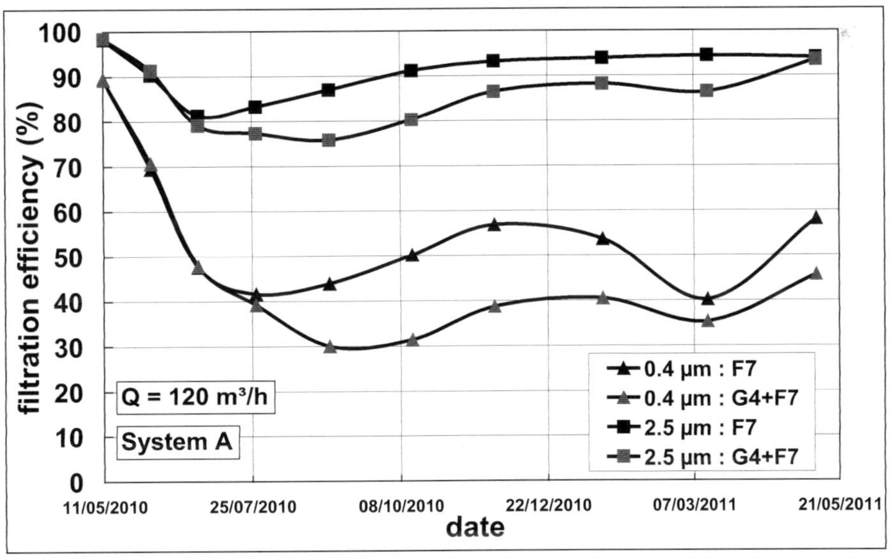

Figure 3. Filtration efficiency of F7 and G4+F7 filters as function of time (system A).

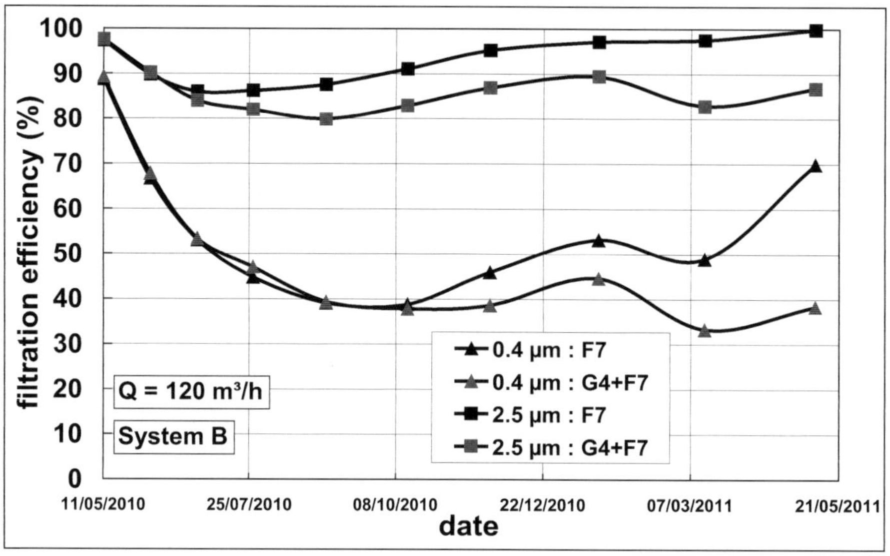

Figure 4. Filtration efficiency of F7 and G4+F7 filters as function of time (system B).

And because during the same time the efficiency tends to increase by mechanical effects, a point where the efficiency is minimum is obtained after a given period of time : 2 months for the F7 filter used alone and 4 months if the F7 filter is protected by a G4 filter in case of system A (Figure 3) considering particles of 0.4 µm ; 4 months for the F7 filter used alone and more than 4 months if the F7 filter is protected by a G4 filter in case of system B (Figure 4) considering particles of 0.4 µm.
The minimum efficiency is obtained more quickly as the particle size is higher because the influence of the electret effect is smaller on the largest particles.

The decrease of the efficiency of the filter configuration is more pronounced when the F7 filter is protected by a G4 filter : the inhibition of the electret effect is produced by the finest particles not retained by the G4 filters and retained by the F7 filters while the increase of the efficiency by mechanical effects is low because the biggest particles are retained by the G4 filters and do not reach the F7 filters.

Pressure drop

The filter pressure drop increase of the different studied filtration configurations is shown on Figure 5.
The pressure drop of the filter configurations continuously increases as function of time. The pressure drop increase of the F7 filters not protected by a G4 filter is linear for around 4.5 months then it increases much sharply as the filters become overloaded by particles especially on their surface (Figures 6a and 6b). When the F7 filters are protected by a G4 filter installed upstream, their pressure drop increase is much smaller and the amont of particles covering their surface is much smaller too (Figures 7a and 7b).
In order to limit the pressure drop increase of the filtration configuration, it appears much better to use an F7 filter protected upstream by a G4 filter instead of the same F7 filter used alone (Figure 5). After 1 year of use, the pressure drop increase was 325 Pa for the F7 filter

used alone and only 41 Pa for the same F7 filter and the G4 filter used in series (system A) ; in case of system B, the pressure drop increase was 611 Pa for the F7 filter used alone and only 20 Pa for the same F7 filter and the G4 filter used in series.

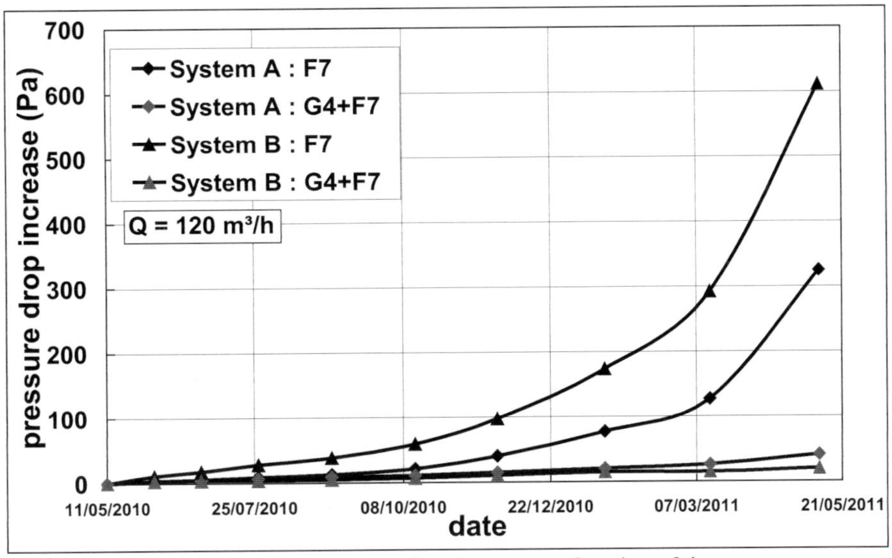

Figure 5. Filter pressure drop increase as function of time.

Figure 6a. F7 filter not protected (1 year), system A.

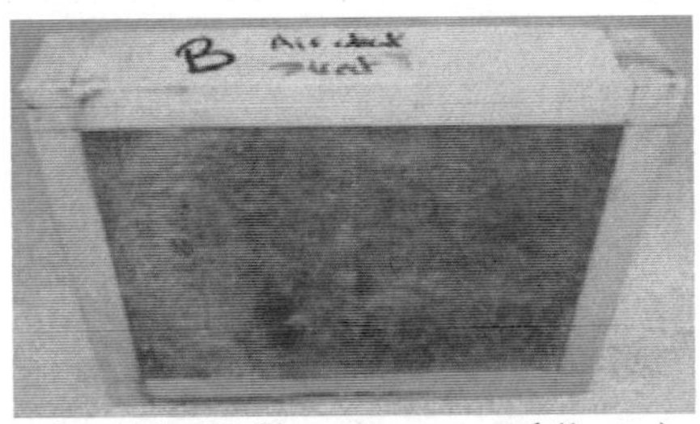

Figure 6b. F7 filter not protected (1 year), system B.

Figure 7a. F7 filter protected (1 year), system A.

Figure 7b. F7 filter protected (1 year), system B.

Dust holding capacity

Compared to the case when a F7 filter is used alone, the increase of the pressure drop in case of a G4 filter is installed upstream is much smaller (Figure 5) because the largest particles are trapped by the G4 filters while the finest fraction is trapped by the F7 filters (the efficiency of the G4 filters at 0.4 µm has always been less than 10 %, data not presented here). In this study, the amount of particles, in mass, retained by the G4 filters (Figure 8) is in the range of 45 to 60 % for system A (Figure 9a) and 65 to 80 % for system B (Figure 9b).

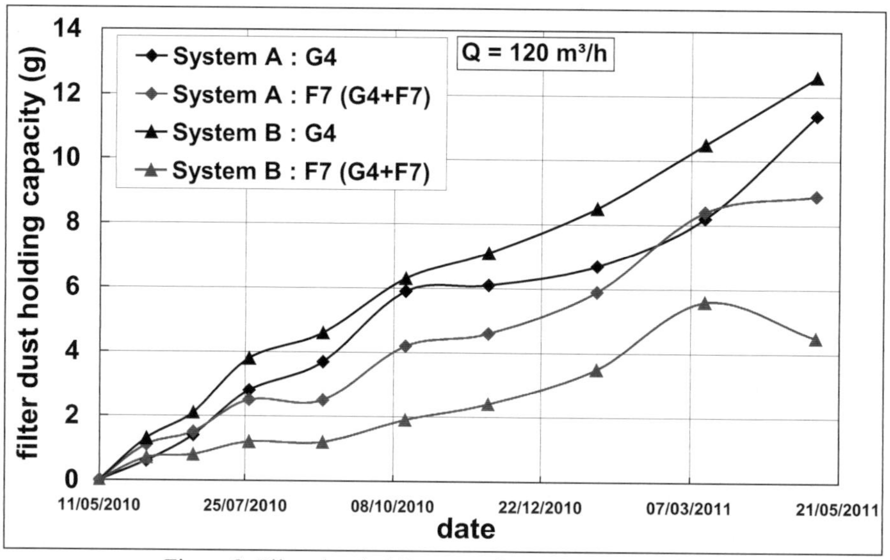

Figure 8. Filter dust holding capacity as function of time.

Figure 9a. G4 filter (1 year), system A.

Figure 9b. G4 filter (1 year), system B.

Quality factor

It is convenient to use the quality factor parameter, qF (calculated by equation (1)), for a filter performance assessment taken into account both pressure drop and filtration efficiency ([4], [5] and [6]) :

$$qF = -\frac{Ln(\frac{100-E}{100})}{\Delta P} \qquad (1)$$

E represents the efficiency of the filter at 0.4 µm (expressed in %) and ΔP represents the pressure drop of the filter configuration (expressed in Pa), with both measured at the same air flow rate. A higher qF factor value means a better filter.

For the 4 studied filtration configurations, the qF value shows a continuous decrease which is due to both efficiency decrease and pressure drop increase (Figure 10).
For system A, at the beginning of the use of the filters, the increase of the pressure drop of the F7 filter used alone is not too high compared to the one of the G4+F7 configuration. Also, the efficiency of the F7 filter used alone becomes higher of the one of the G4+F7 configuration after 2 months of use. For these reasons, the quality factor is higher for the F7 filter used alone during the first months of running then it becomes higher for the G4+F7 configuration when the pressure drop of the F7 filter used alone increases sharply.
In case of system B, the quality factor has always been higher for the G4+F7 configuration (except at the beginning when the filters were new) mainly because the increase of the pressure drop of the F7 filter used alone was very sharp even at the beginning of the use of the filter.

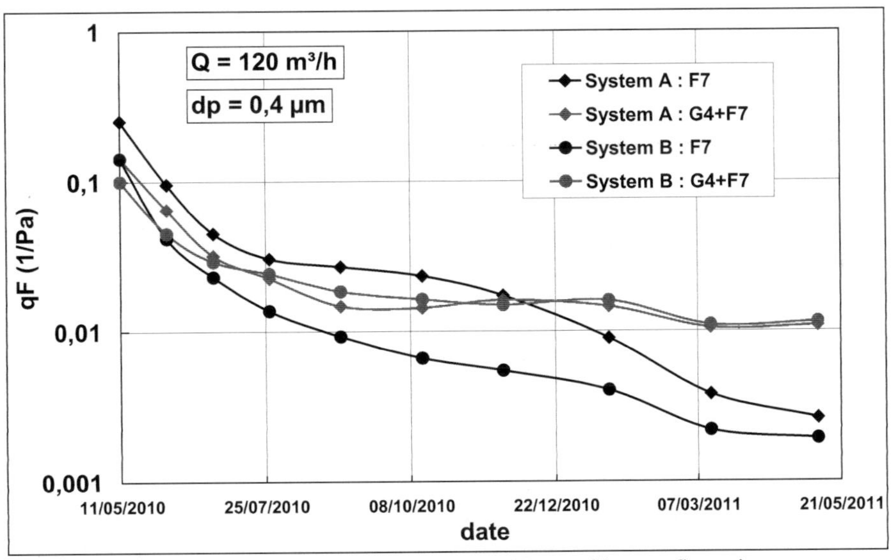

Figure 10. quality factor values for the different filter configurations.

CONCLUSION

On the occasion of long term tests of filters typically used in balanced ventilation systems for single dwellings, it has been shown that regarding the energy consumption point of view (considering systems running on the basis of a constant air flow rate), it is much better to use the F7 filter protected by the G4 filter installed upstream instead of the same F7 filter used

alone because the increase of pressure drop is much lower. This result means that filter life time would be increased and filter maintenance would be reduced in this situation.
In the past it has been shown in laboratory the benefit to use a prefilter installed upstream of a HEPA filter [3] but our results are to our knowledge the first which show in real use the benefit (from the pressure drop point of view) to install a prefilter upstream of a fine filter.
The energy point of view is today a major concern. Anyway, it shall not be done to the detriment of the indoor air quality. The efficiency of the filters has to be high enough to insure that the balanced ventilation systems will provide clean air to the buildings and their occupants. In our study, the pressure drop of the F7 filters protected by a G4 filter is quite low (respectively 41 and 20 Pa after 1 year of use for G4+F7 configurations) and higher values could probably be accepted, but their efficiency has shown a dramatic decrease (from 90 to about 30/40 % at 0.4 µm after few months of use) which cannot be accepted. Engineers have to design filters with acceptable pressure drop increase and filtration efficiency and this compromise is still difficult to achieve. As shown previously, the filter configuration performance may also be analised on the quality factor point of view.

Other types of filters (F7 filters manufactured respectively with synthetic and glass fibres media) are under study since May 2011 in order to provide a more representative set of data.

ACKNOWLEDGEMENTS

This study has been financially and technically (balanced ventilation systems and filters were provided by CETIAT member companies) supported by both the ventilation and filtration CETIAT technical committees.

REFERENCES

[1] Drangsholt, F. 1995. *Long-term tests of filters in a real environment*, STF11 A95052 Report, SINTEF, Trondheim, Norway, 44 pages.
[2] Gustavsson, J., Ginestet, A., Tronville, P. and Hyttinen, M. 2010. *Air filtration in HVAC systems*, REHVA Guidebook No 11, 94 pages.
[3] Davis, W. T. and Kim, G.-D. 1999. *Effect of prefilters on the performance of HEPA filters*, Filtration & Separation, April 1999, 51–56.
[4] Fuchs N. A. 1964. *The mechanics of aerosols*, Pergamon Press Ltd.
[5] Davies C. N. 1966. *Aerosol Science*, Academic Press Inc. (London) Ltd.
[6] Ouyang M. and Liu B. Y. H. 1997. *On the figure of merit for fibrous filters*, The 10th Annual National Technical Conference of the American Filtration & Separations Society, April 29-May 2, Minneapolis, USA. Advances in Filtration and Separation Technology, Vol. 11, 168–179.

ANALYSIS AND IMPLICATIONS OF THE REVISION OF THE SPANISH REGULATION REGARDING VENTILATION AND INFILTRATION.

Salmerón Lissén José Manuel[1], Sánchez de la Flor Francisco José[2], Álvarez Domínguez Servando[1], and Molina Félix Jose Luis[1]

1 University of Seville, Grupo de Termotecnia.
C/ Avda de los Descubrimientos S/N.
41092 Seville,Spain
**Corresponding author: jms@us.es*

2 University of Cadiz, Dpto. De Máquinas y Motores Térmicos.
C/ Chile, 1
11002 Cádiz, Spain

ABSTRACT

The Spanish Technical Building Code is one of the three royal decrees that were approved in Spain as a consequence of the transposition of the European Directive on the energy performance of buildings (2002/91/EU). One basic document of the Technical Building Code deals with the limitations in the energy demand of buildings. Nowadays, due to the recast of the European Directive on the energy performance of buildings (2010/31/EU), a revision process of the current regulations has begun, starting with the Technical Building Code, with its first revision envisaged for 2011. In this article we, as collaborators in this revision process, describe and analyse the main changes regarding ventilation and infiltrations that the "new" Technical Building Code (TBC) is going to introduce.

These main changes can be classified in the next aspects:

1. The calculation methodology.
2. The ventilation technologies that can be considered.
3. The default values and the data that the user can supply to assess its particular case.

Basically, these changes makes possible to move form a previous regulation that consider:

1. One indoor pressure and one outdoor pressure assuming all the indoor spaces as a single-zone (calculation methodology EN 13465:2004).
2. A constant extraction flow 24 hours a day, without the possibility of use a control system.
3. No possibilities of changing the sizing of vents. Default behaviour curve of the vents. Permeability of opaque elements according to the UNE-EN 13465:2004.

To a new regulation that assume:

1. One indoor pressure and one outdoor pressure (calculation methodology EN 15242:2007). But keeping and indoor zoning that allows calculating (using mass conservation) the air flow between zones.
2. The possibility of using a control system allowing different flows each hour for a demand controlled ventilation scheme and pre-heating of supply air flow.
3. The possibility of sizing the vents, using self-regulating vents and changing the permeability of opaque elements, for example if a blower-door test has been carried out.

KEYWORDS

Ventilation and infiltration regulations. Calculation methods of air flow rates in buildings.

INTRODUCTION

This article is going to classify the main changes that appear in the new Spanish regulation regarding ventilation and infiltration in the next topics:

1. Calculation methodology.
2. Ventilation technologies that can be considered.
3. Default values and data requested to the users.

Thus, these topics will be the three main sections of the present paper.
The objective of this paper is then to show how the regulation regarding ventilation and infiltration has been modified in one State Member of the European Union, Spain. Along the paper we will describe the criteria that have been followed to select the new specifications. Finally, we will comment the consequences of these modifications.

CALCULATION METHODOLOGY

Basically, the calculation methodologies regarding air flows implement a multi-zone loop method. Loop methods have been used extensively in the duct networks analysis. They provide an 'exact' analytical approach to size components of natural and hybrid ventilation systems and offer a number of advantages when compared to the node continuity methods [1]. On this basis we are going to describe the particularities of each regulation:

Former Regulation (2006)

The former regulation is based in the EN 13465:2004 [2], thus it implement a loop method with only one indoor pressure and one outdoor pressure by assuming that all the indoor spaces are a single-zone for single family houses and also for blocks of flats.
The Technical Building Code [3] establishes the ventilation rates for indoor air quality in the basic document HS3, and also limits the infiltrations through windows in the basic document HE1.
Following the HS3 the inlet flows –bedrooms and living-room- and exhaust flows –bath-rooms and kitchen- should be calculated separately and consider the ventilation flow as the higher of both. For calculations the next table should be used:

Type of zone	Per occupant	Per square meter	Other
Bed-rooms	5	-	-
Living-rooms and dinning-rooms	3	-	-
Bath-rooms and toilets	-	-	15 per room
Kitchens	-	2	50 per room
Storage rooms	-	0.7	-
Parkings and garages	-	-	120 per place
Waste storage	-	10	-

Table 1. Minimum ventilation flows required (l/s).

The maximum ventilation flow from inlet and exhaust flows is the one considered in a constant schedule of 24 hours a day. Next figure shows this schedule for a building where the maximum air flow is equal to 45 l/s.

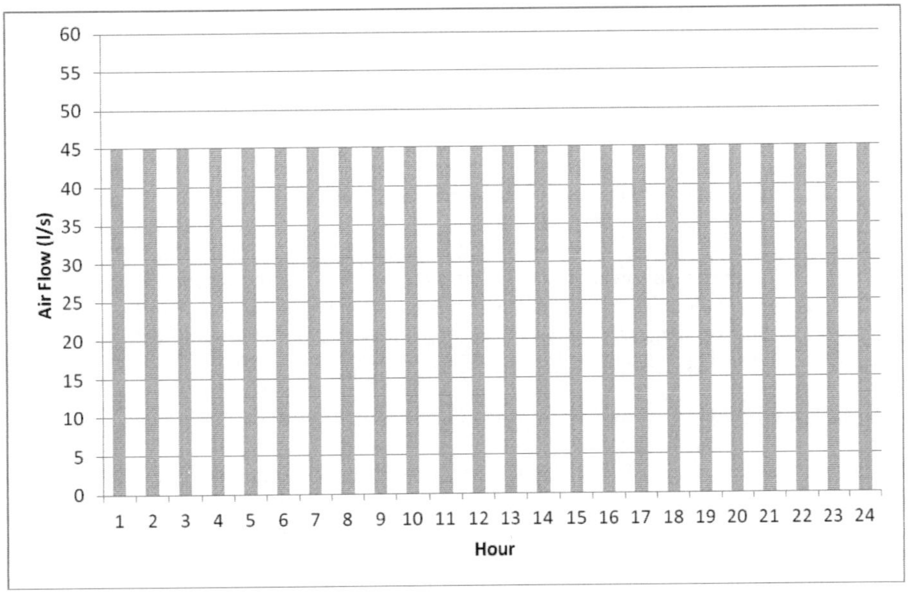

Figure 1. Standard supply air flow schedule.

The air flow of the extraction hood and the eventual air flow requirement of the domestic hot water boiler –or any combustion system- are not envisaged in this regulation.

The equations for the pressure losses in the cracks, windows and vents are the proposed in the European standard.

The default values for the infiltration through opaque elements (cracks) are the next:

- 0.30 air changes per hour at 1 Pa in single-family dwellings. 0.76 at 4 Pa.
- 0.24 air changes per hour at 1 Pa in blocks of flats. 0.61 at 4 Pa.
- 0.10 air changes per hour at 1 Pa in other buildings (tertiary). 0.25 at 4 Pa.

These values are fixed and can not be changed by the building designer.
If we compare these values with the ones that appear in other regulations we can see that these one do not depend on the outdoor or exposed surface of the building like in the French Thermal Regulation [4], on the contrary these are constant values that only depend on the drop pressure between indoor and outdoor, they were obtained from the annex A (not mandatory), tables A.2 and A.3, of the EN 13465:2004.

The permeability of windows is a data that the building designer can modify if the window has a certificate of permeability. The maximum values are 50 m^3/hm^2 at 100 Pa for the south of Spain, and 27 m^3/hm^2 at 100 Pa for the middle and north of Spain. The doors have an invariant permeability coefficient of 60 m^3/hm^2 at 100 Pa due to the absence of a certificate for these elements.

The vents are not defined by the user, they are sized by the official software of the Technical Building Code in order to ensure the air ventilation flow required.

The main assumptions consist on:

- Due to the number of occupants and their position in unknown, it is necessary to suppose that the maximum air flow is supplied 24 hours a day.
- On the other hand, the unizone model implies that the air flow calculated is coming from outdoor to each zone.

Both assumptions have as a consequence an increase in the calculated energy load, the first one because there are a lot of marketable systems that controls the occupancy and tune the ventilation air flow to it, the second one because all the zones does not receive the air from outdoor, on the contrary the air inlets are located in bedrooms and living-room, from there the air flows to the corridor, to end its path in the bathrooms or in the kitchen where the air exhaust vents are installed. So the ventilation thermal load is concentrated in the bedrooms and living-room and it is negligible in the corridors, bathrooms and kitchen if the inlet spaces are thermally conditioned.

New Regulation (2011)

The proposed new regulation uses the calculation methodology of the EN 15242:2007 [5]. This standard is an actualization of the former and derogated EN 13465:2004; it is also based on a loop method and keeps the calculation of air infiltrations trough opaque elements, windows and air vents. The novelty is the indoor zoning that allows calculating (using mass conservation laws) the air flow between zones in a similar way that software SIREN [6] does.

The minimum ventilation air flows are not going to be modified by the moment, but futures modifications are not discarded if the practice shows them to be inappropriated to ensure acceptable indoor air quality levels.

The novelty in the air flows is that the extraction hood in the kitchen is going to run two hours a day in order to take into account the real use of the system, the air flow requirements for combustion (for instance domestic hot water boiler) is going to be considered null because all the new boilers should be airtight.

The default values for the infiltration through opaque elements (cracks) have been changed to fit with the EN 15242:2007 that appeared after the publication of the 2006 regulation. In this standard the Table B.1 of the Annex B (normative) give three values for the permeability of the envelope depending on the air leakage level, in the next point we will justify why we have choose the highest level and thus the default values for the infiltration through opaque elements (cracks) have been fixed to 1.8 m^3/h per square meter of exposed wall surface for single-family dwellings and 1.95 m^3/h per square meter of exposed wall surface for blocks of flats.

As in the former regulation the permeability of windows will be a data that the building designer could modify if the window has a certificate of permeability. The maximum values probably will be more severe than the previous ones.

The software of application of the regulation will size the vents although it will allow to the designer to modify the type -different vent types are listed in the point titled "other energy saving measures"- and probably the nominal air flow in order to give more freedom for the different alternatives.

All these innovations allow to skip the negative effects of the assumptions done on the regulation of 2006: in first term because the calculation of the air flow between zones allows to assign the ventilation energy load to the exact zone that is receiving it, and keep it equal to zero in the zones that only receive the ventilation air from conditioned spaces. A second consequence of calculating the air flow between zones is that it is possible to define the ventilation strategy –demand controlled ventilation, heat recovery ventilation with double flux-. A third consequence is the elimination of a market barrier for the ventilation systems, because now if a designer has implemented a ventilation strategy it can be calculated and it is taken into account in the energy performance of the building, this is very important as the maturity of the ventilation sector is high and the solutions adopted were very basics due to the regulation. Finally, this new regulation open a door for improvements on the tightness of the new buildings as the permeability of the opaque elements is not a default value anymore and now this data that can be given by the designer after a permeability essay, for example a blower door test.

CRITERIA FOR THE NEW SPECIFICATIONS

The criteria for the new specifications have been to keep, when it is possible, the same mean values, updating the reference standards and allowing to the designer a highest degree of freedom in his decisions.

In this sense the main change in the default values have been the change in the opaque permeability value, because is has been moved from a fixed value in air changes per hour – Annex A of EN 13465:2004- to a variable (per square meter of outdoor surface) value from Annex B of EN 15242:2007.
In this section we will explain how we have converted the former values in the new ones keeping the same mean values.

If we compare the values of the EN 13465:2004 with the ones in the EN 15242:2007 there are two main differences: both values are not in homogeneous units -air changes per hour vs. m^3/h per square meter of exposed wall surface-; on the other hand the values of the first standard are referred to the air flow through opaque elements of the building envelope (cracks), while the values of the second one are referred to air infiltration through the whole envelope including opaque elements and window voids.

What we have done is to convert the former regulation values based on the EN13465:2004 into a value comparable to the values of the EN 15242:2007. In order to do that first of all we will multiply the original values with the ratio volume/area of exposed wall surface, thus we will obtain the infiltration through cracks in m^3/h per square meter of envelope surface. After this we will add to this air flow the infiltration through windows using the next equation:

$$Q[m^3/hm^2] = q_{4Pa}^{opaque} \cdot \frac{V}{A^{env}} + perm_{100Pa}^{window} \cdot \left(\frac{4}{100}\right)^{0.67} \cdot \frac{A_{window}}{A^{env}} \quad (1)$$

Where:

$q_{4Pa}^{opaque}[1/h]$: is the infiltration flow through opaque elements at 4 Pa. The normative value following the former regulation is 0.76 for single-family dwellings.
$perm_{100Pa}^{window}[m^3/hm_{window}^2]$: is the window permeability at 100 Pa.
$\frac{V}{A^{env}}[m^3/m^2]$: is the ratio volume/area of envelope surface.
$\frac{A_{window}}{A^{env}}[m_{window}^2/m^2]$: is the ratio window area/area of envelope surface.

Last two parameters depends on the geometry of the building, next tables give representative values of this for typical constructions in Spain.

Type of building	$\frac{V}{A_{cv}^{exp}}[m^3/m^2]$	$\frac{A_{vent}}{A_{cv}^{exp}}[m_{vent}^2/m^2]$
Single-family dwellings	2.13	0.093
Blocks of flats	3.56	0.138

Table 2. Mean values for the ratios in single-family dwellings and blocks of flats.

The resulting air flow Q is comparable to the EN 15242:2007 because now it includes the same infiltration flows and due to the congruency of unit system. Next table give the mean values obtained:

Type of building	$Q[m^3/hm^2]$
Single-family dwellings	2.01
Blocks of flats	2.72

Table 3. Mean values for air infiltration flow in single-family dwellings and blocks of flats.

The conclusion of the comparison is that these values are around the maximum value given by the European standard –for the highest level of leakage- that is 2 m^3/h per square meter of envelope surface. Thus we have assumed this figure as the new air infiltration flow in the new regulation. The permeability of opaque elements at 4 Pa that corresponds to this value is 1.6 m^3/h per square meter of envelope surface, this give 1.8 m^3/h per square meter of exposed wall surface –without window voids- for single-family dwellings and 1.95 in the same units for blocks of flats.

VENTILATION TECHNOLOGIES

The calculation methodology implemented for the new regulations (2011) allows calculating the outdoor air that gets into the building through the walls (infiltrations), through the windows (infiltrations) and through the specific vents designed for the ventilation purposes. Also the methodology allows following the path of the air assessing the air flow from the inlet zones to the corridor and from there to the exhaust zones (bath-rooms).
Thus this methodology is useful for calculating the next systems:

- Demand controlled ventilation: there are two levels of this technology:

- Basic: In this case the maximum air flow would be used when the bath-rooms are in use (3 hours a day), and the rest of the time the required air flow would be equal to the minimum.
- Advanced: In this case the air flow rate would be the necessary for the occupation hour by hour.

- Double-flux system, this system includes a duct network for the inlet and exhaust air flows, so it is possible to pre-heat the inlet air with the exhaust flow. This system includes the installation of a heat recovery unit as exchange unit.

- Combinations of demand controlled ventilation and heat recovery systems.

The technology required for the basic level of demand controlled ventilation consist on presence sensors in the bath-rooms, a double speed exhaust fan and a control device that connect the sensors with the fan. Figure 2 shows the supply air flow hour by hour.

The technology required for the advanced level consist on presence sensors in all the rooms, a multi-speed exhaust fan –it is possible to have more than 6 air flows depending on the number of rooms of the dwelling and the occupancy-, controlled air vents in all the inlet zones – bedrooms and living-room- and a advanced control device that connect all the elements of the system. Figure 2 shows the supply air flow hour by hour.

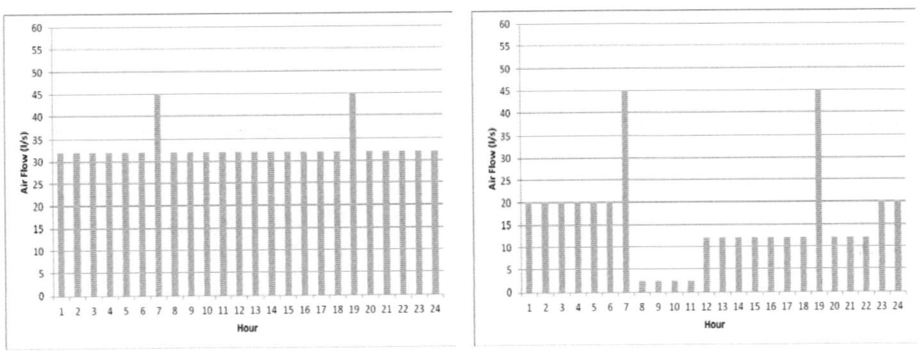

Figure 2. Left: Supply air flow schedule for the basic level of demand controlled ventilation. Right: Supply air flow schedule for the advanced level of demand controlled ventilation.

OTHER ENERGY SAVING MEASURES

Also the new regulation allows calculating the influence of the next energy saving measures in the energy demand:

- Use of self-regulating vents.
- Use of non-return self-regulating vents.
- Use of hygro-regulating vents.
- Improve of the air tightness of opaque elements.
- Improve of the air tightness of windows.

The market of the vents for ventilation is mature and has certain products that could help to improve the energy demand. The former regulation does not give any improved value if the designer chooses one of these products. The new regulation will change the situation giving simulation software that allows calculating the energy consumption and the indoor air quality when using different kinds of vents.

The air tightness of opaque elements is a default value that the designer could not change, but also if a special attention has been put in the construction using vapour barrier and membrane of air tightness for the walls, and using sealants products, silicones, foams or polyethylene tapes for the junction of the walls with the windows or doors openings, the designer could make a permeability test and use the result of this as a input for the software.

The air tightness of windows is a variable value that the designer can move to fulfil with the requirements of the Technical Building Code and also to improve the energy demand or IAQ results.

CONCLUSION

The basic documents of the Technical Building Code have to be updated periodically. These actualizations can be motivated for obsolete standards or for the need of contemplate and assess the effect of using certain technologies that previous versions do not deals with. This is the case of the basic document of energy limitation and it relation to the basic document regarding indoor air quality.

Present document shows how we have updated the references to derogated standards using new ones, and simultaneously keeping the congruence of the default values. Also we have used new simulation methodologies that allows to assess the effect of implement a higher number of ventilation technologies.

Basically we have produced a new regulation that assumes one indoor pressure and one outdoor pressure (calculation methodology EN 15242:2007). But keeping and indoor zoning that allows calculating (using mass conservation) the air flow between zones. The calculation methodology permits the possibility of using a control system allowing different flows each hour for a demand controlled ventilation scheme and pre-heating of supply air flow. And finally the new regulation is flexible to allow sizing the vents, using self-regulating vents or changing the permeability of opaque elements, for example if a blower-door test has been carried out. As a consequence the standard will not be a market barrier and promote the use of technologies of ventilation in order to improve the energy efficiency of buildings.

REFERENCES

[1] Building Ventilation, the state of the art, Santamouris M., Wouters P., Earthscan, London, Sterling, VA, 2006. Chapter 2 by Axley J.

[2] EN 13465:2004. Building ventilation. Methods for the assessment of the air flow in dwellings. European Committee for Normalization. 2004.

[3] Technical Building Code. Basic Document HE-1 Energy limitation. Basic Document HS-3 Ventilation and Indoor Air Quality. Spain Government. Ministry of Housing. 2007

[4] EN 15242:2007. Building ventilation. Methods for the assessment of the air flow rates in buildings, including infiltrations. European Committee for Normalization. 2007.

[5] Logiciel SIREN. Simulation de renouvellement d'air. CSTB.

[6] Reglementetion Thermique, Guide réglementaire. CSTB. 2010.

VENTILATION RATES AND IAQ IN EUROPEAN STANDARDS AND NATIONAL REGULATIONS

Nejc Brelih[*1], Olli Seppänen[1]

[1] *REHVA – Federation of European heating, ventilation and air conditioning associations*
Rue Washington 40, B-1050 Brussels, Belgium
**Corresponding author: nb@rehva.eu*

ABSTRACT

This paper presents some results from the Work Package 5 in the HealthVent project supported by the European Commission. One of the objectives of the project has been to review and critically evaluate the existing requirements on ventilation and IAQ defined in national building codes and European standards. The project's focus has been set on ventilation rates, pollutants, noise, temperature and draft in dwellings, offices, schools and kindergartens. This paper presents a summary of the values given in European regulations and results of comparisons. The up-to-date data in national legislation and codes were collected from 16 European countries with questionnaires, which were sent to project partners and trusted experts on ventilation. The requirements on ventilation rates were found to be given in different units, therefore test cases of real-life design situations were introduced to compare the data on a basis of a common unit. The results show that the ventilation rates given in the regulations are inconsistent and very heterogeneous. The ventilation rates in test cases range from 0.23 to 1.30 h^{-1} in dwellings and 4.2 to 41.7 l/s for local exhaust rates. The ventilation rates per person in test cases of classroom, playroom, and office range from 4 to 25 l/s. Big differences were also found in pollutant levels. Limit CO levels range from 3.0 to 12.5 mg/m^3 and formaldehyde levels from 10 to 100 $\mu g/m^3$. Minimum winter air temperature requirements were found in the range of 15 to 21°C, maximum summer air temperatures from 25 to 28°C, maximum air velocity in summer from 0.15 to 0.30 m/s and in winter from 0.15 to 0.25 m/s. Limit maximum noise levels were also found scattered, from 26 to 40 dB(A) in sleeping rooms and 30 to 50 dB(A) in other investigated room types. In conclusion, the evaluation of the data showed that values in the European local regulations, standards, and those practised locally, are very inconsistent. Moreover, several values in regulations were found to be looser than the recommended values published in European standards and WHO guidelines, thus allowing lower ventilation rates and higher pollutant levels than recommended. Results indicate that there is a considerable need on the European level to harmonize the ventilation and IAQ regulations and adjust them to the values provided in standards and guidelines.

KEYWORDS

ventilation rates, pollutants, indoor air quality, noise levels, thermal environment

INTRODUCTION

Every citizen has a right to indoor air quality that does not endanger his health [1]. Europeans spend on average over 90% of their time indoors – at home, in the office, in school, in kindergarten, etc. In order to assure a healthy and comfortable indoor environment and quality for all citizens as building occupants, the key parameters must be controlled and take into account air pollutants, thermal environment, and acoustic environment.

The building structure and materials as well as other sources in buildings contaminate the indoor air. Besides that, 20 to 100% of the concentrations of outdoor air pollutants are transferred to the inside of the building, adding to the pollution generated by the building itself. Considering the amount of time people spent inside and the concentrations of indoor pollutants, the buildings are the most important factor in air pollution exposure and associated health effects. Ventilation is used to bring outdoor air to the occupied indoor zone and to remove or dilute indoor-generated pollutants. Ventilation rate, as the flow of outdoor air to a space, is one of the most important factors affecting indoor air quality.

In order to achieve a sufficiently comfortable thermal environment in buildings, the following main physical parameters influence a person's sensation of warmth: air temperature, mean radiant temperature, relative air speed, and humidity. In practice, a combination of all listed parameters influences human perception of comfort. Temperature is usually the most important environmental variable affecting thermal comfort. A recommended range of temperature depends mostly on a person's activity and clothing.

An acoustic environment should be free of any unwanted sounds because it causes annoyance to the occupants. The unwanted sound is usually defined as noise. Ventilation and other mechanical systems in buildings must be designed so that the noise level does not cause annoyance to the building's occupants. In a building, a balance must be sought between the noise produced by the building services and the noise coming from the activities taking place within the building. For example, a higher level of noise from the building services may not be disturbing for the occupants in a space with a high level of activity noise. The criteria for the acceptable indoor noise are usually given as maximum A-weighted noise levels or as equivalent continuous A-weighted noise level.

In this paper, we present some of the results from the work performed under Work Package 5 in the HealthVent project[1], supported by the European Commission. The objective of the HealthVent project is to develop health-based ventilation guidelines for the EU. Members of the project group are experts from different disciplines from 9 European countries. One of the objectives of the project was to review and critically evaluate the existing requirements on ventilation and IAQ defined in building codes and European standards. The project's focus was set on ventilation rates, pollutants, noise, temperature and relative air movement in dwellings, offices, schools and kindergartens.

METHODS

The work focused on national regulations and practice in European countries. To overcome the language barriers and to collect data from as many countries as possible, the task of collecting data was performed with a special questionnaire, which was sent to project partners and trusted experts on ventilation in several European countries. The questionnaire comprised of 10 questions and sub-questions. The respondents were asked to provide values of ventilation rates, pollutant limits, noise levels, etc., which can be found in the national regulations. In case if no such values existed in the regulations, they were asked to provide values which are most widely used in practice (from standards, guidelines, etc.). In the responses they had to mark if the provided value is mandatory or voluntary to use.

The returned questionnaires revealed that the ventilation rate criteria are given using various units depending on a country, which do not allow direct comparisons. Criteria are expressed as flow rate per number of persons, flow rate per floor area, flow rate per number of rooms, fixed flow rate per room type, number of air changes per hour, or combination of different units. In order to compare ventilation rates criteria, we developed several test cases, which represent real-world situations. The test cases were developed for two different dwellings, a kitchen, a toilet, a bathroom, a school classroom, a kindergarten playroom, and an office. The

[1] HealthVent project website: www.healthvent.eu

details of the test cases are shown in Tables 1 and 2. Using the details from the tables, we compared the ventilation rates in dwellings on the base of air changes per hour, ventilation of kitchens, bathroom, and toilet as ventilation rate per room, and ventilation of classroom, playroom, and office as air flow rate per person.

Table 1. Properties of the test dwellings

Properties	Dwelling case 1	Dwelling case 2
floor area	50 m^2	90 m^2
ceiling height	2.5 m	2.5 m
main rooms	2: 1 living, 1 sleeping	4: 1 living, 1 sleeping, 2 children
kitchen	1 x 10 m^2 with window and electric stove	1 x 15 m^2 with window and electric stove
toilet	1 x 2 m^2	1 x 2 m^2
bathroom	1 x 5 m^2	1 x 5 m^2
persons	2	4

Table 2. Properties of the test rooms

Properties	Kitchen	Toilet	Bathroom	Classroom	Playroom	Office
area	10 m^2	2 m^2	5 m^2	50 m^2	50 m^2	12 m^2
ceiling height	2.5 m	2.5 m	2.5 m	2.8 m	2.8 m	2.8 m
persons	1	1	1	25	25	1

RESULTS

Respondents in 16 countries returned the questionnaire. A list of the countries and their abbreviations, which are used in the charts, are shown in Table 3.

Table 3. Country abbreviations used in charts

BG	Bulgaria	**DE**	Germany	**LT**	Lithuania	**PT**	Portugal
CZ	Czech Republic	**GR**	Greece	**NL**	Netherlands	**RO**	Romania
FI	Finland	**HU**	Hungary	**NO**	Norway	**SI**	Slovenia
FR	France	**IT**	Italy	**PL**	Poland	**UK**	United Kingdom

Ventilation rates

The results show that values are very inconsistent among European countries. Figures 1 and 2 show ventilation rates[2], which were calculated using the input data from Tables 1 and 2. The lowest ventilation rate in dwellings is 0.23 h^{-1} and the highest 1.3 h^{-1}. Large differences can also be seen in the cases of local exhaust rates, where the highest rates can be up to five times higher than the lowest rates. The ratio is therefore similar to the one of air changes in dwellings, where it is almost 1 to 6. Observing ventilation rates in the cases of classroom and playroom, one can distinguish two groups of countries with similar values. The first group has ventilation rates of around 10 l/s per person. It is formed by the following countries: Finland, Germany, Hungary, the Netherlands, Norway, Slovenia and UK. The second group has rates of around 4 l/s per person and is formed by the following countries: Bulgaria, Czech Republic, France, Greece, Italy, Lithuania, Poland and Romania. Both Nordic countries are in the group with higher ventilation rates, which is predominantly formed by the countries from the North and West of Europe. No countries from the Southern Europe are in that group. The ventilation rates in offices cannot be so obviously divided into two groups, because rates are much more scattered. Two ventilation rates that stand out are from Germany and Hungary, which are calculated according to EN 15251. The region-based conclusions are therefore not possible in the office case.

[2] Ventilation rate in this document is taken as the flow of outdoor air to space

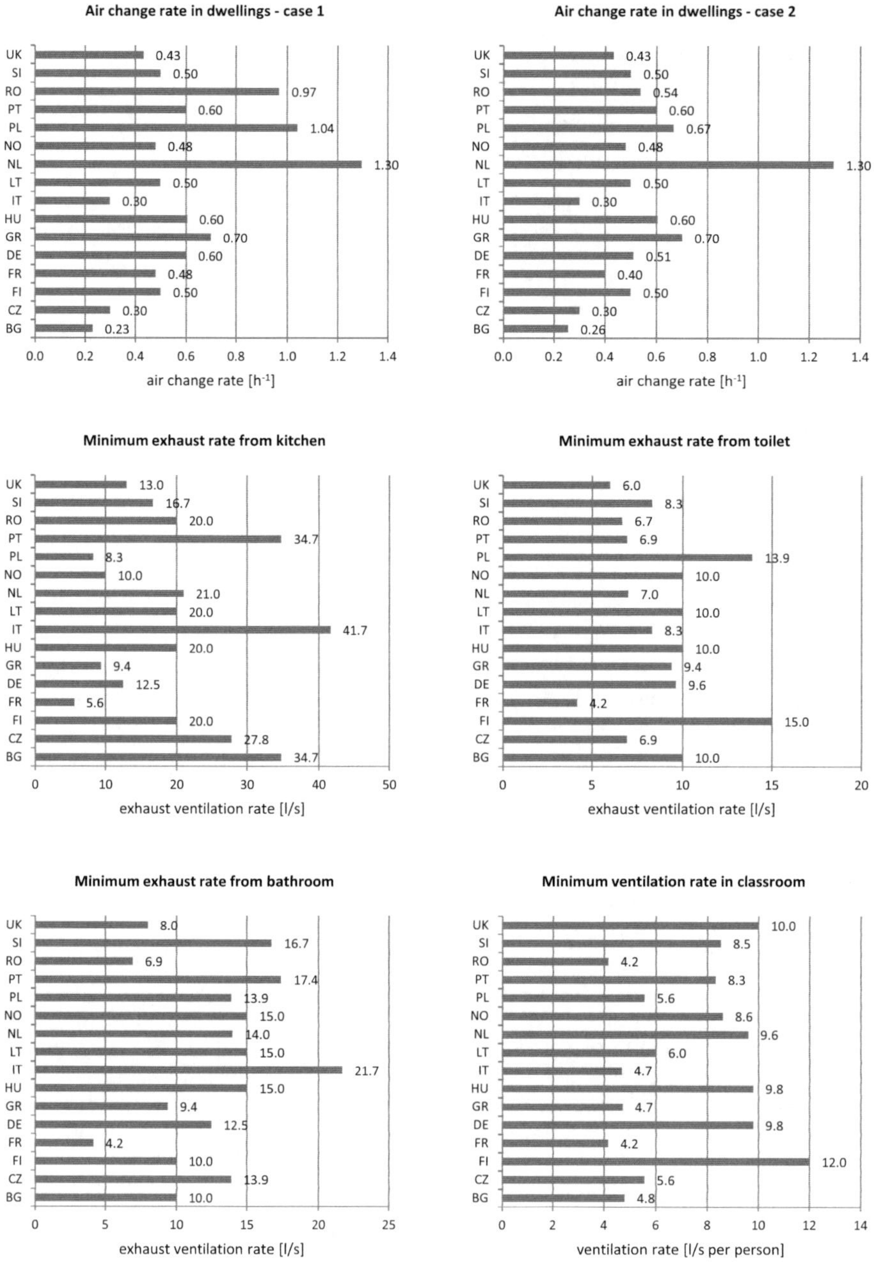

Figure 1. Comparison of ventilation rates in the test cases of dwellings, kitchen, toilet, bathroom and classroom

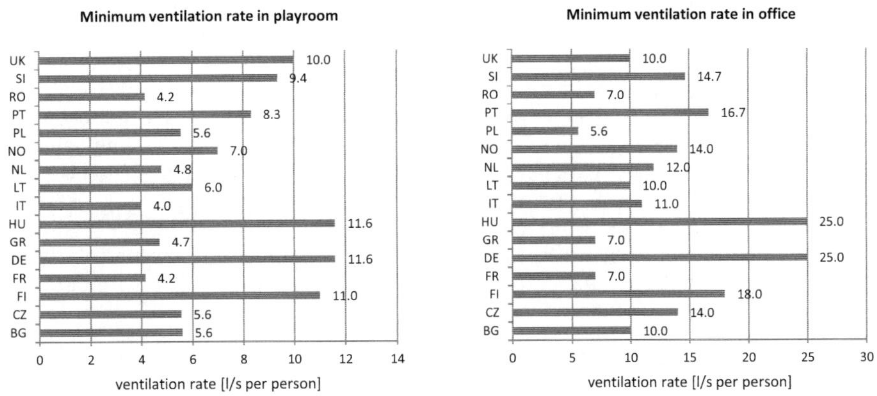

Figure 2. Comparison of ventilation rates in the test cases of kindergarten playroom and office

Indoor pollutants

The required limit levels of selected pollutants are shown in Table 4. The table also includes the WHO suggested values to serve as a comparison [2, 3]. The comparison of all values is difficult, because the limits are given as a maximum or average concentration in a given time. Only 6 out of 16 countries have requirements on limit indoor pollutant levels in non-industrial buildings. Limit levels of only two pollutants are found in the regulations of all the 6 countries: carbon monoxide (CO) and formaldehyde (HCOH). The range of CO limit levels is wide, from 3 to 12.5 mg/m^3. The WHO recommended limit is 7 mg/m^3, therefore the limit of 4 countries exceeds that value. Formaldehyde limit values range from 10 to 100 μg/m^3 and all values are equal to or below the WHO recommended value of 100 μg/m^3. Limit values of other pollutants are not included in the regulations of all countries and their ranges are also wide.

Table 4. Indoor pollutant limit levels

	WHO	FI	LT	NO	PT	RO	SI
Ammonia[μg/m^3]	-	20	40	-	-	-	50
Asbestos	-	0 fb/cm	0.1 mg/m^3	0.1 fb/cm	-	-	-
CO[mg/m^3]	7$^{\#2}$	8	3	10$^{\#5}$	12.5	6$^{\#3}$	10
CO_2[ppm]	-	1200	-	1000	1000	-	1670
Formaldehyde[μg/m^3]	100	50	10	100$^{\#3}$	100	35$^{\#3}$	100
NO_2[μg/m^3]	40	-	40	100$^{\#4}$	-	-	
Ozone[mg/m^3]	0.1$^{\#5}$	-	0.03	-	0.2	-	0.1
PM_{10}[μg/m^3]	20	50	50	-	150	-	100
Radon[Bg/m^3]	-	200$^{\#1}$	-	100	400	140$^{\#6}$	400
Styrene[μg/m^3]	-	1	2	-	-	-	-

#1 – annual average #3 – 30 min average #5 – 8 h average fb – fibre
#2 – daily maximum #4 – 1 h average #6 – instant max

Thermal environment

Figure 3 shows two charts that display the limit values of indoor temperature and relative air velocity in any type of the investigated room types. The temperature limits for summer vary from 25 to 28°C and in winter from 18 to 21°C. Temperature limits that are most commonly found in the regulations are 26°C for summer and 20°C for winter. The lowest low limits of air temperatures are imposed on kindergartens. In the UK, spaces can have a maximum air temperature of 28°C for maximum 1% of the annual occupied hours. Some countries do not have any requirement for limitation of maximum air temperature, but they all have limitation of minimum air temperature. Maximum relative air velocity limits are also inconsistent. They range in summer from 0.15 to 0.30 m/s and in winter from 0.15 to 0.25 m/s. In most of the countries the relative air velocity does not depend on the air temperature. The limit values of air humidity follow the pattern of temperatures and air velocities and are slightly more consistent. Limit levels of humidity are expressed as relative humidity (%) or absolute humidity (g/kg$_{dry\ air}$). Lower limits are constantly at 30% while higher limits are 70% in all cases except one, where it is 75%. The humidity level is given in terms of absolute humidity to limit the highest amount of water in the air and is in both cases the same, i.e. 12 g of water per one kg of dry air.

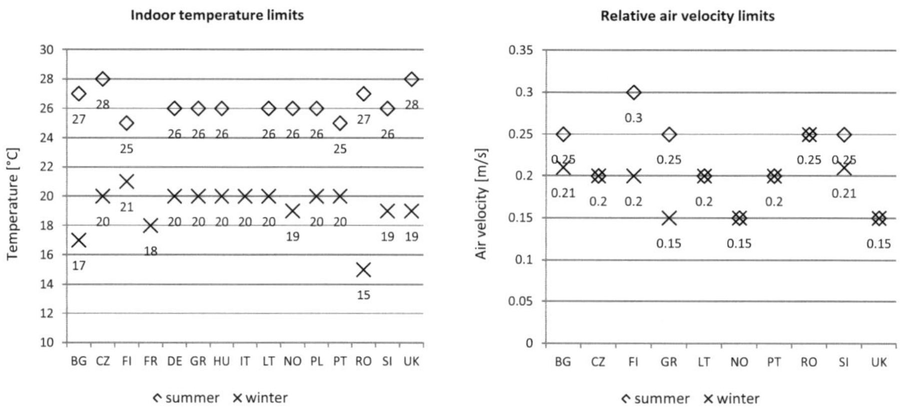

Figure 3. Indoor temperature and relative air velocity limit values

Noise environment

Charts on Figure 4 display limit noise values. Noise limits are expressed with three different criteria which are not directly comparable. The values are most often given as maximum A-weighted noise levels (L_{AFmax}), followed by equivalent continuous A-weighted noise level (L_{eq}), and least often as noise rating (NR) curves. The limit maximum noise levels range from 26 to 40 dB(A) in sleeping rooms and from 30 to 50 dB(A) in other investigated room types. The limit equivalent noise levels range from 28 to 35 dB(A) in sleeping rooms and from 25 to 45 dB(A) in other room types. It seems that in average, the equivalent levels are usually 5 dB lower than the instantaneous levels. Differences min – max are big in all cases, for maximum and equivalent levels. For comparison, European Standard EN 15251 [4] recommends the following limits of maximum noise level: living room 32 dB(A), bedroom 26 dB(A), small offices 35 dB(A), landscape offices 40 dB(A), classrooms 35 dB(A), playrooms 40 dB(A).

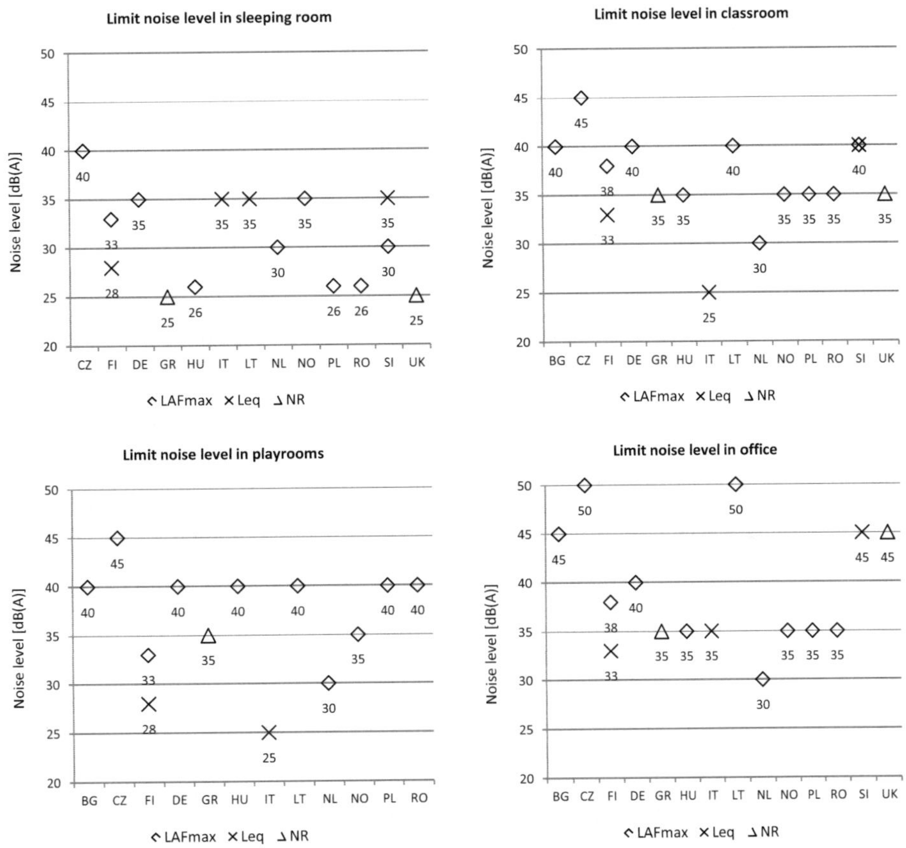

Figure 4. Noise limit levels in sleeping room, classroom, playroom and office

DISCUSSION

The data was collected from 16 countries from all parts of Europe, thus giving a good coverage of regions with different building practice and climate. Although the respondents are experts on ventilation, a certain measure of uncertainty exists regarding the accuracy of the data in the received questionnaires. Due to the language barriers, and limited resources, all data could not be verified. The data presented in this paper are informative and should not be used for the design of ventilation.

Different boundary conditions, which are used to calculate the ventilation rates, show that countries have taken different approaches to define them in regulations. This is further confirmed with the wide range of air change rates and ventilation rates, which were calculated with the data from the test cases. Such wide ranges suggest that countries did not have a common theoretical background for the determination of the required ventilation rates. Approximately one third of countries have requirements for the ventilation of dwellings, which result in air change rate lower than 0.5 h^{-1}.That is in contrast with the health-based recommendations of minimum air change rate of 0.5 h^{-1} [5]. The ventilation rates in classrooms, playrooms and offices are also in contrast with health-based recommendations, because the resulting ventilation rates are often below 10 l/s per person. In the extensive review of studies that investigated the association of ventilation rates with human responses, Seppänen et al. [6] showed that almost all the studies included in the review found that the ventilation rates below 10 l/s per person had been associated with a significantly worse prevalence or value of one or more health perceived air quality outcomes. This is further confirmed by Sundell et al. [5], who in the review of literature shows that the ventilation rates up to 25 l/s per person are associated with reduced adverse health symptoms, and that the number of symptoms increases with lower ventilation rates.

The limit levels of pollutants are often higher than those recommended by the WHO, and missing in the regulations of several countries. The ranges of values are wide, which indicates that the countries do not use common theoretical background to determine the limit values. Minimum requirements for pollutant levels in non-industrial buildings should be included into the regulations of all European countries.

The temperature limit of the lowest winter temperature of 15°C is too low from the health and comfort point of view. The summer maximum temperature limit of 28°C is too high from the perspective of performance, which is higher if the temperatures are lower. All regulations should include temperature limits, which respect comfort and productivity aspects, and may be adapted to local climate conditions. The requirements for relative air velocity usually do not take into account the air temperature. Dissatisfaction due to draught is not only a function of mean air speed, but also of local air temperature and fluctuations of air velocity. The regulations should therefore limit maximum values of air velocity in relation to air temperature.

The noise limit levels are higher than the recommended by the European standards in the majority of the countries. The limits for bedrooms are particularly critical, and significantly too high in almost all the countries. The noise criterion, expressed as equivalent noise level, is more appropriate for the industrial environment than for the environment of non-industrial buildings. If given, it should be supplemented with the limit maximum noise level. The noise limit levels should be harmonized across the national regulations. The maximum noise level should be preferably used as a criterion, supplemented by the equivalent levels if necessary.

CONCLUSION

A review of the European regulations for ventilation rates, indoor air quality and noise showed that the values in regulations are inconsistent and vary greatly according to country. Almost all of the regulated parameters included in the review are already defined in European Standards, which were accepted in the CEN voting process by national bodies. Nevertheless, the values found in the standards and those in the regulations are not harmonized. The inconsistency on the national level between the EN standards and regulations, as well as on the European level from country to country, causes problems to designers and industry, and increases the construction cost. Besides that, the current practice is in contrast with the efforts of unification and standardization of the European common market. Clearly, a common European guideline is needed, which would serve as a basis for national European regulations. The guideline should include ventilation rates, technical properties, and other parameters related to the performance of ventilation.

ACKNOWLEDGEMENTS

The HealthVent project was sponsored by the Executive Agency for Health and Consumers (EAHC) through the grant agreement n° 2009 12 08. The EAHC implements the EU Health Programme, the Consumer Programme and the Better Training for Safer Food initiative.

REFERENCES

[1] WHO Regional Office for Europe Copenhagen, "The Right to Healthy Indoor Air. Report on a WHO Meeting," Bilthoven, 2000.

[2] WHO, "Air quality guidelines. Global update 2005. Particulate matter, ozone, nitrogen dioxide and sulphur dioxide," WHO Regional Office for Europe, Copenhagen, 2006. [Online]. http://www.euro.who.int/__data/assets/pdf_file/0005/78638/E90038.pdf

[3] WHO Regional Office for Europe. (2010) WHO guidelines for indoor air quality: selected pollutants. [Online]. http://www.euro.who.int/__data/assets/pdf_file/0009/128169/e94535.pdf

[4] European Committee for Standardization, "EN 15251: 2007. Indoor environmental input parameters for design and assessment of energy performance of buildings addressing indoor air quality, thermal environment, lighting and acoustics," CEN, 2007.

[5] J. Sundell et al., "Ventilation rates and health: multidisciplinary review of the scientific literature," Indoor Air, vol. 21, no. 3, pp. 191-204, June 2011.

[6] O. Seppänen, W. J. Fisk, and M. J. Mendell, "Association of Ventilation Rates and CO2 Concentrations with Health and Other Responses in Commercial and Institutional Buildings," Indoor Air, vol. 9, no. 4, pp. 226-252, December 1999.

WINDOW OPENING IN HIGH PERFORMANCE BUILDINGS

Piet Jacobs*, Hans Phaff, Bas Knoll, Wim Kornaat, Wouter Borsboom

TNO, Department of Energy and Comfort Systems, van Mourik Broekmanweg 6, the Netherlands
**piet.jacobs@tno.nl*

ABSTRACT
In a typical new housing estate the window use of 64 dwellings have been monitored during a calender year. Questionnaires have been spread and interviews have been held to assess the use of the mechanical ventilation system and the heating energy consumption. Above 2 °C the window opening data show a nearly linear relation with the outside temperature. The heating energy consumption of the town houses shows a moderate correlation (correlation coefficient: 0,25 – 0,34) with the window opening behaviour. On the average the open windows implied an additional gas consumption of about 100 m^3 for the 3 storey town houses and about 300m^3 for the 4 storey houses. In case of the corner house almost no correlation between window opening and heating energy consumption has been found. For this deviating behaviour no explanation has been found.

KEYWORDS
Window opening, balanced ventilation, energy, user behaviour

INTRODUCTION
In the Netherlands energy requirements have stimulated the implementation of balanced ventilation with heat recovery. Up till now about 400.000 dwellings have been built with this system. Especially in high performance buildings, occupant behaviour has a large effect on the indoor climate and energy consumption of dwellings. Research by Santin [1] revealed that only 18-22% of the variation in energy consumption in recently built Dutch housing can be explained with the building characteristics that are used for the Energy Performance Coefficient (EPC). According to Santin two factors explain the remaining variation in energy consumption: actual quality of the construction and actual occupant behaviour. Maybe the evaporation of building moist plays a role too, especially in the first year. Between energy consumption and the use of a mechanical ventilation system Santin found a low correlation. Window and grille ventilation seemed to have a stronger effect on energy consumption than the mechanical ventilation. Andersen [2] has studied the behaviour of occupants with regard to window opening and concluded that the outdoor temperature, the indoor temperature and the indoor CO_2 concentration proved to be the three most important variables in determining the probability of opening and closing a window.
This paper is focussed on how window use in dwellings is connected to the outdoor temperature and whether an effect on energy consumption can be distinguished.

EXPERIMENTAL METHOD
Window opening and natural gas consumption was observed in 64 dwellings, which all were located in the same street in Delft, the Netherlands. The houses were completed in three shifts and were populated in between June 2006 and August 2007. The houses were built in blocks of 8 houses each, of which 2 houses contain four floors (total about 185 m^2) and the other six contain three floors (total about 140 m^2). The houses have almost the same outer dimensions; however the even and the odd numbers were built by different builders.

Description of the houses

The three floor houses complied with an energetic achievement (EPC) of 0,74. For a typical household this would imply a natural gas consumption of about 1100 m^3 per year for the total of cooking (50 m^3), hot tap water (555 m^3) and room heating (508 m^3). The houses are equipped with a balanced ventilation system with heat recovery. The capacity of the ventilation can be controlled with a three way switch. In position 3, with the highest flow, the legal ventilation capacity requirements are just met. However in this position almost all houses suffer from noise generated by the mechanical ventilation system [3]. Table 1 shows the typically amount of operable windows and doors per dwelling. During the years in a number of dwellings extra dormers and roof windows have been placed.

Table 1. Typical amount of operable windows.

	Doors	Windows	roof window
Ground floor	2	1	
First floor		4	
Second floor (some)		4	
Attic		2 – 4	0 - 1

All 64 dwellings use the same tilt and turn windows as shown in Figure 1, 87,5 cm wide and 134 cm high. Tilted, the gap on the upper side is 12 cm. The overhang is 1 m wide and provides good protection against rain. In the summer it provides shading to prevent overheating on the first floor. Windows which are easy accessible are provided with security lock on window knobs (not shown on picture).

Figure 1. tilted window on first floor under the overhang.

Observation of window opening

The window opening was assessed by visual observation from street level. During the period 3 February 2009 – 7 February 2010 each week on Tuesday, between 16.40 and 18.15 hours and on Sunday, between 12.45 and 16.00 hours, these observations were made. A typical inspection round took about 15 minutes. During school holidays no observations have been made as many families were not at home. In total 49 observations were made. Both the front side as on the back side of the houses the windows were observed. How far the windows were opened was not logged. However in general it can be said that during the winter most open windows were tilted. Before and after the visual observations the outside temperature was recorded. The temperature shown in the graphs in this paper is the average of these two temperatures. Due to the daily outside temperature cycle this instantaneous temperature differs from the maximum, minimum and average temperature.

Questionnaire gas consumption and interviews

In september 2010 all residents got a standardized letter in which they could fill in the gas meter position and the date of first occupation of the dwelling. To increase the response the letter contained information about the goal of the investigation and some feedback of the window opening observations. As only 26 of 64 residents returned the letter it has been decided to keep interviews with nearly all other residents. The interviews, which were held between September 2010 and December 2010, increased the gas meter position response to 61. The interviews, 5 to 10 minutes each, were also used to obtain additional information concerning the consumption of natural gas for e.g. cooking, the number of persons present and the use of the balanced ventilation system.

RESULTS

Window opening

The results of the window opening observations are shown in Figure 2. The percentage dwellings with at least one window open is fairly linear with the momentary outside temperature. On Sunday in 5 to 10%point more dwellings the windows are opened compared to Tuesday. The total number of open windows is also fairly linear with the outside temperature. However it seems that at low temperatures, below 2 ºC, the number of open windows remains constant.

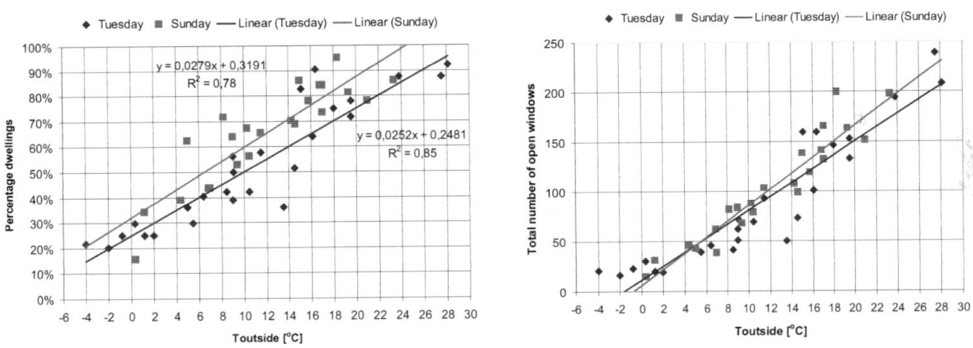

Figure 2. Left: percentage dwellings with one or more windows open; Right: total number of open windows.

Gas consumption

Gas consumption data were obtained from 61 of the 64 dwellings. The gas consumption is averaged over the period between population date and the response date, which varies between 3,1 and 4,5 year. Figure 3, split up in the three different type of dwellings present, shows the consumption of natural gas plotted against the window use. The left diagrams show the counted number of open windows during the whole year, while on the right the window opening data below 12 ºC is shown. This temperature has chosen as it has been observed that below this temperature the central heating is being used. The results of the interviews are summarized in Table 2.

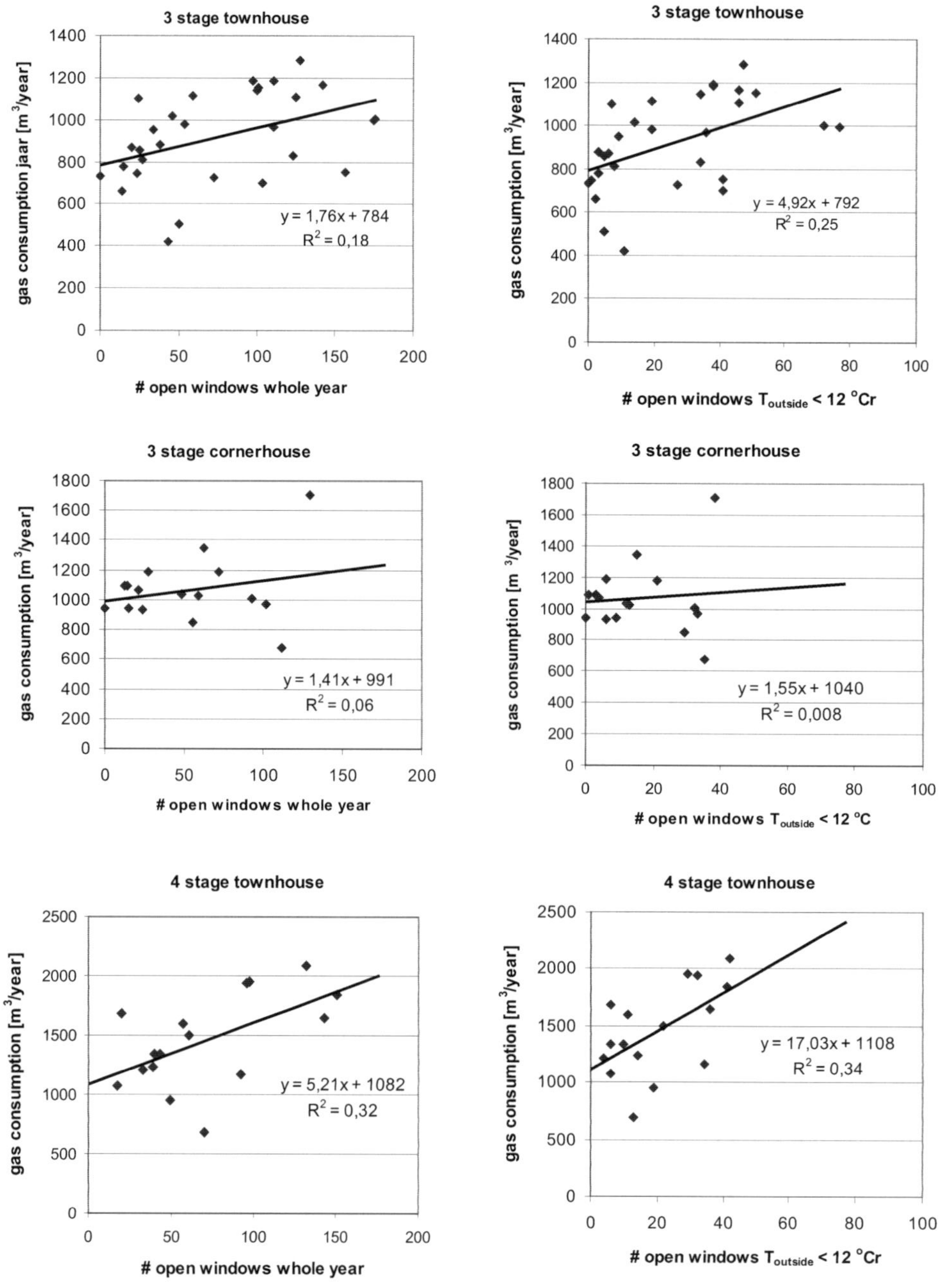

Figure 3. Gas consumption split up in the three different types of dwellings and in two temperature regimes. The number of open windows during the whole year is based on 49 surveys, below 12 °C on 27 surveys.

Table 2. Results of interviews.

# observations	Question	result
	Average gas consumption:	
29	- 3 stage townhouse	918 m^3/year
16	- 3 stage cornerhouse	1065 m^3/year
16	- 4 stage townhouse	1454 m^3/year
35	Percentage cooking on natural gas	94%
45	Door between living room and stairs open	27%
	Average number of persons per dwelling:	
25	- 3 stage townhouse	3,44
15	- 3 stage cornerhouse	3,37
14	- 4 stage townhouse	4,33
30 (total)	Ventilation unit turned off	13%
	Always position 1	27%
	Position 1 and sometimes position 2	20%
	Position 1 and somethimes position 3	30%
	Nearly always position 2	10%
	Nearly always position 3	0%

Figure 4 shows the gas consumption plotted against the number of persons present per dwelling. In the figures linear trendlines and corresponding algebraic relations and correlation coefficients are given.

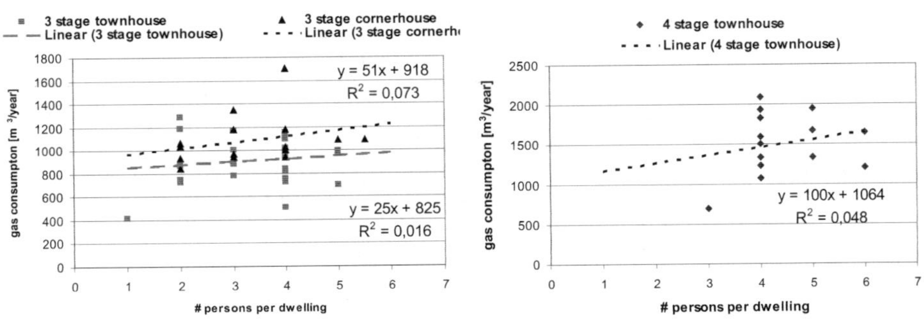

Figure 4. Gas consumption versus the number of persons present.

DISCUSSION

Above 2 °C the window opening data show a nearly linear relation with the outside temperature. The correlation coefficients (R^2) are relative high: 0,78 – 0,85. As the inside temperature is strongly connected to the outside temperature it is conceivable that the window opening is governed by the inside temperature. If the inside temperature becomes too high the occupants open the window. Aside inside and outside temperature according to [2] also the indoor CO_2 level plays an important role in window opening. As the mechanical ventilation system in the 64 dwelling generates high noise levels, up to 40 – 47 dB(a) in the rooms and up to 54 dB(a) on the attic where the ventilation units are located, the ventilation system is in some cases turned off and in most cases only the lowest position is being used. This might be one of the reasons why the occupants for a large part rely on window airing. During winter time probably a trade off is made by the occupants between higher energy consumption and a fresh indoor environment.

Gas consumption

The highest gasconsumption is five times higher than the lowest gas consumption. In general large variations can be seen. In the Energy Performance Coefficient (EPC) the gas consumption is split up in to three groups: cooking (50 m^3), hot tap water (555 m^3) and room heating (508 m^3). Cooking involves a relatively small amount of gas and can be zero in case of electrical cooking. This is the case for 6% of the households. Therefore cooking is not considered to have a large effect. According to the EPC hot tap water preparation has a large effect on the gas consumption. One would expect a strong relation, a high correlation coefficient, with the number of persons present per dwelling. However in this study this can not be confirmed. Probably this is caused by the amount of small children. Room heating can have a larger effect on the variation in gas consumption. An important parameter probably will be the preference for the temperature setpoint in the dwellings. Other occupant behavioiur such as how many hours a week people are present and how much electricity is used will be relevant. However these parameters have not been recorded and can therefore not be taken into account in this study. Another factor is the sun expostion of the dwelling in the heating season. An estimate has been made for each dwelling. This analysis showed a very small correlation coefficient of 0,0004 in combination with a gas saving of 23 m^3 for an optimal exposition. A more important factor, although also with a low correlation coefficient of 0,05, is whether the door between living room and stairs is open in the heating season, an open door increases the gas consumption with 173 m^3.

Among the measured parameters window opening has the largest effect on gas consumption. In case of the town houses the heating energy consumption shows a moderate correlation with the window opening behaviour. The correlation coefficients (R^2) are in the range 0,18 – 0,32 for the number of open windows during the whole year, up to 0,25 – 0,34 for the number of open windows below 12 °C. This indicates that if the occupants are used to open the windows during higher outside temperatures, they still show this behaviour when the outside temperatures are below the heating limit. In case of the corner houses the correlation coefficient is relative low, especially for the number of open windows below 12 °C. For this deviating behaviour no explanation has been found. From the increment in figure 3 it can be derived that each extra counted open window during winter time in case of a 3 storey town house increases the gas consumption with almost 5 m^3. On the average 26 open windows have been counted, this implies an additional gas consumption of about 100 m^3. In case of 4 storey houses the increment is 17 m^3/counted window. On the average 16 open windows have been counted, this implies an additional gas consumption of about 300 m^3.

CONCLUSION

This paper is focussed on how window use in dwellings is connected to the outdoor temperature and whether an effect on energy consumption can be distinguished. Above 2 °C the window opening data show above a nearly linear relation with the outside temperature, with relatively high correlation coefficients. For the town house moderate correlation with the energy use has been found. However for the corner houses this correlation is nearly absent. As this can also be caused by the relatively small sample further research is recommended.

REFERENCES

[1] Santin O.G, Actual energy consumption in dwellings, the effect of energy performance regulations and occupant behaviour, October 2010, *thesis Technical University Delft.*

[2] Andersen R.V., Modelling window opening behaviour in Danish dwellings, 2011, *Proceedings Indoor Air 2011 (Austin).*

[3] Jacobs P, 2008. *Praktijkvoorbeeld van balansventilatie met WTW*, Intech K&S, March, 48-51 (in Dutch).

DUCTWORK AIRTIGHTNESS REQUIREMENTS IN PORTUGAL

Eduardo Maldonado*[1] and Fernando Brito[2]

1 Universidade do Porto
Faculdade de Engenharia, R. Dr. Roberto Frias
4200-465, Porto, Portugal
**Corresponding author: ebm@fe.up.pt*

2 APIRAC
Port. Assoc. of HVAC Industries
R. do Alecrim, 53 -2°
1200-014 Lisboa, Portugal

ABSTRACT

Portugal introduced, for the first time, in the 2006 Building Regulations, a requirement on the airtightness of the ductwork in new HVAC installations. A test is required during commissioning. Data on compliance is however still quite scarce to conclude how effective this requirement is in practice.

KEYWORDS

Duct airtightness; HVAC regulations.

INTRODUCTION

As part of the transposition of the EU Directive 2002/91/EC, on the Energy performance of Buildings (EPBD) [1], new regulations were adopted in Portugal and came into force in 2006. Requirements for new HVAC systems [2] included for the first time a set of mandatory tests that must be carried out during commissioning, before the building receives its use permit. The aim of the tests is to demonstrate that the installation is functioning as designed, in operational terms, but also meeting the minimum energy efficiency and indoor air quality (IAQ) targets set in the legislation.

Proof of the results of these tests, consisting of a detailed report, must be handed to the Qualified Expert (QE) who will issue the Energy Performance Certificate (EPC) for the building, who may ask for further tests if he/she is not satisfied with the report or just for confirmation (random check). Often, the QE is present while the commissioning tests take place. The EPC is required by the local authorities before issuing the building's use permit.

Tests on the ventilation system include:

- Airflow delivered to each room in accordance with design parameters;
- Overall cleanliness of the whole ductwork and other components, such as air handling units and fans;
- Airtightness of the ductwork.

To pass the test on airtightness, ductwork leakage may not exceed 1.5 $l/s.m^2$ under a static pressure of 400 Pa. Airtightness tests should be carried out using the following procedure:

- A 10% random sample of the ductwork is selected and tested. If the measured leakage is below 1.5 $l/s.m^2$, no further testing is required;
- If the first test is not satisfactory, a second test is performed, after the contractor takes corrective measures, again on the initially tested ducts plus an additional randomly selected 20% of the ductwork. If these tests are satisfactory, no further testing is required.

- If the previous test is still unsuccessful, the contractor must take additional corrective measures and the final test(s) must cover the whole ductwork until the required airtightness is met.

MOTIVATION

Up until 2006, there was no check on the quality of the ductwork (most often, building owners did not require the check simply to avoid its cost), and its performance was in general quite poor (high leakage, cheap materials), resulting in significant losses, with consequences in terms of the energy efficiency of the whole installation (more air circulated and treated to compensate for the leakage). Moreover, it was often impossible to meet the minimum fresh air rates in many spaces, resulting in degraded IAQ levels. The new regulation aims at ensuring minimum levels of IAQ and improved energy efficiency during operation of the building, by adopting a life-cycle perspective and moving away from the up-to-then prevailing strategy of lowest possible first cost.

THE NEW REGULATIONS IN ACTION

The new regulations apply to buildings larger than 1000 m^2 that begun their licensing procedure after 2006. Taking into account design and construction, this cycle usually takes, for large buildings, at least 3-4 years before completion. Therefore, there are not yet much data on the success of the new regulations. The first large buildings that had to comply with these new regulations only finished the construction phase late in 2009 and during 2010.

However, there is proof that the market adapted to the regulations. The share of pre-fabricated round ductwork with quality seals between ductwork components increased significantly (from <5% in 2006 to 30% in 2010). For rectangular ducts, the technology evolved to achieve better seals along duct sections and at unions between two consecutive sections, namely at the corners, representing now 20% of the market (extraction ducts carrying air that is not recirculated, e.g., from toilets and wet-zones, are still usually low-quality ducts). Welded and screwed joists disappeared since then. In parallel, "a dozen" specialized companies now offer duct leakage testing services in the market (there were none in 2006).

Although only few EPCs have been issued for large new non-residential buildings so far, there is anecdotal evidence that the required commissioning tests (not just ductwork leakage) resulted, in most cases, in significant delays to the construction phase, with the corresponding negative backlash.

CONCLUSION

It is too early to say if the new regulations have been successful (the number of completed new HVAC installations falling under the new requirements is still rather small) and there are no data regarding the actual performance of the few buildings constructed with the new requirements. But ductwork technology evolved, with better quality components now much more used, and ductwork leakage testing, as well as ductwork cleaning, are now new niche markets that appeared since the new regulations entered into force.

REFERENCES

[1] Directive 2002/91/EC, on the Energy performance of Buildings (EPBD), EU Official Journal, 16 December 2002

[2] RSECE – Regulation on HVAC systems for energy efficiency in buildings, Portuguese Law 79/2006, Portugal Official Journal, April 6, 2006.

HANDS-ON TRAINING COURSES FOR VENTILATION SYSTEMS INSTALLERS WITHIN THE PRAXIBAT INITIATIVE

Anne-Marie Bernard*[1], Roselyne Forestier[2]

1 ALLIE'AIR
4 clos Ballet
01800 Meximieux, France
**annemarie.bernard@allieair.fr*

2 ADEME
Direction Communication et Formation
BP 90406
49004 Angers cedex 01
roselyne.forestier@ademe.fr

ABSTRACT

This paper gives a quick overview of the PRAXIBAT programme led by the French Agency for Energy and Environment to improve building professionals skills to achieve low- or nearly zero-energy buildings. It looks more specifically at the initiatives targeted at ventilation systems installers, with practice-oriented trainings in quasi-real conditions.

KEYWORDS
Ventilation systems, installation, ductwork airtightness

PRAXIBAT CONTEXT

PRAXIBAT is part of the French national framework Grenelle. This initiative aims at implementing the national commission who made the recommendations for building area by training building professions in order to improve their skill.

PRAXIBAT will supply all trainees from the building area (school students, professional training, installers and fitters) studying facilities in 7 main areas:

- energy performance of walls and envelope (inc. air tightness);
- ventilation;
- lighting;
- wood (energy);
- solar (thermal) ;
- solar (PV);
- Heat Pumps.

Within these facilities all trainees should get enough equipment to be able to practice and install as in real situation.

Each administrative area is now preparing the facilities at regional level. In Rhone-Alpes for instance (administrative area near Lyon), almost 20 teaching institutes (inc. highschools, Apprentice schools, private and national training organisms) will get funding to implement the action in their facilities. The total number of funded institutes is not known yet but should be around 150 to 200 for all over France.

OUTLINE OF TRAINING SCHEMES FOR VENTILATION INSTALLERS

The training programme for ventilation systems installers includes:

- Single house equipment (single exhaust and supply and exhaust)
- Ducts and accessories
- Outlets and air inlets
- Measurement equipment (pressure and anemometer)
- Visualization camera
- Documentation
- Access to reference documentation and information
- …

At National level, we prepare the training facilities including:

- Kit for final student (slides, teacher's guide, films, samples, guidelines…)
- The training of reference teachers who will then train themselves in all administrative areas

The training is focused on installation as:

- Today we still have 40% of non compliant installations in new building, which is mainly due to installation difficulties, even with single exhaust systems
- Difficulties with supply and exhaust systems because they need more ducts

The trainees will discover ventilation systems directly in situation by mounting and unmounting a house ventilation system and its components. Installation of ducts and airtightness is an important part of the sessions, including:

- the different kind of ducts and tightness methods,
- how to use mastic and tapes,
- how to connect ducts with seals,
- checking tightness with fumes,
- learning with a video how to proceed with a duct airtightness test….

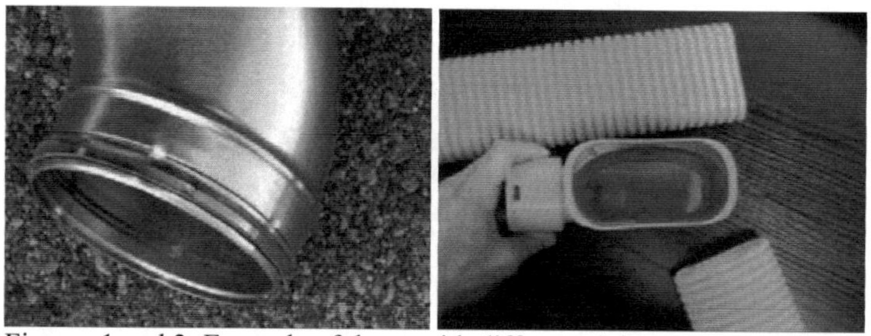

Figures 1 and 2: Example of ducts with different tightness seals

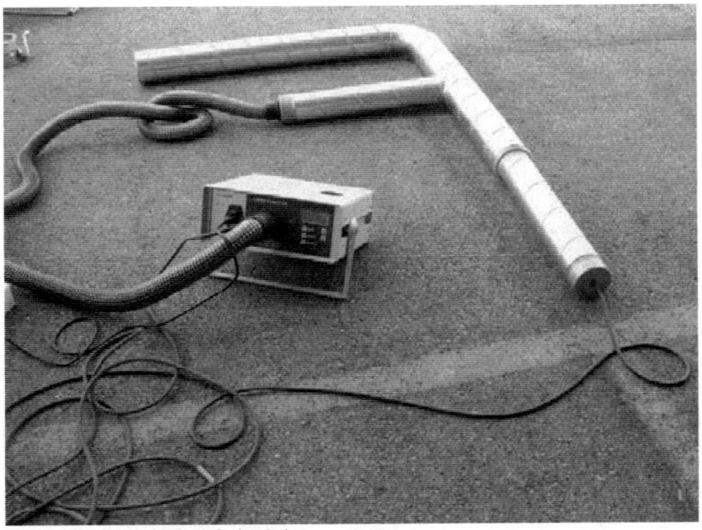
Figure 3 : Example of air tightness test

CONCLUSION AND FUTURE WORK

Although the principle of the training is not new, its originality is to teach only by practical aspect. Students discover the question in situation by practice. They meet afterward to capitalize the teaching and memorise the main points seen on site. But most of the learning is practice-oriented and therefore well adapted to fitters.

The implementation of this system in all areas of France in a large number of institutes should allow to an improvement of practice to fit the objectives and avoid too many references of poor practice on site.

FEASIBILITY STUDY OF VENTILATION SYSTEM AIR-TIGHTNESS

Jeroen Soenens *[1] and Pedro Pattijn[1]

1 Ingenium
Nieuwe St-Annadreef 23
8200 Brugge, Belgium
**Corresponding author: Jeroen.soenens @ingenium.be*

ABSTRACT

The feasibility of good air-tightness in new buildings can be determined based on the obtained air tightness classes as defined in EN 12237. In this paper a model is described which allows to calculate the energy loss caused by leak losses in ventilation systems based on the air tightness class and the feasibility of realising a good air-tightness.

The air-tightness of the ventilation system is building specific as it is preliminary determined by the lay-out of the system and thus by the air-tightness of the used components. Therefore an inventory of the ventilation system components is used to calculate the air losses. If no inventory is available, the model is able to estimate the losses based on square meters floor space and building type.

The total energy cost related to the preconditioning of supply air and the transportation of supply and exhaust air depends on a number of parameters. Some of these parameters are climate-related and therefore fixed, some are case-specific and can be altered by the user (e.g. internal humidity requirements, working hours, type of heat recovery, unity cost energy). The air leak losses have the same unity-cost for transportation and preconditioning as the wanted air.

In this paper the method is applied on three cases: a hospital wing, a rest home and an office building. The simulations show that the total energy consumption related to ventilation can be reduced by over 30% by achieving an airtight ventilation system. Good air-tightness of ventilation systems offers real added value as obtaining good air-tightness lower energy costs and a better indoor comfort. The additional investments to achieve a good air tightness of the ventilation system in new buildings are low compared to the avoided energy losses.

KEYWORDS

ductwork, airtightness, leakage, feasibility, cases

INTRODUCTION

A ventilation system has an important technical function. On the one hand the system moves treated supply air to occupied rooms. On the other hand, used air is extracted from these rooms and the extract air is discharged outside. The ventilation of rooms is required by the (Belgian) energy performance assessment legislation [1] for (new) buildings and also by the (Belgian) ARAB labour legislation.

The energy cost related to the transportation (fan power) and treatment (filtering, heating, cooling, dehumidifying and humidifying) of air is significant compared to the total energy cost when operating a building. Keeping air quality within comfort levels is the main goal of the ventilation system. Uncontrolled air flow losses may lead to unbalances in air pressure in the building, resulting in unwanted infiltration. Therefore, the effectiveness of the applied system is increasingly important in the design of buildings with a low carbon footprint. The desired amount of air has to be transported to the right rooms effectively and in a controlled way. As little ventilation air as possible must be lost. Keeping ventilation air losses to a strict minimum is thus mandatory.

It is important to emphasise that the air-tightness of the ventilation system is determined by the air-tightness of each component. This includes the air ducts themselves, but also all the accessories such as fire dampers, flow-balancing units, silencers and the connections between the elements. The air-tightness of a ductwork is described and quantified in different European standards [2][3][4]. The air-tightness class determines the size of the air leak: air-tightness class C or D indicates a very performing ventilation system, class A or poorer are systems with low air-tightness.

In order to increase the air-tightness class, a ventilation system has to become three times more performant. The leakage flow rate in a type C ductwork is thus three times lower than the leakage flow rate of a type B ductwork. Very poor systems are classified as 3A, 9A, 27A, etc..

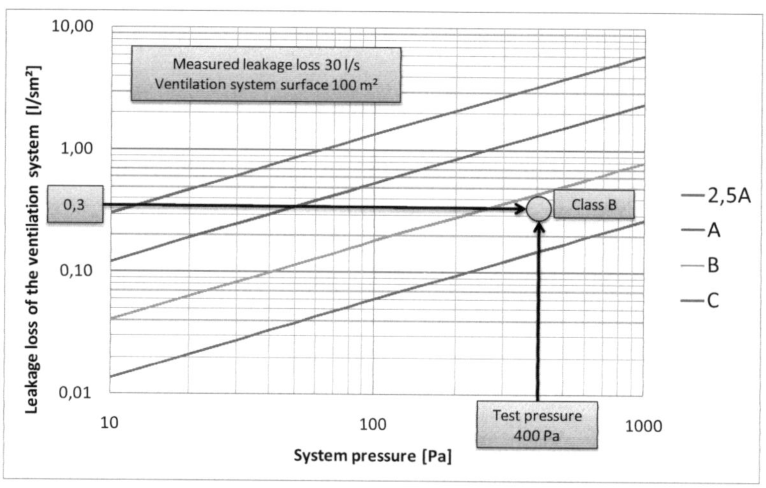

Figure 1. Example of air-tightness class determination, Ingenium, measurement 2010

BELGIUM IS LAGGING BEHIND

Standard Specifications 105 published by the Belgian Buildings Authority [5] is the primary work of reference for designing and installing ventilation systems in Belgium. This report recognises the importance of good air-tightness of the ventilation system as the specifications impose a minimum air-tightness class. The current specifications require class A or B. After revision, the Standard Specifications will impose at least air-tightness class C [6].

In practice, however, it tends to be exceptional that representative air-tightness measurements are carried out. It is thus hardly surprising that the actual air-tightness in Belgium is rather poor. Measurements carried out during the SAVE-DUCT project confirm that the actual air-tightness in ventilation systems in Belgium does not meet the minimum requirements of the Standard Specifications. In many buildings, the air-tightness of the ventilation system is three to nine times worse than class A. Belgian buildings obtain far worse reults than comparable buildings in countries as Sweden [7].

The European standard EN15242 [8] indicates 2,5A as the default value for the air-tightness of a ventilation system. Measurements [7] show that this figure is actually still too optimistic for the current stock of Belgian buildings. In real cases measurements up to 27A are not exceptional.

CALCULATION METHOD

The total airflow is calculated first, both for supply air and extract air. The total air flow is the sum of the wanted air flow and the air losses due to bad air-tightness of the ventilation system. The wanted air is in most buildings well defined and known. The air losses are calculated based on the duct surface, an air-tightness class and a system pressure. AHU casing leakage is not yet included in the model. Default values, based on building type and building surface, are available in the calculation model if no detailed info on the duct system is available.

The model calculates the unity cost to treat the total airflow. This cost, related to the transportation and treatment of air is calculated in five 'energy modules', as shown in figure 2. Two modules calculate the transportation: supply fan, exhaust fan, or both. Three modules calculate the required energy to treat the outside air into indoor air conditions for one specific purpose (heating, cooling, etc.). The preconditions (eg humidity, working hours, type of heat recovery, fuel, unity cost) are related to the building type. For example, in a resthouse, the module 'cooling' is not activated in the default calculation.

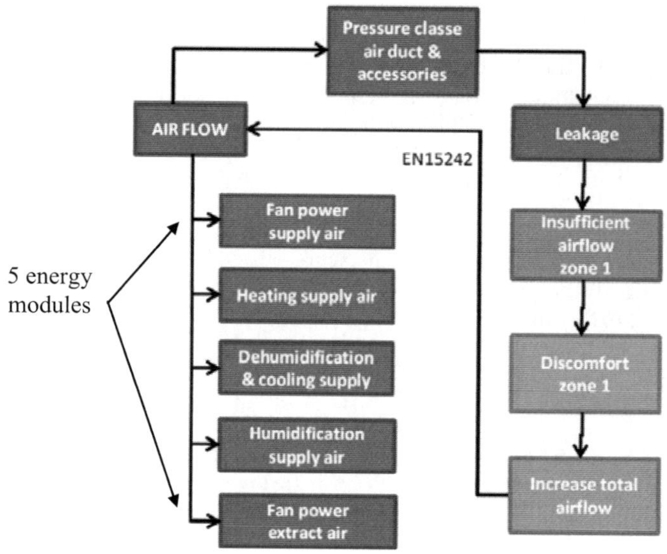

Figure 2. Calculation Method - Energy consumption linked to leakage losses in the ventilation system

The energy gain is supposed to be directly proportional to the avoided leakage loss compared to installations with lower air-tightness.

The applied calculation method incorporates standardised calculation rules from the E-level calculation, as imposed by the Flemish government for office buildings [1]. For non-office buildings, the same methodology is used, buth with other parameters.

All calculations are based on monthly basis.

Building type		Hospital	Resthouse	Office	Other
ventilation rate	(-)	2	1,8	1,5	1,5
average space hight	m	3,1	2,8	2,8	2,8
air flow	/ 1000 m²	6200	5040	4200	4200
duct surface	/ 1000 m³/h	75	65	55	55
number circular ducts	%	0,25	0,25	0,25	0,25
number of circular fire dempers	/ 1000 m³/h	7	6	0,8	0,8
number of rectangular fire dempers	/ 1000 m³/h	3	3	1,6	1,6
average dimension circular fire dempers	mm	160	125	200	200
average dimension rectangular fire dempers	mm	400	400	500	500
heat recovery		60%	60%	60%	no
cooling		yes	no	yes	yes
humidification		yes	no	yes	yes
moisture recovery		no	no	no	no
energy flow for humidification		fuel	elektricity	elektricity	elektricity
speed control fans		no	no	no	no
total fan power (pulsie + extractie)	W / m³/s	3400	2800	3100	3100
unity price electricity	EUR/MWh	120	130	130	140
unity price fuel	EUR/MWh	40	45	45	50
static system pressure	Pa	300	250	200	200
correction factor heating (time factor)	(-)	1	1	0,3	0,6
correction factor cooling (time factor)	(-)	1	1	0,3	0,6

Table 1. Default values calculation

The second step consists of determining the additional investments. The calculation model is based on information obtained from prior studies and from a number of producers (fire dampers).

UP TO 30% SAVINGS ON THE VENTILATION ENERGY

The calculation method was applied to three cases: Case 1 : Renovation of a hospital wing; Case 2 : Rest home; Case 3 : Office building.

Figure 3 shows the results of the simulations. The simulations show that the total energy consumption linked to ventilation can be reduced by over 30% in case 1.

		Hospital wing	Rest home	Office building
Building surface area	m²	11,380	8,830	11,200
Ventilation flow	m³/h	57,450	43,305	53,940
Duct surface (supply and extract)	m²	5,094	2,400	2,480
Percentage of round ducts	-	25%	15%	15%
Fire dampers, number	-	476	496	90
Flow-adjustment units, number	-	527	398	407
Flow adjusters (CAV, VAV), number	-	133	0	77
Silencers, number	-	17	14	19

Table 2. Properties of technical installations

		Case 1 Hospital wing	Case 2 Rest home	Case 3 Office building
Annual energy savings	EUR/y	10,175	6,750	6,750
Investment cost	EUR	14,862	11,068	11,068
Pay Back Time (dynamic)	years	2	2	2

Table 3. Profitability if the ventilation system's air-tightness improves from class A to class C

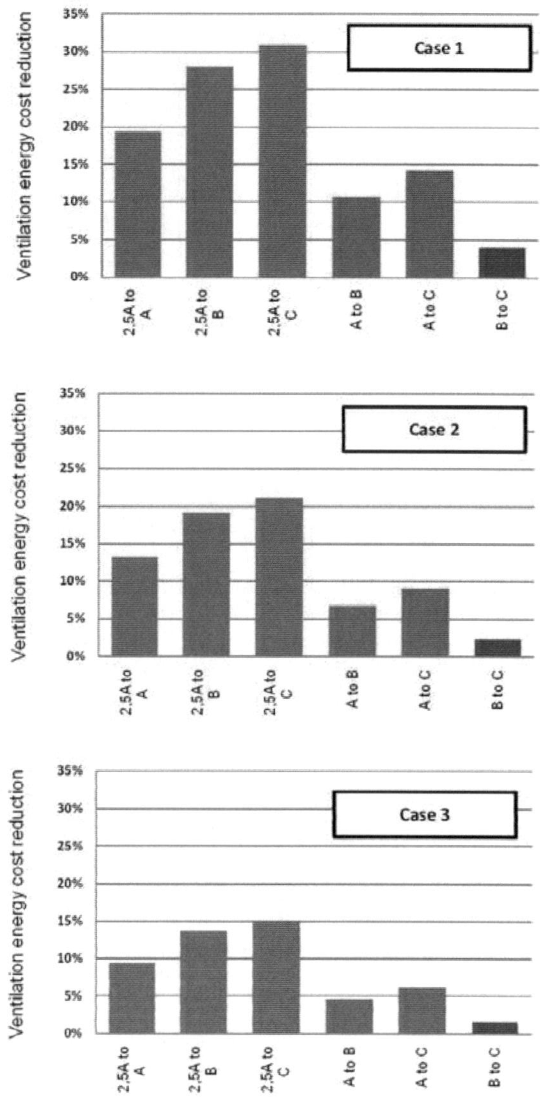

Figure 3. Results

Detailed results are presented in the paper of Peter Stroo, *Class C air-tightness: proven ROI in black and white. Up to 30% savings on the ventilation energy can be defined.*

Detailed calculations can be made for each specific project using the developed calculation model, which is public available on http://www.colorstudio.be/lindab/

CONCLUSION

Paying attention to high air-tightness of the ductwork is worthwhile. The energy related to the ventilation system can be reduced by up to 30%. Detailed calculations can be made for each specific project using the developed calculation model.

The importance of good air-tightness is also acknowledged as the Belgian Buildings Authority will impose class C. In view of the pioneering role of the Standard Specifications, we can expect the Belgian installation world to catch up so that (at least) air-tightness class C will soon be standard.

The Flemish government can also support the air-tightness of ductwork, amongst other things by revising the EPU calculation method for offices and schools. In the current calculation method, air-tightness of ventilation systems, in contrast to air-tightness of the building's shell, has not been taken into account. Revision of the EPU calculation method and thus a reward with a lower E-level for buildings with a (measured) airtight ventilation system would be an interesting incentive to support the air-tightness of ventilation systems in practice.

Nevertheless, good air-tightness of ventilation systems offers real added value. The reduction of the energy use related to ventilation will result in lower energy bills and a better indoor comfort. This reason alone justifies the limited additional cost of achieving an airtight ventilation system [9].

REFERENCES

[1] EPB regulations Flanders Decision by the Flemish government on 11 March 2005 to determine the requirements in the area of energy performances and the inside climate of buildings – Annex II – method to determine the level of primary energy consumption of office and school buildings

[2] EN 12237:2003 Ventilation for buildings - Ductwork. Strength and leakage of circular sheet metal ducts

[3] EN 1507:2006 Ventilation for buildings - Sheet metal air ducts with rectangular section Requirements for strength and leakage

[4] EN 1751:1999 Ventilation for buildings - Air terminal devices – Aerodynamic testing of dampers and valves

[5] Belgian Buildings authority *Standard Specifications 105* Central heating, ventilation and air conditioning (1990)

[6] Belgian Buildings authority Standard Specifications 105 - final draft 2010 Article E5 PAR5 measurement of the air-tightness of ducts; Article C14 Air ducts

[7] SAVE-DUCT Duct leakage in European buildings: status and perspectives; François Rémi Carrié et al.; 1999

[8] EN 15242:2007 Ventilation for buildings - Calculation methods for the determination of air flow rates in buildings including infiltration.
[9] TUV - 2008 *TUV Rheinland - Advisory opinion regarding the comparative test and* evaluation of the overall installation costs of various installed air systems

CASE STUDY: EFFECT OF EXCESSIVE DUCT LEAKAGE IN A LARGE PHARMACEUTICAL PLANT

David F. Dyer

Auburn University
270 Ross Hall
Auburn, AL 36849
USA

ABSTRACT

A study of excessive air leakage in the ductwork of a large pharmaceutical plant located in the Southeast United States is executed in order to determine the energy loss associated with the excessive ductwork leakage. Much of the air supplied by the ductwork is delivered to clean rooms. The analysis requires the development of a model that is used to predict the increased energy costs. The model is applied for each 15 minute interval over the entire year (approximately 35,000 data points). The results are broken down into extra energy used to heat, humidify, and also cool/dehumidify the air over each 15 minute period throughout the year. The results are presented which show that the excessive duct leakage results in more than $1,000,000 loss over the life of the system. Also the additional makeup outside air dramatically increases the dust loading on the hepa filters used to clean the air before introduction into the clean rooms.

KEYWORDS
Duct Leakage, Energy Efficiency

INTRODUCTION

The situation being considered is an actual pharmaceutical manufacturing plant located in the Southeast United States. The HVAC duct work being investigated is used to supply clean rooms used in the manufacturing process. The duct work in question has approximately 3,800 square meters of surface area and air is delivered at a constant rate through the duct work. The installation of the duct work did not meet SMACCNA standards [1] in terms of leakage. For this ductwork the SMACCNA standard calls for a maximum leak rate of approximately 81.6 CMM. The duct work as installed had a leakage rate of 406.6 CMM. The excessive leak rate of 325 CMM requires that the plant's chillers and boilers provide additional chill water and steam to condition the outside air that must be supplied to replace the excessive leaked air from the system. In addition, the HEPA filters dust loading as well as fan HP is increased. This paper examines only the energy penalty associated with conditioning additional outside air due to leaks as compared to return air. The paper is divided into the following sections: (1) thermodynamic model, (2) results from application of model, and (3) conclusions.

1. THERMODYNAMIC MODEL

1.1 Introduction to Model

The air that leaks out of the duct system must be replaced with outside air rather than return air. For many hours/year, the energy level in outside air requires significantly more energy (either as heat or cooling) to reach project supply conditions. Thus, the excessive leak rate is considered to be a flow of outside air which must be conditioned to the return state.

1.2 Control of Temperature and Humidity

The following section describes the model used to determine whether heating or cooling and humidification is required. If the outside air humidity level is less than the desired value (in this case, .008291 kg water/air), the outside air is humidified to this level. Conversely, if the outside air humidity is higher than the required level, the outside air must be dehumidified to this level.

The "trigger" point between heating and cooling is 23 °C, i.e. if the outside air temperature is less than 23 °C, heating is required, and conversely, if the outside air is greater than 23 °C, cooling is required.

1.3 Governing Equations

The following equations, taken from reference [2] were employed in determining costs of heating and cooling of excessive leakage.

Relative humidity $\equiv \emptyset = P_v/Pg$ **[1]**

Where Pg is saturation pressure of water at air temperature and Pv is partial of water vapor in air all measured in bars

Specific humidity $\equiv \omega = .622\ [P_v/(.98 - P_v)]$ **[2]**

Where Pv is measured in bars

Enthalpy per kg of dry air in kJ/kg $= h = T + \omega\ [2501.3 + 1.86T]$ [3]

Where T is dry bulb T in °C

SensibleE_{total} = **Energy required to heat or cool air in kJ/kg =**
$\text{Abs}\ [h_{supply} - h_{outside}]$ **[4]**

Where h_{supply} = supply air enthalpy from Eq [3]
$h_{outside}$ = outside air enthalpy from Eq [3]

LatentE_{total} = **Latent energy required to humidify air in kJ/kg =**
$568(\omega_{supply} - \omega_{outside})$ **[5]**

Where ω_{supply} = specific humidity of supply air
$\omega_{outside}$ = specific humidity of outside air

Annual Cost of Cooling $= (\text{SensibleE}_{total}/\text{COP})(\text{CMM}_{excess})(\text{Time})(\rho)(\text{cost})$ **[6]**

Where E_{total} is given by Eq [4]
COP is the ratio of cooling energy achieved/input electrical energy
CMM_{excess} is the excess leak rate in cubic meters per minute
Time is the number of minutes of cooling over one year
ρ is the density of air in kg/cubic meter
cost is the cost of electricity in $/kJ

Annual cost of heating $= (\text{SensibleE}_{total}/\eta)(\text{CMM}_{excess})(\text{Time})(\rho)(\text{cost})$ **[7]**

Where E_{total} is given by Eq [4]
η is the efficiency of transforming natural gas energy to heating air
CMM_{excess} is the excess leak rate in cubic meters per minute
Time is the number of minutes of heating over one year
ρ is the density of air in kg/cubic meter
cost is the cost of natural gas in $/kJ

Annual cost of humidifying = (LatentE$_{total}$/η)(CMM$_{excess}$)(Time)(ρ)(cost) **[8]**

Where LatentE$_{total}$ is given by Eq [5]
η is the efficiency of transforming natural gas energy to heating air
CMM_{excess} is the excess leak rate in cubic meters per minute
Time is the number of minutes of humidifying over one year
ρ is the density of air in kg/cubic meter
cost is the cost of natural gas in $/kJ

2. APPLICATION OF EQUATIONS

Weather data for the location of the plant were recorded for every fifteen minute interval for the year 2009. The properties recorded include termperature and relative humidity. The calculation of energy use is made for each fifteen minute interval based on these weather data.

Table 1 summarizes the sequence of steps used to calculate the energy used to condition outside air as opposed to return air on a KJ/Kg basis:

Table 1. Sequence to Calculate Energy Used to Condition Outside Air

Step	Action	Equation Employed
1	Use weather data for temperature T and relative humidity ω, to calculate partial pressure of water vapor in air, Pv	Eq[1]
2	Calculate the specific humidity of outside air using value of Pv from step 1	Eq [2]
3	Calculate the enthalpy of outside air using the temperature from weather data and specific humidity from step 2	Eq [3]
4	Calculate the return air enthalpy based on a temperature of 23 °C and sepcific humidity of .008291 kg water/kg dry air	Eq [3]
5	Calculate the heating or cooling required	Eq [4]
6	If the specific humidity of outside air is less than .008291 kg water/kg air, calculate the energy required to humidify air	Eq [5]
7	Calculate the annual cost of energy to cool excess leakage	Eq [6]
8	Calculate the annual cost of energy to heat excess leakage	Eq [7]
9	Calculate the annual cost of energy to humidify excess leakage	Eq [8]

CONCLUSIONS

The governing equations were applied according to the sequence listed in Table 1. Table 2 lists the results of this calculation based on 2009 weather data. In table 1, the average cost of electrical energy is assumed to be \$.10/kwh and the cost of natural gas is assumed to average \$10/million kJ. These average data are assumed to be for a 30 year life of the plant. In addition the boiler efficiency is assumed to be 80% in converting natural gas energy to steam and system losses in converting steam to heat the air are taken to be 5%. Hence, the true heating to fuel energy input rate, η, is 75%. Finally, the actual ratio of cooling to electrical energy input, COP, is considered to be 5 including all auxiliary inputs such as cooling tower fans and condenser water pumps.

Table 2. Summary Costs of Excessive Duct Leakage

Month	Amount of Energy to Cool Excess leakage in Kw-hr	Amount of Energy to Heat & Humidify Excess Leakage in millions of kJ	Cost of Energy to Cool Excess Leakage in \$	Cost of Energy to Heat and Humidify Excess Leakage in \$	Total Cost to condition Excess Leakage in \$
January	1425	329.9231	28.5	4398.975	\$ 4,427.48
February	4987.5	205.2855	99.75	2737.14	\$ 2,836.89
March	1425	329.9231	28.5	4398.975	\$ 4,427.48
April	24937.5	51.32138	498.75	684.285	\$ 1,183.04
May	88350	6.28425	1767	83.79	\$ 1,850.79
June	78375	5.236875	1567.5	69.825	\$ 1,637.33
July	86925	4.1895	1738.5	55.86	\$ 1,794.36
August	59850	20.9475	1197	279.3	\$ 1,476.30
September	15675	163.3905	313.5	2178.54	\$ 2,492.04
October	0	372.8655	0	4971.54	\$ 4,971.54
November	0	594.909	0	7932.12	\$ 7,932.12
December	0	703.836	0	9384.48	\$ 9,384.48
Total	361237.5	2789.16	7224.75	37188.8	\$ 44,413.55

The data shows that over a 30 year life the energy penalty associated with excessive duct leakage is more than 1.3 million dollars. The cost of the leakage far exceeds the marginal increase in the initial cost to install duct work that meets industrial standards. It should be pointed out that proper testing of the duct work installation before acceptance is critical because it usually is not cost effective to remedy leak problems after the plant is in operation. This is true in this particular case. The lack of initial proper testing has led to a situation where the leak rate penalty that exists must be accepted and the production plant is at a disadvantage in comparison to the equivalent plant with duct work with minimum leakage. Of course, the direct energy loss is only part of the loss due to excessive duct leakage. Poorer air quality leading to poorer quality of product, extra fan HP, and HEPA filter maintenance are further areas of economic loss.

REFERENCES

[1] No Author. 2010. *Architectural Sheet Metal Manual - SMACNA, 6th Edition*; Sheet Metal and Air Conditioning Contractor's National Association, Washington, DC.

[2] McQuiston, F.C., Parker, J.D., and Spither, J.D 2005. *Heating, Ventilating, and Air Conditioning, Sixth Edition*, John Wiley and Sons, New York.

Wednesday 12 October 2011

15.30 – 16.30 Parallel Session 2A with short oral presentations and posters – Assessment of ventialtion system performance

- Ventilation rates and indoor air humidity depending on local climate – Simulations and measurements of 9 European countries (Rainer Pfluger, Germany)
- Whole year simulation of humidity based demand controlled hybrid ventilation in multiapartment building (Jerzy Sowa, Poland)
- Method to assess the performance of ventilation systems in dwellings considering the influence of uncertainties (Zhiming Yang, Netherlands)
- Evaluation of some DCV control strategies based on building types (Ke Xu, Norway)
- Demand-controlled Ventilation: an outline of assessment methods and simulations tools (Jean-Luc Savin, France)

15.30 – 16.30 Parallel Session 2B with short oral presentations and posters - Airtightness of buildings and ductwork

- Behavior of leakages exposed to dynamic wind loads. A numerical study using CFD on a single zone model (Dimitrios Kraniotis, Norway)
- Influence of Air Leakage on Indoor Air Quality in Low Energy Buildings: a case study (Juslin Koffi, France)
- The use of building own ventilation system in measuring airtightness (Timo Kauppinen, Finland)
- The use of a sampling method for airtightness measurement of multi-family residential buildings – an example (Jiri Novak, Czech Republic)
- Application Of Airtightness To Healthcare Buildings (William Booth, UK)
- Class C air-tightness: proven ROI in black and white (Peter Stroo, Belgium)
- The Power of Quality (Christophe Debrabander, Belgium)
- Quality Management Approach to Improve Buildings Airtightness, Requirements and Verification (Valérie Leprince, France)

16.30 – 17.00 Room Change and coffee break

VENTILATION RATES AND INDOOR AIR HUMIDITY DEPENDING ON LOCAL CLIMATE – SIMULATIONS AND MEASUREMENTS OF 9 EUROPEAN COUNTRIES

Rainer Pfluger

University of Innsbruck, Unit Energy Efficient Buildings
Technikerstr. 13
6020 Innsbruck, Austria

ABSTRACT

Most European standards and national regulations about ventilation rates are based on indoor air quality assumptions in terms of contamination. On the other hand, indoor air humidity is important for human health as well. In case of high flow rates during the heating seasons in cold climates, the indoor air humidity tends to low values.
This presentation gives an overview of the influence of the local climate (ambient temperature and humidity) in 9 European countries at given flow rates and moisture sources. In case of low indoor air humidity, ventilation systems with humidity recovery might be an useful option. The simulations show which humidity recovery rate is necessary at which climate. Besides the simulations, real measurement data of a project in Innsbruck (Austria) are presented.

KEYWORDS
Ventilation, flow rate, indoor air humidity, climate

INTRODUCTION
The design of ventilation in climate zones which are characterized by a heating season, have to consider both, air quality as well as the humidity balance.
The ventilation flow rate should be adjusted to the moisture sources in space and occasionally reduced, especially in case of very low outside air humidity. In cold climates during heating season, the absolute external humidity is about 1 to 5 g/kg whereas the indoor absolute humidity is around 7 to 9 g/kg. Each cubic meter of air change indoor- by outdoor air takes some 2 to 8 g of water vapour out of the building. Reducing the ventilation rate and accounting for the air quality at the same time is one of the most appropriate ways of raising the indoor humidity level. If the flow rate would become to low in terms of air quality, moisture recovery helps to increase in the indoor air humidity.
Only in extreme cases humidifying action is necessary and recommended, taking into account the hygiene requirements.
This paper gives an overview of the ambient humidity and the consequences for the indoor air humidity as well as some examples of measured data.

MEASUREMENT DATA INNSBRUCK LODENAREAL (AUSTRIA)

"Measurements have been conducted since the fall of 2009 as part of the research projekt titled "Passive House residential complex in Lodenareal – indoor air quality, building services losses and household electricity consumption in Passive House rental apartments" and commissioned by the State of Tyrol and Innsbruck's Kommunalbetriebe (Municipal Utilities Company)." [1]

Figure 1. Passive House Innsbruck Lodenareal (Austria), Monitoring in 18 dwelling units (6 dwellings in detail).

The Passive House Lodenareal (Innsbruck, Austria – see contribution of Kapferer, R. in this proceeding) consists of different apartments. In this contribution the 81 m^2-type apartment is investigated. A detailed measurement evaluation was done concerning air quality and thermal comfort. In this contribution, the measurement data for CO2 and indoor air humidity is evaluated.

Figure 2: Floor plan of the apartment (PH Lodenareal, Innsbruck) under investigation

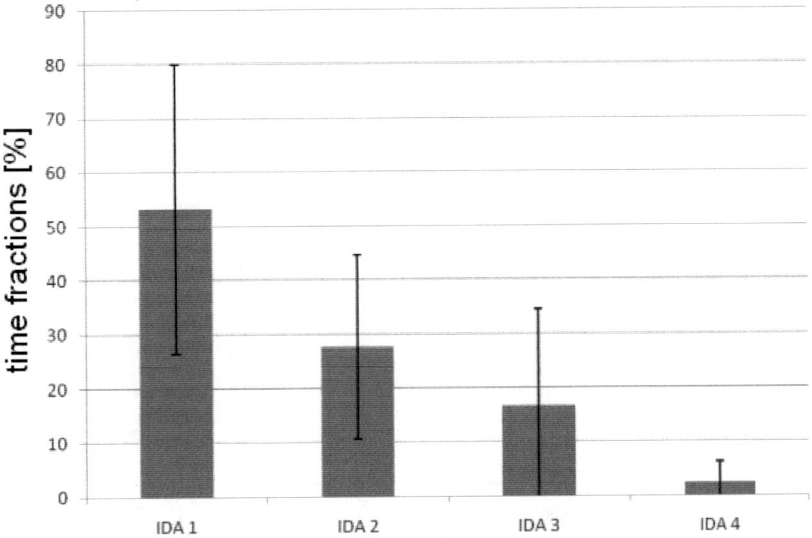

Figure 3. Measurement data of indoor air quality in living rooms of 18 dwellings (0-24 h), only Dec. 1st-Jan. 31st 2010/11

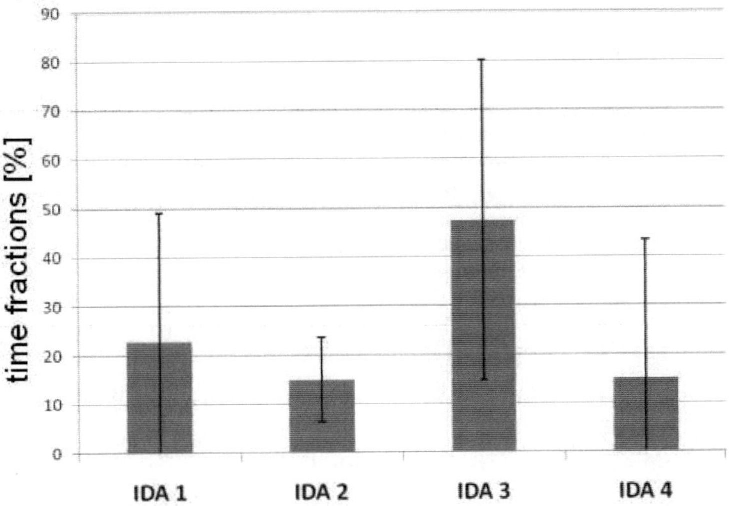

Figure 4. Measurement data of indoor air quality in bedrooms of 6 dwellings (23-7h), only Dec. 1st-Jan. 31st 2010/11

As shown in Figure 3 and Figure 4, the air quality (indicated by CO_2-concentration of the indoor air) is very good in the living rooms and rather good in the bedrooms. There is a large variation in between the individual dwellings, pointing out, that the air quality strongly depends on the number of inhabitants and their behaviour. In general, the ventilation rate of around 80 m^3/h per dwelling is low, but it shows to be adequate from air quality point of view.

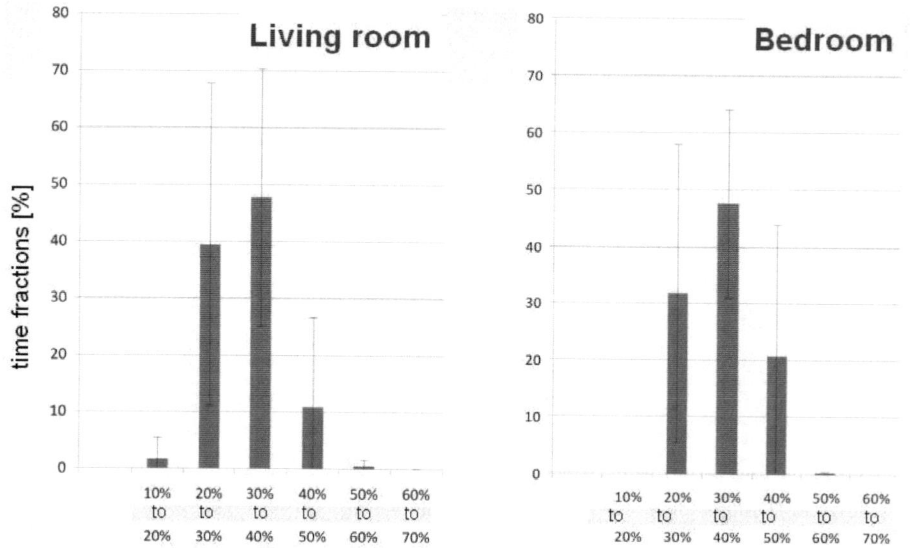

Figure 5. Measurement data of indoor relative humidity in living rooms (18 dwellings, left figure, 0-24 h) and bedrooms (6 dwellings, right figure, 23-7 h), only Dec. 1st-Jan. 31st 2010/11

The relative humidity measured in the living rooms and bedrooms (see Figure 5) was rather low during the core heating season. Higher flow rates without humidity recovery would not be suitable. As seen in case of the CO2-measurement data, also the humidity data show a high variation depending on the number of inhabitants and the user behaviour (humidity production and indoor air temperature). In most cases low relative humidity is also due to high indoor air temperature of around 23 °C – 24° C.

VENTILATEN FLOW RATE AND INDOOR AIR HUMIDITY BALANCE

As pointed out by the evaluation of the measurement data, there is a wide range of humidity production depending on the user behaviour.

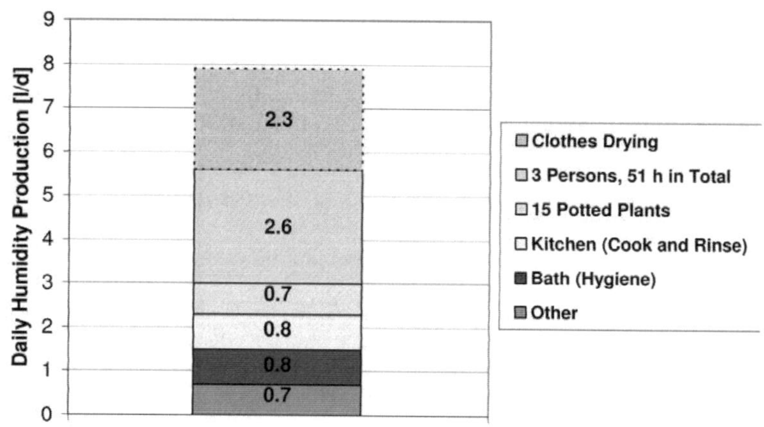

Figure 6. Daily humidity production for a family of three according to Hartmann 2001 [1]

Hartmann evaluated the daily humidity production as shown in Figure 6. Humidity is produced within the dwelling by persons, potted plants, cook and rinse, bath and others. The most important user dependant variation is the drying of clothes. According to Hartmann, this humidity production is around 2.3 l/d if the drying of clothes is done within the dwelling. For the following evaluations a mean daily humidity production of 7 l/d was used for a family of three.

As design value for the flow rate, a value of 30 m³/h per person was applied, this is found to be appropriate both in terms of air quality as well as in terms of indoor air humidity. Moreover this value is recommended in standards and guidelines.

COMPARISON OF DIFFERENT CLIMATE DATA

In order to show the climate dependancy of the indoor air humidity, some examples of ambient conditions mostly in middle and northern Europe where used. The locations of the meteostations (data from METEONORM) are illustrated in Figure 7. The absolute humidity of the ambient air not only depends on the latitude but also on the distance to the sea (e.g. the climate in Dublin and Nantes are strongly influenced by the sea whereas Kiev is a location representing continental climate).

Figure 7. Distribution of locations (hourly climate data from METEONORM), map source: google earth

In oder to illustrate the differences in absolute humidity, hourly data from Innsbruck (Austria) and Dublin (UK) are plotted versus the ambient temperature in Figure 8 and Figure 9 respectively. The red line represents the minimum value of ambient absolute humidity (2 g/kg) at which (under the given boundary conditions and assumptions, see next chapter) an indoor air humidity of at least 30 % RH can be achieved.

It can be seen, that in Dublin, there is no problem with dry air during the heating season, whereas in Innsbruck there are data points below the threshold value. This finding is in good accordance with the measurement data from Innsbruck Lodenareal presented in this paper.

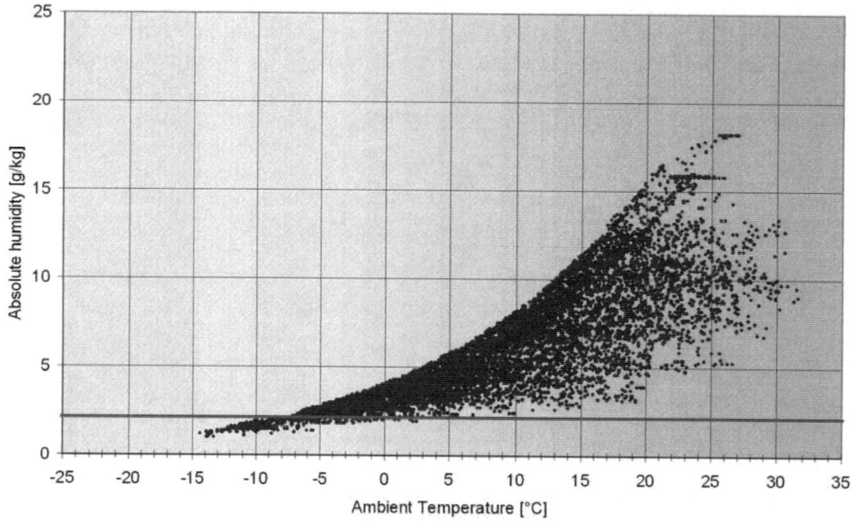

Figure 8. Hourly data of absolute humidity (METEONORM climate data Innsbruck, Austria)

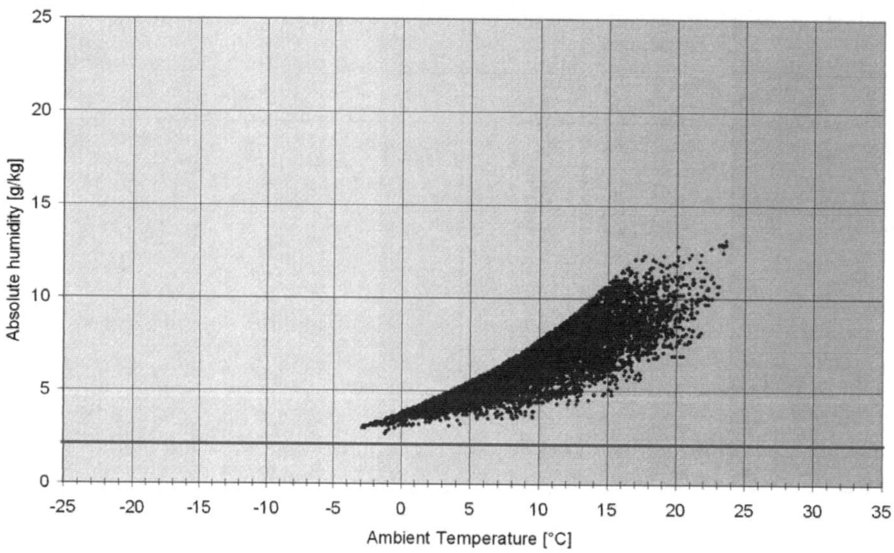

Figure 9. Hourly data of absolute humidity (METEONORM climate data Dublin, UK)

EVALUATION AND RESULTS

The climate data as described above was used as boundary condition for a simplified (without humidity buffering elements, air well mixed) humidity balace calculation for a dwelling with tree inhabitants. The assumed humidity source is 7 l of water per day, whereas the flow rate was set to a constant value of 30 m³/h per person. The relative humidity was calculated assuming an air temperature of 21 °C.

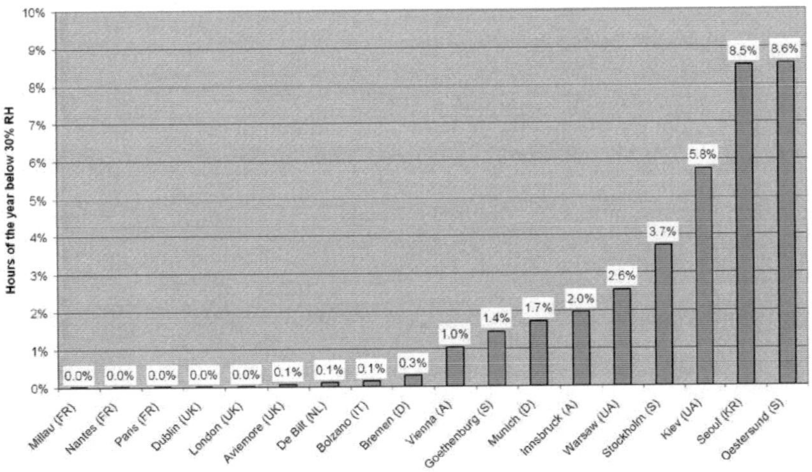

Figure 10. Hours of the year with relative indoor air humidity below 30 % (assumed ventilation rate: 30 m³/h per person)

The indoor relative humidity calculated from the humidity balance for each of the 18 climate data sets where used to evaluate the percentage of hours below 30 % RH plotted in Figure 10. In case of Milau, Nantes, Paris, Dublin, London Aviemore, De Bilt, Bolzano and Bremen, this percentage is negligible, whereas for the other locations, a certain number of hours during the heating seasons occur. For example in Stockholm the calculated result is 3.7 %, which is 324 hours (round about 2 weeks) per year with a relative humidity below 30 %.

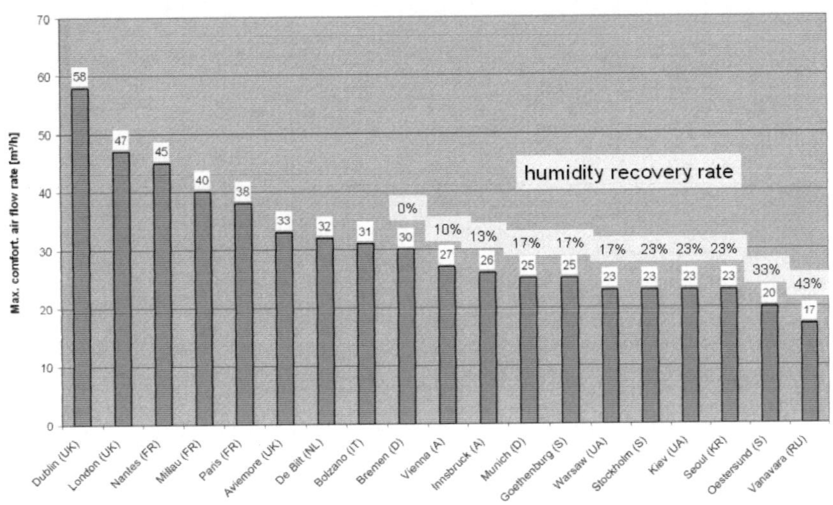

Figure 11. Maximum air flow rate per person [m³/h] with less than 1 % of the hours per year below 30 % RH and minimum necessary humidity recovery rate in case of constant flow rate of 30 m³/ per person (see labels above)

If the flow rate of 30 m³/h per person should be maintained for air quality reasons, an equivalent humidity recovery rate of 23 % would be enough to keep the percentage of hours with a relative humidity below 30 % lower than 1%.

CONCLUSION

In case of a flow rate of 25 up to 30 m³/h per person, both, indoor air quality and relative humidity is within an acceptable range for most of the European countries. In case of cold climates, humidity recovery might help to enhance the humidity level even at a constant flow rate of 30 m³/h per person. According to the simulation results, a humidity recovery rate of around 30 % is mostly sufficient to keep the indoor air humidity within the adequate range during most of the time throughout the heating season. Only in extreme cases (very low humidity sources) active humidification is necessary.

ACKNOWLEDGEMENTS

- State of Tyrol
- Innsbruck's Kommunalbetriebe (Municipal Utilities Company).
- Building owner: Neue Heimat Tirol (NHT)

REFERENCES

[1] Kapferer, R., Pfluger, R., Suschek-Berger, J. *Evaluating measurements from the Lodenareal Passive House residential complex in Innsbruck* in Feist, W. 2011.: Conference Proceedings, 15th International Passive House conference Innsbruck, 191-196

[1] Hartmann, Thomas, Ralf Gritzki, Jens Bolsius, André Kremonke, Alf Perschk, and Wolfgang Richter: *Bedarfslüftung im Wohnungsbau.* Stuttgart, Fraunhofer IRB Verlag, 2001.

WHOLE YEAR SIMULATION OF HUMIDITY BASED DEMAND CONTROLLED HYBRID VENTILATION IN MULTIAPARTMENT BUILDING

Jerzy Sowa*, Maciej Mijakowski

Warsaw University of Technology
Faculty of Environmental Engineering
Nowowiejska 20
00-653 Warsaw, Poland
**Corresponding author: jerzy.sowa@is.pw.edu.pl*

ABSTRACT

The paper presents the whole year simulation of humidity based demand controlled hybrid ventilation in multiapartment building. The simulation was performed for NAPE (National Energy Conservation Agency) multifamily residential reference building. This allowed the authors to compare obtained results with earlier investigated behaviour of the NAPE building with passive stack ventilation and mechanical exhaust ventilation.
Simulations were performed using computer programme CONTAM (developed by NIST). In CONTAM environment the building together with analysed ventilation system has been idealized as 127 zones and 884 flow paths. Simulations were performed with 5 min time step (results were stored with 1 h time step). Huge set of results allow the authors to compare behaviour of the analysed ventilation systems depending on the number of storey, size of the apartment, type of space etc. CONTAM does not allow users to perform thermal analysis of the buildings, but thermal and energetic analyses can be performed separately at second step (using airflows calculated by CONTAM). The 6R1C model developed at Warsaw University of Technology was used for this purpose.
Performed simulations presented the possibilities of utilisation computer programme CONTAM for modelling behaviour of humid air in buildings, even when they are huge and complex. Additionally obtained results indicated once again that humidity based demand controlled hybrid ventilation systems can reduce substantially amount of energy in residential buildings.

KEYWORDS

Simulation, CONTAM, Demand controlled ventilation

INTRODUCTION

Humidity based demand controlled hybrid ventilation systems create many problems during modelling. If the system is analyzed in residential or office building with many rooms, the applied model has to allow researcher to perform multizone simulations. Hybrid nature of the system causes that natural as well as mechanical forces driving airflows have to be taken into account. Utilisation of humidity based demand controlled strategies calls for simultaneous calculations of humidity ratio in each of the building zone and calculations of airflows between zones. Moreover software used for simulations has to allow users to model control elements.
The paper presents the whole year simulation of humidity based demand controlled hybrid ventilation in multiapartment building. The simulation were performed for NAPE (National Energy Conservation Agency) multifamily residential reference building. This allowed the

authors to compare obtained results with earlier investigated behaviour of the NAPE building with passive stack ventilation and mechanical exhaust ventilation.

The simulations were performed in two steps. Airflows were analysed using CONTAM with 5 min time step (results were stored with 1 h time step) while energetic assessments were performed using 6R1C model (hurly calculations).

ANALYSED BUILDING

The National Energy Conservation Agency NAPE reference building is a virtual, residential building with 8 storeys. The building has the total volume of Ve = 5865 m^3, surface of envelope Ae=2028.5 m^2 (shape ratio Ae/Ve = 0.35) and usable area Af = 1634 m^2. All assumed parameters (e.g. size of windows, thermal resistances of walls etc.) fulfil minimum requirements for new buildings described in Polish building codes and related ministerial ordinances.

There are 3 different small flats at each storey with 1, 2 and 3 rooms respectively (fig. 1). Altogether there are 23 flats, occupied by 47 persons.

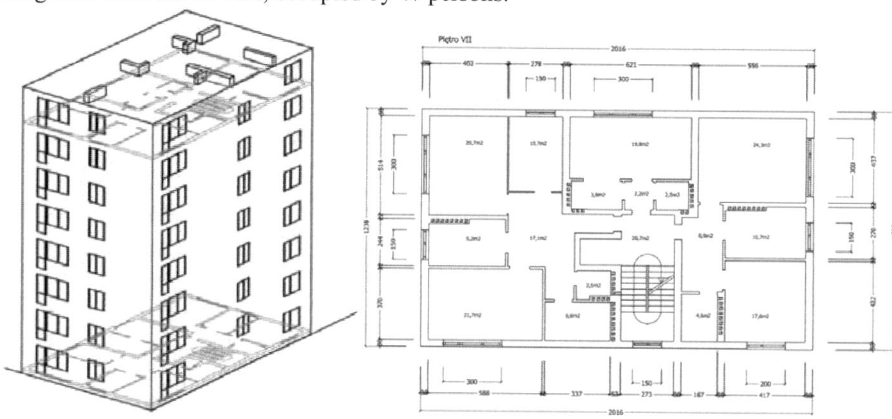

Figure 1. The NAPE reference building - view and plan of the typial storey.

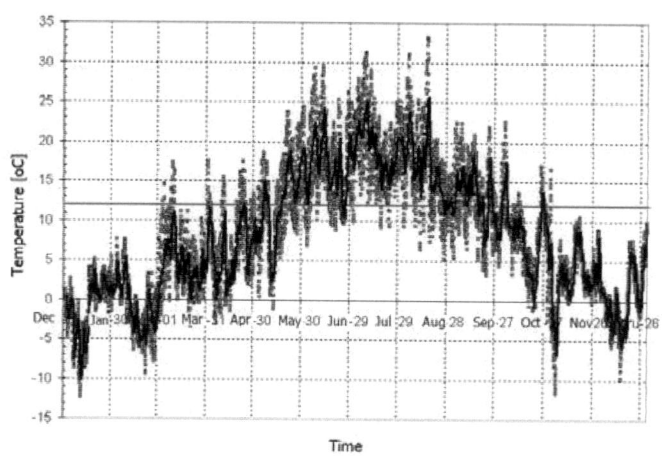

Figure 2. Ambient temperature for Warsaw- dots hourly data, black line daily average.

The building is virtually located in Warsaw. Any airflow or energetic calculations for this building are performed using typical meteorological year published by the Polish Ministry of Infrastructure http://www.mi.gov.pl/2-48203f1e24e2f-1787735-p_1.htm . This file has been prepared according to EN ISO 15927-4, [1]. According to these data in Warsaw ambient temperature varies from -12 °C to +34 °C, while span of daily average temperature is -7 °C to +24 °C (fig. 2).

The NAPE reference building is equipped with 2 optional systems of ventilation. Option 1 is passive stack ventilation. Exhaust grills connected to individual stacks with cross-section 14x14 cm are located in kitchens, bathrooms, toilets. Air is supplied to the flats through air vents installed in each window. In test conditions at 10 Pa pressure difference these vets should provide 50 m^3/h [4].

Option 2 is mechanical exhaust ventilation that meets minimal requirements of Polish standards. Air is exhausted from kitchens (70 m^3/h for kitchens equipped with gas cooker), bathrooms with or without toilet (50 m^3/h) and separate toilets (30 m^3/h) [4]. Air supply is through air vents which in case of mechanical exhaust ventilation provides 30 m^3/h of air when tested with 10 Pa pressure difference [4].

In this study the building was equipped with humidity based demand controlled hybrid ventilation. Air vents used in that system has variable characteristics influenced by relative humidity. For given pressure drop air flow is proportional to relative humidity (in range 30÷70 %). Characteristics of exhaust grills also depend on relative humidity. Additionally exhaust grills mounted in bathrooms and toilets are equipped with presence sensors that force opening of a control damper when users are in a space (delay for switching off is 20 min). Exhaust fans mounted on a roof above collecting ducts are equipped with pressure sensors and can reduce fan speed when needed.

AIRFLOW SIMULATIONS

Air flow simulations were performed using well known and verified computer programme CONTAM (developed by NIST [5]). In CONTAM environment the building together with analysed ventilation system has been idealized as 127 zones and 884 flow paths. Additionally in case of humidity based demand controlled hybrid ventilation systems the model takes into account controls (in analysed case humidity influences characteristics of air vents, exhaust grills and exhaust fans). It should be pointed out that real control network is more complex than presented on a sketchpad as some nodes can be so called superelements that in fact are control subnetworks.

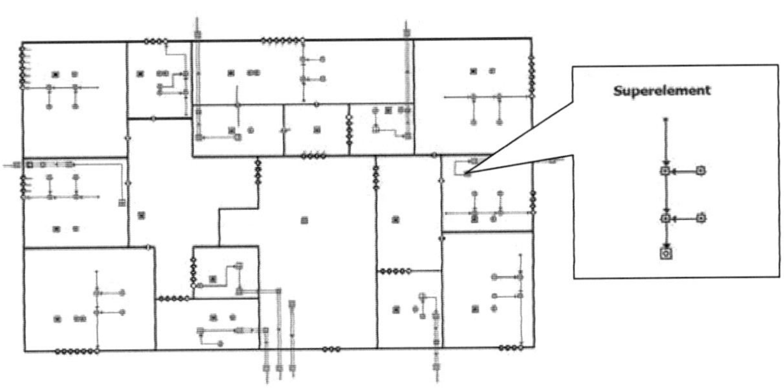

Fig. 3. Sketchpad (CONTAM) presenting 8 floor of analyzed building for analyzed variant.

Figure 4 presents the difference between performance of exhaust grill (example: kitchen in M3 flat at 8 floor) used in regular mechanical exhaust ventilation and grill used in humidity based demand controlled ventilation. One may observe that in case of very high relative humidity analysed demand controlled system is able to ventilate moist space more effectively.

a) Mechanical exhaust ventilation

b) Humidity based demand controlled ventilation

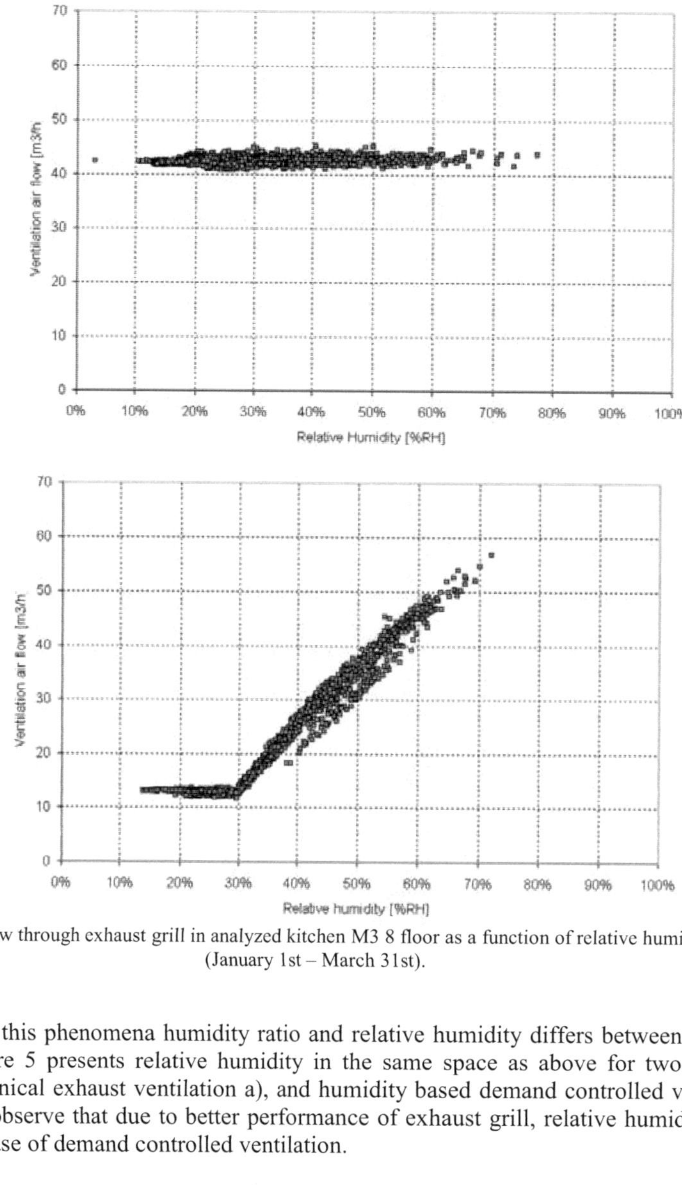

Figure 4. Airflow through exhaust grill in analyzed kitchen M3 8 floor as a function of relative humidity of air (January 1st – March 31st).

As a result of this phenomena humidity ratio and relative humidity differs between analysed systems. Figure 5 presents relative humidity in the same space as above for two variants: regular mechanical exhaust ventilation a), and humidity based demand controlled ventilation b). One may observe that due to better performance of exhaust grill, relative humidity peaks are lower in case of demand controlled ventilation.

b) Mechanical exhaust ventilation

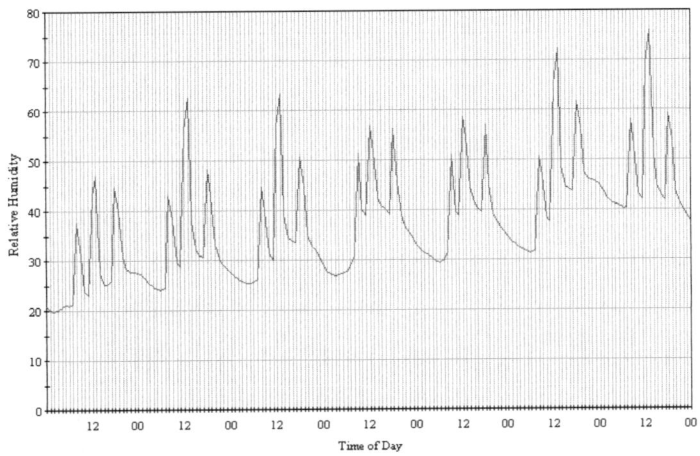

b) Humidity based demand controlled ventilation

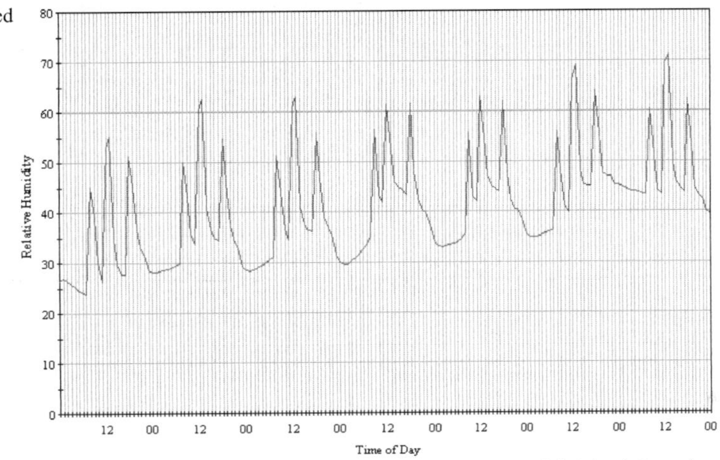

Figure 5. Relative humidity in kitchen on last floor in M3 dwelling during one week in March for variant 1 (a) and 2 (b).

As during summer buildings are not heated and users strongly change thier bahavuior related to opening windows the calculations were performed only for heating season. It has been assumed that heating season starts in autumn when the daily average temperature drops below +12 °C and the heating period stops when daily avereage temparature exceeds +12 °C at spring. Meteorological data corrected in such a way gives for Warsaw 3855 degreedays that is in good agreement with commonly used in simplified methods value 3885 degreedays.

As CONTAM [5] does not allow users to perform thermal analysis of the buildings, thermal and energetic analyses were performed separately at second step (using airflows calculated by CONTAM at first step). Energetic aspects were analysed using lumped capacitance building 6R1C heat exchange model (developed at Warsaw University of Technology, [3]). The model is the further development of simple hourly method described in ISO FDIS 13790:2007, [2]. The most important modification is related to splitting air flows between outdoor and indoor into controlled airflow (with known or calculated supply temperature) and uncontrolled infiltration/exfiltration. The model, similarly as 5R1C, allows supplying the heat energy to three nodes – to interior of building construction, to the internal surface of building construction and to indoor air. Developed model has been successfully verified with the BESTEST [3].

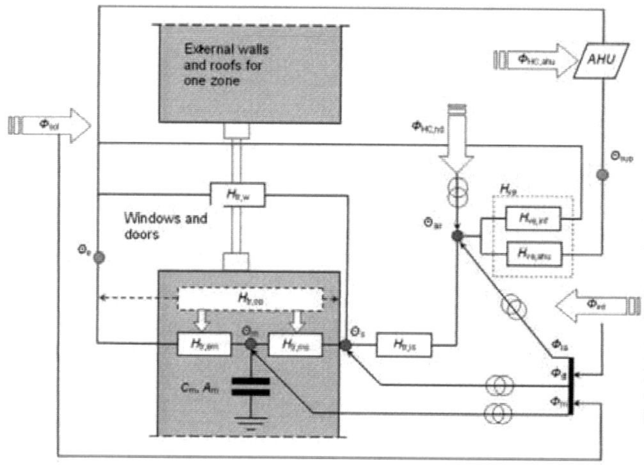

Figure 6. Lumped capacitance building 6R1C heat exchange model

RESULTS

Detailed simulations provided huge set of results that allowed the authors to compare behaviour of the analysed ventilation systems for whole building, as well as for different flats located at different storey.

Figure 7 presents ventilation rate for whole building during the heating period in case of humidity based demand controlled hybrid ventilation. It should be pointed out that obtained ventilation rates are much below values required by Polish Standard PN-82/B-03430/Az3:2000 [4] and commonly used rough indicator equal 1 air change rate per hour.

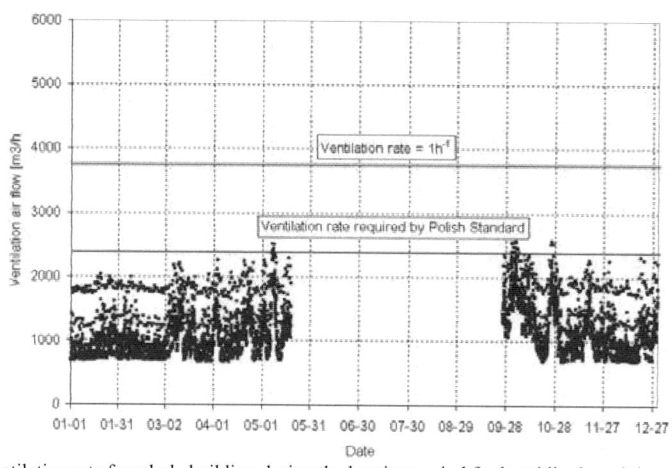

Figure. 7. Ventilation rate for whole building during the heating period for humidity based demand controlled hybrid ventilation.

Table 1 presents comparison of air ventilation rates for different flats and different storey. Simulation indicated that humidity based DCV system works with substantial differences in ventilation rate over time but without important differences between storey. Average air volume is ~ 40 % of maximum value that is approximately equal required ventilation rate.
The mechanical exhaust ventilation (table 2) works very stabile over time and also without differences between flats. Performance of passive stack ventilation presented in table 3 shows that there are huge differences in ventilation rate not only over time but also between similar flats located at different storey. Maximum values 2-times exceed requires ventilation rates.

	Ventilation rate. m^3/h						
	Total	**Flat M1 8 storey**	**Flat M2 8 storey**	**Flat M3 8 storey**	**Flat M1 2 storey**	**Flat M2 1 storey**	**Flat M3 1 storey**
Average	1133	38	36	66	39	37	67
Minimum	667	24	21	34	25	22	36
Maximum	2566	102	90	151	102	90	148
Standard deviation	402.4	14.5	14.7	25.5	14.9	14.9	25.6

Table 1. Summary of airflows analysis for humidity based demand controlled ventilation.

	Ventilation rate. m^3/h						
	Total	**Flat M1 8 storey**	**Flat M2 8 storey**	**Flat M3 8 storey**	**Flat M1 2 storey**	**Flat M2 1 storey**	**Flat M3 1 storey**
Average	2160	87	68	112	91	75	120
Minimum	2069	83	65	107	87	68	110
Maximum	2314	101	78	122	102	86	130
Standard deviation	37.2	1.5	1.6	1.9	2.1	2.7	3.5

Table 2. Summary of airflows analysis for mechanical exhaust ventilation.

	Ventilation rate. m^3/h						
	Total	**Flat M1 8 storey**	**Flat M2 8 storey**	**Flat M3 8 storey**	**Flat M1 2 storey**	**Flat M2 1 storey**	**Flat M3 1 storey**
Average	3253	54	88	119	120	167	230
Minimum	1086	3	28	32	45	41	48
Maximum	5590	261	289	360	243	270	349
Standard deviation	615.8	29.1	36.1	45.7	24.6	30.3	41.9

Table 3. Summary of airflows analysis for passive stack ventilation.

These differences influences also energy performance of the whole building. The primary energy use for heating and ventilation (mechanical exhaust system), calculated using reference climatic data for Warsaw and heating system efficiency 0.9 and primary energy conversion factors 0,8 for heat (produced in cogeneration) and 3,0 for electricity is 106 681 kWh/year (65.29 kWh/(m^2year)).
Application of the humidity based demand controlled hybrid ventilation system reduces primary energy use by ~25% to 79 712 kWh/year (48.78 kWh/(m^2year)).
Benefits are even more obvious when comparing this system with passive stack ventilation which has primary energy use 132889 kWh/year (81,33 kWh/(m^2year)). In such a comparison benefits resulting from application demand controlled hybrid ventilation may reach 40% of primary energy use.

CONCLUSION

Performed simulations presented the possibilities of utilisation computer programme CONTAM for modelling behaviour of humid air in buildings, even when they are huge and complex and building systems contains control elements. Additionally obtained results indicate once again that humidity based demand controlled hybrid ventilation systems can reduce substantially amount of energy in residential buildings.

REFERENCES

[1] EN ISO 15927-4:2007 – *Hygrothermal performance of buildings -- Calculation and presentation of climatic data -- Part 4: Hourly data for assessing the annual energy use for heating and cooling*

[2] ISO-FDIS 13790:2007 *Energy performance of buildings - Calculation of energy use for space heating and cooling.*

[3] Narowski P., Mijakowski M., Panek A.D., Rucińska J., Sowa J., *Integrated 6R1C Energy Simulation Method – Principles, Verification And Application*, Preceedings of the 10th Rehva World Congress "Sustainable Energy Use in Buildings" CLIMA 2010, May 9-12 2010, Antalya, Turkey

[4] PN-83/B-03430/Az3:2000, *Ventilation in dwelling and public utility buildings Specifications (in Polish)*

[5] Walton G.N., Dols S.W., *CONTAM – User Guide and Program Documentation*, NISTIR 7251, October 2005

METHOD TO ASSESS THE PERFORMANCE OF VENTILATION SYSTEMS IN DWELLINGS CONSIDERING THE INFLUENCE OF UNCERTAINTIES

Zhiming Yang[1], J.J.M. Cauberg[2], M.J. Tenpierik[3]

1 Delft University of Technology
Julianalaan 134
Delft, the Netherlands
zhiming.yang@tudelft.nl

2 Delft University of Technology
Julianalaan 134
Delft, the Netherlands
**cauberg@chri.nl*

2 Delft University of Technology
Julianalaan 134
Delft, the Netherlands
m.j.tenpierik@tudelft.nl

ABSTRACT

Normally, the design of a ventilation system in a dwelling is based on national regulations, related design rules, building tradition and general knowledge about healthy indoor air quality, ventilation and air handling units. In practice, the actual performance of ventilation systems is determined by ventilation components, building properties, outdoor environment and occupant behavior. Unspecified items in the design rules and uncontrollable items in the design stage will bring uncertainties which may cause the actual performance deviating from the designed performance. In this research, an assessment method considering the influence of such uncertainties is proposed and developed. First, the method for defining criteria for assessing the performance of ventilation systems in houses is described. The basic idea is that the criteria should be defined based on the national ventilation regulations. Then, the process for estimating the uncertainties in four aspects (including ventilation components, building properties, outdoor environment and occupant behavior) is introduced based on the definition of 5 main uncertainty sources. The relevant parameters in each aspect and the main uncertainty sources for each parameter are figured out. A point which may be interesting is that we propose to explore the reaction of the system performance to certain occupant behavior rather than predicting the occupant behavior pattern. Later on, the uncertainty analysis techniques including uncertainty propagation technique and sensitivity analysis technique are introduced as Monte-Carlo simulation (with Latin-hypercube sampling) and Morris factorial sampling respectively. The uncertainties in ventilation components, building properties and outdoor environment are involved in both uncertainty propagation and sensitivity analysis while the occupant behavior is only analyzed via a sensitivity analysis. A pilot case study using the described method is given afterwards. The conclusions are that there is a necessity to integrate the influence of uncertainties in the assessment of ventilation systems in houses and that the introduced method could give a useful framework and approach for such assessment.

KEYWORDS

Uncertainties, performance assessment, ventilation system

1 INTRODUCTION

A common awareness regarding ventilation is that the proper amount of air change should be maintained while unnecessary air change should be minimized. For this aim, people realized that in a house a properly designed ventilation system was required and that this needed to be regulated. In order to obtain a properly designed ventilation system, two questions have to be answered, i.e. how much air flow is required and how to realize this amount of air change by available measures. To answer these two questions, relevant required air flow rates and design solutions have been developed in different countries, including the required *performance requirements* and *design rules* (including prescriptive design requirements and generally accepted design rules) for ventilation systems. Many different types of ventilation systems for dwellings have been designed based on the same regulations and design rules within a country. But in practice, performance deviations can be found in two aspects, i.e. the performance difference between the same type of system equipped in the same or a similar type of building and the performance deviation between the actual and designed performance. A possible explanation is that such performance deviations could be partly due to the uncertainties existing in the development process of a ventilation system in dwellings which have not been well addressed in the design rules.

Several studies were already carried out for investigating the uncertainties in building performance analysis/simulation. Macdonald (2001) [1] compared and summarized the different uncertainty analysis techniques. De Wit (2002) [2] focused on modeling uncertainties in thermal performance predictions. Wouters et al. (2004) [3] gave a rough discussion about which parameters related to the performance of ventilation systems in houses should be considered as uncertain, focusing on the uncertainties in building properties and on scenarios (mainly including climate and occupant behavior). Generally speaking, in such researches, uncertainties related to the physical properties of building and the modeling were well addressed and the analysis techniques were also well illustrated. But deficiencies also exist, which were our main initiatives for conducting the current study, including the following aspects:

- The discussion of uncertainties mainly focused on the uncertainties in physical data, modeling uncertainties and scenarios (design boundaries, including e.g. climate and occupant behavior) based on a definitively designed building or system; but the uncertainties caused by a lack of specification in the design rules have not been well discussed.
- Researches were mostly conducted under specific conditions. But in different ventilation projects, different uncertainties may emerge. A general method for defining and estimating the relevant uncertainties in the development process of ventilation systems deserves attention.
- Although the uncertainties in occupant behavior and treatment method were discussed by several researchers, predicting future occupant behavior is still difficult because it involves a stochastic process. In our current research, we propose to explore the reaction of the system's functions to occupant behavior instead of predicting the uncertainties in the system's performances by predicting the uncertainties in occupant behavior.

Indoor air quality is generally determined by three factors: the functions of ventilation systems, indoor emissions (materials) and occupant behavior. Different from the traditional assessment of indoor air quality, which focuses on the concentrations of indoor air pollutants, in the current research we focus on only one factor, i.e. the functions of the ventilation system. More specifically, we focus on the functions required in ventilation regulations for a ventilation system in a house. In order to give a more clear explanation of our ideas, the structure of the current research is explained in below figure 1.

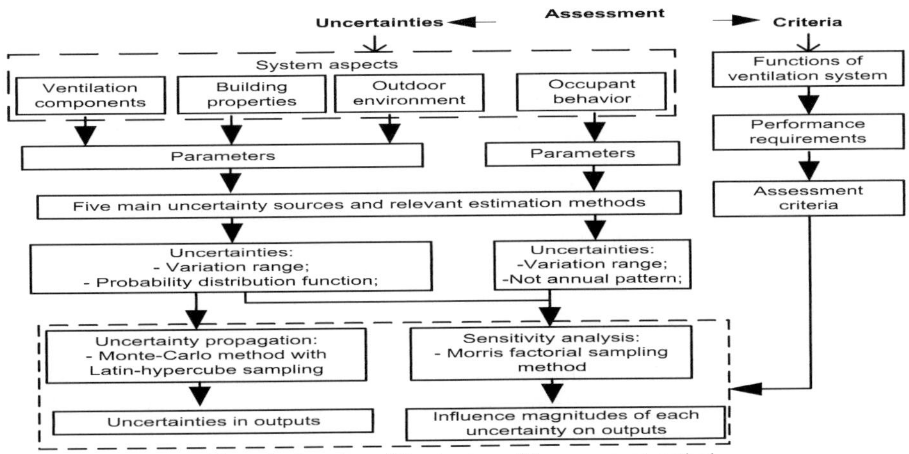

Figure 1 : Overview of the structure of the assessment method

2 ASSESSMENT CRITERIA

The required functionality or performance of a ventilation system in a dwelling is actually defined using performance criteria and required values derived from the national ventilation regulations in each country. Thus, assessment criteria and required values should also be defined based on specific national ventilation regulations. Various different performance criteria can be derived from different national regulations. A review of the ventilation regulations and standards of four countries, including Canada [4], Denmark [5], UK [6] and the Netherlands [7] was carried out to figure out the key criteria and aspects to be used generally for expressing the required performance of ventilation systems in houses. Consequently, the most relevant criteria to be considered in the current method are summarized with short description as:

- Minimum ventilation capacity;
- Air flow direction;
- And Controllability.

2.1 Minimum ventilation capacity

Ventilation capacity includes exhaust capacity and supply capacity. The corresponding criteria should be defined for both aspects based on local or national regulations. Defining criteria for minimum exhaust capacity is simple because they mostly can be straightforwardly derived from the relevant ventilation regulations. In ventilation regulations, such criteria are normally specified based on room type, and/ or room volume or space.

Criteria for minimum capacity should be divided into room/ space level and whole house level. Furthermore, four types of air flow elements could be considered as supply air, i.e. air flow from designed openings or grilles, infiltration from cracks, overflow from internal rooms, and recirculation flow. In different ventilation regulations, there are different considerations of the contribution of each of these air flow elements to the indoor air quality. For defining criteria, such contributions should be considered based on the relevant ventilation regulations and the researcher's interests.

Although normally not specified in regulations, the total designed exhaust rate and supply rate should be equal to each other. Besides, the whole house exhaust rate or supply rate should not be less than the higher value of the required minimum total exhaust and supply rate.

2.2 Air flow direction

Not every ventilation regulation defines the air flow direction as a performance criterion, but the importance of avoiding unwanted ventilation directions is obvious. Generally speaking, among the regulations in different countries, the unwanted air flow directions can be summarized into the following parts:

- Air flow from polluted/ wet rooms to habitable space;
- Air flow from combustion appliance or chimney into the habitable space;
- Back flow crossing the exhaust ducts/ grilles;
- Reverse air flow crossing the facade air inlets/ outlets (not for cross or single-sided natural ventilation systems).

2.3 Individual controllability

Individual controllability is typically not defined in ventilation regulations and is an indicator proposed by us. The employment of individual controllability is to describe the reactions of the performance of a ventilation system to certain individual occupant behavior. The following aspects are considered:

- The influence on the air flow in other rooms by a change in the position of the air inlet in one room;
- The influence on the air flow in other rooms by a change in the position of window openings in one room;
- The influence of a change in the position of internal doors on the overall air flow in the house.

If the individual behavior of an occupant could strongly affect the system performance, for example strongly reduce the supply air flow in other rooms, certain measures may be required to deal with such effects.

3 UNCERTAINTIES

The ventilation performance not only depends on ventilation components but also on other external factors. In each of these factors, uncertainties could be produced. We generally consider the uncertainties in the following aspects:

- Ventilation components;
- Building properties;
- Outdoor environment;
- Occupant behavior.

For identifying the uncertainties in such aspects, five uncertainty sources are defined. For an uncertainty analysis, certain information of each uncertainty may be required, including the variation range and probability distribution. The description of these five uncertainty sources and proposed estimation methods for each of them is given below:

- Specification uncertainty;
- Design alternative;
- Production deviation;
- Modeling uncertainty;
- Stochastic process.

Specification uncertainty means that not all relevant parameters which are needed as inputs for an assessment process or calculation model are specified in the design or required in the design rules. For example, in the selection of a fan, only the working point is considered but not the whole fan curve which also influences the practical performance of the fan. This uncertainty source is related to the lack of specification in design rules. To estimate this specification uncertainty, it is first important to know which parameters are not specified and then to ascertain the variation ranges of the values of such parameters. The most straightforward way is to use measurement data of such unspecified parameters.

Design alternative is the design choice which the designer has without conflicting with the design rules. For example, only the minimum requirements on the designed air flows are defined, meaning the designer could use the values higher than the minimum values. For the designer; if the designer has no preferred design choice yet and wants to know the influence of different design choices, the design alternative could be an uncertainty source. This type of uncertainty can also be considered as a kind of lack of specification in the design rules, but is different from specification uncertainty. The variation range should be estimated or defined based on the possible alternatives resulting from the design rules. For example, the air tightness level is specified; the designer may have alternatives from a high air tightness level to a medium air tightness level.

Production (construction) deviation is the difference between the practical properties of the products with the nominal/ theoretical properties of the products caused by the realization process. The reasons can be summarized as two aspects: nominally identical products that are produced or constructed by different machines or workers, and the random error that the same process may result in different results. In practice, it is difficult to completely separate these two aspects. Anyway, such uncertainties could be reflected in measurement data. Estimations are best made directly through measurement. For example, for a certain type of construction, like a window, the air leakage value could vary in a certain range. If measurement data is missing, estimation may be made based on the requirements in relevant standards and assumptions.

Modeling uncertainty involves the employment of various models to represent reality. Models are developed usually with simplifications and approximations. Arbitrarily speaking, no one model can 100% represent the reality. During the assessment of a ventilation system, many inputs not measurable are derived from models. In [2], the methods for estimating modeling uncertainty were discussed and proposed to be used. For a rough estimation, the scatter between different models could be used. If more a strict estimation is required, costly methods like expert judgment, experiments and measurements may be needed.

Stochastic process is related to the factors which actually require prediction, including e.g. future climate, occupancy and occupant behavior. Such stochastic process factors are normally assumed to be certain default values or profiles in the design. Therefore, uncertainties are introduced. Different treatments for climate and occupant behavior are used. For climate, we recommend to used different typical patterns of annual weather data to represent the climate variations; for example, typical average weather year, typical hot year and typical cold year. Except the indoor heating temperature, the other defined occupant behaviors are treated as part of a sensitivity analysis which only requires the variation range of the parameters but not the whole annual behavior pattern.

The relevant parameters, uncertainties, dominant uncertainty sources and proposed estimation methods are summarized in table 2. The list in table 2 is considered to cover most of the uncertainties but is not exhaustive.

Parameters	Main uncertainty sources	Possible estimation methods
Ventilation components		
Geometry properties	Design alternatives; Production deviation;	Specified probable variations in design rules; Estimation from measurement data;
Flow coefficients & Flow exponents	Production deviation; Modeling uncertainties	Measurement or products data; Measurement data and interactions;
Duct leakage	Specification uncertainty; Production deviation;	Probable duct type or airtight levels; Measurement data;
Exhaust/supply grilles	Construction deviation;	Accuracy requirements on fittings; Measurement data;
Heat recovery efficiency	Production deviation;	Product information and measurement data;
Internal leakage	Production deviation;	Product information and measurement data;
Building properties		
Building direction	Specification uncertainty;	Probable directions and design limitations;
Building air tightness	Specification uncertainty; Construction deviation	Specified air tightness level; Measurement data.
Outdoor environment		
Climate data	Specification uncertainty; Stochastic process;	Probable climate regions; Using different typical weather data;
wind reduction factor	Modeling uncertainty	Estimation from calculation models; Expert judgment;
Wind pressure coefficients	Modeling uncertainty	Estimation from calculation models; Expert judgment;
wind direction	Modeling uncertainty	Estimation between different climate data; Expert suggestions;
Local temperature	Modeling uncertainty	Estimation from literature or data;
Occupant behavior		
Indoor setting temperature	Specification uncertainty	Estimation from survey data;
Control of air inlet	Stochastic process	Variation range of the position of air inlets;
Control of windows	Stochastic process	Variation range of the position of windows;
Control of doors	Stochastic process	Variation range of the position of doors;

(Note: Assumption is always a possible estimation method when other methods are not available.)

Table 2: parameters, uncertainties, dominant uncertainty sources and relevant estimation methods

4. ANALYSIS TECHNIQUES

4.1 Basic uncertainty quantification techniques

With uncertainty quantification studies, there are generally two different analysis purposes, i.e. quantification of the uncertainties in output results from uncertainties in inputs, and figuring out the individual influence or importance of each uncertainty on the outputs. The respective analysis techniques can be generally categorized as uncertainty propagation and sensitivity analysis.

For uncertainty propagation, there exist many techniques including a black box technique (Monte-Carlo method), a statistical method (Factorial regressions), and an internal method (integrating uncertainties into the model equations). Among such techniques, we propose the Monte-Carlo method with Latin-hypercube sampling to be used for uncertainty propagation. The main reason for this selection is that the Monte-Carlo method does not require particular knowledge of statistics and is suitable for calculations with complex models or software packages. Latin-hypercube sampling is a kind of improved stratified sampling method which divides each of the parameters into N disjoint intervals with equal probability mass. This technique could provide a good coverage of the parameter samples space with relatively small number of samples compared to random sampling. More detailed information about the Latin-hypercube sampling method can be found in Mckay et al. (1979) [8].

Sensitivity analysis can be divided in different levels, from addressing the important or influential parameters to addressing the interaction between parameters and the quadratic influence of a single parameter. For uncertainty quantification studies in engineering, normally a parameter screening process will be carried out first to address the influential parameters from a large set of parameters. For parameter screening, the Morris factorial sampling method is recommended to be used. The Morris method indicates which factors are important, and also gives information on the directions of the main effects and on the severity of these interactions or nonlinear effects. A more detailed introduction into the Morris sampling method can be found in [9].

4.2 Treatment of occupant behavior

Above, we have stated that the research is not going to predict the whole occupant behavior pattern during the usage stage but is going to explore the reaction of the system's performance to certain occupant behavior, as we defined the criteria of individual controllability. For the parameters related to occupant behavior we defined in section 3 that the variation ranges are all considered as (0, 1). A parameter screening analysis (sensitivity analysis) among such parameters is proposed to be carried out. Then the main influence on the performance and the interaction among such parameters can be estimated.

5. PILOT CASE STUDY

A pilot case study is carried out to show the application of above described assessment method. The case dwelling is assumed to be located in Delft in the Netherlands, identical to the reference Dutch single family house from Agentschap NL (Tussenwoning) which can be found in [10]. The ventilation system equipped in this reference house uses a central mechanical exhaust and natural supply of air (standard Dutch system C). The process of the assessment is described below.

Identification of assessment criteria. According to the Bouwbesluit 2003 (Dutch regulation), relevant assessment criteria are (for air flow rate, expressed in dm^3/s): the room exhaust rates for kitchen, bathroom and toilet, room supply air for living room and bedroom 1, 2, 3, whole house supply rate (we consider two conditions: with infiltration and without infiltration) and unwanted air flow direction, including the air flow from bathroom and toilet room to habitable space.

Estimation of uncertainties. The uncertainties estimated are related to the following components or parameters: facade air inlets, overflow components, positions of exhaust grilles, building air tightness, wind pressure coefficient, wind reduction factor and use of air inlets and internal doors.

Calculation process. Calculations are based on the model built in the simulation tool TRNFLOW (a combination of TRNSYS and COMIS). A whole year simulation is carried out, while the typical annual climate data for De Bilt is used. 130 runs were executed for uncertainty propagation of 26 parameters (the probability distribution of each parameter is divided into 130 non-overlapping intervals with equal probability mass and then each interval is sampled once) and 148 runs were executed for parameter screening among 35 parameters (each parameter is sampled 4 values; use of air inlets and internal doors are calculated separately from the other three aspects).

Results. Part of the results is shown in figure 2. The value in figure 2a is annually averaged hourly flow rate for bedroom1 (main bedroom), including ventilation and infiltration (when all air inlets are fully opened) which is even below the designed pure supply value 14.7 dm3/s (without infiltration). Figure 2b shows that the ventilation and infiltration rate of bedroom1 is actually highly determined by the position of the air inlet in the living room. This means the individual controllability of the air inlet in the living room is limited. Other results can be displayed in the same forms as shown in figure 2 a/b.

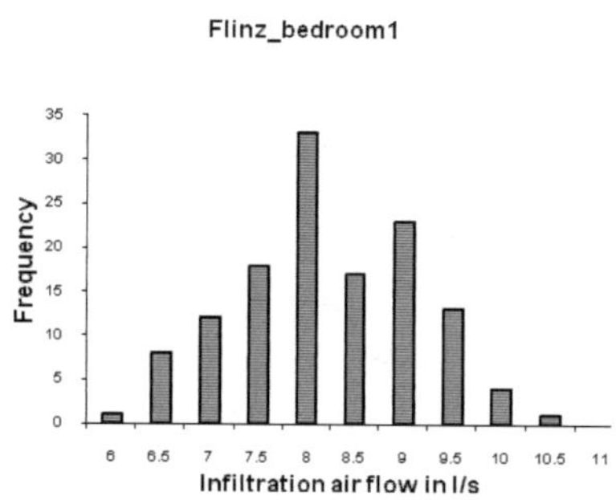

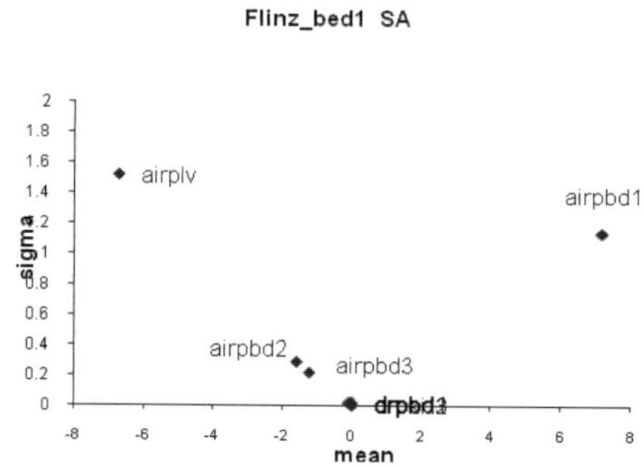

(Note: "airplv" and "airbd1" mean the positions of facade grille in living room and in bedroom1 respectively.)
Figure 2 a/b. Uncertainties in total ventilation in bedroom1 and influential factors

6. CONCLUSION

In this article, we introduced an assessment method of ventilation systems in dwellings considering the influence of uncertainties. The core objective of the method is to explore the performance of ventilation systems in houses expressed by performance criteria defined in ventilation regulations with considering the uncertainties in 4 categories (ventilation components, building properties, outdoor environment and occupant behavior). The relevant methods for defining criteria and estimating uncertainties are given. The pilot case study shows that uncertainties could have significant influence on the performance of ventilation systems in houses and the method described could give a useful framework and routine for carrying out such analysis.

REFERENCES

[1] Macdonald, I. A. 2002, *Quantifying the Effects of Uncertainty in Building Simulation*, Doctoral thesis, University of Strathclyde.

[2] De wit, M. S. 2001, *Uncertainty in Predictions of Thernal Comfort in Buildings*, Doctoral thesis, Delft University of Technology.

[3] Wouters, P., Heijmans, N., and Loncour, X. 2004. *Outline for a General Framework for the Assessment of Innovative Ventilation Systems*, RESHYVENT, WP 4, public report.

[4] Residential Mechanical Ventilation System CAN/CSA-F326-M91, Canada.

[5] The 2010 Danish Building Regulations, Denmark.

[6] The Building Regulations 2010 Approved Document F: F1 Means of Ventilation, United Kingdom.

[7] Bouwbesluit 2003, the Netherlands, available via: www.boubesluitonline.nl.

[8] Mckay, M. D., Beckman, R. J., and Conover, W. J. 1979, A Comparison of Three Methods for Selecting *Values of Input Variables in the Analysis of Output From a Computer Code, Technometrics*, 42, 55-61.

[9] Morris, M. D. 1991, Factorial *Sampling Plans for Preliminary Computational Experiments*, Technometrics, 33, 161-174.

[10] Referentiewoningen nieuwbouw, Agentschap NL (SenterNovem), 2006, available from:http://www.rijksoverheid.nl/documenten-en-publicaties/brochures/2010/11/24/referentiewoningen-nieuwbouw.html

EVALUATION OF SOME DCV CONTROL STRATEGIES BASED ON BUILDING TYPES

Ke Xu, Bjørn R. Sørensen

Narvik University College
P.O. Box 385, N-8505 Narvik
Norway

ABSTRACT

During the recent decades, energy consumption of buildings, together with the costs for operation, has gained increasing concern. HVAC systems stand for a significant share of the total energy consumption in buildings. Demand-controlled ventilation (DCV) has proved to be an efficient system that gives opportunity to strongly reduce energy consumption, especially when contamination loads or temperature load vary during the operating hours. 30-60% energy reduction can be expected by applying proper DCV. However, while focusing on energy savings, it is important to simultaneously maintain acceptable indoor air quality (IAQ) and thermal comfort. There is thus a need for a balance of emphasis between different IAQ factors such as CO_2 concerntration, volatile organic compounds (VOC), etc. when the DCV control strategies are assessed and implemented.

The aim of this paper is to provide advices for choosing DCV control strategies by evaluating existing ones such as CO_2 concentration control, Multi-zone control, dynamic-occupancy-detecting control etc. by matching them with users' preferences according to different building types for residential buildings, office buildings and schools.

KEYWORDS

DCV, evaluation, control strategies, building type preferences

1 INTRODUCTION

The aim of ventilation in a building is to provide users with good and comfortable indoor air environment. It should be noticed that 'good' and 'comfortable' are subjective conceptions, which depends on not only objective composition of the air handling components, but also users' individual experiences. To achieve this aim of ventilation, there are mainly two approaches. One is to regulate indoor air handling components by using technical means, diluting or removing contaminants and improving air 'freshness'. The other is to determine IAQ preferences of different user groups through the investigation of different building types or research about building regulations and standards. The first one is to achieve more efficient and more economical methods through engineering, such as better control behavior, new detection technology or sensors with higher quality etc. The latter belongs to a certain extent to a more sociological related category.

After more than 20 years of research and application, demand-controlled ventilation (DCV) has proved to be an efficient mean that provides opportunity to strongly reduce energy consumption when contamination loads or temperature load vary during the operating hours. However, as researched further, a lot of control strategies are developed which take more and more parameters into account simultaneously. This leads to more complex systems, and increased number of components. Growth in the complexity of DCV systems directly results in growth in possibility of malfunction or error in measurements and control. There is thus a need to emphasize a balance of different IAQ factors such as CO_2 concentration, volatile organic compounds (VOC), etc, namely to chose the most representative indicator when the DCV control strategies are assessed and implemented and the complexity of DCV system should be taken into consideration while evaluated. 'All in one' solutions are not always the best choice.

The aim of this paper is to provide some basis for selecting DCV control strategies by reviewing and evaluating existing ones such as CO_2 concentration control, Multi-zone control, dynamic-occupancy-detecting control etc., and further by matching them with users' preferences according to different building types for residential buildings, office buildings and schools.

2 REVIEW OF SOME DCV CONTROL STRATEGIES ACCORDING TO BUILDING TYPES

2.1 Residential buildings

In many countries ventilation with constant air volume (CAV) is mostly chosen type in dwellings. Due to fewer fluctuations in the occupancy level, there is less expected energy saving potential caused by change in number of occupants in dwellings compared with other building types like schools or offices. However, DCV can still be a good solution for energy saving in dwellings. On one hand the contaminants from users' activities vary during the day and the houses are unoccupied for a long time (usually the working hours) every day. It is hence possible to reduce energy consumption considerably by implementing proper DCV. On the other hand, because of small number of occupants and relatively regular (more predictable) activity patterns, DCV in dwellings requires smaller systems and simpler control strategies.

2.1.1 Description of control strategy

Besides CO_2 concentration, moisture is also important to dilute or remove by ventilation in residential houses. Research on different DCV control strategies shows that it is possible to achieve energy savings by using CO_2 or humidity controlled ventilation to reduce flow rate while maintaining acceptable IAQ [1]. A control strategy was developed by two Danish researchers to reduce energy use without large change in IAQ and moisture concentration compared with CAV. To simplify the system and reduce the investment costs of DCV, this developed control strategy is based only on measurements in the air handling unit that control the speed of the fans. There are only two ventilation rates: either a high rate designed to maintain acceptable IAQ while the occupants are present in the house or a low rate designed to remove contaminants due to building materials or indoor equipments while the house is unoccupied. The movement of occupants (entering or leaving the house) is based on measured difference in CO_2 concentration between the exhaust air and outdoor air. The difference in absolute humidity between exhaust air and outdoor air is used to assure that the high ventilation rate is maintained until the humidity in the dwelling is below a certain threshold [2].

This control strategy was tested in a single family house with a floor area of $140m^2$. Occupants were two adults and two young children, and all occupants were out of the house during working hours for work or school. The measurements were focused on CO_2 concentration and humidity in the living rooms. The chosen flow rates for the test house were

216m^3/h for the high rate and 80m^3/h for the low rate, based on the requirements in Danish building regulations and indoor air quality standards [2].

While choosing different threshold levels for difference in CO_2 concentration and threshold for difference in absolute humidity between the exhaust air and outdoor air, the aforementioned control strategy was tested in four different periods. The results are summarized in Table 1.

Table 1. Results from four different time periods [2].

Series	Measurement period	Threshold for difference in CO_2 concentration [ppm]	Threshold for difference in absolute humidity [g/kg]	Fraction of time on low fan
1	20/2-26/2 2009	100	1	13%
2	18/3-26/3 2009	200	2	31%
3	28/3-13/4 2009	100	2	34%
4	15/4-24/4 2009	150	2	37%

The best result was obtained with a 150ppm limit on difference in CO_2 concentration and 2 g/kg limit on difference in absolute humidity. In this case, the ventilation was running with the low ventilation rate 37% of the time. By assuming that electric power depends on air flow rate to the third, electricity consumption is theoretically reduced by 95% while reducing the flow rate from 216m^3/h to 80m^3/h. Thus, operating on the low flow rate 37% of the time resulted in 35% lower electricity consumption compared to CAV.

It is shown in Fig. 1 that the implemented control strategy does cause significant change in IAQ. Both CO_2 concentration and indoor moisture level were kept within the designated limit most of the time.

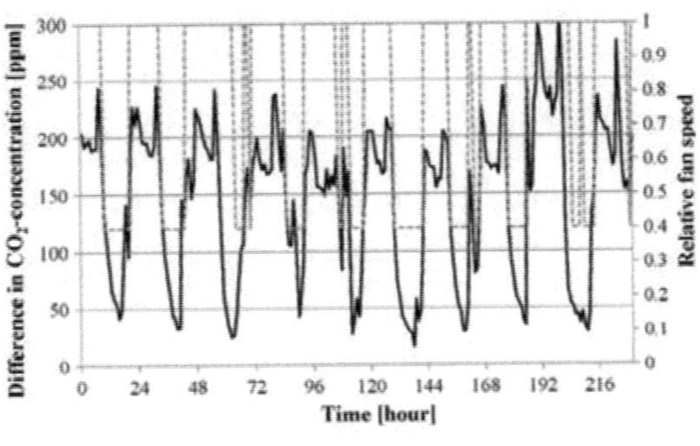

Fig. 1. Difference in CO_2 concentration between extracted air and outdoor air, and relative fan speed for series 4 (150 ppm, 2 g/kg). Time 0 h is midnight [2].

2.1.2 Evaluation of control strategy

The control strategy has following advantages:

1. Simple. There are only four control parameters need to be calculated: the high ventilation air flow rate and the low one, the limit of difference in CO_2 concentration between extracted air and outdoor air and the limit of difference in absolute humidity between extracted air and outdoor air. And only the two differences above need to be measured during operating period.

2. Inexpensive. All components used for the control are located in the air handling unit. That suggests very limited investments on implementing DCV in single family house compared with other building types.
3. Reliable data. The control strategy is tested in field experiment which provides more reliable data than only in simulations.

This control strategy satisfies requirements for DCV in residential buildings: simple, inexpensive, and easy to operate. However, in order to achieve the benefits of this control strategy, there are two issues that need further research:

1. The high and low air supply rates are the key parameters of this control strategy; they are calculated based on the relevant Danish residential building regulations. When applied in other countries or regions, the control strategy needs to be adapted to the respective building regulations and indoor air quality standards. The difference between the air supply rate when the house is occupied and the minimum air supply rate when the house is unoccupied directly affects the efficiency of this control strategy.
2. The relationship between length of low air flow rate duration and electric power saving potential is influenced by a series of parameters such as fan type, the efficiency of the heat recovery etc. Although the field experiments verify that this control strategy is able to achieve low air supply rate for nearly 40% of the operating hours, the energy saving capacity need to be confirmed by taking the actual technical parameters of components in the DCV system into consideration

2.2 Schools

The pollution loads pattern in schools is quite different from other building types. People play a more important role as pollution source than in office- or residential buildings. Thus the pollution loads concentration (mainly CO_2) can vary significantly during occupied period in local places such as class rooms or auditoriums due to strongly changing occupancy.

In order to study the pattern of occupancy change in school, records of occupancy changes are taken for three different classrooms in a college in two weeks for each classroom. By taking records for a large auditorium in two working days as an exsample, it is shown in Fig. 2 that occupancy changes significantly each day and the occupied period is not of the same pattern or magnitude for different days. That means occupancy pattern in schools is not so fixed as in office- or residential buildings but more unpredictable.

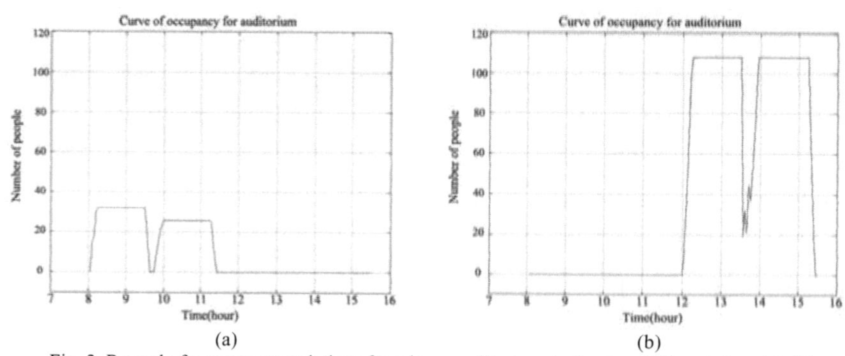

Fig. 2. Record of occupancy variations for a large auditorium during two different days (a), (b).

In order to operate DCV effectively in schools under the condition of such unpredictable occupancy changes, detecting current–time population accurately is the weighting factor

2.2.1 Description of control strategy

Dynamic-occupancy-detecting control was developed to obtain a reliable calculated current–time population to determine the minimum required outdoor air flow rate [3].

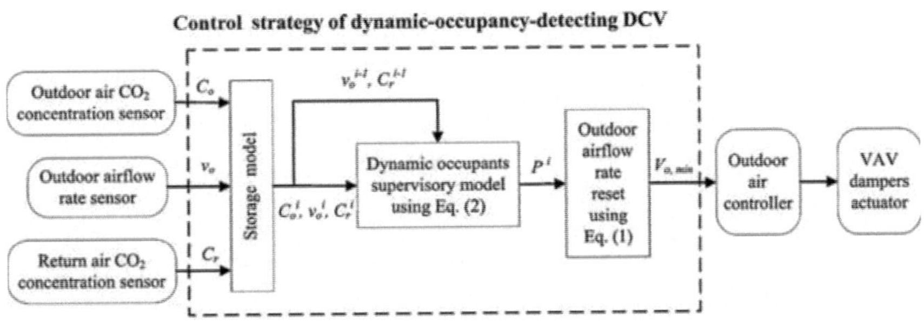

Fig. 3 Control strategy scheme of dynamic occupancy detection [3]

As shown in Fig. 3, three parameters are measured by sensors and saved in a storage model: the indoor CO_2 concentration, the outdoor CO_2 concentration and the outdoor air flow rate. The dynamic occupants in controlling zone is detected based on the difference between indoor and outdoor CO_2 concentration, and the minimum outdoor air flow rate is calculated by Eq. (1).

Required outdoor air flow rate is determined by the combination of occupant-related and area-related contaminant (odor) concentrations. The equation for required outdoor air flow rate according to ASHRAE Standard [4] can be depicted as:

$$v^i_{o,\min} = R_P \times P^i + R_a \times A \tag{1}$$

Where

$v^i_{o,min}$ is the total minimum outdoor air flow rate at time i,
R_P is the outdoor airflow rate required per person,
P^i is the total number of occupants at time i,
R_a is the outdoor airflow rate required per unit area,
A is the total occupied floor area.

The current–time population P^i can be calculated by Eq. (2) [5]. The values at time *i-1* and time *i*, which represent the previous and current sampling time, are measured and used as inputs for the dynamic occupant supervisory model.

$$P^i = \frac{(v^i_o + v^{i-1}_o) \times (C^i_r - C^i_o)}{2S} + \frac{V \times (C^i_r - C^{i-1}_r)}{S\Delta t} \tag{2}$$

Where

v_o is total outdoor air volume,
C_r is the CO_2 concentration for indoor air,
C_o is the CO_2 concentration for outdoor air,
S is the CO_2 emission rate per person,

V is the total occupied floor volume,
Δt is the sampling interval.

Fig.4 shows that the calculated results from the dynamic occupancy detection model are very close to the actual current–time population when tested in an office room. Accurate indoor dynamic occupancy is calculated by using this control strategy to determine the minimum required outdoor air flow rate.

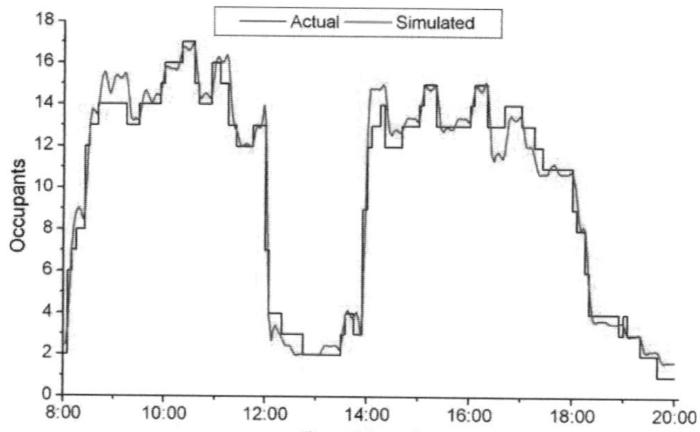

Fig. 4 Comparison of population between actual and dynamical detection of occupants for an office [3].

2.2.2 Evaluation of control strategy

The aforementioned control strategy has the following advantages:

1. Simple calculation models. Only three parameters need to be measured: the indoor CO_2 concentration, the outdoor CO_2 concentration and the outdoor air flow rate. Only two equations are applied in calculation.
2. Acceptable accuracy. Experiment shows calculated current–time population from the dynamic occupancy detection model are very close to the actual value to determine the minimum required outdoor air flow rate.

The dynamic-occupancy-detecting control strategy adapt to unpredictable occupancy in schools by providing a calculated current–time population with exceptional accuracy, but there are some factors which impact the actual accuracy of this control strategy:

1. Sensors. Due to there are only three parameters measured to calculate current–time population, the detecting accuracy at each sensor will strongly impact the final calculated result. In order to guarantee reliable measurements, assessment of number of sensors, location of sensors etc. must be performed.
2. Further test. As this control strategy only solve the problem to produce a detected occupancy by using a quite simple calculation model, its applicability to actual DCV system in schools and potential of energy saving need to be further tested in specific situations.

A similar occupancy calculation procedure was also decribed in [6].

2.3 Office buildings

Similar to residential buildings, office buildings have relatively regular change over time in the number of users since they are occupied during the working hours only. However, in practical situations, it is rare that all the rooms in office buildings are occupied at the same time. In order to determine the potential energy savings and optimize the capacity of a DCV system, it is necessary to know the occupancy level. Occupancy factor, which is defined as the actual number of occupied rooms, divided by the total number of rooms, is often used to describe the occupancy level. Early study shows that OF (occupancy factor) is about 50% most of the time, and the peak OF is 70%. Typical fluctuation of OF in an office building is shown in Fig 5 [7].

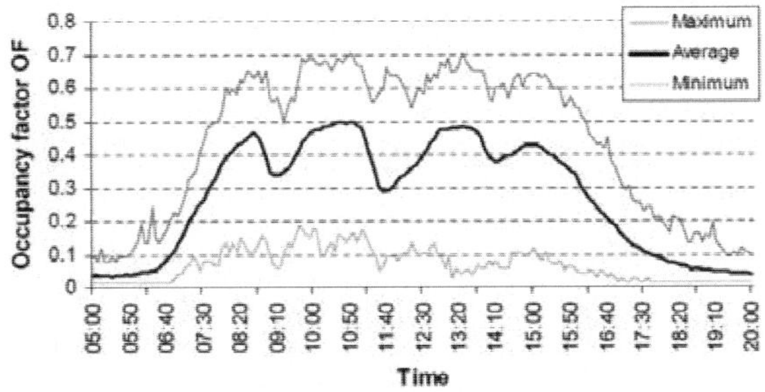

Fig. 5. Occupancy factor for offices for all working days over the one year measurement period for a case study in an office building [7].

The fluctuation of OF for offices is caused by occupants moving from offices to break rooms at noon, or to meeting rooms at any time of the day. Due to this movement, some zones in the office building become over-ventilated or under-ventilated although the total number of occupants does not change significantly. Under such conditions, multi-zone control has potential to provide proper IAQ.

2.3.1 Description of control strategy

Based on dynamic-occupancy-detecting control, a multi-zone DCV control strategy is developed with consideration of local IAQ satisfaction in each zone. As shown in Fig. 6, total required outdoor air flow rate is first calculated based on detection of total occupancy [8]. Required outdoor air flow rates in different zones are then calculated based on their own occupancies. By detecting the critical zone (or room), total outdoor air fraction is corrected by outdoor air fraction of the critical zone. At last the corrected outdoor flow rate is calculated. A more precise formulation of the calculation model can be found in [8]. This is a type of distributed DCV that works efficiently. Distribution is not dependent on the measurement of pressure, and more accurate demand based flow rates to the zones can be obtained.

2.3.2 Evaluation of control strategy

Validated in both simulation and field test [8], the multi-zone DCV control strategy can help to reduce the energy consumption and running cost without significant IAQ reduction (compared to the conditions normally provided by a CAV system). Special attention must be directed to local IAQ when this control strategy is implemented. To enable online control and ultimately achieve better local IAQ, this control strategy needs an Intelligent Building Management (IBM) and Integration platform as overall communication platform for operating the DCV strategy efficiently. Establishing IBM systems may lead to increased investment costs, but it is generally believed that operating costs will be lower, especially for large buildings. In [8], investigations were performed for a 500m high skyscraper of about 321,000 m^2 floor area.

The balance between the control accuracy and system size as well as initial investment on equipment to achieve this control strategy needs to be further evaluated when implementing in office buildings of different sizes.

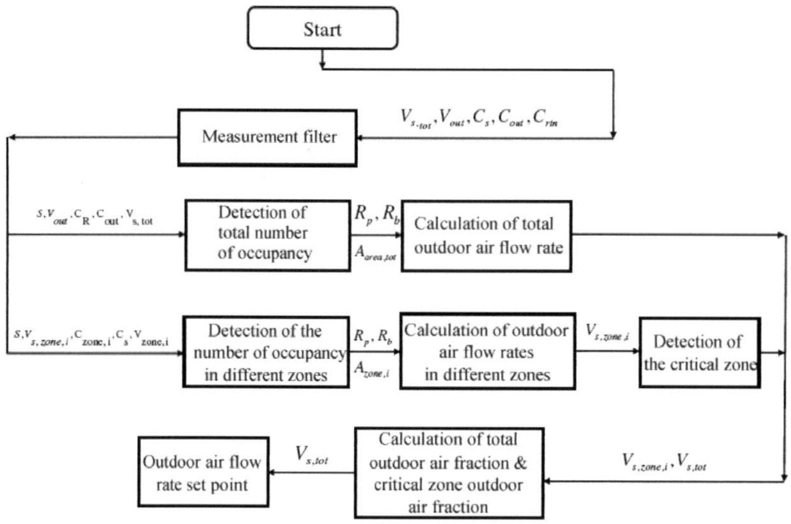

Fig. 6 Flow chart of multi-zone DCV [8]

SUMMARY AND CONCLUSIONS

A few relevant DCV control strategies have been reviewed in this paper. Each type of building has its own load patterns, and therefore also a corresponding building type-related preference for DCV control strategy. Although CO_2 concentration is a commonly chosen control objective and in many cases used as the main parameter, there are other factors according to building type-related preference for residential buildings, schools and office buildings that need to be assessed.

1. Residential buildings. The expected energy savings potential is not so high as for other building types due to the relatively fixed occupant behavior and small size of houses/apartments/dwellings. DCV can however still be a good solution for energy savings in dwellings as they are unoccupied in long periods every day. These characteristics also suggest a simplified control strategy, which is easy to operate and not too expensive to establish.

2. Schools. Compared with that in other building types, indoor contaminants in schools are to a great extent occupancy-related due to the large number of occupants and significant variations both during one day and between different days. To address an efficient DCV strategy for schools, that is able to maintain IAQ conditions at acceptable levels at all times, detecting the occupancy of the different rooms may be crucial. However, the fluctuations in the number of occupants make it difficult to predict the occupancy accurately, and a dynamic-occupancy-detecting control strategy can be a solution. This also enables possibility for direct flow control (not using duct static pressure or fan differential pressure as control objectives of the fans).

3. Office buildings. Similar to residential buildings, office buildings have relatively regular change in contaminant load over time; just they are occupied during the working hours. However the total occupancy has not so large change as in schools, the local IAQ should be paid attention to due to that not all the rooms in the office building are occupied simultaneously. Over-ventilation or under-ventilation is caused by the movement of people within the building and unsatisfied local IAQ is formed. Both lower energy use and better local IAQ should be the target for implementing DCV in office buildings.

ACKNOWLEDGEMENT

Thanks to the financially support for the work presented in this paper from Narvik University College.

REFERENCES

[1] Vitalijus Pavlovas, Demand controlled ventilation: *A case study for existing Swedish multifamily buildings*, Energy and Buildings 36 (2004) 1029–1034

[2] Toke Rammer Nielsena, Christian Drivsholm, *Energy efficient demand controlled ventilation in single family houses*, Energy and Buildings 42 (2010) 1995-1998

[3] Xue-Bin Yang, Xin-Qiao Jin, Zhi-Min Du, Bo Fan, Xiao-Feng Chai, *Evaluation of four control strategies for building VAV air-conditioning systems*, Energy and Buildings 43 (2011) 414-422.

[4] Tullie Circle NE, Atlanta,GA 30329. *ASHRAE. Energy standard for buildings except low-rise residential buildings. ANSI/ ASHRAE/IESNA standard* 90.1-2004.: *American Society of Heating, Refrigerating and Air-conditioning Engineers*, Inc.; 2004.

[5] S. Wang, X. Jin, *CO_2-based occupancy detection for on-line outdoor air flow control*, Indoor Built Environment 7 (3) (1998) 165-181.

[6] Sørensen Bjørn R., *A decipol predictive controller for VAV systems*, Proceedings of the 18th AIVC conference, 1997 Greece.

[7] Mari-Liis Maripuu, *Demand control ventilation (DCV) systems in commercial buildings-Functional requirements on system and components*, Doctoral thesis 2009

[8] Zhongwei Sun, Shengwei Wang, Zhenjun Ma, *In-situ implementation and validation of a CO_2-based adaptive demand-controlled ventilation strategy in a multi-zone office building*, Building and Environment 46 (2011) 124-133

DEMAND-CONTROLLED VENTILATION: AN OUTLINE OF ASSESSMENT METHODS AND SIMULATIONS TOOLS

Jean-Luc Savin[1] and Jelle Laverge[2]

1 Aereco SA
9 allée du clos des charmes
Collégien
77615 Marne la Vallée cedex 3
France
**jeanluc.savin@aereco.com*

2 Ghent University
Research Unit Building Physics, Construction and Climate Control
Department of Architecture and Urban Planning
Faculty of Engineering
Jozef Plateaustraat 22
B-9000 Gent
Belgium

ABSTRACT

Enhancing the energy efficiency of buildings imposed by global warming and by the perspective of fossil fuel dwindling requires new technical solutions, more efficient. The race for efficiency directly affects ventilation and air tightness of buildings, the main potential causes of heat loss in homes. If heat recovery is emerging as an effective solution to meet energy performance and indoor air quality in climates with harsh winters, some other solutions appear to be very efficient in moderate climates. The unbalanced demand-controlled ventilation, by adjusting the airflow as to the needs, provides a particularly effective alternative to heat recovery on climatic zones where the difference in indoor-outdoor temperature may not be enough to balance the excess power consumption generated by the heat recovery. Used with humidity control (automatic control of the airflow according to humidity level through a mechanical sensor), unbalanced demand-controlled ventilation system has met a large success due to its simplicity of implementation and operation, especially in the field of social housing where the investment and maintenance constraints are often dominant.

Born in France in the early eighties already in a context of an energy crisis, humidity controlled ventilation has since spread to various countries where it now benefits from a real technical assessment in the regulation. This is particularly the case in Belgium, Spain, Germany and Poland. Different evaluation methods, different tools and assumptions are used according to countries, leading to substantially different results from one country to another. This variety of performance, whether on the energy side or IAQ, finds its origins in various parameters such as regulated airflow rate for reference, outdoor climate, and occupancy modes. This article aims to provide a state of the art of simulation tools, assumptions and evaluation procedure used in different countries to assess DCV. Reference to several studies complement this presentation, such as that conducted by the Fraunhofer Institute in 2009 to compare heat recovery and unbalanced demand-controlled ventilation or the one by Air.H in 2009 to assess the reliability of the simulation tool used in France. This software has been directly compared with actual measurements made on site during a large-scale monitoring in Paris (2007-2009), with very convincing results on the relevance of the calculation algorithm and on the assumptions.

Several tools such as SIREN from CSTB, CONTAM from NIST and WUFi® from Fraunhofer IBP have proven their reliability to assess DCV. Various evaluation methods are now used in different countries, moving DCV and humidity controlled ventilation from innovative to standard. The availability of simulation tools and assessment methods remains essential for all countries wishing to exploit the energy gain potential offered by this technology.

KEYWORDS

Demand, controlled, ventilation, assessment, procedure, model, simulation

INTRODUCTION

Demand controlled ventilation is considered today as a particularly relevant alternative to other efficient ventilation systems such as heat recovery, especially in residential application. With its growing diffusion increases the need for assessing this technology, which requires reliable simulation tools as well as defining precise hypothesis concerning the occupancy, the dwelling characteristics, the weather data and the ventilation components. Different procedures and tools are today used in Europe to assess the energy and IAQ performance of DCV systems. This paper aims at presenting some of the most representative ones, with the application examples of France, Belgium and Germany.

FRENCH APPROACH AND SIREN SIMULATIONS

Regulation and ventilation systems

The French regulation for residential ventilation is defined by a bylaw ("arrêté du 24 mars 1982 modifié le 28 octobre 1983"). The regulation text requires a ventilation system for all new dwelling, natural or mechanical, with a permanent airflow that can be constant or variable according to the demand. Air inlets must be located in the main rooms (bedrooms and living rooms), exhaust is placed in the technical rooms (kitchen, bathroom and toilets). Specific airflows are defined according to the type of dwelling, characterised by the number of main rooms. The regulation requires a minimum value for maximum airflows (see Table 1), and a minimum value for the total airflow (see Table 2) that is agreed only if the demand controlled system is automatic such as the humidity controlled MEV system[1] for example.

Number of main rooms	Airflows (m3/h)				
	Kitchen	Bathroom w/wo WC	Other bathroom	WC	
				One	Several
1	75	15	15	15	15
2	90	15	15	15	15
3	105	30	15	15	15
4	120	30	15	30	15
5 and more	135	30	15	30	15

Table 1. Minimum value for maximum airflow according to the number of main rooms.

	Number of main rooms						
	1	2	3	4	5	6	7
Minimum value for total Airflow (m3/h)	10	10	15	20	25	30	35

Table 2. Minimum value for total airflow according to the number of main rooms when using automatic demand controlled ventilation system.

[1] The humidity controlled MEV system commonly used in France is composed of humidity controlled air inlets and humidity controlled (mixed with presence detection or switch) extract units. The humidity sensor is mechanical, made of polyamide strips.

Assessment procedure and model

The distribution of demand controlled ventilation systems in France is submitted to availability of technical agreements ("avis techniques") delivered by CSTB. This document not only supplies all the technical requirements, applications, constraints and limits for the system; it validates as well its performance for indoor air quality and energy. The procedure uses the equivalence principle to assess the demand controlled ventilation system: the energy performance of the system is expressed in terms of a constant airflow for each type of dwelling, that is equivalent with the variable one for energy losses. These value can be directly used in the EPB software to assess the global energy performance of the dwelling.

SIREN software has been created by the CSTB more than 15 years ago to assess the IAQ and energy performance of DCV and other innovative ventilation systems in the framework of the Technical Agreement procedure. First used in France, it is currently used to assess demand controlled ventilation in Spain. This dynamic simulation tool offers a calculation method to characterise the aeraulic behaviour of ventilation systems as well as occupants' exposure to pollutants according to various impacting parameters (see Figure 3).

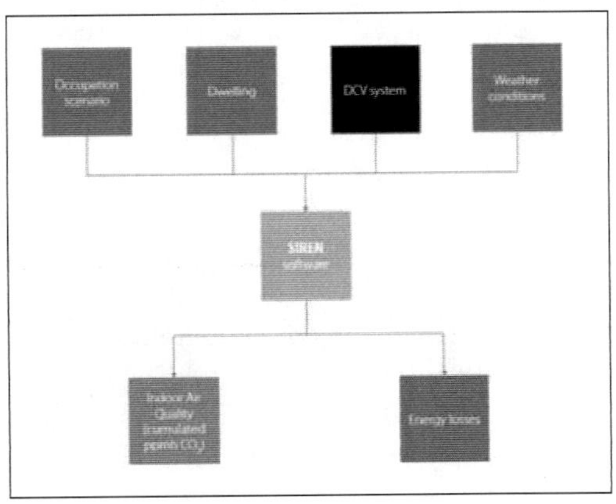

Figure 3. Input and output data for SIREN simulations

The software takes into account numerous factors such as orientation of the dwelling (influence of wind pressures), air leakage, infiltration, location of ventilation components, activity of the inhabitants (through occupation scenario - see Figure 4), adsorption and desorption phenomena, etc. to best mimic the real physical behaviour. A stratification of internal temperatures is also taken into account. The tolerance of the ventilation components must be precisely described to enable the calculation for IAQ (product at the lowest airflow in the tolerance) and for energy savings.

	Number of living spaces						
	1	2	3	4	5	6	7
For IAQ testing	2	2	3	3	3	4	4
For average flow rates	1	1,5	2	2,5	3	3,5	4

CO_2	awake	16 l/h
	sleeping	10 l/h
H_2O	awake	40 g/h
	sleeping	

Figure 4. Base input data for occupation scenarii: Number of occupants according to type of dwelling (left); Metabolism CO2 and H2O production (right).

SIREN computes the indoor air quality (expressed in cumulated CO2 values [ppm.h.]) and the energy losses due to ventilation (thermal losses), expressed as a constant airflow equivalent for energy losses. The ventilation system is validated when the number of hours when CO2 concentration exceed 2000 ppm multiplied by the CO2 concentration during this period does not exceed 500 000 ppm.h. for all the heating season. Several additional parameters such as the number of hours with condensations on windows, numbers of hours with varied humidity ranges, etc. are calculated too, to give a complementary indicative performance result. A data for air sections of demand controlled air inlets is as well given by the simulation tool to be used in the EPB software (influence on the cross-flow ventilation).

Validation of the simulation software

SIREN predictive model has been checked through a large scale experiment in Paris ("Performance de la ventilation et du bâti") in 2009. This monitoring study, carried out on a total of 29 occupied dwellings in two buildings, has confirmed the simulation results with a discrepancy lower than 10%, as showed on Figure 5. The measured results have demonstrated a fairly good reliability of the dynamic tool for energy and aerodynamic simulations. Although SIREN seems to over-evaluate condensation risks (this may be due to some assumptions on room surfaces, doors and windows opening… etc), the software has been proved to be particularly relevant for the evaluation of humidity and demand controlled ventilation systems.

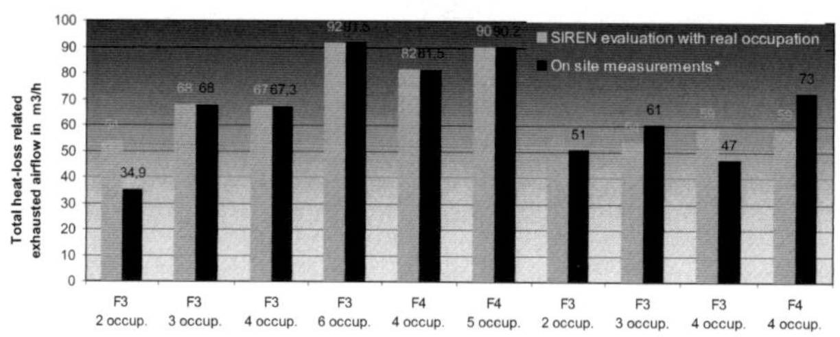

Figure 5. Comparison between SIREN simulations and in-situ measured airflows in the framework of a large scale monitoring ("Performance"). Yearly average heat-losses related airflow per dwelling / occupation type.

From a mono-zone dwelling calculation exclusively designed for MEV, SIREN software is now able to simulate a multi-house building, communicating with an external aeraulic simulation tool to take into account the variable pressure at the terminals. Next developments will allow the calculations in natural and hybrid ventilations; user licences can be bought from CSTB.

BELGIAN APPROACH AND CONTAM CALCULATIONS

Regulation and ventilation systems

In the competence structure of the different state levels in the federal state Belgium, the implementation of the EPB-directive is a regional competence. The Belgian residential

ventilation requirements are set forward in the Belgian Standard NBN D 50-001, which is annexed to the EPB-decrees of the different regions. This standard dates back from 1991 and in it, the ventilation systems are presented in a descriptive manner. 4 standard systems are described, ranging from natural ventilation (system A), over simple exhaust (system C) or supply ventilation (system B) to fully mechanical ventilation (system D). In the market, systems A, C and D are dominant, while system B is virtually inexistent.
The standard requires the air supply and return components of the systems to be sized according to the function of the room in which they are located. These required design flow rates are listed in Table 6. Non-mechanical components such as trickle ventilators and transfer grilles are to be sized to this flow rate at a reference pressure drop of 2 Pa.

Since the standard only mentions demand control as a possible extension of the reference systems without any detail as to how this demand control should be achieved, an equivalence approach is used to rate the performance of demand controlled ventilation systems. This assessment is done by UBAtc, a technical approval agency in Belgium and is valid for all the regions. In order to be deemed equivalent to the reference systems in the standard and therefore acceptable under the building code, the performance of the demand controlled system can not be inferior to the worst performance obtained by the application of each of the reference systems.

room	minimum	Limited to:
living	1 l/s/m^2 > 21 l/s	42 l/s
Sleep/office	1 l/s/m^2 > 7 l/s	20 l/s

room	minimum	Limited to:
Kitchen, storage, bathroom	1 l/s/m^2 > 14 l/s	21 l/s
WC	7 l/s	
Open kitchen	> 21 l/s	

Table 6. Design flow rates in the Belgian residential ventilation standard NBN D 50-001

Simulation tool and parameters

A demand controlled residential ventilation system is assessed through numerical simulations with the multi-zone airflow model Contam, developed by NIST. The uncertainty associated with different implementations of the system under review in specific dwellings is taken into account by applying a Monte-Carlo algorithm, varying such parameters as occupancy, environmental conditions and building orientation. The distributions of these parameters are based on available data from the Belgian statistics institute. 100 sample sets are used. Other parameters such as system characteristics, building leakage level and geometry and indoor temperature are fixed at different intervals. This amounts to over 1000 simulations for each system or variant of a system. The simulations are run over the typical heating season in Belgium for a moderately insulated dwelling. Both the system under review and the 3 reference systems that have a reasonable market share are assessed.

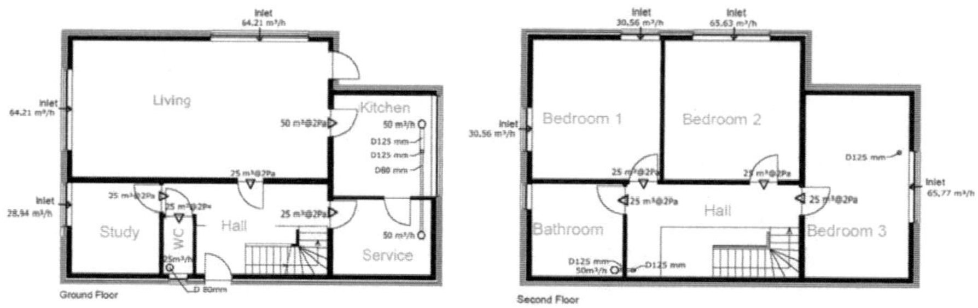

Figure 7. Geometry of the reference building used in the equivalence procedure.

The performance of the demand controlled system is assessed on 3 IAQ parameters, namely humidity level, exposure to odours (tracer) and perceived indoor air quality (carbon dioxide). If performance of the system under review is equal to that of the worst performing reference system for each of these parameters, it is accepted as equivalent and an energy saving coefficient is determined.

The level of IAQ that is attained by the application of the system is taken into account to determine the energy saving coefficient. The coefficient is defined as the ratio of the heating season integrated ventilation heat loss with the exclusion of infiltration losses and a reference. This reference is determined by interpolating between the heat loss levels of the 2 reference systems from the standard (A, C and D) that produce an IAQ level just higher and lower respectively than the system under review, based on the IAQ level attained by the latter system. The process for the determination of the reference and the energy saving coefficient for a system 'X' demonstrated in Figure 8.

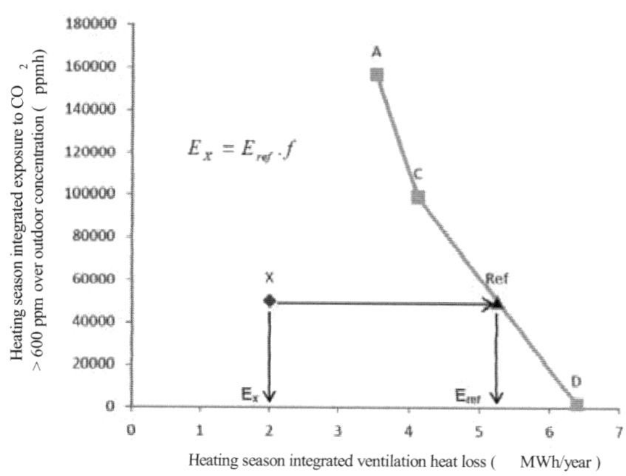

Figure 8. Reference and energy saving coefficient calculation for a ventilation system X

The infiltration heat losses are treated separately in the EPB-calculation method. Nevertheless, should the building leakage to infiltration heat loss ratio obtained under operation of the demand controlled system be different from that of the reference systems, an additional correction coefficient for this is calculated. So far, however, no system reviewed presented such behaviour.

COMPARED PERFORMANCES OF A HUMIDITY CONTROL MEV AND A BALANCED HEAT RECOVERY IN GERMANY USING "WUFI® PLUS" SIMULATION TOOL

Study

If Germany does not propose technical agreement for assessing DCV as it is done in France, Spain and Belgium, several studies have been realised using comparable simulation tools and methodology. A study conducted in 2010 by the Fraunhofer Institut Bauphysik IBP has consisted in simulating the working of two ventilation systems on a single-family house: on one hand, a demand controlled (humidity controlled) exhaust MEV system which varies the air changes depending on the relative humidity in the room and, on the other hand, a standard supply and exhaust system with a constant air change rate of 0.4 ACH and an integrated heat recovery. The heat recovery efficiency of this unit is of 93% according to its manufacturer with additional electrical preheating device. The comparative calculations has been carried out for both ventilation systems for a new built house with high insulation (according to DIN V 4108-6:2003 and DIN EN 12831-Bbl. 1) with a heated and ventilated living area of 205.6 m2 and a ventilated volume of 534.6 m3. The heat transfer coefficient of the outside walls is 0.25 W/m2K, with the roof at 0.18 W/m2K and the ground floor at 0.7 W/m2K. The windows have a U-value of 1.1 W/m2K. The reference cold city of Hof (Germany) has been selected as for the weather data. The indoor temperature is 20° C.

Simulation software

The evaluation of the various ventilation systems has been made possible by the newly-developed hygrothermal indoor climate simulation model WUFI® Plus. This last belongs to a larger software family which allows realistic calculation of the transient coupled one- and two dimensional heat and moisture transport in multi-layer building components exposed to natural weather. It is based on the newest findings regarding vapour diffusion and liquid transport in building materials and has been validated by detailed comparison with measurements obtained in the laboratory and on IBP's outdoor testing field. The sofware is available at Fraunhofer IBP.

The study, that has simulated two years working, has supplied a complete comparison between the different ventilation systems in termes of energy performance, separating heat losses from electricity consumed by the fan(s) and for the preheating (see result on Figure 8). It has demonstrated –notably- the low difference between humidity controlled MEV and balanced system (HR) with 93% heat recovery. WUFI® Plus enabled also to check the indoor air quality through the recording of CO2 concentration, according to different occupation scenarii.

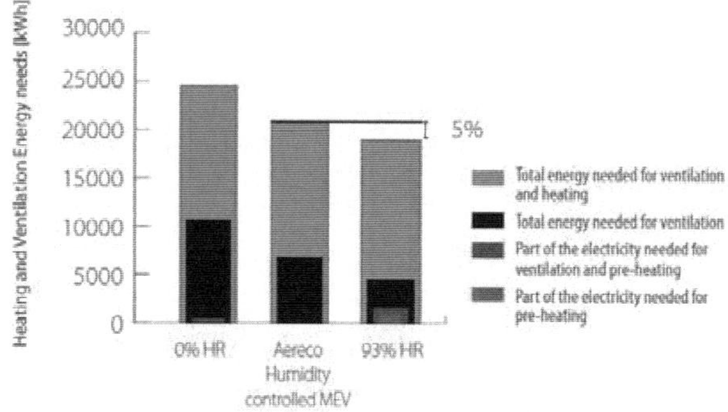

Figure 8. Total energy consumption of various compared ventilation systems

CONCLUSIONS AND PERSPECTIVES

Defining adapted assessment procedures and identifying reliable simulation tools is a crucial point to take into account demand control ventilation systems in the EPB regulations. Several procedure and simulation tools such as SIREN (CSTB), CONTAM (NIST) and WUFI® Plus (Fraunhofer IBP) for example are today used in different countries such as in France, Spain, Belgium and Germany to assess the energy and IAQ performance of DCV systems. We could cite as well studies realised in Switzerland and in Poland (through NAPE), demonstrating the strong interest and need for proposing an assessment of DCV systems.

If parameters and hypothesis must be adapted to the software, to the regulation and to specific conditions, all available simulation tools have proved their reliability, notably by comparison with laboratory or in-situ measurements. From MEV only, new developments are today being carried out to widen the application field, for hybrid and natural ventilation notably.

REFERENCES

[1] Air.H. 2009. *"Performance de la ventilation et du bâti" Final report.* Ademe convention n.0504C0114

[2] CSTB, EvalIE (Evaluation des Installations Energétiques). 2009. *Règles de calculs «siren» pour l'instruction des avis techniques sur les systèmes de ventilation mécanique hygroréglable.*

[3] J.L. Savin and A.M. Bernard, 2009. *"Performance" project: Improvement of the ventilation and building air tightness performance by implementing a quality approach when building and monitoring of results in occupied dwellings in France.* AIVC Conference 2009.

[4] A. Holm, M. Krus and D. Rösler, 2010. *Calculation of the energy demand of a supply and exhaust system with heat recovery compared to a demand controlled (humidity controlled) exhaust system in a single-family house.* IBP Summary Report RK 032/2010/292

Behavior of leakages exposed to dynamic wind loads. A numerical study using CDF on a single zone model

Dimitrios Kraniotis*, Thomas Kringlebotn Thiis, Tormod Aurlien

Department of Mathematical Sciences and Technology, Norwegian University of Life Sciences
Drøbakveien 31, P.O. box: 5003, IMT, N-1423, Ås, Norway
**Corresponding author: dimitrios.kraniotis@umb.no*

ABSTRACT

Wind is a potential dominant factor regarding the air infiltration through building envelopes. Due to its dynamic characteristics, quite complex aerodynamic phenomena arise around a structure or through cracks and openings. Energy perfomance is influenced by the climate conditions and thus it should be much more researched. Despite the fact that steady state measurements of infiltration rates offer a simple and easy way of estimating an enclosure's airtightness level, a supplement to those methods might be imposed. In this context, computational fluid dynamics could be a useful tool, generating interesting results and helping understand the role of flow mechanisms.

The current numerical study deals with the influence of wind gust on the infiltration rate of a bluff body ('box') on which cracks of varied size are located on different sides. Different internal volumes are also researched. On discussing the results, the dynamic leakage rates (infiltration and exfiltration) are under investigation, while the airflow patterns inside the 'box' are pointed out as well. Frequency characteristics of the wind gust and their correlation with the internal volume are also studied.

The results suggest that gust frequency is the dominant factor for air flows through cracks on the windward side, while the internal volume also plays an important role on the instantaneous leakage rates under unsteady wind conditions. The impact of the enclosure's volume seems to be of importance for relatively small cracks. In addition, there is a difference among the leakage rates through the studied cracks and openings depending on their location or switching from a small to a big volume. Comparisons with the respective behavior of the cracks-openings under steady state enclosure's pressurization at 50Pa are shown as well.

Since the studied enclosure represents a airtight single zone, where air flows only through the opening(s), a parallelism to the internal space of a high level airtight building arises. Taking into account the results of the current study, the influence of the frequency of the gust, the internal volume of a relatively airtight building and the location of the leakage area seem to be critical factors under climate conditions. Having ensured that leakages through the internal walls are very low, a further research should be proceed towards the importance of the location and airtightness of the internal walls. On the same mode internal volume's control and exterior local climate might be also considered.

KEYWORDS

Leakage rate, air flow, air infiltration, crack, dynamic wind load, wind gust, computational fluid dynamics, shear-stress-transport, single zone model

INTRODUCTION

Air infiltration is important towards the optimal design of energy efficient buildings. Moreover, being an uncertain phenomenon, affected by various factors, it is recognized as one of the major losses in residence buildings [1]. Steady state measurements of infiltration rates offer a simple and easy way of estimating an enclosure's airtightness level. The dynamic characteristics of air infiltration have been pointed out by Hill and Kusuda [2] and therefore challenges arise upon that field. The role of the climate parameters and location characteristics on average infiltration rates has also been studied by Sherman [3]. Turbulence causing wind gustiness is recognized as one major factor that affects infiltration [4]. In

addition, building aerodynamics contributes to air infiltration too. In that context, modelling approaches have been presented by Haghighat et al. [5, 6], while the theoretical background of the flow equations through cracks has also been studied [7, 8, 9].
Computational fluid dynamics (CFD) could be a useful tool, generating interesting results and helping understand the role of flow features under unsteady conditions. Numerical studies could contribute to the prediction of potential leakage areas and the evaluation of their contribution to leakages rates. Therefore, a supplement based on numerical methods to the flows through cracks should be imposed.

CASE STUDY

The current numerical study deals with the influence of wind gust to the infiltration rate of a bluff body ('box') on which cracks of various size are located on different sides. Three cracks/openings are studied; (i) of 3mm height (referrred as A), (ii) of 6mm height (referrred as B) and (iii) of 30mm (referrred as C). The lenght of the cracks/openings is 1,5*m* in all cases. Three different internal volumes are also researched (fig. 1). The volumes V_2 and V_3 are equal and double the cubic volume V_1. In addition, three different locations of cracks are examined; (i) the first case refers to one crack/opening across the windward side, 10*cm* below the roof of the 'box' (notation: F), (ii) the second case refers to two cracks of the same size, both situated across the windward side, one 10*cm* below the roof of the 'box' and the other 10*cm* above the ground (notation: FF), (iii) the last case refers to two cracks of the same size, one situated across the windward side, 10*cm* below the roof of the 'box' and the other across the leeward side, 10*cm* above the ground (notation: FB). Four cases of wind load (wind gusts of different frequencies) are used as inlet boundary conditions.

(a) (b) (c)
Figure 1. The three different volumes V_1 (a), V_2 (b) and V_3 (c) simulated and tested.

In total 3*3*3*4 = 108 cases are studied. The notation adopted follows the general rule:
(size of the crack)(volume of the 'box')(location of the cracks)-(frequency of the wind gust).
The table 1 presents the notation for the crack A (3*mm*), the cubic volume (V_1) and the wind gust frequency $\omega_1 = 1/2\pi$. In a similar way, a notation of 2 or 3 or 4 at the end of the notation shows the other frequencies ω_2, ω_3 and ω_4 respectively.

Case	Size of crack-opening	Volume	Location
AV_1F-1	3*mm*	V_1	one crack on the windward side (10*cm* from roof)
AAV_1FF-1	3*mm*	V_1	two cracks on the windward side (one 10*cm* from roof and the other 10*cm* from ground)
AAV_1FB-1	3*mm*	V_1	one crack on the windward side (10cm from roof) and one on the leeward side (10cm from ground)

Table 1. Example of the notation followed. Here, the notations of the studied cases for the crack A (3*mm*), the cubic volume (V_1) and the wind gust frequency ω_1 (= $1/2\pi$) are presented.

METHODOLOGY

The CAD model was developed in ANSYS Design Modeler™ 12.1. The CFX-mesh method of the ANSYS Mesh program (involved in ANSYS Workbench) was used to mesh the domains. The fluid dynamic package ANSYS CFX 12.1 was used as solver for the numerical simulations. Pressure distribution around a building is important to get correct prediction of the pressure gradients and consequently of the air infiltration through the envelope. Among the available turbulence models, the Shear-Stress-Transport (SST) model, a two equation k-ω based model [10], was imposed. The reason for that is the inclusion of transport effects into the formulation of the eddy-viscosity. This results in a major improvement in terms of flow separation predictions [11]. In addition, other relevant studies have shown a good agreement between SST model and full scale data, better rather than compared with standard k-ε and RNG k-ε models [12]. A period of 30*sec* was assumed to be the total time per run, while the timestep was selected to equal to 0,5*sec*. At the inlet of the domain, a fully turbulent velocity profile was assumed based on the equation:

$$u = 7 + 4 * \sin(\omega * t) * \left(\frac{y}{y_{ref}}\right)^{\frac{1}{7}} \quad (2)$$

where u is the wind velocity, the factor $4 * \sin(\omega * t)$ governs the frequency of the wind gust and $\left(\frac{y}{y_{ref}}\right)^{\frac{1}{7}} y$ is the 1/7 power-law factor that describes the wind profile. The exponent factor v = 1/7 corresponds to a widely applicable regarding low surface roughness and well exposed sites from which conventional National Climatic Data Center (NCDC) data are available [13]. At y = 2,5m (the height of the studied box, while y_{ref} = 10m is the reference height), the velocity of the inlet boundary is according to (1):

$$u = 7 + 3{,}82 * \sin(\omega * t) \quad (3)$$

As said above, the factor ω formulates the frequency of the wind gust. Four different frequencies – cases of wind gusts were simulated. As shown in the figure 1, ω_1 = 1/2π[*rad/s*] and T_1 = 2π[*sec*], ω_2 = 1/4π[*rad/s*] and T_2 = 4π[*sec*], ω_3 = 1/6π[*rad/s*] and T_3 = 6π[*sec*], ω_4 = 1/8π[*rad/s*] and T_4 = 8π[*sec*]. Figure 2 shows u against t at y = 2,5m.
The instantaneous mass flow rate Q_m and thus the instantaneous volumentric flow rate Q_v across the crack(s)/opening(s) are calculated during the run interval time (30*sec*) for every case (figure 3). The infiltration phase is assumed to have positive values. In contrast, the exfiltration period has negative values. In figure 3 (as well as done in the rest cases), the first 4*sec* have been excluded, because during that time period the domain was not yet under fully turbulent conditions. Both the infiltration and exfiltration rates follow a sinusoidal function, similar to the velocity of the inlet boundary condition does. Based on the Q_v, accepting its sinusoidal form and assuming that the mathematical behavior of the later does not change within an hour, the equivalent air change rate ΣACH_i extrapolated over time t_{tot} = 1h is calculated (the superscript *in* declares the infiltration phase:

$$\Sigma ACH_i^{in} = \frac{3600}{V}\left(\int_0^{t_{tot}} Q_v dt\right) \quad (4)$$

The annual average infiltration rate n is also calculated in order to be compared with ΣACH_i [14]:

$$n = \frac{ACH_{50}}{20} \quad (5)$$

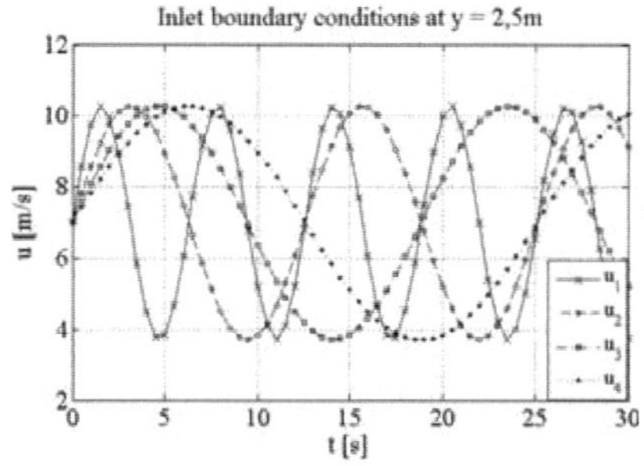

Figure 2. Inlet boundary conditions at $y = 2{,}5m$.

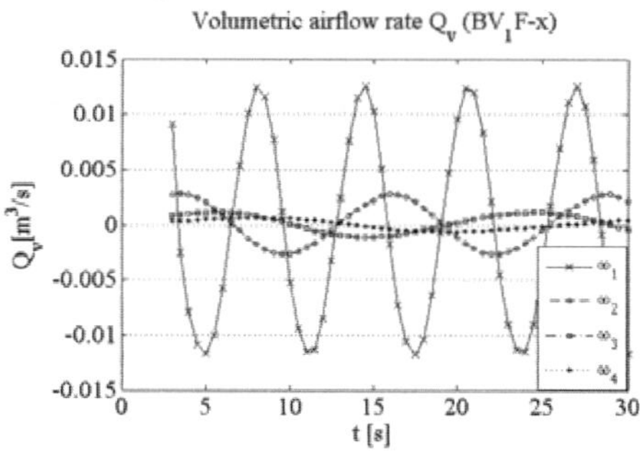

Figure 3. Variation of volumetric airflow rate Q_v through crack B ($6mm$) against time for each gust frequency ω.

RESULTS

The equivalent air change rate ΣACH_i is plotted against gust frequency ω_i and is shown in figures 4, 5 and 6. The annual average air change rate is also shown. The leakage rate increases with ω_i for the cases of (i) one crack on the windward (fig. 4) and (ii) two cracks on the windward side (fig. 5). The impact of ω_i seems to become much more significant in higher frequencies (slightly parabolic form) for those cases. It plays an important role for a single zone model with leakages located perpendicular to the main wind flow direction, while it becomes the dominant factor for high frequencies. Again, ω governs the wind-induced pressure difference across the envelope of the 'box', which is the driving force of the air infiltration (and exfiltration) through leakage areas of such size. In addition, the ΣACH_1 for the highest gust frequency ω_1 is higher compared to annual average infiltration rate n, for the small crack(s) A and B ($3mm$ and $6mm$) (fig. 4a, 4b and 5a, 5b).

In contrast, the other set of cases (one crack on the windward and one crack on the leeward side), shown in figure 6, seems to be slightly influenced by ω_i. The flow pattern of cross ventilation is totally affected by the volume and the size of the crack(s)/opening(s).

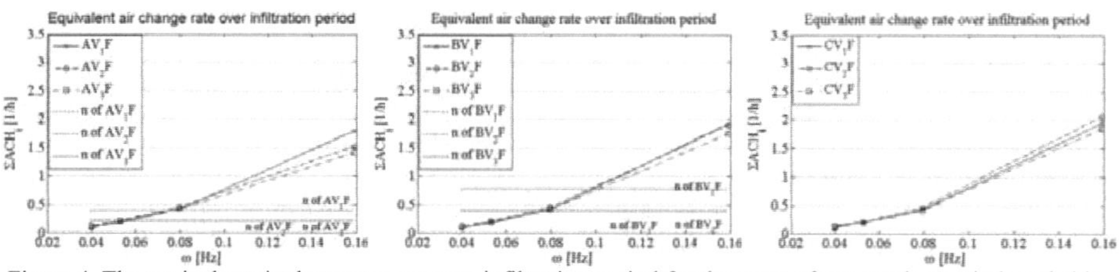

Figure 4. The equivalent air change rate over an infiltration period for the cases of one crack on windward side.

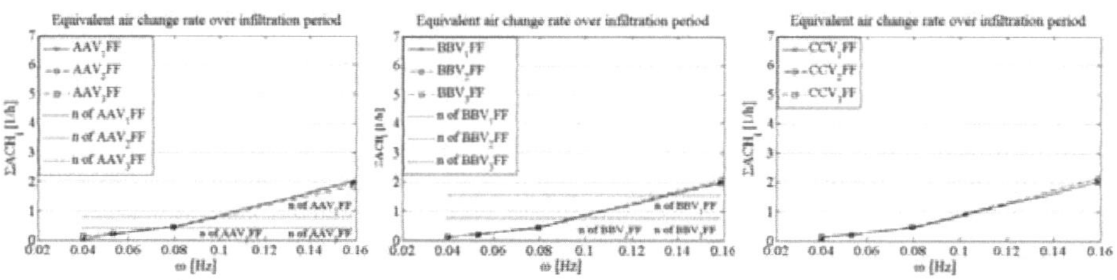

Figure 5. The equivalent air change rate over an infiltration period for the cases of two cracks on windward side.

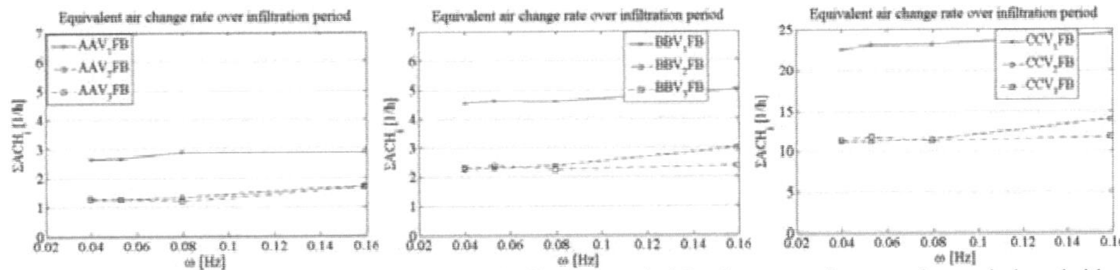

Figure 6. The equivalent air change rate over an infiltration period for the cases of one crack on windward side and one crack on the leeward side.

The equivalent ΣACH_2 for the gust frequency ω_2 against the size of the crack(s)/opening(s) d is presented in figure 7. For the cases of one crack on the windward side (no outlet), the ΣACH_2 of volume V_2 is higher compared to the cubic volume V_1 ($V_2 = 2V_1$); the crack A (3*mm*) of V_2 gives higher equivalent ΣACH_2 than the crack of 6*mm* or even of 30*mm* of V_1 (fig. 7a). That fact is also valid for the cases of two cracks located on the windward side. The lower inertia forces (compressibility) of V_2 seems to affect the flow rate through the crack(s) (fig. 7b). The influence of d to the infiltration rates appears to be negligible for such sizes.

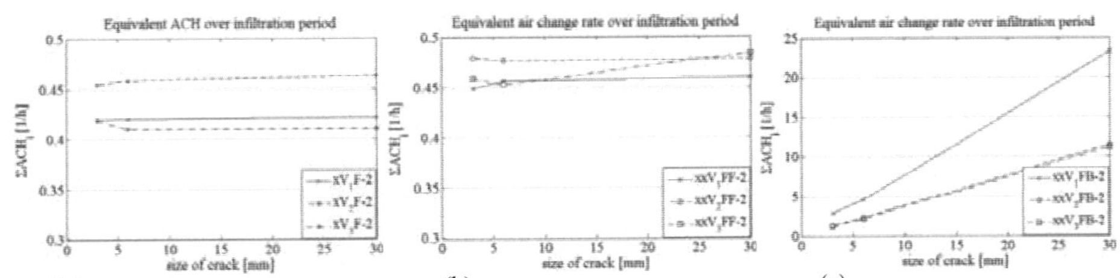

(a) (b) (c)

Figure 7. The equivalent air change rate against the size of the crack. (a) one crack on the windard side, (b) two cracks on the windward side and (c) one crack the windward side and one crack on the leeward side.

In contrast, the V_3 ($V_3 = V_2$) does not 'behave' as V_2 does but like V_1 in the case of one crack on the windward crack (fig. 7a); the inertia forces (compressibility) of V_3 appears to be approximately the same compared to V_1. It could be reasonable to claim that the difference is due to the elongated side. In V_2, the later is along the wind direction, while in V_3 it is perpendicular to that, increasing the resistance of the volume. For the cases of two cracks-openings on the windward side of V_3, the ΣACH_2 increases with the d (fig. 7b). Indeed, in the later case, the driving forces caused by the existence of cracks seem to be more important, explaining also the weaker impact of the volume.

Regarding the cross air flow (fig. 7c), the ΣACH_2 seems to be completely connected to size of the cracks-openings for all the cases, while the volumes behave in analogy to a steady state case and the role of the volume of the enclosure is not important anymore.

CONCLUSION

A single zone model ('box') was simulated and studied numerically under unsteady wind conditions. Four wind gust frequencies were used to describe the inlet boundary conditions. Three sizes of cracks-openings located on the 'box' in three different ways were tested (no outlet, inlet-outlet on the same windward side and cross air flow). In addition, three enclosures were studied, in order to explore the role of the volume. In total 108 cases were solved with the shear-stress turbulent model (SST). The equivalent air change rate ΣACH_i, extrapolated over time $t_{tot} = 1h$, was calculated and was shown against (i) the gust frequency

ω_i and (ii) the size of the crack(s). Comparisons with relevant annual average infiltration rates were also presented. The influence of the frequency of the gust, the internal volume of a relatively airtight building and the location of the leakage area seem to be critical factors under climate conditions. The ΣACH_i increases with ω_i for the cases of one or two cracks located on the windward side. The impact of ω_i seems to become much more significant in higher frequencies (slightly parabolic form). The gust frequency seems to be the dominant factor for a single zone model with leakages located on the same side perpendicular to the main wind flow direction. In contrast, in case of cross air flow the infiltration rates seems to be slightly influenced by ω_i. The internal volume of the enclosure plays also an important role on the leakage rates under unsteady wind conditions in cases of no outlet or of both inlet and outlet being on the windward side. The inertia forces (compressibility) of the volumes is a major parameter when air infiltrates the envelope across the windward, while the size of the cracks does not seriously affect the rates. The location of the cracks is finally of high importance; in case of cross air flow, the gust frequency slightly affects the leakage rates and the enclosure 'behaves' in analogy to the steady state case.

REFERENCES

[1] Shaw C.-Y. 1981. *A correlation between air infiltration and air tightness for houses in a developed residential area*, National Research Council Canada, Ottawa.

[2] Hill J.E., Kusuda T. 1975. *Dynamic characteristics of air infiltration*, ASHRAE Trans. 81 (1) 168–185, ASHRAE, New York.

[3] Sherman M. H. 1986. *Estimation of infiltration from leakages and climate indicators*, Energy and buildings Vol. 10, 81–86.

[4] Haghighat F., Brohus H., Rao J. 1999. *Modelling air infiltration due to wind fluctuations – a review,* Building and environment Vol. 35, 377–385.

[5] Haghighat F., Rao J., Fazio P. 1991. *The influence of turbulent wind on air change rates – a modelling approach*, Building and environment Vol. 26, 95–109.

[6] Haghighat F., Rao J., Riberon J. 1992. *Modelling fluctuating airflow through large openings*, 13th AIVC conference, Nice, France.

[7] Baker P.H., Sharples S., Ward I.C. 1987. *Airflow through cracks*, Building and environment Vol. 22, 293–304.

[8] Walker I.S., Wilson D.J., Sherman M.H. 1997. *A comparison of the power law to quadratic formulations for air infiltration calculations*, Energy and buildings Vol. 27, 293– 299.

[9] D.W. Etheridge 1977, *Crack flow equations and scale effect*, Building and environment, Vol. 12, 181 – 189.

[10] Menter F.R. 1994. *Two-equation eddy-viscosity turbulence models for engineering applications*, AIAA Journal, Vol. 32, no 8. 1598–1605.

[11] ANSYS Inc. 2009. *ANSYS CFX-Solver Modeling Guide,* ANSYS CFX 12.1 help manual.

[12] Guha, T.K., Sharma, R.N., Richards, P.J. 2009. *CFD modeling of wind induced mean and fluctuating external pressure coefficients on the Texas Technical University building*, EACWE 5.

[13] Elliott D.L. 1979. *Adjustment and Analysis of Data for Regional Wind Energy Assessments*, Workshop on Wind Climate, Asheville, North Carolina.

[14] Sherman M.H. 1998. *The use of blower-door data*, Lawrence Berkeley National Laboratory.

Innovative Ventilation systems for Low Energy Buildings: a studied case

Juslin Koffi, University of La Rochelle, France
Francis Allard, University of La Rochelle, France

ABSTRACT

Demanded-controlled ventilation (DCV) systems can be an alternative for achieving the goals of ventilation in new buildings. Such ventilation systems are likely to be more suitable for low-energy buildings, as far as these buildings should be energy efficient. Due to higher envelope insulation, reducing or adapting the airflow rates to the occupancy schedule or to the pollution level appears as an imperative for guaranteeing good performances of ventilation systems. The control of the ventilation should nevertheless carefully manage pollution risks without increasing the energy consumption due to air renewal, while this energy is expected to be reduced.

However, the regulations on ventilation do not so far specifically deal with low-energy buildings. Most of the time, engineers use existing ventilation systems. In this context, the French "QUAD-BBC" project aims to determine, by simulating various local climate, occupancy and pollution conditions, the suitable ventilation systems according to the type of building, i.e. a single-family house, an apartment in a residential building, an office building and a school building. The studies aim to assess the performance of various innovative ventilation systems operating in these buildings. The examination of the impact of these parameters on the indoor air quality (IAQ) would help bringing out solutions for the management of ventilation systems in low-energy buildings.

This paper presents the first results of the simulations carried out in the residential low-energy buidings. The simulations are perfomed using three demanded-controlled ventilation (DCV) systems respectively based on: 1°) CO_2 concentration, 2°) occupancy and 3°) both CO_2 and humidity. The DCV systems are compared to permanent mechanical exhaust and balanced ventilation systems. We analyse the influence of the control of the ventilation systems on the air quality, using a newly built IAQ index associated to the guideline values of the usual building pollutants.

In the studied cases, the results tend to show some different concerning the performance of the ventilation systems. In most of the cases, better indoor air quality is obtained when using presence DCV as far as the air supply is adapted to the fresh air need. The energy consumptions are in adequacy with the energy requirements, however, this factor depends a lot on the local climate and the control strategy.

THE USE OF BUILDING OWN VENTILATION SYSTEM IN MEASURING AIR TIGHTNESS

Timo Kauppinen*[1], Sami Siikanen[2], Erkki Vähäsöyrinki[3], Markku Seppänen[4]

1 VTT Technologies and Services for Buildings
PO Box 1100
FI-90591 Oulu, Finland
timo.kauppinen@vtt.fi

2 VTT Optical Measurement Technologies
PO Box 1199
FI-70211 Kuopio, Finland

3 VTT Expert Services
PO Box 1100
FI-90591 Oulu, Finland

4 Oulu University of Applied Sciences
Kotkantie 1
FI-90250 Oulu, Finland

ABSTRACT

The improvement of energy efficiency is the key issue after the energy performance of buildings directive came into the force in European Union countries. The city of Kuopio in Finland participate a project, in which different tools will be used and tested to improve the energy efficiency of public buildings. In this project there were pilot buildings e.g. schools. The other pilot school consumed much more heating energy than the other same type of school. Air tightness was measured using the own ventilation system of the building and by remote control from the central operation room. In this paper the procedure and some results are presented. There is also one example about a kindergarten, a health center, about a swimming hall and a new administration building in City of Oulu, in which the air tightness was measured using Blower-Door method and building's own ventilation system. Thermography, air tightness test and other supporting measurements can be used together to solve energy loss problems – if these measurements will be carried out by proper way. The ventilation system of the building can be used to test air tightness within certain limitations.

KEYWORDS

Air tightness, Blower Door, Energy efficiency of buildings, Building performance

1. INTRODUCTION

Air tightness of a building or part of the building can be measured by specific device or set of devices (s.c Blower Door). The own ventilation system of a building can also be utilized with certain limitations. Tracer gas-method, such as concentration decay method can also be used, especially in case of uniform and continuous large volume buildings, like sport halls. In ongoing ENEFIR-project [1] the objective is to develop new measurement and instrumentation concepts both analyzing tools to improve total energy efficiency of buildings. In the project one sub targets was to verify the performance of building envelope by thermography and supporting tools. Three municipalities participate the project, one of them is Kuopio. The energy consumption of the public building stock of Kuopio has been relatively low compared with the cities same size, in the year 2008 the weather corrected consumption of heating energy of the building stock was 31,5 kWh/m^3 and electricity 14,1 kWh/m^3. The average consumption of heating energy of school buildings in the whole country was 45, 4 kWh/m^3 and the consumption of electricity 14, 6 kWh/m^3.

City of Kuopio chose 2 schools, 2 kindergartens and an art museum for the project. The buildings were selected based on their specific energy consumption – the other school and the other kindergarten consumed relatively much energy compared with their counterparts. The aim was to analyze the possible reasons for that and to know which factors caused the differences in energy consumption.. School 1 (building volume 13 210 m^3) was renovated in the year 2007 and the specific consumptions in 2009 were: Heating 43 kWh/m^3 and electricity 21, 8 kWh/m^3. School 2, renovated in the year 2005 (the cubic content 30 012 m$^{3)}$ consumed in 2009 25, 6 kWh/m^3 of heating energy and 10, 9 kWh/m^3 electricity. The specific consumption of energy in the school 2 was only 56 % of the consumption of the school 1. Both schools were connected in district heating system. In school 1 the most problematic space was the library. In the winter 2009 - 2010 additional heaters were used to keep the indoor temperature in acceptable level. The complaints related to decreased thermal comfort. There were no reclamations from school #2.

2. MEASURING AIR TIGHTNESS BY BUILDING'S OWN VENTILATION SYSTEM

2.1 Preconditions

There have not been requirements of numerical values dealing with air tightness in Finland before 2008. Requirements of energy performance calculations caused changes in building codes 2008 and then 2010. According the new energy performance code from July 2012 [2], air tightness number q_{50} cannot be more than 4 m^3/ ($h*m^2$). Better air tightness can be shown by measurements. The air infiltration must be calculated in compensation calculations based on air tightness number 2.0 m^3/ (h*m2). Air tightness measurement standard SFS-EN 13829 is presented in the building code [3].

Measurements has been done using blower-door or self-made equipments. Also tracer-gas method has been used. The ventilation system of a building has sometimes been used to determine air tightness – mainly in connection of research projects or occasional experiments. There have been two main factors which makes the use of ventilation system difficult in air tightness measurements: The fans of air supply units have been two-step type devices; it means that the machine has run either by 1/1 or by ½ -power. The other obstacle has been difficulties to measure air flow in primary air ducts. Normally the ducts have been equipped with measurement connections, that Pitot-tube or thermoanemometers can be used. In some cases ducts has been equipped with measuring devices. Modern systems have frequency controlled fans, often remote controlled from operation room. Also pressure difference over than fan is measured, or there are at least fittings to do that. Pressure differences can be converted to air flow units, and in the wall of the machine an installed indicator/meter can be found. If the fans are frequency controlled and the primary air flow is measurable, air tightness test can be done in principle.

The practical problems in the use of the own ventilation system is:

1. Do you measure the whole building or only part of it?
   - In that case one must secure that the area what is going to be measured can be separated tightly enough from the other part of the building
2. If there are many units, one must choose the unit effective enough (in general the biggest unit) and verify if that particular device can be used. In addition, one must reserve another fan if the test cannot be carried out with one fan.
3. Is it possible to operate with wan exhaust fan without probles to stop other fans?
   - control system can include forced connection and some manual operations should be used (discontacting the alarm functions, sometimes removing fuses, to shut manually dampers, to disconnect blast regulators)
4. Sealing and tightening:
   - Is it possible to ensure that all the potential lead-troughs which could disturb to have correct measurents are sealed?
5. Measurement of air flow
   - Can we use reliably the measuring options of the system (either from the machine and/or in the control room)?
   - the pressure-flow chart and specific curves of the fan which is used must be available
6. Measurement of pressure conditions

- Is it possible to measure pressure conditions from each facade and also from different levels, if needed?

2.2 Course of the procedure

Preliminary works:

Sealing and tightening:

- Air inlets (outdoor air) must be closed and covered.
- All the air exhausts of the air supply units must be closed and covered, except the one which will be used in the test.
- All fans and roof exhaust fans must be stopped and roof exhaust fans must be covered (plastic foil, inneer tubes of football, volley ball or specific devices).
- If tightening can not be done in the machine, the vents must be taped in service areas of the fan.

Measurements:

Measurement of air flow (depressurization)

- Pressure drop over the fan must be measured both from the unit and in the control room (if possible)
- Use calibrated meters (the displays, pressure transducers and indicators of the system can give false readings) in parallel of existing measuring devices
- the pressure-flow chart and specific curves of the fan which is used must be available
- normally the fan frequencies are from 0 Hz up to 60 Hz – using 10 Hz steps
- In each stage the pressure difference between outdoor and indoor must be measured
- drive the fan at least two times from the minimum value to maximum value and back

Measurement of pressure conditions

- at least from two facades
- the measurer of pressure conditions must have on-line connection (cell phone or SW phone) with the operator/machine room

Also the pressurization can be carried out using the procedure above. In that case one must use one air supply fan and the other devices must be stopped. Typically, the most time consuming operation is preliminary works. It will take 2-4 hours, depending on how much assistance is available and, of course, on the size of the building. Measurements can be done in one hour.

3. MEASUREMENT OF AIR TIGHTNESS IN TWO SCHOOLS

Both schools were connected in remote monitoring system of the city. The heating and ventilation system could be controlled and consumption monitored from control room which situates in the center of the city. The in-situ operations involved indoor and outdoor thermal scanning in prevailing conditions, pressure drop measurements between indoors and outdoors, and indoor temperature and relative humidity (RH) measurements in all workspaces.

Measurements were focused on the library, where the indoor environment problems occurred. The library was depressurized. To the same service area of ventilation system belonged also the gym and the canteen. The canteen was not directly connected with the library –other room space was between them. The other spaces of the service area of same air supply unit were separated from the tested area but there was no full certainty about that; it might be possible that there will be some leaks through the ducts from the other areas, too. The breakthroughs, vents, blast controls and dampers were shut, and the incoming air unit was closed by plastic film. The exhaust fan was driven by frequency controller in 25 % steps from 0 % to 100 % capacity and the air flow was measured from the pressure difference units over the fan – the pressure difference can be converted to volume flow using the specific curve of the fan. By the same the pressure difference between indoors and outdoors was measured on each capacity range. The measured air flow of the exhaust fan represents air leaks through building envelope. The measured air flow at 50 Pa pressure drop is divided by the volume of measured area. The result represents the air leakage number and gives a conception of the air tightness of the building (in this case the air tightness of measured part of the building). If 50 Pa pressure difference cannot be reached, an estimated value based on the calculations and leakage curve equation can be used. In this case only negative pressure difference was used because of practical reasons.

The structures were scanned again by thermal imager then under the maximum pressure difference reached. By comparison of the thermal images in actual normal conditions and images under negative pressure drop one can evaluate the thermal performance of the building envelope and locate air leak patterns. By thermal scanning in normal operating conditions it's also possible to monitor the performance of ventilation system (supplied air and exhaust air temperatures) and the functioning of heating system (surface temperature of radiators).

The pressure condition in the normal situation in the library was -4 Pa, which can be considered as normal; the balancing of the ventilation system was properly done. Air flow rates were not measured. The canteen was under highly negative pressure difference (-30 Pa) compared with outdoors. The problem was caused by the kitchen which had a separate ventilation system but which was connected with canteen. Also the gym had -30 Pa negative pressures drop. The kitchen was under -40 negative pressure drop compared with outdoors. The ventilation of kitchen was totally unbalanced which also caused problems to the canteen. The first conclusion was that the ventilation system of the kitchen should be checked – there was no possibility to isolate the canteen from the kitchen area, so the kitchen ventilation affected the indoor conditions of the canteen. After repairs carried out in the summer 2010 (based on the results of the first measurements) the measurements were repeated in the school 1.

4. RESULTS OF AIRTIGHTNESS OF TWO SCHOOLS

Indoor thermography showed the problems in the library – the surface temperature of ventilation ducts all over the way were low (4 – 5 °C) and the temperature of supplied air was 14 °C (18 °C recommended as minimum). Figures 1 and 2 show the situation in the ceiling – the surroundings of the supplied air duct were cold and also some extremely low surface temperatures in rear part of the library, where the works space of librarians are situated. In the shower room the surface temperatures were below 0 °C at lowest (the outdoor temperature had been very low in the night), which means that there was a risk of condensation and structural damages. The radiators seemed to work properly, and except some leaks from window weather-strip and spotty low surface temperatures in the junction of exterior wall and ceiling. There was no possibility to reach -50 Pa pressure difference using the exhaust fan – by the maximum capacity (100 %) the negative pressure drop was -31 Pa. This indicated leaks. Figure 3 shows the air leak curve. Air leak number must be evaluated using the equation of the curve. N50 was approximately 7 - 8 1/h (changes/hour) which is a very poor result. The acceptable value, taking into account the age of the building and structures could be 2 – 3 1/h at 50 Pa. The eligible value should be 0, 5 – 1 1/h at 50 Pa. It was very probable that without very exhaustive repairs it would be difficult to achieve a proper level of air tightness. The actual level of air tightness caused significant uncontrolled leakage air ventilation which will increase the consumption of heating energy. In generally, the library part of the building was very leaky – one reason for that could be the renovation which was carried out in 2007 was not properly done.

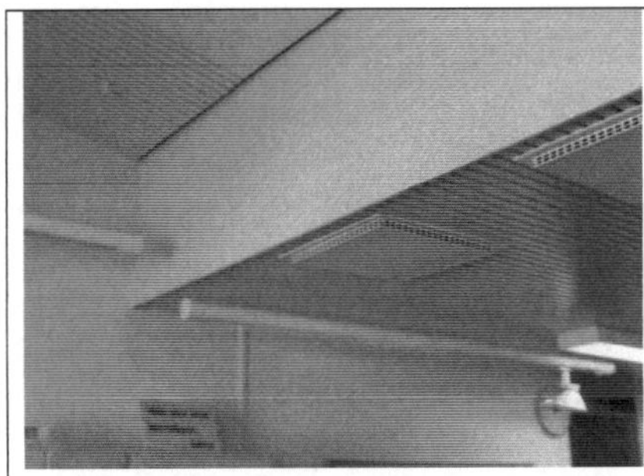

Figure 1. Ventilation duct, inlet to the library

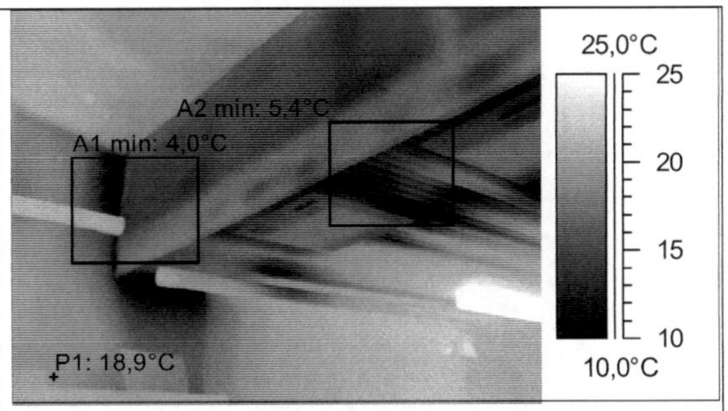

Figure 2. Thermal image of the same point

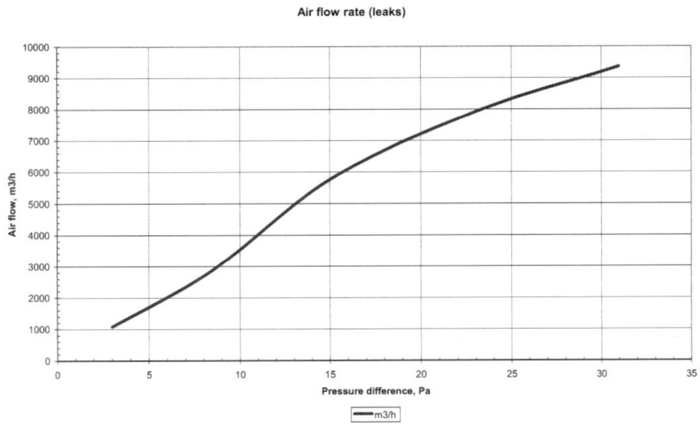

Figure 3. Air leak curve

5. OTHER CASE STUDIES

5.1 Case study: Kindergarten

Except 2 schools, in ENEFIR-project two kindergartens were studied and compared, built 1991 and 2001. The structures were more or less equal, but the older kindergarten consumed significantly more energy than the point of comparison. The results of measurements showed the reason for higher energy consumption in the older kindergarten. The lowered thermal comfort was caused by thermal bridges of structures, and air leaks of windows and doors. Figures 4 and 5 show one example of window leaks – the surface temperatures were extremely low (winter conditions) because of narrow metallic window frames and air leaks. In kindergartens the occupation zone of the children is close to floors and also close to wall-floor junctions. In these facilities special attention must be paid to structural details and must try to avoid cooling air leaks and cold surfaces.

In the older kindergarten the ventilation system was two-step operated; no frequency control and no flow measurement options. In this case the exhaust air flow was measured both by Pitot-tube and by thermo anemometer according to valid measurement standards. The results are presented in table 1. The measured air flow rates were surprisingly close to each other. The measurement point was the same. The goal of 50 Pa negative pressure difference was not reached. Interference distance requirements were fulfilled, which is not always possible in small machine rooms with short straight part tubing. In the other kindergarten it was possible to carry out measurements using frequency controlled fan and pressure difference-based exhaust air flow measurements. Also 50 Pa pressure difference was reached.

Meter type	Pitot	TA	
Capacity	flow	flow	diff.
	m3/h	m3/h	%
50 % (1/2)	1435	1452	1,2
100 %/(1/1)	3897	4087	4,6

Table 1. Results of two different measurements

Figure 4. Window at normal conditions

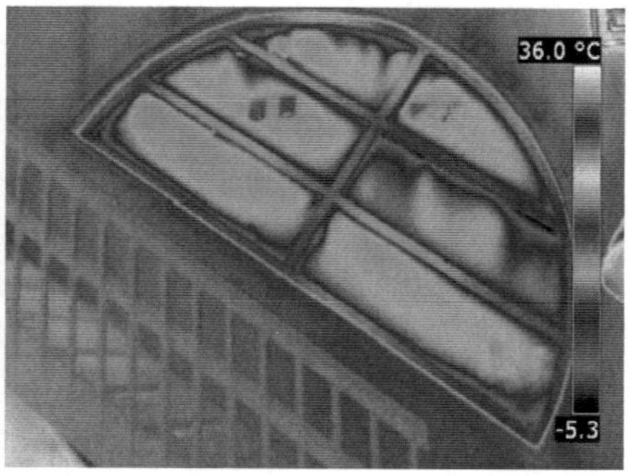

Figure 5. Window under depressurization

5.2 Case Study: Health Center

One pilot building was a health center and it`s sickroom department. The air tightness was determined causing just a service area of one air supply unit, which covered on floor of the building. In this case the measurements could be done using the own ventilation system after certain adjustments of the control unit; it was not possible to switch off the air supply fan without doing some manual operations. In many systems exhaust and supplied air fans are operating with forced connection.

Figure 6 shows the air tightness curve and results. Air leakage number n_{50} was 1, 9 1/h (changes/hour), which is excepted result but not acceptable if the question would be about a new hospital. The main leaky points were the windows (figures 7 and 8). Because the patients in health centers are mainly elderly people (in this case old long-term or chronic patients), the indoor conditions should be best possible; if there is draft in the rooms, it will be compensated by increasing indoor temperatures, which will increase also energy consumption.

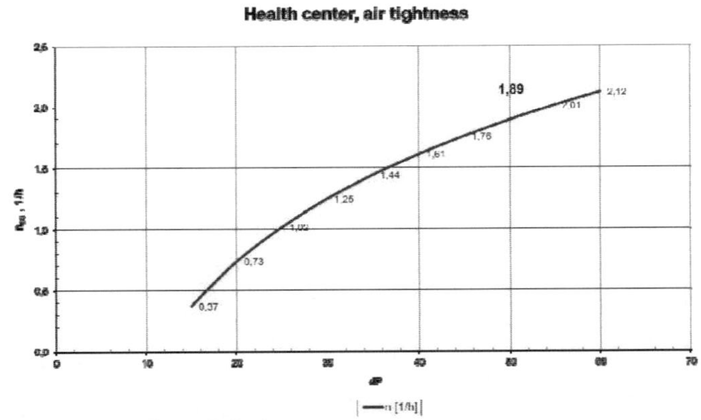

Figure 6. Leakage curve

Figure 7. Windows of a sickroom

Figure8. Windows at negative pressure difference

5.3 Case Study: Swimming Hall

A swimming hall in Northern Finland was tested in aim to have information for planning renovation measures. The ventilation system of the hall was renovated 2002-2003 and the hall was built in 80`s. The test was limited concerning the pool and dressing room department. The strongest moisture load is against the structures in the pool section and also in showers and saunas. In normal operating conditions the pool section had -10 Pa negative pressure drop at height of 1, 5 m compared with outdoors. It means that the pressure difference against roof structures is very small or even positive. The indoor temperature was +30 °C at the same level. The air tightness of the space to be studied was determined with the own ventilation system of the hall by closing all other ventilation fans, shutters and using exhaust fan of the pool section. The designed capacity of the fan is 7 m^3/s. Air flow of the exhaust fan was controlled by frequency converter. The pressure drop over the fan was measured by calibrated pressure difference meter. When compared the measured readings with readings given by pressure difference transmitter/indicator, it came out that the values did not match with measured from fittings – it means that the operator will have false readings and false air flow values during operation. This is not so unusual when talking about the local measurements an devices – in general, sensor and detectors can be inappropriately or wrongly installed, pressure tubes could be blocked, pressure difference signal (or other measuring signal) could be interfered or incorrectly adjusted etc.

Thermal scanning was made both indoors and outdoors before pressurization and during depressurization. Figures 8 – 11 show some of the located leak patterns under depressurization.

Especially air flow routes in the roof structures can cause several damages, because the diffusion can drive water vapor into the structures, even the pressure difference across the roof were slightly negative.

Figure 8. Entrance door of pool unit

Figure 9. Door at 50 Pa depressurization

Figure 10. Ceiling of spa

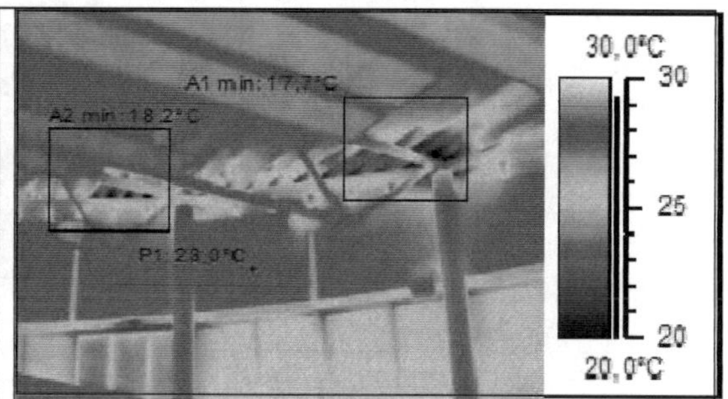

Figure 11. Ceiling under negative pressure difference

5.4 Case study: Office Building

A new office building was taken into the use in the summer 2011. The built volume is 28 000 m^3. It was designed according to strict demands. The requirements for the building are presented in table 2. The set air-tightness value n_{50} was **1, 0** 1/h. Air tightness was measured both by the own ventilation system and by blower-door. The pressure difference between indoors and outdoors during the test was measured in the middle from both facades (the building is approximately 20 m high). There were 3 air supply units in the machine room and the unit with highest capacity was used, the others were stopped and taped. Also the roof exhaust fans were stopped and taped. Figure and table show the results. The results (table 3) diverged from each other. The mean value n_{50} of air leak number by blower-door under pressurization was 0,41 1/h and q_{50} 1,86 $m^3/h*m^2$. The result of depressurization, which can be compared with the result of the own ventilation system was 0,39 1/h (1,78 $m^3/h*m^2$). The measured result using ventilation system was 0,32 1/h. The tolerance of blower door result is ± 3%, according to manufacturer. The blower-door results are typical, generally pressurized value is a little bit higher than depressurized ones, depending on the structural details. The theoretical tolerance of the exhaust fan of ventilation system under controlled testing conditions according to manufacturer is ± 5 %. In prevailing conditions one can assume that the real margin of error is higher, because the flow profile probably changes compared with fan test conditions.

Structural component	U-Values SRMK C3-2010 (building codes)	TeVi(Tech. center) (city requirements)
-Wall	**0,17**	**0,15**
-Base floor	**0,16**	**0,14**
-Roof	**0,09**	**0,09**
-Window	**1,0**	**1,0**
Structural component adjacent to the ground (Basement wall)	**0,16**	**0,14**

Table 2. Energy efficiency requirements

The results were close enough and the result using building's own ventilation system gave a good approximation for air tightness. The problem becomes obvious in case that the measured value is close to the requisite value. The difference between blower-door and ventilation system measured results was 18 % calculated from the blower-door value.

Comparison of air-tightness results from an office building		
Test method	**Result, 1/h; $m^3/(h*m^2)$**	**Estimated air-flow, m^3/h at 50 Pa**
measured by blower-door (depressurization)	**0,39**/1,78	10948 (+/- 2.7 %)
measured by blower-door system(pressurization)	0,41/1,86	11517 (+/- 2.8 %)
n50, measured by ventilation system(depressurization)	**0,32**/1,46	8980 (+/- ? %)

Table 3. Measured results

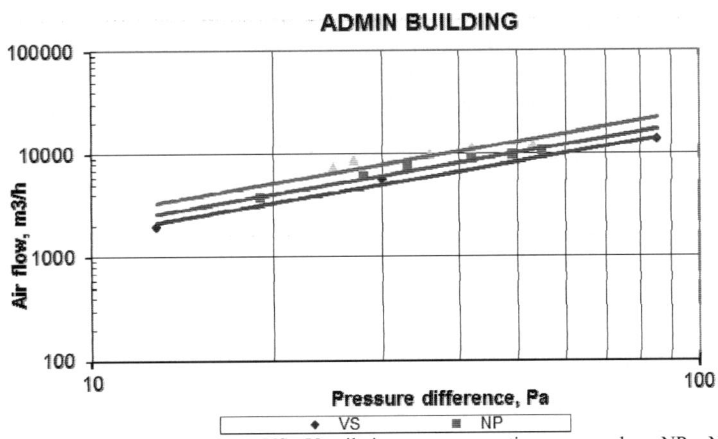

Figure 12. Air tightness test. VS= Ventilation system, negative pressure drop, NP = Negative pressure drop, blower-door, PP = Positive pressure drop, blower door

The result, 0,4 1/h was below the passive-house value (0,6 1/h). During the tests all the windows and breakthroughs were checked, and some windows were not properly adjusted, and some electric tubing lead-ins were not tightened; these works were made after the test. Thermography was not used during depressurization for locating the leaks, because of summer conditions. The building will be tested again in colder climate conditions. The air tightness level was good.

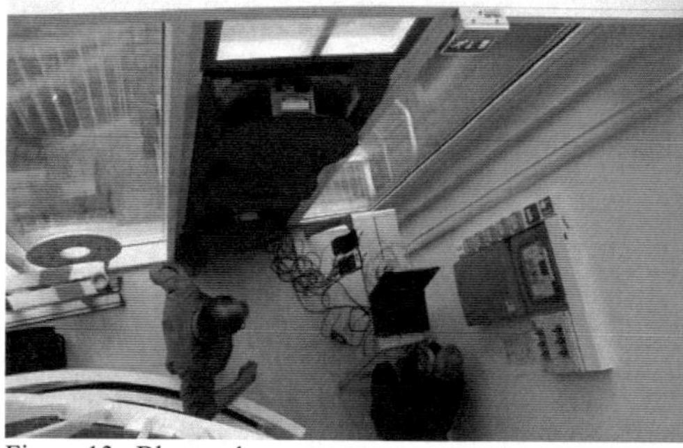

Figure 13. Blower-door

Figure 14. Measurements over the exhaust fan

6. CONCLUSIONS

The tests, which were made in the schools were typical measurements when solving existing problems. By thermography it is possible to determine the thermal performance of building envelope and also the functioning of heating and ventilation system within certain limitations and preconditions. Air tightness test will give a good conception of the performance of the building envelope. The results can be compared with the other buildings of same type. Two-stage thermography helps to separate the thermal bridges from air leaks and to find leak patterns which maybe could not be found in normal conditions. Pressure conditions are necessary to measure in connection of thermography. The normal negative pressure difference in cold climate conditions should be < 10 Pa (if the question is about mechanical ventilation system). If big pressure differences exist, there are some problems in balancing or in use of the ventilation system. The studies proved that the air tightness test is possible to make using the building's own ventilation systems – if the fans are frequency controlled and the main air flow can be measured. Based on the experiences of the project a guideline has been written into the use of maintenance staff of the city. In the future they can do air tightness tests by themselves. The main problem is when the measured value is very close to the required value but remains above it; the accuracy of measurements will vary and the tolerance is difficult to determine. The accuracy of blower-door may be better compared with method using building's own ventilation system. Repeatability of the tests using the own ventilation system was good in all tests.

ACKNOWLEDGEMENTS

Results reported in this work were carried out in a research project funded by European Union European Regional Development Fund and TEKES Finnish Funding Agency for Technology and Innovation. We thank the staff of the Kuopio City for co-operation.

REFERENCES

[1] Kauppinen, Timo, Siikanen, Sami. Improvement of energy efficiency – the use of thermography and air-tightness test in verification of thermal performance of school buildings. SPIE Proceedings of Thermosense 2011. SPIE, Bellingham 2011.

[2] D3. Suomen rakentamismääräyskokoelma. Ympäristöministeriö. Rakennusten energiatehokkuus. Määräykset ja ohjeet 2012. *(D 3, Finnish building codes. Ministry of Environment. Energy efficiency of buildings. Requirements and instructions 2012. Ministry of Environment 2011)*

[3] SFS-EN 13829

THE USE OF A SAMPLING METHOD FOR AIRTIGHTNESS MEASUREMENT OF MULTIFAMILY RESIDENTIAL BUILDINGS – AN EXAMPLE

Jiri Novak[*1]

Faculty of Civil Engineering, Czech Technical University
Thakurova 7
166 29 Prague 6, Czech Republic
jiri.novak.4@fsv.cvut.cz

ABSTRACT

Large buildings can not always be tested as a single pressure zone. In Europe, different approaches have been proposed concerning the choice of representative parts of the building (sampling method) and the compliance check in situations, when several parts of the building have to be tested separately. The preliminary Czech standard TNI 73 0330 defines a sampling method, as well as subsequent treatment of results and compliance check procedures for multifamily residential buildings. This contribution reports the results of a trial test of TNI 73 0330 method. The estimation of the whole building n_{50} value by means of this sampling method was found questionable, namely due to the potential presence of internal air leakage. Alternative method based on measurements of building parts using so called guard zone technique was proposed. Inconsistency between the compliance check based on the whole building test results and the sampling method test results was identified.

KEYWORDS

airtightness measurements, blower door test, multifamily residential buildings

INTRODUCTION

The airtightness measurements of large buildings can become complicated or even impossible, since in some situations the whole building can not be measured as a single pressure zone. The common reasons are well known: (i) limited performance of the measuring device fan, (ii) absence of the internal openings in building partitions making impossible to connect all the parts of the building into one pressure zone, (iii) impossibility to interconnect all the building parts for organisation reasons (e.g. access refusal). In such situations, different parts of the building can be tested separately, each one as a single pressure zone. Then, the whole building airtightness can be estimated based on the results of these individual tests.

Until now, no generally approved rules have been formulated concerning the choice of representative parts of the building tested (sampling method), the estimation of the airtightness of the building envelope and the compliance check in situations, when several parts of the building were tested separately. Different approaches were proposed e.g. in Germany, France, Belgium and UK [1]. In the Czech Republic, the preliminary standard TNI 73 0330 [2] defines a sampling method, subsequent evaluation of results and compliance check procedures for multifamily residential buildings. The practical applicability of this method was investigated by means of a trial testing of selected multifamily residential building. The results are reported in this contribution.

SAMPLING METHOD ACCORDING TO TNI 730330

TNI 73 0330 defines following three airtightness measurement procedures of low energy and passive multifamily residential buildings for the purpose of compliance check with airtightness requirements. However, the method can be used regardless the energy performance of the building tested. All measurements should be performed using the fan pressurization method according to [3], test method B, by means of standard blower door devices.

.

Procedure 1

If possible, the building is measured as a single pressure zone. If the resulting n_{50} value complies with the limit value, the building is supposed to fulfil the requirements.

Procedure 2

If the whole building can not be measured as a single zone, then the residential part is measured separately. If the n_{50} value of the residential part complies with the limit value, the building is supposed to fulfil the requirements. If not, the test can be repeated using an appropriate method allowing exclusion of air exchange between the tested residential part and adjacent spaces through the leaks in the internal partitions (internal air leakage).

Procedure 3

If even the residential part can not be measured as a single zone, than a relevant number of residential part units (e.g. flats) has to be tested separately. The building is supposed to fulfil the requirements if: (i) the n_{50} value of each tested unit complies with the limit value, or (ii) the mean $n_{50,m}$ value of all tested units complies with the limit value and simultaneously the particular n_{50} values of a majority of tested units comply with the limit value (the internal air volume of this "majority" has to be at least 2/3 of internal air volume of all the tested units). If the building does not fulfil the requirement, the tests can be repeated using an appropriate method allowing exclusion of internal air leakage.

The "mean $n_{50,m}$ value of all tested units" is calculated as weighted average of n_{50} values of particular tested units over their internal air volumes. Furthermore, TNI 73 0330 sets rules for the choice of the "relevant number of residential part units" which have to be tested separately. Priority is given to units facing exterior. The units should be chosen so as all types of building envelope elements (walls, ceilings, roofs, etc.) would be subject of the test.

TRIAL TEST OF THE SAMPLING METHOD

Building tested

A multifamily residential building built in 1978 was selected for the purpose of this work (figure 1). The building has 6 floors. The 1st floor contains only garage, technical and storage rooms. The 6th floor contains a boiler room. The residential part contains 16 identical flats located from the 2nd to the 5th floor.

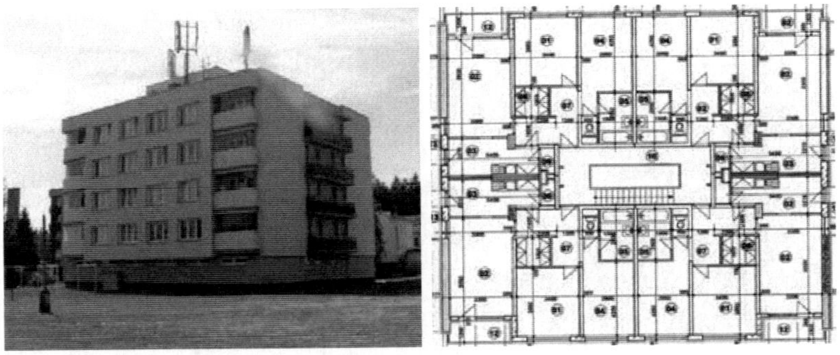

Figure 1.Measured building – frontal view and plan of a residential floor

Method

The airtightness of the building was tested using each of the three measurement procedures according to TNI 73 0330. In addition, the airtightness of one flat in the 5th floor was tested using the so called guard zone technique (GZT) [4] in order to limit the internal air leakage to adjacent rooms. In order to create the guard pressure zone around the flat tested, entrance doors of all the neighbouring flats were kept open during the test as well as all internal doors in these flats. A second blower door fan was fitted into the entrance door of the guard zone. Its speed was automatically controlled in order to maintain zero pressure difference between the tested flat and the guard zone. Only pressurization test was performed using the GZT.

Based on the results, reliability of the estimation of the whole building airtightness from the sampling method results and reliability of the compliance check procedure according to TNI 73 0330 was examined. The magnitude of the internal leakage was evaluated based on the results of procedure and the results of the measurement using the GZT. Finally, an alternative method was proposed for estimation of the whole building n_{50} value from the air permeability of segments of the building envelope measured by means of the GZT.

RESULTS

Overview

Measured part of the building	Test procedure	Internal volume V [m^3]	Depressurization t.		Pressurization test		Average	
			V_{50-} [m^3/h]	n_{50-} [h^{-1}]	V_{50+} [m^3/h]	n_{50+} [h^{-1}]	$V_{50-/+}$ [m^3/h]	$n_{50-/+}$ [h^{-1}]
whole building	1	3 909	3 001	0.77	2 828	0.72	2 915	0.75
residential part	2	3 492	2 953	0.85	2 971	0.85	2 962	0.85
flat 2.1	3	189	573	3.03	777	4.12	675	3.57
flat 3.1	3	189	545	2.88	710	3.76	628	3.32
flat 3.3	3	189	317	1.68	369	1.95	343	1.82
flat 4.3	3	189	1 068	5.65	1 126	5.96	1 097	5.81
flat 5.1	3	189	379	2.01	566	3.00	473	2.50
flat 5.2	3	189	660	3.50	849	4.49	754	3.99
flat 5.4	3	189	479	2.54	605	3.20	542	2.87
flat 5.1+GZT		189			271	1.44		

Table 1. Overview of individual test results – air flow rate at 50 Pa V_{50} and air change rate at 50 Pa n_{50}.

Procedure 3

In case of the building tested, the resulting mean $n_{50,m}$ value should be calculated as weighted average of individual n_{50} values of at least the 5 flats selected according to the rules of TNI 73 0330. Since finally 7 flats were tested, the mean $n_{50,m}$ value can be calculated in different manners from 5, 6 or 7 n_{50} values of different flats. The n_{50} values of flats 2.1, 5.1, 5.2 and 5.4 should be always included in the $n_{50,m}$ calculation, since these flats have a higher proportion of envelope area facing to the external environment. The results are summarised in table 2 (based on TNI 73 0330, all these results are correct and equivalent).

Calculation manner		A	B	C	D	E	F	G
	flat 2.1	3.6	3.6	3.6	3.6	3.6	3.6	3.6
	flat 3.1	3.3	-	-	3.3	3.3	-	3.3
measured	flat 3.3	-	1.8	-	1.8	-	1.8	1.8
$n_{50-/+}$ [h^{-1}]	flat 4.3	-	-	5.8	-	5.8	5.8	5.8
	flat 5.1	2.5	2.5	2.5	2.5	2.5	2.5	2.5
	flat 5.2	4.0	4.0	4.0	4.0	4.0	4.0	4.0
	flat 5.4	2.9	2.9	2.9	2.9	2.9	2.9	2.9
calculated	$n_{50,m}$ [h^{-1}]	3.3	3.0	3.7	3.0	3.7	3.4	3.4

Table 2 Procedure 3 test results calculated in different manners from the particular test results of the flats tested.

Measurement of a flat using the guard zone technique

The flat 5.1 (under the roof) was tested using the GZT with $n_{50+} = 1.4\ h^{-1}$ as the result. The air permeability of the flat envelope $q_{50,5.1,ext+}$ was calculated dividing the measured air flow rate at 50 Pa by the area of the surface of the flat envelope facing exterior. Since the air leakage trough the internal partitions was supposed effectively eliminated thanks to the use of the GZT, this value represents the air permeability of building envelope around the flat 5.1.

COMMENTS, DISCUSSION

Procedure 1 - influence of the building shape and size

The procedure 1 test result, $n_{50-/+} = 0{,}75\ h^{-1}$, seems to be surprisingly good. However, the air permeability of the building envelope $q_{50-/+} = 1{,}56\ m^3/(h.m^2)$, is substantially higher than the values usually achieved in case of single family houses equipped with a high performance air barrier system and having a similar air change rate $n_{50-/+}$. Hence, the satisfactory result was not achieved thanks to the quality of building envelope but it is rather a consequence of a favourable building shape and size (ratio of the envelope surface and internal volume A_E/V).

Procedure 2

The air flow rates measured during the pressurization test are higher than in case of the whole building tests. It is probably consequence of a measurement error (e.g. unfastened temporarily sealing of the intentional openings in the building envelope).

Procedure 3 and measurement of the flat 5.1 using the GZT - internal air leakage

For each flat tested, theoretical, expected $n_{50-/+}$ values were calculated from the measured air permeability of the whole building $q_{50-/+}$ assuming that this value is constant over the building envelope and the internal partitions delimiting the flats are perfectly airtight. These expected values were compared to the measured ones (figure 2, left). The difference between the expected and measured n_{50} values can be caused by either the internal air leakage or variations of the air permeability over the building envelope. However, the magnitude of the difference points out a very significant internal air leakage. The difference between the n_{50} values of flat 5.1 measured with and without the use of the GZT confirms the high significance of the internal leakage (figure 2, right). The difference between the expected n_{50} value and the value measured with the use of GZT can be a consequence of a local deterioration of the building envelope air permeability (compared to its average level) or a consequence of a measurement error.

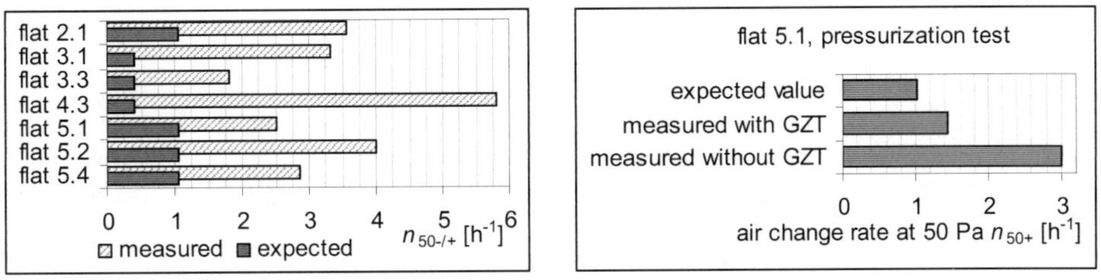

Figure 2 Magnitude of the internal air leakage. Left – measured n_{50} values of flats tested and expected values assuming no internal leakage. Right – flat 5.1 expected n_{50} value and values measured with and without the GZT.

Procedure 3 and measurement using the GZT – estimation of the building n_{50} value

Obviously, the mean $n_{50,m}$ value depends on the choice of the flats included into the assessment (table 2). Regardless the calculation manner, the mean $n_{50,m}$ values are substantially higher than the whole building n_{50} value. The difference may be the consequence of (i) the different shape of the building and flats (different ratio A_E/V), (ii) internal air leakage (which is the dominant factor – see above), (iii) variation of the air permeability over the building envelope. Hence, the procedure 3 test result, $n_{50,m}$, can not represent a correct estimate of the whole building n_{50} value. In case of the building tested, such a simplified estimation would lead to un error ranging from 300% to 400% (depending on the manner of $n_{50,m}$ calculation).

Based on the assumption that the variation of air permeability over the building envelope is negligible, an estimate of the whole building n_{50} value ($n_{50,\text{building}}$) can be calculated from the air permeability of the flat 5.1, measured using the GZT ($q_{50,5.1,\text{ext}}$):

$$n_{50,\text{building}} = \frac{A_{E,\text{building}}}{V_{\text{building}}} \cdot q_{50,5.1,\text{ext}} \quad (1)$$

where $A_{E,\text{building}}$ and V_{building} are the whole building envelope area and internal volume respectively. In the studied case, such estimation of the whole building n_{50+} value would lead to an error of 39 %. Furthermore, the air leakage through the slab on the ground could be neglected in the studied case. Hence, in order to correctly estimate the whole building n_{50} value, only the area of surfaces facing the air should be accounted into $A_{E,\text{building}}$. Then, the estimation error of n_{50+} would be 24%.

CONCLUSIONS

Obviously, the TNI 73 0330 procedure 3 test result ($n_{50,m}$) can neither be directly compared to the procedure 1 test result (whole building n_{50}), nor used as its estimate. More generally - when using a sampling method in order to estimate the whole building airtightness from test results of its parts, the differences in the size and shape of the whole building and the parts tested should be taken into account as well as a potentially significant effect of the internal air leakage. Alternatively to the procedure 3 of TNI 73 0330, the whole building n_{50} value can be estimated more accurately from the air permeability of the external surface of the envelope of the separately tested building parts. Then, the use of methods allowing exclusion of internal air leakage is required (e.g. the guard zone technique). This study proves applicability of this approach despite higher time and equipment requirements. However, the accuracy of this method seems to be limited and should be investigated more deeply.

The mean $n_{50,m}$ value depends on the choice of number of the building parts tested separately and their position inside the building. Therefore, it seems reasonable to introduce certain tolerance when checking the compliance of $n_{50,m}$ with the limit value. The same approach would be followed when evaluating the results of the alternative method.

When several parts of the building are tested separately according to the rules of TNI 73 0330 procedure 3, the majority of them has to comply with the limit n_{50} value (valid for the whole building), so as the building could be declared to fulfil the airtightness requirement. Namely in case of small flats tested and severe limit n_{50} values, the fulfilment of this condition might become difficult. Due to potentially significant impact of internal air leakage, it would require very airtight execution of the envelope of flats, including its internal surfaces.

Furthermore, in case of the same building the procedure 1 test result can comply with the limit value (e.g. due to the favourable shape or size) whilst the procedure 3 test result may not (e.g. due to internal air leakage). Such unclear situation could lead to legal disputes.

ACKNOWLEDGEMENTS

This outcome has been achieved with the financial support of the Ministry of Education, Youth and Sports and Ministry of Industry and Trade of the Czech Republic, projects No. 1M0579, and No. 1221420507. The measurements were performed in close cooperation with TZÚS a.s. Ceske Budejovice.

REFERENCES

[1] Walther, W., Rosenthal, B. 2010. *Airtightness testing of large and Multi-family Buildings in an Energy Performance Regulation Context.* Stimulating increased energy efficiency and better building ventilation INIVE, Brusel, 2010

[2] TNI 73 0330 *Simplified calculation assessment and classification of residential buildings with low energy demand for space heating - Multifamily residential buildings.* Czech Office for Standards, Metrology and Testing. 2009.

[3] EN 13829 *Thermal performance of buildings – Determination of air permeability of buildings – Fan pressurization method.* CEN. 2000.

[4] Roulet, C.A., Vandaele, L. 1991. *Air flow patterns within buildings, measurement techniques, AIVC TN 34*, Air infiltration and ventilation centre, Coventry

APPLICATION OF AIRTIGHTNESS TO HEALTHCARE BUILDINGS

William Booth*[1], Tom Jones*[1] Blanca Beato Arribas*[1],

[1]BSRIA Ltd
Old Bracknell Lane West
Bracknell
Berkshire
RG12 7AH
UK

ABSTRACT

The thrust of airtightness specification and testing is derived from energy considerations. The application to healthcare buildings and specialist laboratory facilities embodies the same principles but derives the appropriateness of the criteria with reference to [a] producing controlled and controllable cascading pressure zones and [b] specifying or quantifying the potential exposure in the event of failure of mechanical ventilation.

The paper discusses the application of airtightness testing to two full scale physical models of isolation rooms. It will be shown that important information can be readily obtained to allow the commissioning of such facilities by determining component flows through closed doors, pressure stabilisers and the like. The modifications in approach necessary for application to Category 3 and above facilities are presented with discussion of the findings from field measurement data.

KEYWORDS
Airtightness, ventilation strategy, health care buildings, airborne pathogens.

INTRODUCTION

The development of airtightness standards now common in commercial and domestic building codes[1] has been driven by energy considerations – reducing the amount of unwanted and uncontrolled infiltration of external air. Low energy buildings require a high quality of design and construction. In the field of healthcare buildings and specialist laboratory facilities, conventional practice has been based on the priority of cascading pressure zones with the emphasis on the magnitude of the static pressure differences between adjacent zones – and by implication with doors closed. A secondary importance has often been assigned to the magnitude of the air volume flow rates – the priority being the cascading negative pressure. However, the dilution of airborne contaminants and pathogens is determined to a large extent by the provision of an appropriate quantity of air coupled to an effective ventilation strategy. The ventilation strategy is generally a well- mixed process in the context of isolation rooms but could involve displacement ventilation or a uni-directional (piston effect) regime.

Efficient design and effective construction of this type of facility can only be carried out by appropriate cognisance of the airtightness standards and subsequent representation and expression as inter-zonal flows.

A further consideration is that a component such as a (passive) pressure stabiliser regulates the static pressure differential between two spaces by changing the resistance to air flow. This can be done only over a limited dynamic range – once the damper is fully open or closed it will behave as a fixed opening. Typically, this dynamic range may only be 5 to 10 Pa with a near linear volume flow rate versus pressure differential characteristic. Under door flows (assuming intentional) can be estimated but in practice are difficult to adjust.

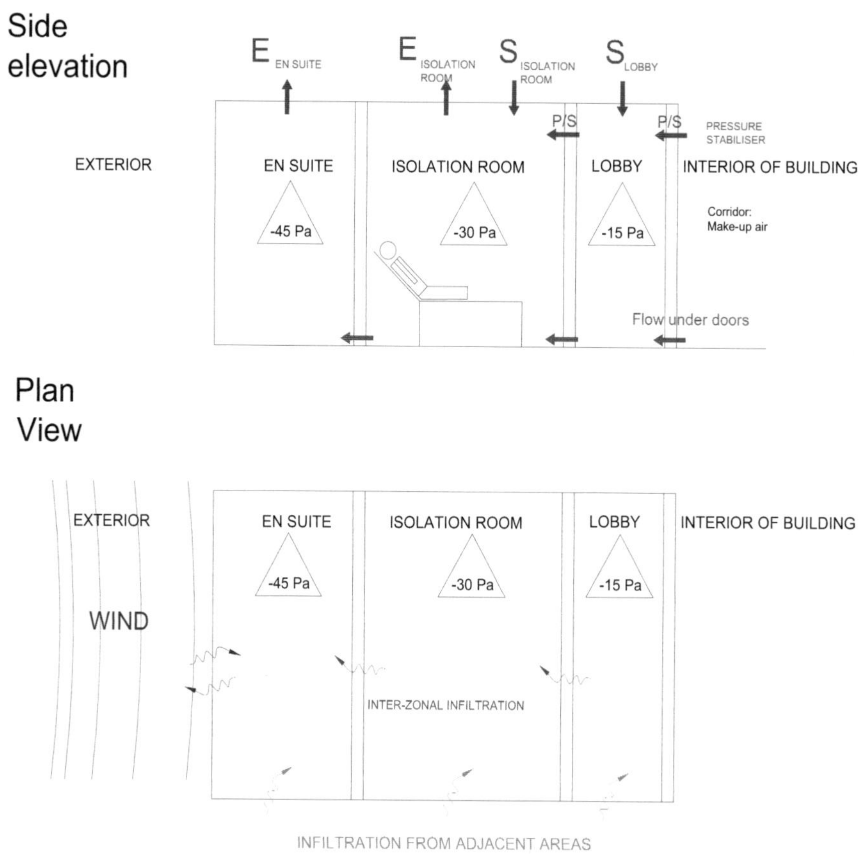

Figure 1. Schematic of component flows

Figure 1 illustrates in schematic form the key component flows for a (UK) typical negative pressure isolation room with en suite bathroom and entrance lobby. In the idealised situation with a perfect enclosure and no external wind effects, the relationship between the flows is readily defined. (NB There is an implicit assumption that sufficient air is available to be drawn into the lobby.) At commissioning and subsequent maintenance, the two mechanical extracts and two mechanical supplies can be adjusted to produce the desired regime although this assumes that the pressure stabilisers and under-door flows are appropriate. If either of these are not appropriate, the consequences of adjusting the mechanical volume flow rates to achieve the desired cascading pressure differentials can cause practical problems such as poor room air movement and mixing (supply much lower than design compromising diffuser selection), draught and thermal discomfort risk (supply much higher than design). Lower than design dilution rates are also a probability. In the non-idealised situation (i.e. when constructed), leakage into (NB outward for positive pressure) the controlled spaces may reduce the mechanical supply requirement. It follows from the expression of the inter-zonal flows that practical bounds can be set on acceptable levels of infiltration (and hence minimum airtightness requirements) by realising that the undesirable flow rates must be less than the dynamic range of control offered by the pressure stabilisers and the maximum design variances in mechanical supply and extract volume flow rates. It is prudent to include the potential uncertainty in under door flows within the overall allowance.

In reality, the situation becomes more complicated with the effect of wind on external walls being one factor to be considered. The effect of the rest of the building or hospital on the corridor also has an influence.

PHYSICAL MODEL INVESTIGATIONS

BSRIA has conducted a programme of investigations[2] using full size physical models. The objectives were to validate two different isolation room designs in England and Northern Ireland. Understanding the component flows and the potentially significant impact of airtightness criteria became one of the major findings. The first study, commissioned by DH, is described in the Guidance HBN4[3] and it consisted of an entrance lobby, isolation room and en suite. The second study, a proposed Intensive Care Unit(ICU) isolation room for the Royal Victoria Hospital of Belfast, commissioned by DHSSPS, consisted of an entrance lobby and the patient isolation room but no en suite. The ICU had a much higher provision of medical equipment and therefore heat load than the HBN4 design.

Both rooms were tested under a cascading negative pressure configuration and as neutral isolation rooms with a Positively Pressurised Ventilated Lobby (PPVL). A PPVL room provides protection from infections originating in the room (equivalent to a negative pressure isolation room) and from infections originating in the corridor (equivalent to a positive pressure type).

In the HBN4 design, air was supplied into the lobby through a 4-way square ceiling diffuser and then entered the isolation room via a pressure stabiliser and gaps around and below the lobby- to- isolation room door. The pressure stabiliser was designed to open when the ΔP between the lobby and the room was 10 Pa.

The air mixed in the room, diluting the contaminants concentration and went through a door grille into the en suite, where it was extracted. The lobby was kept at a ΔP of +10 Pa with respect to the isolation room and the hospital corridor, the room was kept at Neutral pressure with respect to the corridor and the en suite at -10 Pa. The air change rate (ACR) in the room was 10 ACH (nominal).

To achieve a cascading negative pressure design, the room was modified and an extract was built in the isolation room and only the top blade of the pressure stabiliser was used. The air change rate was still 10 ACH and the pressures were -4 Pa in the lobby, -18 Pa in the isolation room and -18 Pa in the en suite.

The PPVL design of the Intensive care unit at the Royal Victoria Hospital, also consisted of a pressurised lobby at +10 Pa and a 10ACR in the neutral isolation room. In this case, the air passed into the Isolation room through two pressure stabilisers above the door and it was extracted from the room into an adjacent plant room where the air was treated (filtered) before being released outside the hospital.

In order to achieve a cascading negative pressure configuration in the ICU, the supply and extract flows were modified, the doors were changed so that they opened into the corridor, and the pressure stabilisers were set so they opened at a differential pressure of 15 Pa. The ACR was now 18 ACH instead of 10ACH in the patient's room. The heat load in the room was not modified. The lobby was kept at –15 Pa and the room at -30 Pa with reference to the hospital corridor.

FINDINGS

The investigations carried out by BSRIA demonstrated that the effectiveness of an isolation room does not depend solely on pressure differentials, but also on achieving good mixing and dilution within the space. Protection levels of 10^3 were obtained inside the room (outside the 1 m patient breathing zone) and of 10^5 outside the room. [4]

Air tightness tests were carried out in these rooms for two reasons. Firstly, in order to commission the room, the airflows through the structure derived from the pressure differentials between the rooms, (in addition to the flows under and around doors and through the pressure stabilisers) needed to be quantified and minimised. (i.e in a fan failure mode resulting in altered pressure differentials in the room , or anomalous room operation-leaving doors open) the walls offer a degree of protection, this term is known as passive protection.

Unknown or unaccounted for air leakage though the structure of an Isolation room could translate into risk of contamination in adjacent areas where it would not be expected. Secondly, air tightness tests were used to determine the flow through key design components (pressure stabilisers, doors).

Table 2 shows the study of air leakage through the envelope. HBN4 suggests a test pressure of double the operating pressure for isolation rooms. For a given Air leakage index (ALI) the maximum test flow rate is calculated. Table 2 shows that the air leakage through the walls for a room at 20 Pa is considerably much higher than the same room at 10Pa (45% for Model 1 and 50% for Model 2).

Model	Working P	Test P	ALI	ALI (used)	Max Test Flow Rate	Max. Test Flow Rate / unit volume
	(Pa)	(Pa)	$m^3/(h.m^2)$	$m^3/(h.m^2)$	$(l.s^{-1})$	$(l.s^{-1}$ per $m^3)$
1	10	20	3.16	3	72.5	1.01
	15	30	3.87	3.9	94.3	1.31
	20	40	4.47	4.5	108.8	1.51
	25	50	5	5	120.8	1.68
2	10	20	3.16	3	99.2	0.75
	15	30	3.87	3.9	128.9	0.98
	20	40	4.47	4.5	148.8	1.13
	25	50	5	5	165.3	1.25

Table 1. Air leakage maximum values

The experimental work clearly demonstrated that determination of airtightness enables component flows to be readily determined thus providing a straightforward route for commissioning and in-service maintenance.

FACILITIES UP TO CAT3

The traditional view of facilities such as isolation rooms has been to focus on the relative de-pressurisation (negative) or pressurisation (positive). The quality and quantity of airflow within the space has generally not been accorded the same attention – the dilution of potential pathogens is the mechanism by which personnel within a space are afforded protection. The provision of all three parameters – pressure steps, air change rate and effective ventilation – has been shown to be necessary.

The integrity of the building fabric forming isolation and other specialist facilities subject to cascading pressure differentials is expressed in the form of appropriate airtightness specifications. The primary driver for specification is to ensure the pressure cascades can be achieved and controlled. The secondary driver is to ensure there are no adverse effects on delivered air change rate and effective ventilation.

For facilities up to CAT2, some consideration should be applied to the challenge arising from a first order failure. For example, for a negative pressure facility, loss of the mechanical extract could result in positive pressurisation of the all or part of the facility with the consequent potential hazard of leakage of airborne pathogens to surrounding spaces or the external environment. *(NB It is worth noting completely different considerations apply when a door is opened between adjacent spaces. Only in very specialist applications will there be a high enough inward volume flow rate for negative pressure cascades (outward for positive). Generally, this is regarded as a transient situation with some potential for interchange of air between the adjacent spaces, possibly compounded by entrainment of air in the wake of a walking person, for example.)* A risk analysis approach could involve identification of the maximum acceptable leakage of pathogens from the isolation room under normal operation (e.g. potential cross-transfer due to air interchange from isolation room to lobby and lobby to corridor resulting from personnel movement) and under a single mode failure (e.g. extract fan failure leading to positive pressurisation of the isolation room). Relative pressurisation of adjacent spaces such as flanking corridors, plant rooms and service voids must be taken into account. It is not realistic to prescribe a zero leakage. It may be helpful to view this target in more prosaic terms such as a permissible "*bugs per minute*" transmission.

This target can then be converted by designers into specifications that in turn can be validated at commissioning and throughout the operational life of the isolation facility.

EXTENSION TO CAT3 AND ABOVE FACILITIES

When dealing with extremely hazardous situations, the rationale for setting performance based targets is shifted to reflect the overriding concern to avoid escape of airborne pathogens under failure modes as the priority. This leads to much high airtightness standards and directly to lower actual flow rates and changes to the measurement techniques applied.

The method of measuring the air leakage of rooms using a pressurisation (or de-pressurisation) technique is well documented with various standards giving the basic test method (for example)[5] [6]. These involve supplying (or extracting) air to an enclosure and measuring the resulting pressure differential. This is repeated for a number of different flow rates, the air flow rate and resulting pressure differential being related by the following equation:-

$$Q = C_L.\,(\Delta P)^n \qquad 1$$

Where Q	= air flow rate supplied to the building	($m^3.s^{-1}$)
ΔP	= pressure differential across building	(Pa)
C_L	= the air leakage coefficient	($m^3.s^{-1}.Pa^{-n}$)
n	= an exponent normally between 0.5 and 1.0	

A regression analysis is carried out on the data and the results expressed as an air flow rate at a specific pressure differential, and normally normalised by either surface area ($m^3.(h.m^2)^{-1}$, or volume (ach^{-1})).

Where the air leakage through the door is small compared with the remaining air leakage the use of blower door fan sets can be used, as the air flow rate required to test an enclosure reduces, the air leakage through a normal door fan set-up could become a significant proportion of the air leakage, in these circumstances an alternative method of supplying (or extracting) will be required. This can be either a purpose made attachment or utilise existing apertures within the enclosure such as the balancing dampers. In addition the air flow rate range of this type of equipment may not be appropriate or provide the required pressure differential. In these circumstances it may be preferable to use an in-line flow meter rather than the 'inlet flow meter' which is regularly used to measure the air leakage of buildings. The in-line flow meter could be, for example, a laminar flow element or a venture nozzle. The air flow rate being determined by measuring the pressure differential across the flow meter. In addition, the information on the airflow rate through the access door will be required.

As the room air leakage decreases the time taken for the room pressure to stabilise after the air flow rate is adjusted can increase, and some circumstances can be in excess of 45 minutes. In these circumstances a different method has to be adopted. The method used has been to set the air flow rate at the required air leakage, and allow the room pressure differential to stabilise (some adjustment of the air flow rate may be required during this period). The room pressure attained should be greater than required by the air leakage criteria to indicate compliance. Once the room has attained its maximum pressure differential, the fan is left to operate for a minimum of 15 minutes, with measurements recorded every 2 minutes to confirm that stability has been reached.

Table 2 lists some of the performance targets for some of the test work carried out by BSRIA over the last four years. To facilitate comparison, the criteria have been converted to an equivalent air tightness expressed as a conventionally reported air permeability at 50 Pa.

Room description	Air tightness criteria	Equivalent air tightness expressed as an air permeability $m^3.h^{-1}.m^{-2}$ @ 50 Pa
Clean room with close control of humidity required	0.2 $m^3.h^{-1}.m^{-2}$ @ 50 Pa	0.2
Clean rooms & labs within a hospital	2.5 $m^3.h^{-1}.m^{-2}$ @ 50 Pa	2.5
Cat 2 & 3 labs	0.6 air changes per hour at ±60Pa	0.45 – 0.5
Cat 3 labs within a hospital	1 $l.s^{-1}$ @ 100 Pa	0.012
Cat 3 labs within an animal research facility, (temporary building)	2 $l.s^{-1}$ @ 200 Pa	0.015
Cat 4 Rooms within an animal research establishment	0.009 $m^3.h^{-1}.m^{-2}$ @ 200 Pa	0.0025

Table 2. Recent air tightness criteria

CONCLUSIONS

The extension of energy based airtightness specifications to healthcare and specialist laboratory facilities has been illustrated. The rationale for performance criteria based has been explained via component flows identified from an idealised scenario. The need to characterise pressure stabilisers and under door flows has been shown to be essential for design selection and in use commissioning and maintenance.

Application of blower door techniques to two types of isolation room has been reported. These techniques can yield data on component flows (e.g. pressure stabilisers) as well as the actual standard of airtightness achieved at construction. These techniques can be utilised routinely.

The modification in measurement technique necessary when extending to high integrity structures (i.e. for CAT3 and above) has been described.

It has been shown how high risk applications can use the component flow approach to specify airtightness standards based on hazard assessment in failure modes. This may require buffer zones in extreme applications to decouple protected areas from the external environment and the effects of wind in particular.

In summary, the application of appropriate airtightness standards to healthcare buildings and specialist laboratory facilities goes beyond the energy-based origins of airtightness. Confidence in design and construction can be achieved by specification, effective construction and good commissioning practice. The underlying theme is the optimisation of desirable air flows and the minimisation of undesirable flows.

ACKNOWLEDGEMENTS

Part of the work carried out by BSRIA reported here was supported by The Department of Health (England) and DHSSPS, Northern Ireland.

REFERENCES

1 Building Regulations, England:(2010) *"Approved Document L2A"*.

2 William Booth and Blanca Beato Arribas:(2007). *"Characterisation of Contaminant Distribution In A Neutral Pressure Isolation Room"* Roomvent 2007, Helsinki. Paper 1156.

3 NHS Estates, England:(2005). *"HBN4 Supplement 1: Isolation facilites in acute settings"*. Efm-standards. ISBN 0-11-322711-6.

4 W. Booth, B. Beato Arribas, C. Noakes, L. Fletcher, A. Sleigh. and N. Tomlinson (2009), *"Characterisation of the protection provided by the ventilation strategy in hopsital isolation rooms"*, Healthy Buildings Conference Syracuse USA.

5 ATTMA: (2010) *"Technical Standard L2 Measuring Air Permeability of Building Envelopes"*.

6 BS EN 13829:2001 *"Thermal Performance of Buildings – Detemination of Air Permeability of Buildings – Fan Pressurisation Method"*.

CLASS C AIR-TIGHTNESS: PROVEN ROI IN BLACK AND WHITE

Peter Stroo

Rf-Technologies
Lange Ambachtstraat 40
9860 Oosterzele
Belgium

ABSTRACT

Class C air-tightness: proven ROI in black and white

At the end of 2010, two manufacturers have commissioned an independent engineering firm to carry out a cost-benefit analysis of air-tightness in ventilation. The study report uncovers the clear return of investment in class C air-tight ventilation systems in Belgium.

The study comprises:

- The composition of a detailed spreadsheet with the result of the energy cost savings and additional investment cost;
- The application of the spreadsheet to three Belgian cases (the renovation of a hospital wing, a nursing home and an office building);

The simulations show that the total energy consumption linked to ventilation can be reduced by as much as 30% with a break-even-point between 2 and 3 years.

KEYWORDS

air-tightness, ventilation, class C, ROI, fire dampers

INTRODUCTION

A **ventilation system** has an important technical function. On the one hand the system moves treated air to rooms where there are people present (supply air). On the other hand, used air is extracted from these rooms and discharged outside (extract air).

The transportation (fan power) and treatment (filtering, heating, cooling, dehumidifying and humidifying) of air is a **significant energy cost when operating a building**. It is therefore important to get the desired amount of air in the right rooms effectively and in a controlled way. On the way in and out, as little ventilation air as possible must be lost; and this is only possible by building an airtight ventilation system.

This paper offers an answer to a number of legitimate questions regarding air-tightness of ventilation systems: What impact do investments in ventilation air-tightness have on the energy consumption of a building? How can we calculate the **break-even-point of such investments**?

AIR-TIGHTNESS OF VENTILATION SYSTEMS

The air-tightness of a ventilation system is determined by the air-tightness of each component of the system: this includes the air ducts themselves, but also all the accessories such as fire dampers, flow-balancing units, silencers, etc.

The air-tightness of ductworks is described and quantified in European standards (including EN12237, EN1507, and EN1751). The air-tightness class determines the size of the air leak: air-tightness class C or D indicates a very performing ventilation system, class A or lower (3A, 9A) are awarded to systems with minimum air-tightness.

In order to go up an air-tightness class, a ventilation system must become three times more efficient: so the leakage flow rate in a type C ductwork is three times lower than the leakage flow rate of a type B ductwork.

AIR-TIGHTNESS STUDY

At the end of 2010, a manufacturer of ventilation ducts and a manufacturer of fire dampers have commissioned an independent engineering firm to carry out a cost-benefit analysis of air-tightness in ventilation in Belgium.

The study comprises the composition of a detailed spreadsheet [1] which summarizes the energy cost savings and additional investment cost related to air-tightness; as well as the application of the spreadsheet to three cases, which is the subject of this paper.

The calculation method was applied to following (existing) sites:

- Case 1 – Renovation of a hospital wing, Antwerp
- Case 2 – Nursing home, West-Flanders
- Case 3 – Office building, Flemish Brabant

		Case 1	Case 2	Case 3
Building surface area	-	11,380	8,830	11,200
Ventilation flow	m³/h	57,450	43,305	53,940
Duct surface (supply and extract)	m²	5,094	2,400	2,480
Percentage of round ducts	-	25%	15%	15%
Fire dampers, number	-	476	496	90
Eq. surface area dampers	m²	461	395	183
Flow-adjustment units, number	-	527	398	407
Dynamic flow adjusters (Constant Air Volume, Variable Air Volume), number	-	133	0	77
Silencers, number	-	17	14	19

Table 1 Case study – properties of technical installations

UP TO MORE THAN 30% SAVINGS

The simulations show that the total energy consumption linked to ventilation can be reduced by as much as 30% with a break-even-point between 2 and 3 years.

The table and figures below show the results of the simulations. The simulations show that the total energy consumption linked to ventilation can be reduced by over 30% (case 1).

		Case 1	Case 2	Case 3
Annual energy savings	EUR/y	10,175	6,750	2,335
Investment cost	EUR	14,862	11,068	7,212
Pay Back Time (dynamic)	years	2	2	3

Table 2 Profitability if the ventilation system's air-tightness improves from class A to class C

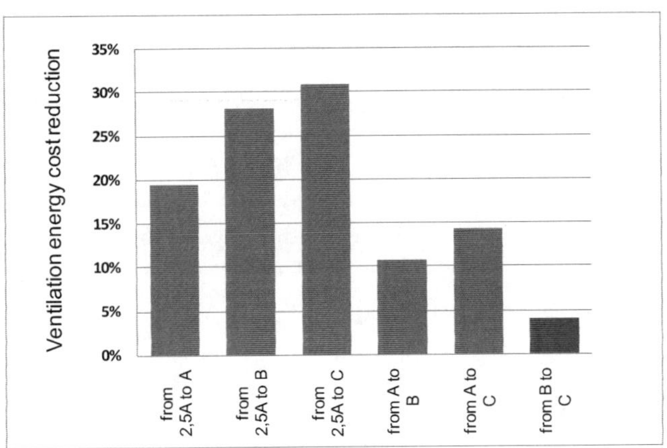

Figure 1 - Results simulation case 1

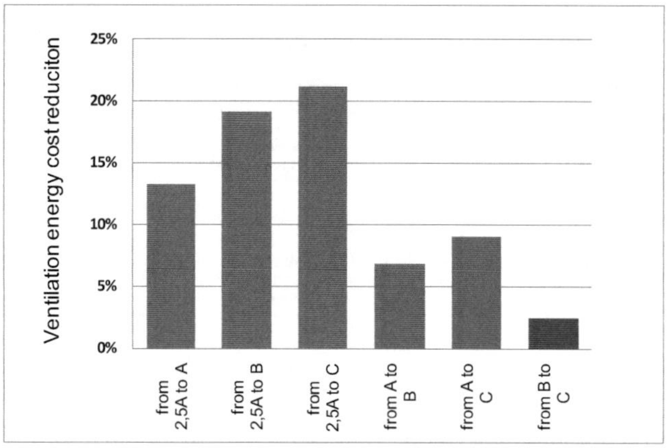

Figure 2 - Results simulation case 2

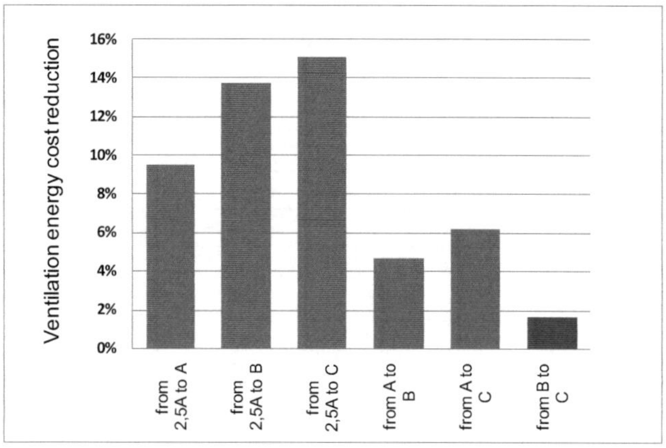

Figure 3 - Results simulation case 3

CONCLUSION

Optimising the air-tightness of the ductwork is worthwhile. The total amount of energy linked to the ventilation system can be reduced by as much as 30%.

The importance of air-tightness is also acknowledged by the Belgian Buildings Authority, which is expected to impose class C very soon in its Standard Specifications. In view of the pioneering role of the Standard Specifications, we can expect the Belgian installation world to catch up so that (at least) air-tightness class C will soon be standard.

REFERENCES

[1] the spreadsheet is covered in another paper in this congress.

THE POWER OF QUALITY

Arch. Christophe Debrabander, Ir. Arch. Kristof Neutens

Bostoen
Koninginnenlaan 2-3
Drongen, Belgium

ABSTRACT

Through the experiences gained by building a sufficient number of air-tight buildings, the author will illustrate the ease of detailing and constructing an air tight building.
Using parallels to conventional building typologies, the methods of making an air-tight building enveloppe will be explained. The presentation will be divided into following chapters:

1. Precicious building methodology.

   When everything is detailled and executed with precision and according to art and best practices (e.g. finishing the interior of the windows with plaster in stead of using wood; using windows 'better' than class 4; correct placement of the vapour barrier to prevent internal condensation; using correct materials in the correct application; …), the n50 value will already be astonishingly low.

   All of these precautions taken to improve the quality of the building, benefit the airtightness. So the measures taken are a logical choice.

2. Quantity.

   Our tests prove that applying these logical choices lead to a great number of succesful tests (e.g. 10 tests a week, all of them lower than 0,6/h. Examples: 0,50/h; 0,43/h; …).

3. Example.

   The best examples are 2 of the employees of Bostoen who build their own passive zero-energy house. They took into account the quality precautions and reached immediately the n50 value of 0,47/h.

4. Quality through mentality and motivation.

   The example and our experience prove that you can have a good blower-door score without making any special and/or expensive interventions. The quality of the materials, the mentality and the motivation of everyone involved suffices to be able to build and design air-tight buildings. One of the only prerequisites for buildings with an n50 that's smaller than 0.6/h, is that all the people working on it, from architect to construction worker are aware of what they're doing.

5. Other areas of application.

   When we keep in mind that a cooler for fruit has a n50 value of around 0.05, we know that we still can do better.

KEYWORDS

Air-tightness, precicious building, detailling, building methodology, passive houses, zero energy building, blowerdoor

INTRODUCTION

Through the experiences gained by building a sufficient number of passive houses and zero energy buildings, the author will illustrate the ease of detailing and constructing an air tight and well ventilated building.
This year (2011) already 45 passive houses and zero energy buildings were finished, each of them with a possitive ventilation report and each of them with an n50 value < 0,6/h.

INSURING AIR-TIGHTNESS

Detailling

Using parallels to conventional building typologies, the methods of making an air-tight building enveloppe will be explained.

The first important element is precicious building methodology.
When everything is detailled and executed with precision and according to art and best practices, the n50 value will already be astonishingly low. We will demonstrate this through a number of examples.

Figure 1.

The example shown in Fig. 1, shows one of the triangles beneath the roof that won't be visible after the rooms are finished. So strickly for the functionallity and the appearance of the room, they normally don't have to be covered with plaster, yet to immediately do so while plastering the wall, requires only little extra effort. Otherwhise, when we perform a blowerdoor test, air is going to infiltrate between the bricks. This infiltration, depending on how many triangles are present, can cost a few thenths of a point for the n50 value.

Figure 2.

Figure 2 shows the connection between a steel I beam that carries the hollow core slabs. When the open space above the slabs isn't filled with cement or plaster, there will be a considerable amount of air infiltration. (cfr Figure 3a)
So there are 2 option, either the space is filled up with concrete or cement when the construction workers fill up the holes between the slabs as shown in Figure 3b, or the workers who plaster the walls also fill up the I-beam on both sides.

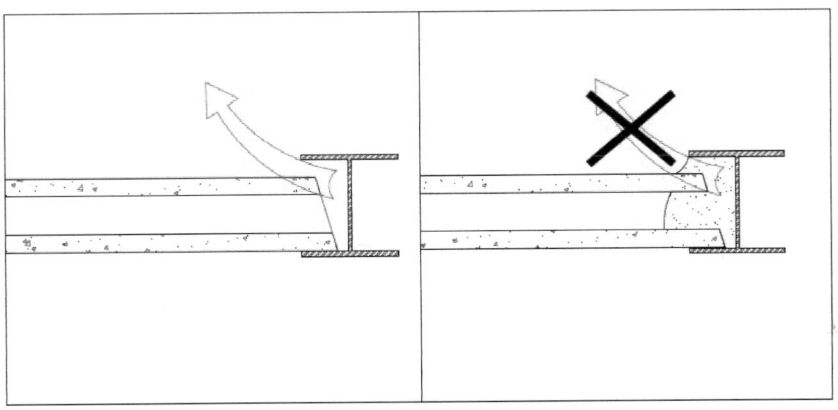

Figure 3a. Figure 3b.

Figure 4.

In figure 4 we can see the holes in the hollow core slab. The edges of the slabs have to be filled with concrete even if it is not strictly necessary for stability purposes. This prevents air from flowing through the canals and entering the house where holes are drilled in the slabs to allow ventilation and electricity to pass (cfr. Figure 5).

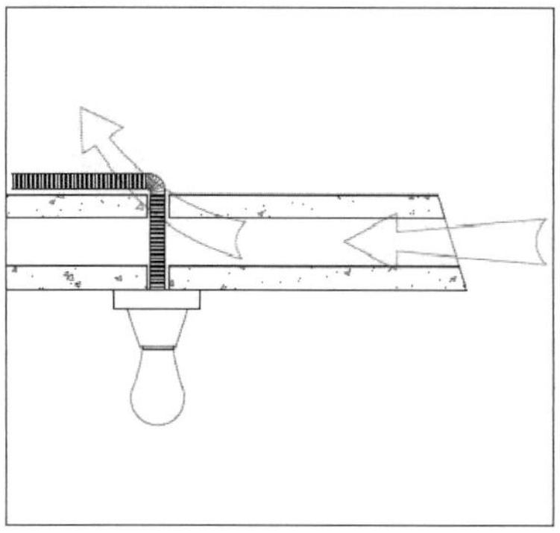

Figure 5.

Figure 6.

The situation as depicted in Figure 6, shows a way to insure that there are no air leaks through the connection between the bricks and the hollow core slab. When applying cement to the base of the wall, one can make sure that there's no air infiltration between the concrete and the brick. By choosing plaster to finish the walls on the inside, we make sure that the entire wall is air tight.
Again, this measure doesn't require a lot of time or money, but has a big inpact on the blower door test.

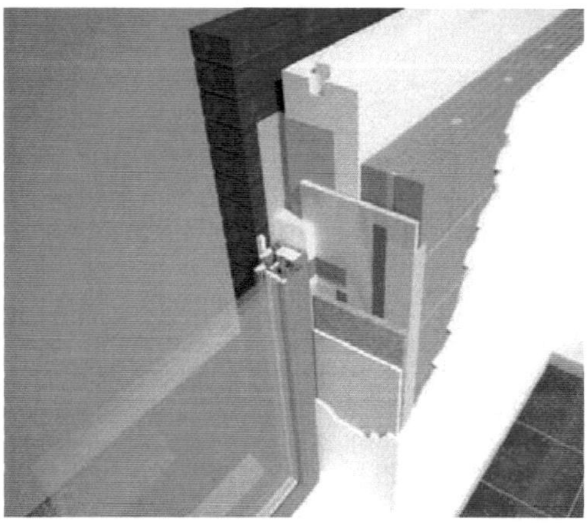

Figure 7.

When detailing the connection between the windows and the wall, one should keep in mind that the air-tightness is of the utmost importance.
In this render we can see the various air-tight foils (pink in picture) and their connection to the plywood and the brick wall. Everything is done so no air will infiltrate and to see to it that the best thermal resistance is achieved.

To insure the air-tighness of the building, an excellent detailing isn't the only step one should take. The use of windows that have an air-tightness 'better' than class 4 is also recommended. This way you allow yourself some room for errors, be it in the execution of the details or little faults in the materials. It speaks for itself that air-tight detailing is of little use, when using windows from class 3, 2 or 1.

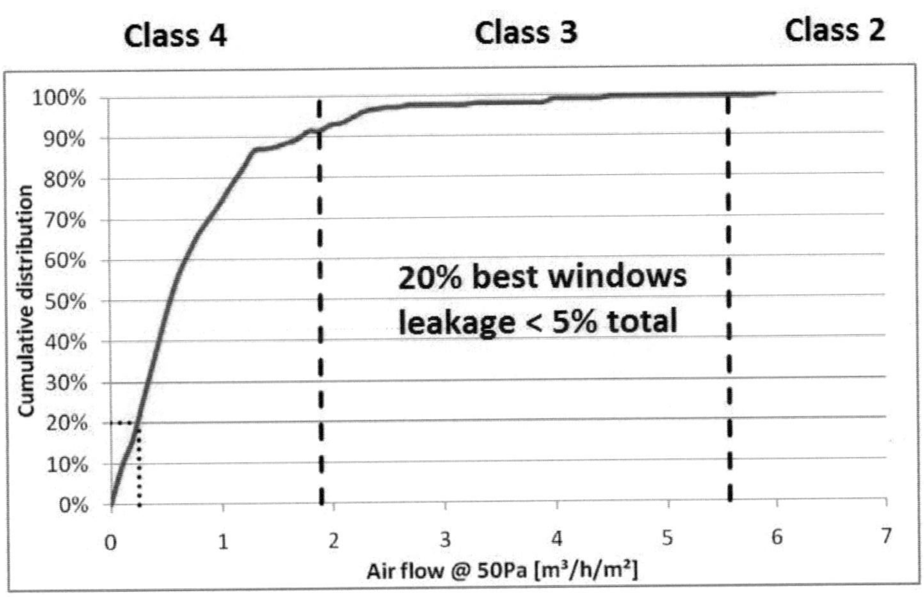

Graph 1.

All of these precautions shown in Figures 1 - 7, are taken to improve the quality of the building and benefit the airtightness. The measures taken are a logical choice and demand little extra effort.

The 3 M's

Quality through mentality and motivation is another important issue in this matter.
The quality of the **materials**, the **mentality** and the **motivation** of everyone involved suffices to be able to build and design air-tight buildings. One of the only prerequisites for building with an n50 value that's smaller than 0.6, is that all the people working on it, from architect to construction worker are aware of what they're doing.
This can be achieved by providing adequate (materials, mentality and motivation) training for everybody involved and allowing feedback from every level of the chain of command.

Figure 8.

The proof

Our tests prove that applying these logical choices (as shown in Figures 1 - …) and a correct mindset (as illustrated above), lead to a great number of succesful tests (e.g. 10 tests a week, all of them lower than 0,6 cfr. Appendix I - IV).

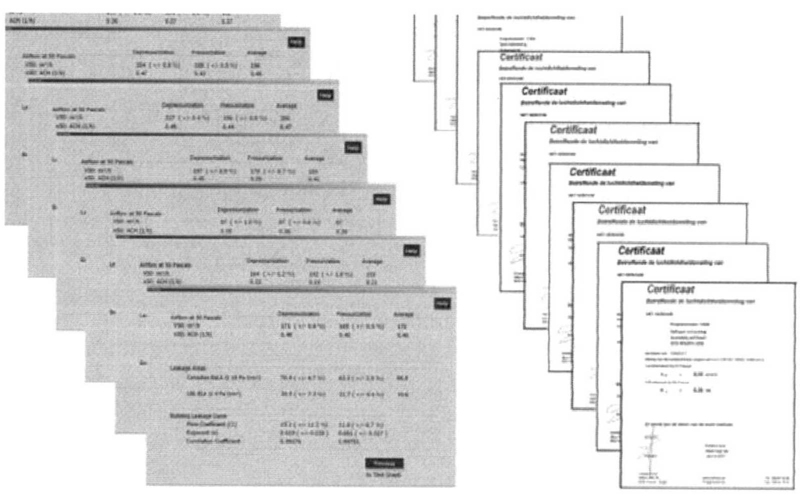

The best examples are 2 of the employees of Bostoen who build their own passive zero-energy house. As we can see in the pictures, they took into account the quality precautions and immediately reached the n50 value of 0,47/h.

Figuur 10: The ends of the hollow core slabs have been filled with concrete

Figuur 11: Air-tight finishing of the wooden roof

Figuur 9: Insulation and installation of the windows

These examples and our experience prove that you can have a good blower-door score without making any special and/or expensive interventions.

CONCLUSION

When one strives to build with an n50 value < 0.6, there are a couple of things that can help to achieve that goal.

The building should be designed with the utmost care as not to allow any in- or exfiltration of air. This does not only concern construction details, but the entire building concept. All the steps of the building process, from the construction of the walls to the installation of the ventilation, have to be in tune with each other.

To insure a good execution, the construction workers and everyone involved should be properly trained and have to be aware of what they are doing. An investment in the schooling of the personel rapidly repays itself.

APPENDIX I

Housing project Temse
Net Volume: 430.40 m^3
n50 Value : 0.37

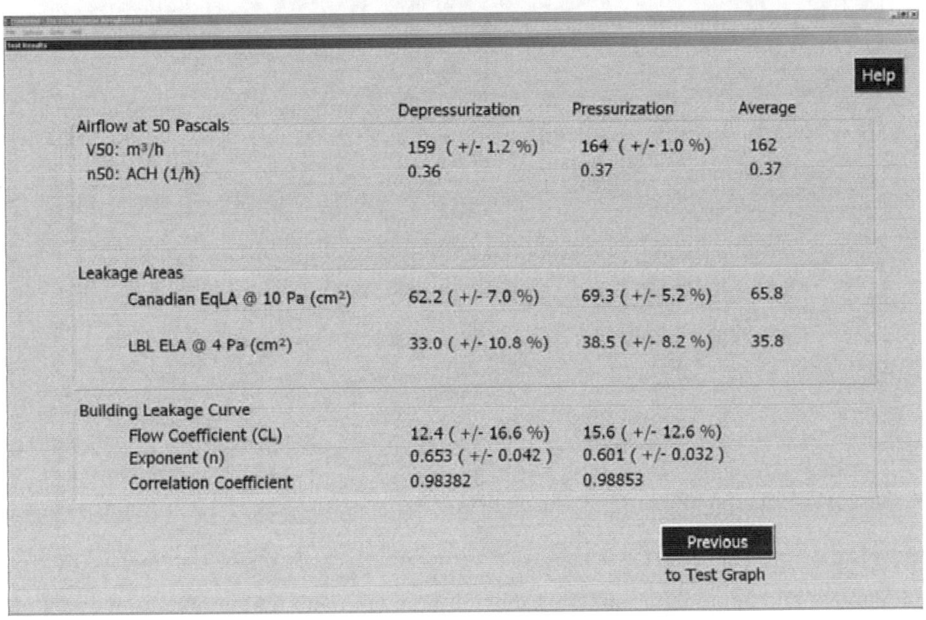
Help

	Depressurization	Pressurization	Average
Airflow at 50 Pascals			
V50: m³/h	159 (+/- 1.2 %)	164 (+/- 1.0 %)	162
n50: ACH (1/h)	0.36	0.37	0.37
Leakage Areas			
Canadian EqLA @ 10 Pa (cm²)	62.2 (+/- 7.0 %)	69.3 (+/- 5.2 %)	65.8
LBL ELA @ 4 Pa (cm²)	33.0 (+/- 10.8 %)	38.5 (+/- 8.2 %)	35.8
Building Leakage Curve			
Flow Coefficient (CL)	12.4 (+/- 16.6 %)	15.6 (+/- 12.6 %)	
Exponent (n)	0.653 (+/- 0.042)	0.601 (+/- 0.032)	
Correlation Coefficient	0.98382	0.98853	

Previous
to Test Graph

APPENDIX II

Housing project Eke (16179)
Net Volume: 426.93 m^3
n50 Value : 0.40

Airflow at 50 Pascals	Depressurization	Pressurization	Average
V50: m^3/h	171 (+/- 0.8 %)	169 (+/- 0.5 %)	170
n50: ACH (1/h)	0.40	0.40	0.40
Leakage Areas			
Canadian EqLA @ 10 Pa (cm^2)	70.4 (+/- 4.7 %)	63.3 (+/- 2.8 %)	66.8
LBL ELA @ 4 Pa (cm^2)	38.5 (+/- 7.3 %)	32.7 (+/- 4.4 %)	35.6
Building Leakage Curve			
Flow Coefficient (CL)	15.1 (+/- 11.2 %)	11.8 (+/- 6.7 %)	
Exponent (n)	0.619 (+/- 0.028)	0.681 (+/- 0.017)	
Correlation Coefficient	0.99176	0.99751	

Help

Previous
to Test Graph

APPENDIX III

Housing project Oostende
Net Volume: 263.81 m³
n50 Value : 0.38

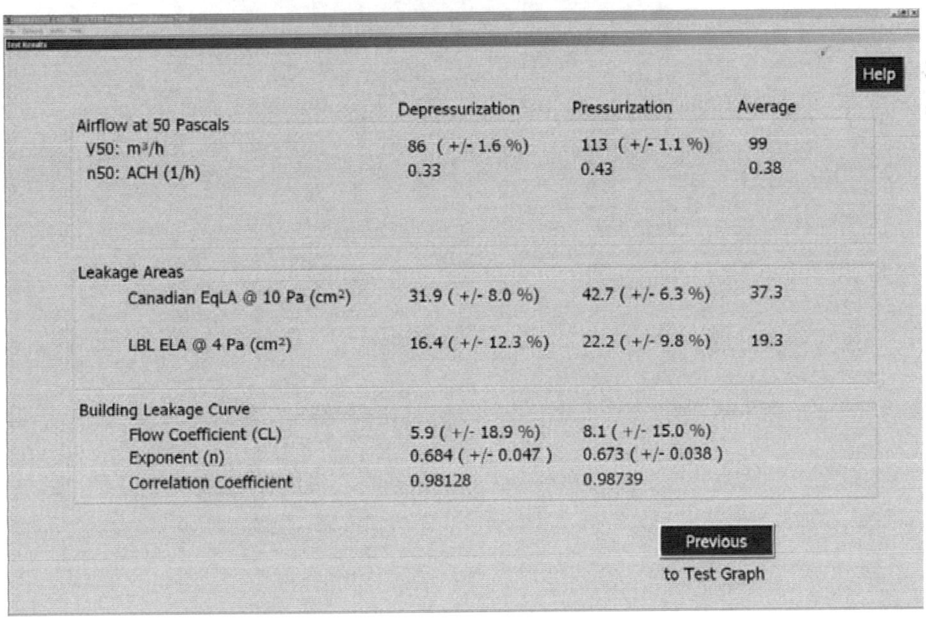

Help

	Depressurization	Pressurization	Average
Airflow at 50 Pascals			
V50: m³/h	86 (+/- 1.6 %)	113 (+/- 1.1 %)	99
n50: ACH (1/h)	0.33	0.43	0.38
Leakage Areas			
Canadian EqLA @ 10 Pa (cm²)	31.9 (+/- 8.0 %)	42.7 (+/- 6.3 %)	37.3
LBL ELA @ 4 Pa (cm²)	16.4 (+/- 12.3 %)	22.2 (+/- 9.8 %)	19.3
Building Leakage Curve			
Flow Coefficient (CL)	5.9 (+/- 18.9 %)	8.1 (+/- 15.0 %)	
Exponent (n)	0.684 (+/- 0.047)	0.673 (+/- 0.038)	
Correlation Coefficient	0.98128	0.98739	

Previous
to Test Graph

APPENDIX IV

Housing project Hove
Net Volume: 738.44 m³
n50 Value : 0.21

Help

Airflow at 50 Pascals	Depressurization	Pressurization	Average
V50: m³/h	164 (+/- 1.2 %)	142 (+/- 1.0 %)	153
n50: ACH (1/h)	0.22	0.19	0.21

Leakage Areas			
Canadian EqLA @ 10 Pa (cm²)	55.9 (+/- 7.2 %)	47.9 (+/- 6.4 %)	51.9
LBL ELA @ 4 Pa (cm²)	27.5 (+/- 11.1 %)	23.4 (+/- 9.9 %)	25.5

Building Leakage Curve		
Flow Coefficient (CL)	9.2 (+/- 17.1 %)	7.8 (+/- 15.3 %)
Exponent (n)	0.735 (+/- 0.043)	0.742 (+/- 0.039)
Correlation Coefficient	0.98651	0.98928

Previous
to Test Graph

Quality Management Approach to Improve Buildings Airtightness

Requirements and Verification

Valerie Leprince*[1], Joris Biaunier [1], Rémi Carrié [2] and Myriam Olivier [1]

1 Centre d'Etudes Techniques de l'Equipement de Lyon.
46 rue Saint Théobald. F – 38081 L'Isle d'Abeau Cedex, France,
**Corresponding author*
valerie.leprince@developpement-durable.gouv.fr

2 INIVE,
Lozenberg 7
BE-1932 Sint-Stevens-Woluwe, Belgium

ABSTRACT

In France, starting on January 1st, 2013, a minimum airtightness value for all residential building will be required by the energy performance regulation (RT 2012). It will be compulsory to justify for any new residential building that its airtightness is below 0.6m³/h.m² at 4 Pa (Q_{4Pa_surf}) for single-family houses and 1 m³/h.m² for multi-family buildings.

The new regulation specifies two ways to prove the building airtightness compliance: either a State certified technician measures the building airtightness at commissioning, or the builder implements an approved quality management approach.

Firstly, this paper discusses the requirements for approved quality management approaches. Then, since this process started in 2008, results obtained by companies who implemented this quality management have been analysed.

Secondly, in order to check if the airtightness of constructions built by those companies is in line with expectations, State technicians perform controls. Each company quality management approach is evaluated through measurements and files analysis. This paper describes the process of these controls, and gives a preliminary analysis of the verifications performed.

The key result is that measurements ordered by the builders show significant improvement in envelope airtightness. This trend merits to be confirmed with evaluations by state technicians that are underway, but whose preliminary results already show the relevance of state control namely to avoid competition distortion.

KEYWORDS

Airtightness, Quality management, Control, Thermal regulation

INTRODUCTION

Building airtightness became a key subject in the nineties for low-energy buildings labels such as Passivhaus and Minergie. In the 2000's, airtightness is confirmed as a prerequisite to design low-energy buildings.

The 2005 version of the French energy performance regulation (RT2005) included the possible adoption of an approved quality management approach to justify the airtightness level. However, justifying airtightness treatment was compulsory only for the low-energy building label "BBC-Effinergie" [3].

In the 2012 version of the French energy performance regulation (RT2012), which imposes the low-energy level to all new construction, a minimum requirement for the envelope air tightness of residential buildings is included, with two options to justify its treatment: a) measurement at commissioning or b) adoption of an approved quality management approach.

On January 1st 2013, when the RT2012 will come in force for residential building, the quality management will become a key approach to justify airtightness level.
This paper describes first the requirements to obtain an approved quality management approach. Then, since the first companies were approved in October 2008, the impact of measures taken by those companies on airtightness performance can be analysed and compared to conventional buildings airtightness.
Finally, the paper describes the control process set by State technicians to check on randomly selected buildings the correct application of the quality management and performance.

REQUIREMENTS TO SET AN APPROVED QUALITY MANAGEMENT

Procedure to file a quality approach

The RT2005 introduced the possibility to claim for a lower than the default airtightness value in the EP-calculation, without performing a test, provided that an approved quality management approach would be applied.
This possibility is maintained with the 2012 version with strengthened and clarified requirements, because justification of airtightness level will become compulsory. Every 2005 version approved applicant will have to apply for a new approval in 2012.
A candidate submits an application to a national committee. Each application is evaluated by two independents experts who either approve the applicant, or request additional documents, or else reject the applicant.
Every year the approved applicant must provide a yearly report of its quality approach. The yearly report includes measurements on a sample (at least 5% of the production) and the last version of every quality documents.

Key elements of the quality approach

The quality application includes basic requirements for quality management approach, measurements on sample and training documents on airtightness.
The quality management basic requirements to be approved are:[1]

- Identification of "who-does-what" and when;
- Trace of each step of the approach;
- Proof of the approach effectiveness based on sample measurements;
- Proposal of a scheme to ensure that the approach will remain effective with time, based on measurements on a sample.

The process should also include training and education of all craftsmen on site. This requirement should lead to a better airtightness knowledge dissemination in the building community including awareness raising and treatment of air-leakage.
In addition, the RT 2012 application version will request to provide not only measurement to prove the effectiveness but also all documents linked with the quality approach for several randomly selected buildings.

ANALYSIS OF RESULTS OBTAINED BY APPROVED COMPANIES

Analysis' procedure

As of September 2011, the committee set by the Ministry has approved twenty RT2005 quality management approaches, 2 are in process and 2 were rejected.
The first appliance for RT2012 is already in process.

Table 1 summarizes all approved applicants with their approval date, the number of buildings produced each year and the average $Q_{4Pasurf}$ measure in the last yearly report (5% of the production).
As we can see on the table the first builders were approved by the end of 2008, which means that they have experience with the quality approach for more than 2 years.

	Approval date	Type of applicant	Estimated production for the coming year	$Q_{4Pasurf}$ ($m^3/h.m^2$) average in the 2010 production
1	20/10/08	Builder	98	0.36
2	25/11/08	Builder	48	0.57
3	25/11/08	Builder	405	0.50
4	24/4/09	Builder	Not Available	0.53
5	26/10/09	Manufacturer	13	0.27
6	24/4/09	Builder	89	0.40
7	5/8/09	Builder	291	0.37
8	5/8/09	Builder	83	0.46
9	30/11/10	Builder	Not Available	Not Available
10	2/4/10	Builder	127	Not Available
11	30/6/10	Builder	89	Not Available
12	2/4/10	Builder	184	Not Available
13	2/4/10	Builder	508	Not Available
14	31/8/10	Builder	110	Not Available
15	27/9/10	Builder	Not Available	Not Available
16	5/11/10	Builder	Not Available	Not Available
17	6/9/10	Builder	Not Available	Not Available
18	25/11/10	Builder	Not Available	Not Available
19	2/2/11	Builder	Not Available	Not Available
20	30/6/11	Builder	Not Available	Not Available

Table 1: Approved applicants and production

Besides, those results can be compared with almost 2000 other measurements. Indeed in France a quality framework has been set for airtightness measurement [1] which includes the annual recovering of all measurements done by each authorized measurer.

Obtained results

Figure 1 compares results obtained on buildings made with (8 applicants) and without quality approach. It shows the effectiveness of the quality approach as 100% are under 0.8 $m^3/h.m^2$ when it is only represent 81% of other buildings. In 2005, the requirement for the quality approach was to build single-family houses with $Q_{4Pasurf} \leq 0.8$ $m^3/h.m^2$; in 2012 the threshold will be lowered to 0.6 $m^3/h.m^2$.

The 1792 measurements referred to in Figure 1 were extracted from the measurement databases of "technicians authorized to perform pressurization tests" in low-energy (BBC-Effinergie certified) buildings. In fact, the authorization process described by Carrié et al (2010) [1] requires for each authorized tester to send an annual report that includes all of his air leakage measurements results. Therefore, the sample is heavily biased towards low-energy buildings: 47% of the buildings tested were involved in a BBC-Effinergie certification process, whereas this certification has a market share of only 7% of all new constructions. As a result, the distribution "without approved QM approach" represented in Figure 1 is certainly quite optimistic.

If in the two samples the average is not so different (0.42 $m^3/h.m^2$ with quality approach, 0.48 $m^3/h.m^2$ without), the standard deviation goes from 1.17 without quality approach to 0.15 with quality approach. The most efficient applicant who obtain an average of 0.27 $m^3/h.m^2$ even have a standard deviation of only 0.09, cf. Table 1.

Results with the quality approach consistently show values well bellow the required limits, but the main interest of the quality approach is the reliability of results with very low standard deviation.

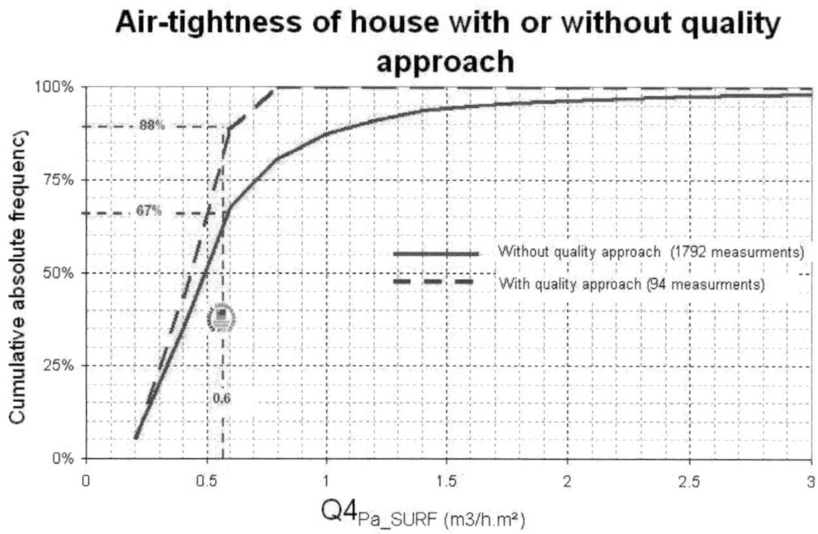

Figure 1: Distribution of measured airtightness of houses a) with implementation of an approved QM approach (dotted in green) and b) without approved QM approach (solid, in red)

Besides, based on discussions with applicants, they seem really satisfied with the benefits of the implementation of quality management approaches for various reasons. First, although it is expensive to start, it gives a positive image to the customer. Second, it requires measurement on a limited sample (typically 10% of the yearly production). Third, some mention that it has an impact on the overall building quality, which implies significant savings on customer service.

These convincing results obtained by several companies, lead to a growing confidence of the QM approach although this framework needs a careful independent evaluation.

CONTROLS CARRIED OUT BY STATE TECHNICIANS

The underlying philosophy of the quality approach was that it was better to think airtightness from the beginning than to cure at the end till obtaining the required value. The other interest was to disseminate good practice.

The results presented above are based on measurement usually performed by "authorized measurers" in ISO 9001 bodies. In the RT2005 version the measurers are not necessarily independent of the contractor, which does not guarantee for example that buildings are randomly selected.

However, because it gives significant benefits to the applicants without independent control, this QM approach was quite controversial and therefore needed an independent evaluation to ensure its credibility. (Note that RT 2012 will require independence between the builder and the tester but state control may still remain necessary to avoid biased quality control.)

Procedure

The committee has validated the process of those controls.

Every year a state technician carries the process. He contacts every applicant approved for more than a year and asks for the list of all constructions whose delivery is expected in the coming year. The list should include the estimated date of commissioning, address, name and phone number of the future inhabitant as well as of the construction superintendent. At the end of the year, the committee checks the consistency of the list when the applicant sends his yearly report.

If not cooperative, the applicant is warned that his agreement may be suspended.

The percentage of control depends on the number of buildings and the availability of state technicians all over France. The first set of controls will cover 5% of the buildings, Table 2. Most applicants will be controlled by different state units as they build in various locations, the distribution of controls is automatically determined by an Excel solver.

	Approval date	Type of applicant	Estimated production for the coming year	Number of building controlled by state technicians in 2011
1	20/10/08	Builder	98	5
2	25/11/08	Builder	48	2
3	25/11/08	Builder	405	20
5	26/10/09	Manufacturer	13	1
6	24/4/09	Builder	89	4
7	5/8/09	Builder	291	14
8	5/8/09	Builder	83	4
10	2/4/10	Builder	127	6
11	30/6/10	Builder	89	4
12	2/4/10	Builder	184	9
13	2/4/10	Builder	508	25
14	31/8/10	Builder	110	5

Table 2: Number of building controlled

The control consists in two different parts. First of all 5% of the production of each applicant is randomly chosen and measured at commissioning by trained and certified state technicians. Secondly, the application of the quality management approach is checked while asking the following element for 2% of the building production:

- A brief description including at least: the kind of construction (masonry, brick, wood, etc.), the type of insulation (inside, outside, spread), the number of level and kind of floor (wood, concrete...)
- The list of companies working on the site
- The contract or subcontract signed by all companies working on the site (which should include an airtightness statement)
- Craftsmen's certificate of training
- Site supervision documents (each step of the quality approach)
- Actions in case of non-compliance (corrective treatment, reports of treatment meetings...)
- Report if a measure was carried out during construction

The documents will be requested just twice. If the applicant is unable to produce them within two months, the building is considered non-compliant with the quality management approach.

First results – 4 houses

Measurements began in June 2011 and the checking of the quality management approach will begin in September 2011. As of September 2011, four houses of two different applicants have been measured.

For the first applicant, results are in line with our expectations ($Q_{4Pasurf}$= 0.47 and 0.57 $m^3/h.m^2$).

But for the second applicant one result was much less satisfactory with $Q_{4Pasurf}$ = 1.26 and 0.51 $m^3/h.m^2$. In fact, this applicant is accustomed to deliver unfinished home, according to the future occupant wills. In the first building the screed layer was not cast and sanitary equipments weres not installed. It is likely that when the house will be finished airtightness will much improve, but as far as the applicant is supposed to deliver houses respecting $Q_{4Pasurf}<0.8m^3/h.m^2$, works made by the occupants should not be necessary to comply with the requirement. Therefore, this is considered as an non-compliance with requirements.

CONCLUSION

Airtight construction generalisation is a challenge, however, it is compulsory to address this issue given the objective to generalize near-zero energy buildings in 2020. The definition of quality frameworks for airtight envelopes achievement is one of the proposed solutions in the French regulation. Interesting lessons arose from the preliminary evaluation of this framework, whose first applicant was approved in November 2008. The wide scope of applicants, from small builder to large ones or manufacturers, is quite encouraging, especially in view of RT2012, so are the results presented in the yearly reports. No doubt that it will lead to a better dissemination of airtightness knowledge.

Nevertheless, given the first state-control results, it is clear that external checking has to be implemented in order to ensure the credibility of the approach and to value those who best meet requirements.

ACKNOWLEDGEMENTS

The contribution of CETE de Lyon to the implementation of these quality frameworks is funded by the ministry of ecology, sustainable development, transport and housing (MEDDTL).

The sole responsibility for the content of this publication lies with the authors. It does not necessarily reflect the opinion of the Ministry

REFERENCES

[1] Carrie, F.R., Charrier, S. and Leprince, V. 2010. *Implementation of measurement and quality frameworks in the French regulation for achieving airtight envelopes*, 31th AIVC Conference, 26-28 October 2010, Seoul Korea

[2]*Arrêté du 26 octobre 2010 relatif aux caractéristiques thermiques et aux exigences de performance énergétique des bâtiments nouveaux et des parties nouvelles de bâtiments*, JO 27 octobre 2010

[3]*Arrêté du 24 mai 2006 relatif aux caractéristiques thermiques des bâtiments nouveaux et des parties nouvelles de bâtiments*, JO 25 mai 2006

Wednesday 12 October 2011

17.00 – 18.30 Parallel Session 3A - Ventilation and cooling

This session is organized within the scope of a starting AIVC-TightVent project that deals with various aspects of ventilation for cooling.

- Air/Ground heat exchangers for heating and cooling: dimensioning guidelines
  (Pierre Hollmuller, Switzerland)
- Future climate effect on building refurbishment using ventilation for cooling: a case study
  (Maria Kolokotroni, UK)
- Ventilation solutions in net zero energy buildings, the Elithis Tower case study
  (Oscar Hernandez, France)
- Low-energy buildings with night and air-to-air heat exchangers – case studies and analysis
  (Jens Pfafferott, Germany)

17.00 – 18.30 Parallel Session 3B - Development of air leakage databases

There are several national initiatives to collect air leakage data from field measurements. The objective of this session is to begin structuring communication between some of these initiatives. It falls within the scope of a AIVC-TightVent project.

- Preliminary analysis of U.S. Residential Air Leakage Database Update v.2011 (Wanyu R. Chan, USA)
- U.S. Commercial Building Airtightness Requirements and Measurements (Andrew Persily, USA)
- The Web@set project: reasons behind, objectives and on-going developments (Andrés Litvak, France)
- Experience with the development of an air leakage database in Germany (Oliver Solcher, Germany)

18.30 – 19.30 Poster and industry exhibition

19.30 End of the first day

AIR/GROUND HEAT EXCHANGERS FOR HEATING AND COOLING: DIMENSIONING GUIDELINES

Pierre Hollmuller

Groupe Energie
Institut des Sciences de l'Environnement
Université de Genève
7 route de Drize - CH 1227 Carouge

pierre.hollmuller@unige.ch

ABSTRACT

This paper deals with air-soil heat exchangers used for heating or cooling airflows used for ventilation of buildings. Basing on a previously published analytical solution concerning heat charge and discharge around an array of buried pipes, it is shown how convective air/pipe and diffusive pipe/soil heat exchange are combining, and how the characteristic exponential amplitude-dampening along the pipe is achieved. The main result consists in dimensioning guidelines, in terms of relations between airflow and pipe length necessary for complete dampening of yearly or daily amplitude. Latter guidelines are finally illustrated and validated with numerical simulation

KEYWORDS

passive heating and cooling, buried pipes, daily and seasonal heat storage

SYSTEM DESCRIPTION

As considered throughout this article, an air-soil heat exchanger consists of an array of horizontal pipes, possibly a unique pipe, buried underneath or next to a building and situated at the inlet of the ventilation system (fig. 1). Purpose is to take advantage of the soil's thermal inertia, so as to absorb and dampen the winter/summer or the day/night meteorological oscillation carried by the airflow, hence cutting off the cold or hot thermal peaks which would otherwise be transferred to the building. In principle this technique may be used as well for winter heating as for summer cooling purposes, although it was shown that under Mid-European climates both these modes differ as well in terms of the effect that is to be striven for (yearly versus daily amplitude dampening), as of synergy with the rest of the ventilation system and with the building itself [1].

As is schematically depicted here (fig. 1), and as will be discussed further down in all details, the main working characteristics of an air-soil heat exchanger are as follows:

- Winter preheating and summer cooling characterize as well by daily amplitude-dampening (the day/night meteorological extremes contracting around the daily average) as by yearly amplitude-dampening (the daily average approaching the yearly meteorological average).

- While yearly heat storage propagates approximately 3 m around the pipes, daily heat storage does so on approximately 15 cm. Daily amplitude-dampening hence always surpasses yearly amplitude-dampening, which requires a wider storage volume and is limited by more in depth heat diffusion.

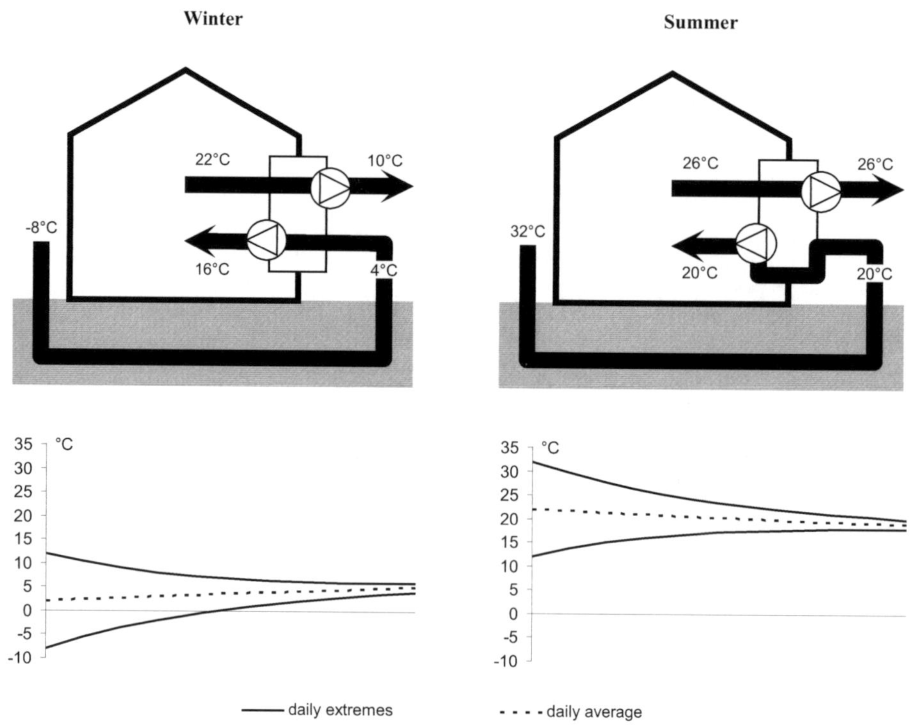

Fig. 1: System schematic with typical daily and yearly amplitude dampening along the pipe.

GOVERNING PRINCIPLES AND PARAMETERS

So as to establish appropriate dimensioning guidelines, we will start by characterizing the governing principles and their link to amplitude-dampening along the pipes. We therefore will base on the simplified and analytically resolved case of a constant airflow subject to sinusoidal temperature input, with explicit treatment of diffusive heat storage into the soil [2], which yields following main results.

Heat penetration depth

As long as the soil layer around each pipe is at least that thick, heat charge and discharge around the pipes naturally extends over a penetration depth δ, which depends on the oscillation period:

$$\delta = \sqrt{\frac{\lambda_s}{c_s \rho_s} \frac{\tau}{\pi}}$$

In the case of a typical soil (conductivity: 1.9 W/K.m ; capacity : 1.9 MJ/K.m3), which will be taken as a reference, the daily heat oscillation hence extends on approximately 15 cm around the pipes, against 3 m for the yearly oscillation:

$$\delta_{day} \approx 15 \text{ cm}$$

$$\delta_{year} \approx 3 \text{ m}$$

Oscillation dampening

Thanks to the diffusive heat storage in the soil, a sinusoidal temperature input carried by the airflow:

$$T_{in} = \theta_0 \sin(\omega t)$$

dampens exponentially along the pipe exchange surface *S*:

$$T_{out} = \theta_0 \exp\left(-\frac{Sh}{c\dot{m}}\right) \sin(\omega t)$$

Strictly speaking, and as for all type of diffusive heat storage, this amplitude-dampening phenomenon actually goes along with delaying or phase-shifting of the input signal. It is however shown that, as long as heat storage may extend over its natural penetration depth δ, phase-shifting remains secondary, reason why it is ignored here.

The amplitude-dampening coefficient h, which will govern the dimensioning guidelines, basically results from serial linking between the convective exchange coefficient h_a (air/pipe) and a diffusive exchange coefficient h_s (pipe/soil):

$$h \approx \frac{h_a h_s}{h_a + h_s}$$

The diffusive exchange coefficient h_s relates to heat charge and discharge in a soil layer of thickness δ and can be approximated by the corresponding static conduction coefficient (fig. 2).The convective exchange coefficient h_a can be computed by way of a simplified form of the Gnielinski relation [3].

GEOMETRICAL LAYOUT

As comes out of preceding relations, guidelines for dimensioning of air-soil heat exchangers relate to careful examination of convective and diffusive heat exchange coefficients, which themselves depend on geometrical parameters:

- The convective air/pipe heat exchange increases almost linearly with air velocity, with a lesser dependence on pipe diameter, reaching values between 4 and 16 W/K.m2 for velocities between 1 and 4 m/s.
- The diffusive heat exchange on its turn depends on the available soil layer around the pipes and the way heat diffusion can actually take place.

We therefore will consider two distinct geometrical layouts, with quite different behaviours (fig. 2):

- The first case concerns an extensive geometry, made of pipes buried deep and apart from each other (or possibly a unique deep buried pipe), at around 3 m depth and with 6 m inter-axial distance, so that both daily and yearly heat diffusion may fully propagate in radial mode (fig. 2, left). For both these frequencies, the diffusive coefficient takes values in the same order of magnitude as the convective coefficient, although more important in daily than in yearly mode. As a consequence, in such a geometry daily and yearly amplitude-dampening always go on par, although latter somewhat less effectively.
- The second case concerns a compact geometry, made of pipes buried close to the surface and to each other, with approximately 15 cm soil around each, so that the daily heat oscillation carried by the airflow may be charged and discharged in the soil in the same radial way as before (fig. 2, right). To the contrary, charge and discharge of the yearly heat oscillation saturates the immediate vicinity of the pipes and can hence only take place in plane mode, downwards into the ground. As a consequence, such geometry enables to reach daily amplitude-dampening with the same pipe length as before, but almost without dampening of the yearly amplitude.

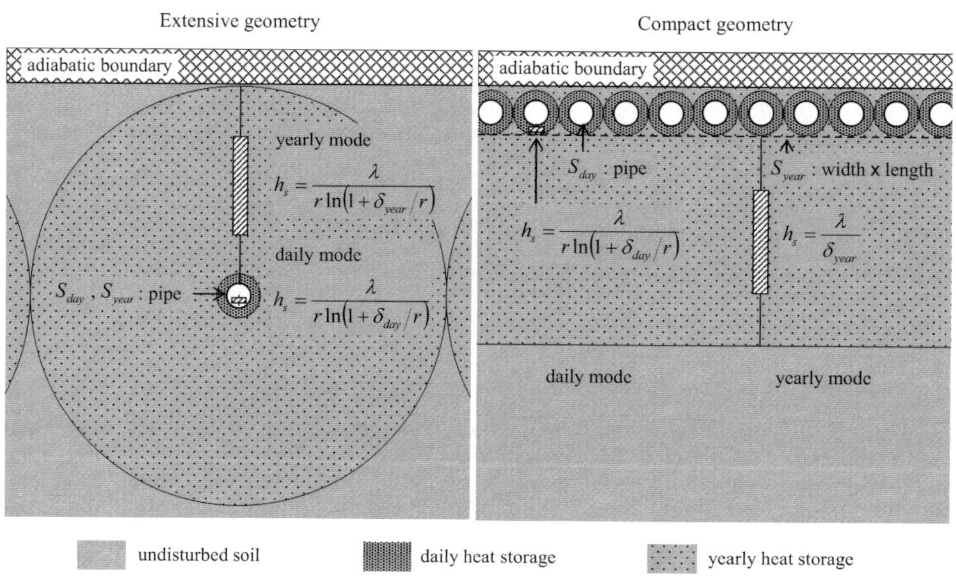

Fig. 2: Basic geometries and associated storage volumes

PIPE LENGTH

For both the above defined geometries it is now possible, via the exponential form of the temperature dampening, to give dimensioning rules in terms of necessary pipe length for "complete" amplitude-dampening, which we define as a residual amplitude of $e^{-2} \sim 15\%$. The resulting guidelines are as follows (fig. 3):

- If the point of interest concerns dampening of the yearly oscillation, it is necessary to work with distant of each other and deeply buried pipes, with approximately 3 m soil around each. In this case, a rough dimensioning rule for complete amplitude-dampening is 30 m pipe per 100 m3/h airflow (fig. 3, left), for diameters yielding an air velocity in the 1 to 4 m/s range. As pointed out above, the daily amplitude will have vanished on approximately half that distance, with the same rules as hereafter.
- If the point of interest concerns dampening of the daily oscillation, only with slight or negligible dampening of the yearly component, it is sufficient to work with close to each other and shallow buried pipes, with approximately 15 cm soil around each. Typical required length is roughly 15 m per 100 m3/h airflow (fig.3, right), a relation which however isn't as linear anymore and slightly depends on the pipe diameter and associated air velocity.

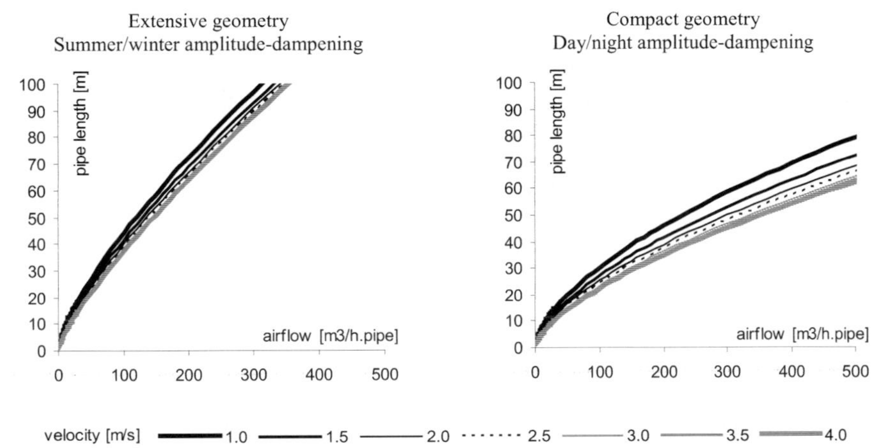

Fig. 3: Guidelines for dampening of daily or yearly amplitude

Note that preceding guidelines for pipe length concern residual amplitude of e^{-2}, but may be extrapolated by way of the exponential form of the temperature dampening. As an example, the guidelines for the compact configuration indicate that a 200 m3/h airflow in a 22 cm diameter pipe (~ 1.5 m/s) needs 55 m for complete daily dampening (fig. 3, right); a 20 m pipe hence would yield a residual amplitude of $e^{-2\cdot(20/55)} = 48\%$.

It should finally be stressed that the model on which these guidelines have been established does not take into account perturbation of the storage mechanism by thermal drive from upper surface (ambient or building), nor the possible effects of transient airflow or latent heat exchanges. As a complement to the basic guidelines elaborated in this study, these effects can however be investigated by way of numerical simulation.

VALIDATION

Validation as well as illustration of these behaviours and guidelines will now be given on hand of an extensively validated finite element numerical model for buried pipe systems [4], which accounts for fully three dimensional heat diffusion in soil and flexible border conditions, as well as for sensible and latent heat exchanges (latter not used within this study).

Both the considered configurations consist of a layer of 20 cm diameter pipes, each swept by a 200 m3/h dry airflow, with hourly input given by the standard annual meteorological temperature for Geneva. The pipes are buried in a soil with same thermal characteristics as above, in one case at 310 cm depth (center of pipe) and with 620 cm inter-axial distance, in the other case at 25 cm depth and with 50 cm inter-axial distance. Superior border conditions are in both cases taken as adiabatic, whereas a sufficiently important soil layer is taken into account for the heat wave to expand as deep as necessary (adiabatic border conditions 15 m below pipes). The arrays are supposed to consist of a sufficient amount of pipes for lateral border effects to be negligible, so that the system may be described by way of a unique pipe (adiabatic conditions at inter-axial distance).
Simulated hourly temperatures are depicted in form of daily minima and maxima profiles at 0, 20 and 50 m distance (fig. 4, top), and confirm preceding analysis: deep and wide apart buried pipes induce combined daily and yearly amplitude-dampening, eventually reaching the constant annual average, whereas the compact pipe configuration essentially allows for daily amplitude-dampening, with a constant output over the day/night period but an almost unaltered seasonal trend.
The hourly data at every 10 m is Fourier analyzed, so that residual yearly and daily amplitudes may be compared to the values given by the above dimensioning rules (fig. 4, bottom): a very good correlation is manifest in either case, theoretical guidelines only slightly overestimating the residual amplitudes given by numerical simulation

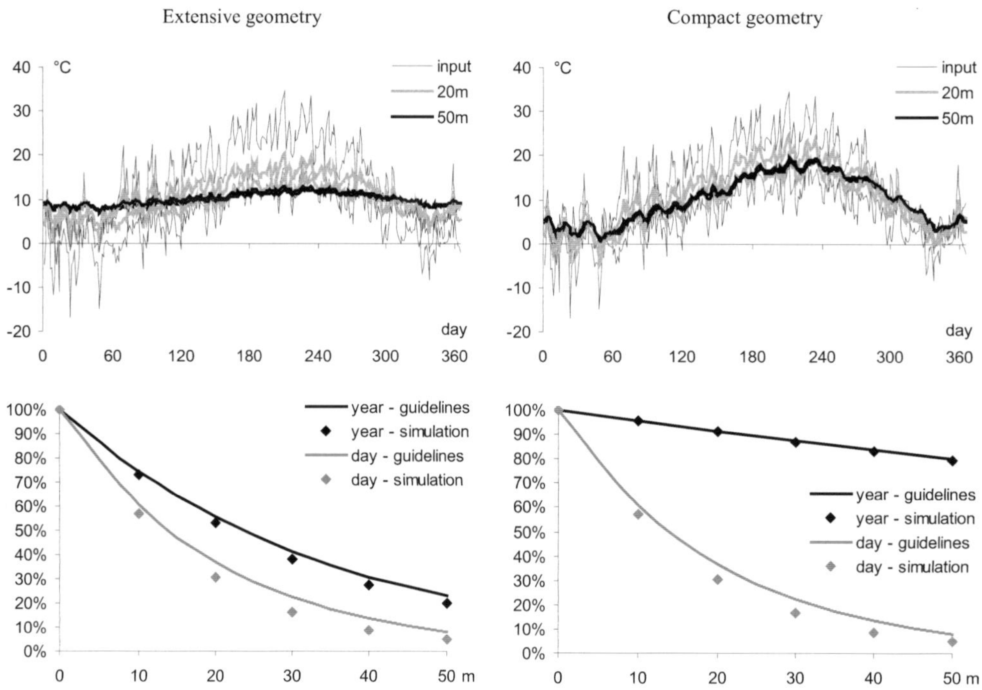

Fig. 4: Dynamic of daily minima and maxima (top) as well as resulting amplitude dampening (bottom).

REFERENCES

[1] Hollmuller P., Lachal B. (2001). *Cooling and preheating with buried pipe systems : monitoring, simulation and economic aspects*. Energy and Buildings, 33(5), p. 509-518.

[2] Hollmuller P. (2003). *Analytical characterisation of amplitude-dampening and phase-shifting in air/soil heat-exchangers*. International Journal of Heat and Mass Transfer, vol. 46, p. 4303-4317.

[3] Gnielinski V. (1975) Neue Gleichungen für den Wärme- und Stoffübergang in turbulent durchströmten Rohren und Kanälen, Forsch. Ing.-Wes., Vol.41(1), pp. 8-15.

[4] Hollmuller P., Lachal B. (2005). *Buried pipe systems with sensible and latent heat exchanges : validation of numerical simulation against analytical solution and long-term monitoring*, in : Building Simulation, proceedings of the 9th International Building Performance Simulation Association, 15-18 August 2005, Montréal, Québec, École Polytechnique de Montréal, Vol.1, p. 411-418. http://www.unige.ch/cuepe/html/biblio/detail.php?id=362

FUTURE CLIMATE EFFECT ON BUILDING REFURBISHMENT USING VENTILATION FOR COOLING: A CASE STUDY

Feras M Al-Mukhtar and Maria Kolokotroni*

Brunel University, Mechanical Engineering
Kingston Lane
Uxbridge UB8 3PH, UK
**Corresponding author: maria.kolokotroni@brunel.ac.uk*

ABSTRACT

This paper presents the effect of specific future climate changes scenarios on the resilience of the refurbishment of a 1960s office building in suburban London. A model of the building was created and simulated using IESVE to predict current energy consumption calibrated with operational energy data. Energy efficient improvements were incorporated which mainly consist of improving the insulation and air-tightness of external envelope, reducing solar and internal gain and utilising natural ventilation during the day and night for improving thermal comfort in the summer. Climate change scenarios were investigated by using future climates weather files available for the UK. It was found that thermal comfort in the summer can be significantly improve and provide acceptable condition in most areas of the building at present and in the future. Energy consumption can be improved by 80% of present consumption. This can be compared with projections that office CO_2 emissions from office building will increase two to three-fold in the future due to increased cooling requirements.

KEYWORDS

Future climate, building simulation, refurbishment, passive design.

INTRODUCTION

In recent years, global temperature has increased significantly [1] and it is probable that the average annual temperature will increase by several degrees during this century. In the UK, work on climate change projections are carried out under the umbrella of UKCIP (UK Climate Impact Programme) [2].

Literature review has revealed that there exist significant recent work on the effect on climate change on energy consumption by building; for example [3-5] which examine the impact of temperature changes on the building energy consumption for heating and cooling in the US, [6] which examines the case in the UK, [7] for Hong Kong, [8] in Australia and in the UK for office buildings [9,10]. In addition, work is under progress in many countries to construct weather files suitable for building energy consumption simulation taking into consideration future climate change scenarios. In the UK, the creation of future climate weather files for building simulation (termed 'morphing') is currently based on the methodology developed by [11] and implemented by [12] based on UKCIP02 scenarios. Recently more detailed and probabilistic-based UKCIP09 climate projections are made available [2], and researchers have constructed weather files based on these projections [13].

Recent work by the authors using UKCIP02 weather files to examine energy consumptions and CO_2 emissions by typical office buildings has revealed that with no passive design intervention, overheating hours will increase in the future and therefore active cooling is likely to be added to existing buildings. A comparison on the environmental impact (CO_2 emissions) was carried out between non-cooled office of today (2000) versus comfort cooled offices of the future (2050) [14] indicating that CO_2 emissions can increase between 230% to 340% in existing offices in suburban London. Another study [15] has revealed that in non-cooled offices night ventilation coupled with thermal mass is very effective in reducing internal temperatures. The cooling potential of night cooling is expected to be even greater in the future due to the greater diurnal temperature range.

This paper presents a study of passive design refurbishment of an existing building and how present refurbishment proposals will perform in the future. The study used the commercial software ISEVE for the simulations and UKCIP09 future weather files.

BUILDING DESCRIPTION

The case-study presented in this paper of the Howell Building on Brunel University Campus which was constructed in the 1968. The building has a square planform with a large central well. As originally conceived, there were three floors with the lowest supported on 'stilts' allowing full access to an open central court. In the 1990s, the central court was replaced by a large lecture theatre and additional teaching rooms were inserted under the original lowest floor, providing a full four storey building. The ground floor consists of the lecture theatre and a number of teaching and meeting rooms, together with a plant room. Floor 1 and 2 principally contain individual cellular offices linked by a central corridor. Floor 3 contains teaching rooms, two large open-plan postgraduate offices and a number of individual offices. The floor area of the building is 4650 m^2 and it consists of 35% of offices, 18% of teaching rooms, 31% of circulation areas and the rest of computer/meeting rooms and plant rooms.

Figure 1 shows external views of the North and South facades of the building and Figure 2 is a view of the model of the building developed for analysis and shows the general layout, including the inner well and, within the well, the roof of the lecture theatre situated at Ground Floor level. A geometrical and thermal model of the building was created within the simulation model IESVE which is used by many researchers and building services consultants for building design. It should be noted that IESVE is listed as one of the approved tools for compliance with the current Building Regulations in the UK and therefore comply with the national implementation of EPBD.

A survey was carried out to determine construction details as close to the existing building as possible and also internal heat gains based on observed use of the building due to occupancy levels, equipment and lighting. These were inputed in the model for each room and use and a simulation for energy cosumption was carried out using weather files as will be described in the next section.

Gas and electricity actual consumption of the actual building were available as used for DEC (Display Energy Certificate) and the simulated results for the building as is currently were compared with the actual consumption favourably. The current energy consumption on DEC is 426 $kWh/m^2/year$ and the emissions 640 tonnes CO_2 (137 $kgCO_2/m^2/year$).

Figure 1: Howell Building, North and South Facades

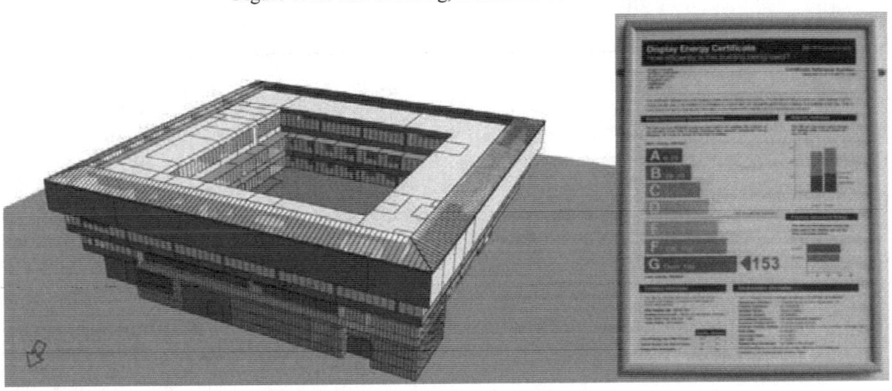

Figure 2: Computer model of the Howell Building and the current DEC.

CLIMATE CHANGE SCENARIOS AND WEATHER FILES FOR BUILDINGS

The weather files used for simulation are derived from the 'Emissions Scenarios' used in the UKCP09 from the IPCC SRES (2000) [16] report; are A1FI (High), A1B (Medium) and B1 (Low), each projects a global mean temperature range.

What does this mean in terms of external weather parameters? As an example the following projections of weather variables are based on the 2050 (Medium Emissions), 50% Probability (Central Estimate), this corresponds to the Howell building being renovated now to handle climate change during its 40-50 year (typical) life and renovated again in the 2060s. This provides a qualitative insight of what weather variables to expect in the London region for the 2050s under the medium-emissions-scenario, 50% probability, relative-to the 1961–1990 (Baseline).

Key findings for London, 2050s (medium emissions scenario), central estimate;

- Increase in winter mean temperature is 2.2°C; it is very unlikely to be less than 1.2°C or more than 3.5°C. A wider range of uncertainty is from 0.9°C to 3.8°C.

- Increase in summer mean temperature is 2.7°C; it is very unlikely to be less than 1.3°C or more than 4.6°C. A wider range of uncertainty is from 1.1°C to 5.2°C.
- Increase in summer mean daily maximum temperature is 3.7°C; it is very unlikely to be less than 1.4°C and more than 6.6°C. A wider range of uncertainty is from 1.2°C to 7.4°C.
- Increase in summer mean daily minimum temperature is 2.9°C; it is very unlikely to be less than 1.3°C and more than 5°C. A wider range of uncertainty is from 1.2°C to 5.7°C.
- Change in annual mean precipitation is 0%; it is very unlikely to be less than –5% and more than 5%. A wider range of uncertainty is from –5% to 5%.
- Change in winter mean precipitation is 14%; it is very unlikely to be less than 2% and more than 32%. A wider range of uncertainty is from 0% to 35%.
- Change in summer mean precipitation is –19%; it is very unlikely to be less than –41% and more than 7%. A wider range of uncertainty is from –43% to 16%. [17].

From these scenarios a weather generator was constructed which can provide seven weather variables at an hourly signal for; precipitation, temperature, vapour pressure, relative humidity, sunshine fraction and direct & diffuse radiation for a given location, time and future emissions scenario. However, in order to facilitate dynamic building thermal modelling, weather variables for; wind speed, wind direction, air pressure and cloud cover also need to be generated in a consistent manner with the rest of the (UKCP09 WG) weather signal and needs to be in the same file format as the reference years used currently by building simulation programs [18]. This study used these weather files to simulate projected energy consumption of the studied building as shown in Table 1[available from 19]:

Weather File Name	Location	IPCC SRES	CDF%	Date Range
WG_COMBINED_cntr_5100180_DSY.epw (Baseline)	London/Heathrow	N/A	N/A	1961-1990
LondonDSY05.fwt (Current)	London/Heathrow	N/A	N/A	1983-2004 Current CIBSE
WG_2050_5100180_a1b_50_percentile_DSY.epw	London/Heathrow	Medium (A1B)	50% Central Est.	(2050s) 2040-2069
WG_2050_5100180_a1fi_50_percentile_DSY.epw	London/Heathrow	High (A1FI)	50% Central Est.	(2050s) 2040-2069
WG_2080_5100180_a1b_50_percentile_DSY.epw	London/Heathrow	Medium (A1B)	50% Central Est.	(2080s) 2070-2099
WG_2080_5100180_a1fi_50_percentile_DSY.epw	London/Heathrow	High (A1FI)	50% Central Est.	(2080s) 2070-2099

Table 1. Simulation weather file names, location, cumulative distribution percentage and date range.

ADAPTED BUILDING

Guidelines have been published on how to adapt buildings in the UK for future climate change [20, 21]. The investigation comprised 11 examples of various building types (dwellings, offices and schools) and a single climate scenario (UKCP02 Medium-High) for his analysis, the purpose was to show how buildings with particular features respond to climate change and what can be done to obtain a more acceptable performance in the future. The results showed that buildings with a high thermal mass combined with an intelligent ventilation strategy performed best, new modern buildings performed better than older buildings, due to greater insulation and air-tightness from build quality. It was also evident that, with the exception of air-conditioned buildings, all unadapted buildings showed instances of overheating (1% occupied hours > 28°C) after the 2020s. From the results [21] a simple design philosophy termed 'adaptation strategy' was proposed that can reduce the effects of the risk of thermal discomfort (overheating) and simultaneously reduce energy consumption, these principles are as follows;

i. Switch off - Reduce unnecessary internal heat gains, and use solar shading,
ii. Absorb - increase thermal mass (new builds only), or effectively use thermal mass,
iii. Blow away - introduce an intelligent ventilation strategy i.e. night cooling then mixed-mode if required,
iv. Cool - introduce active cooling only when all passive options have been assessed, minimise active cooling use to times when thermal discomfort are probable, this is termed 'peak lopping', [21].

Changes to the building were made inline with the overall adaptation strategy as follows:

- Office equipment load has been reduced by 30% over the base case building to reduce internal gains,
- Lighting has been reduced to 9 W/m2 for all offices to reduce internal gains,
- Shading has been added on the east, west & south facade to minimise solar gain/cooling load, the atrium roof also offers shading to the central well and offices,
- Glazing has been changed to reduce direct & diffuse solar radiation, by using tinted glass and/or reflective films & using triple glazing (xenon gas) with a thermal break in the frame,
- Cool Roof (reflective paint coating) has been used to reduce heat gain to the top floor offices and thermal storage to the roof slab to improve thermal comfort,
- Improvements to the building fabric were aimed at reducing the wall elements thermal transmittance (U-value) to reduce heat flow, which was achieved by using vacuum insulation panels,
- Building air tightness & infiltration has been improved and a value of 0.25 ACH has been used to simulate a well sealed envelope.
- Natural ventilation with night cooling has been used to pre-cool the structure during the summer months, designed to allow air to cross flow across the floor slab and exhaust out through the atria,
- Mixed mode ventilation system employed to minimise energy use and only switch on the A/C system (generic convective) when thermal comfort thresholds are breached 25°C,
- Heating system (generic system) has been replaced with efficiencies representing those of a modern condensing boiler and new pipe distribution.

Ventilation Strategy: The current Howell geometry lends itself well to applying cross flow ventilation with a passive stack, with relatively minor alterations to the building i.e. no structural changes. It is well known that cross flow ventilation relies on establishing an unimpeded air flow path between the inflow and outflow air streams, which should pass through the zone of occupancy; calculations shown in Figure 3. There is a practical limit for cross ventilated air penetrating the building, i.e. cross flow maximum depth 5*height = (5*2.56m = 12.8m), the Howell building is within limits as the cross section for the 1st, 2nd & 3rd floor are 9m, 10m & 11m respectively.

The operation of the external window openings in the offices are separated for summer and winter conditions as follows:
<u>Summer</u>: Apr to Sept (The controller is ON, IF)

- During Night; 23:00 - 07:00, (Inside DB Air Temperature > Outside DB Air Temperature AND Inside DB Air Temperature > 20°C)

- During Day; 07:01 - 22:59, (Inside DB Air Temperature > Outside DB Air Temperature AND (Inside Air Temperature > 23°C, Step function 3°C) OR CO2 > 1200ppm)

<u>Winter</u>: Oct to Mar (The controller is ON, IF)

- All Day; 00:00 - 23:59, (Outside DB Air Temperature > 12°C AND Inside DB Air Temperature > Outside DB Air Temperature AND (Inside DB Air Temperature > 25°C, Step function 2°C) OR CO2 > 1200ppm)

Due to these modification the floor area of the building is increased to 5275 m^2 because of the addition of the atrium.

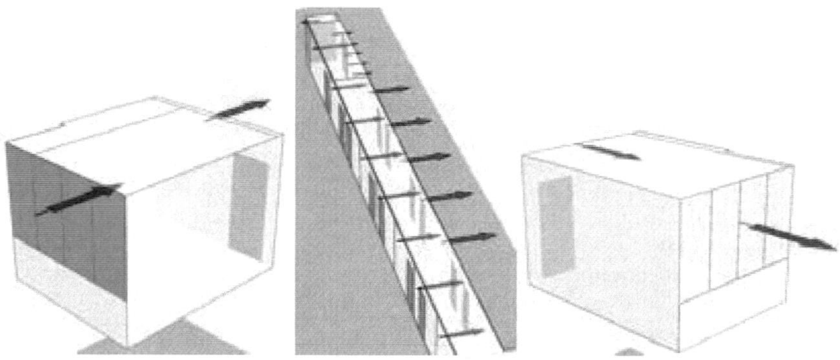

Figure 3: IES Cross flow ventilation through Office 119, Corridor 102a, Office 116 (July CIBSE 05 DSY). Inflow (Blue), Outflow (Red).

An atrium (glass covered internal void for the Howell building) has been added to allow air flow to vent up and out through the atrium openings (see Figure 4). Natural ventilation can be applied by just using the atrium itself as a passive stack, where the atrium is extended above the 3rd Floor Roof.

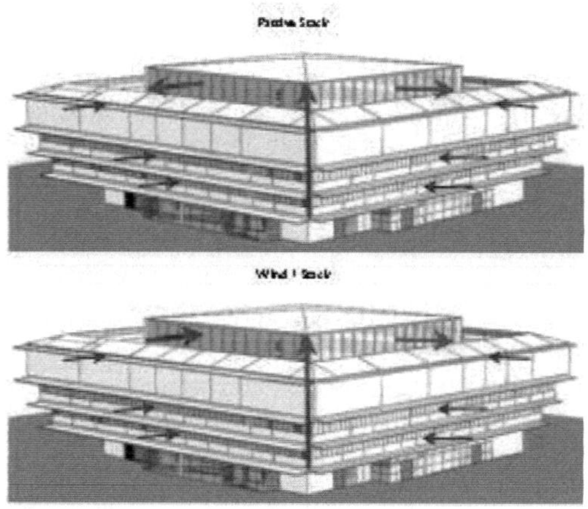

Figure 4: (a) Howell building atrium (passive stack) and (b) wind & stack. Inflow (Blue), Outflow (Red).

RESULTS: EXISTING BUILDING

Simulations were carried out with the building as is and summarised results are presented in Figures 5-7. Figure 5 shows the overheating hours on the top floor of the building (which suffered worse overheating than the other floors. There are some overheating problems presently and as expected these will be increased in the future. Figure 6 shows the specific energy consumption which is mainly due to heating requirements - these will reduce in the future so the building will perform better in terms of energy consunption and CO_2 emissions (Figure 7).

Simulations were carried out with the adapted building and summarised results are presented in Figures 8-10. Figure 8 shows the overheating hours on the top floor of the building; it can be seen that temperatures above 28 °C have been eliminated at present although there might be some overheating problems in the future that might require some cooling - however this is very small proportion of the total required energy as shown in Figure 9. This is because of the refurbishment according to energy efficient design principles and the use of ventilation for cooling during the day and utilising night time ventilation.

DISCUSSION

The passive adaptation strategy (no active cooling) demonstrated that there were reductions in the yearly overheating hours in both the >25°C & >28°C thresholds for the mid-floor offices and for all scenarios the >28°C threshold was reduced by 100%. This showed that for the future climate scenarios, the mid-floor adapted case is resilient to the climate change, and a required change, would be to extend the current night cooling strategy into October for the future climate.

However, the adapted case top-floor only showed reductions in overheating hours in the baseline and current (CIBSE DSY '05) weather scenarios for the dry-resultant temperature >25°C for 5% of occupied hours. In all the future scenarios the >25°C is above 5% of the occupied hours, and the solution to extend the Night Cooling strategy into October (and even November for the top-floor) was seen to reduce overheating hours, but was insufficient to lower them below the 5% of occupied hours threshold. The improvement in the overheating risk by extending night cooling to October and November is shown in Figure 11.

The Cool-Roof was successful in reducing the dry-resultant temp. >28°C for 1% of occupied hours. This was achieved in all weather scenarios for the top-floor except for the 2080(High), which is marginally above the limit at 1.17%. This showed that part of adaptation strategy had been successful at lowering the internal dry-resultant temperature below the upper range of thermal comfort levels and the internal dry-resultant temperature is maintained within the 25-28°C temperature range.

The cooling loads in the adapted case were determined to be an increase over the base case as there is no current cooling system in the base case, but its operation was designed to be less energy intensive than a conventional fully air-conditioned office by employing a mixed-mode system. Therefore, the cooling load progressively increases through each scenario relative-to the baseline period. This indicates that, there will be an increase in cooling demand to maintain the offices at the lower comfort threshold >25°C (5% of occupied hours) in the future weather scenarios.

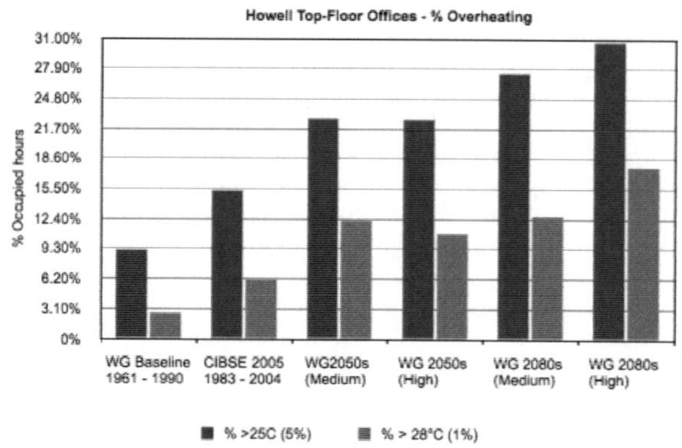

Figure 5: Base Case, Top-Floor Offices overheating hours for defined comfort thresholds.

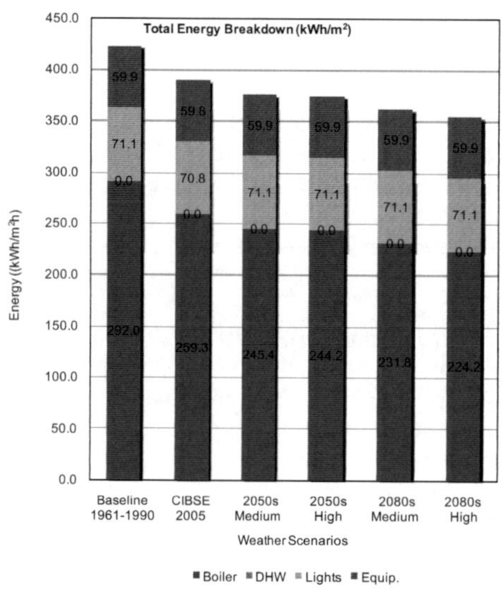

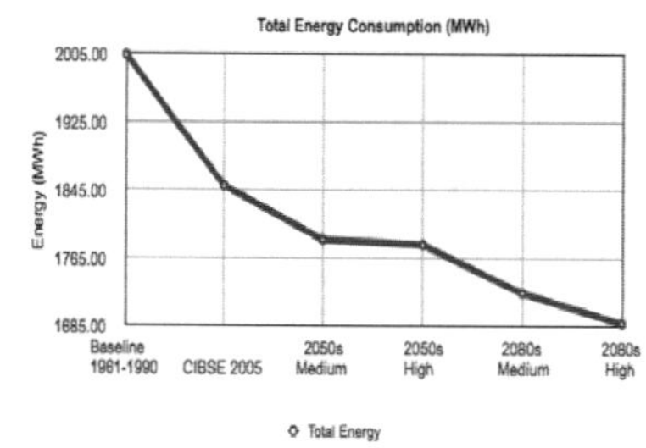

Figure 6: Base case, normalised energy breakdown and total energy per scenario

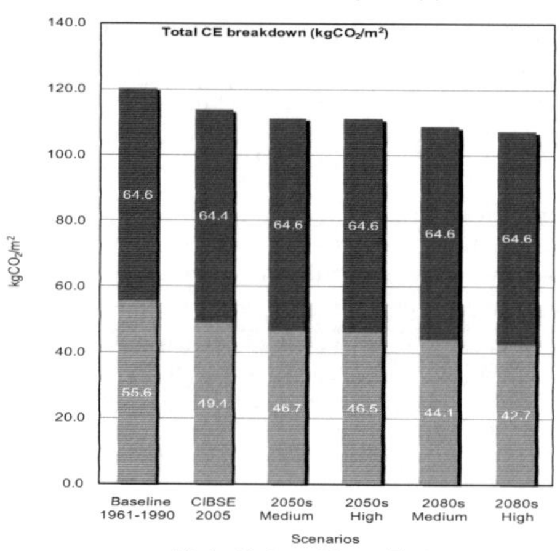

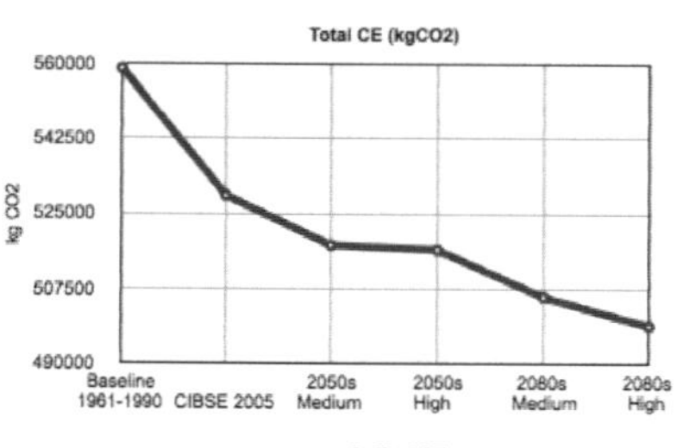

Figure 7: Base case, normalised carbon emissions breakdown and total carbon emissions per scenario

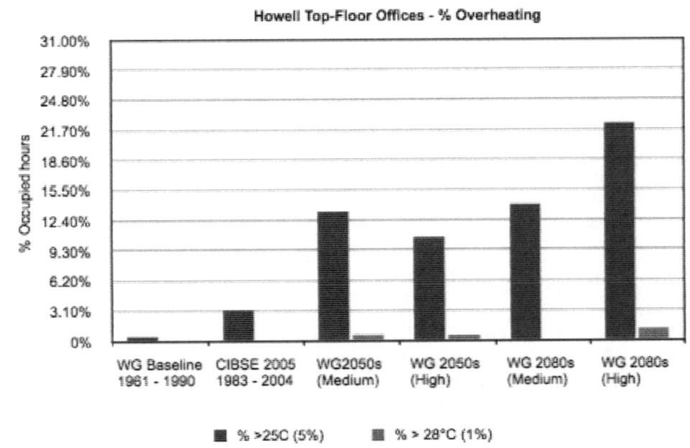

Figure 8: Adapted Case, Top-Floor Offices overheating hours for defined comfort thresholds.

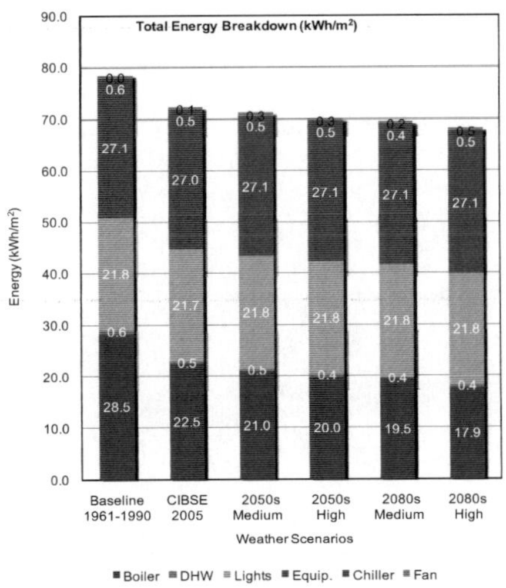

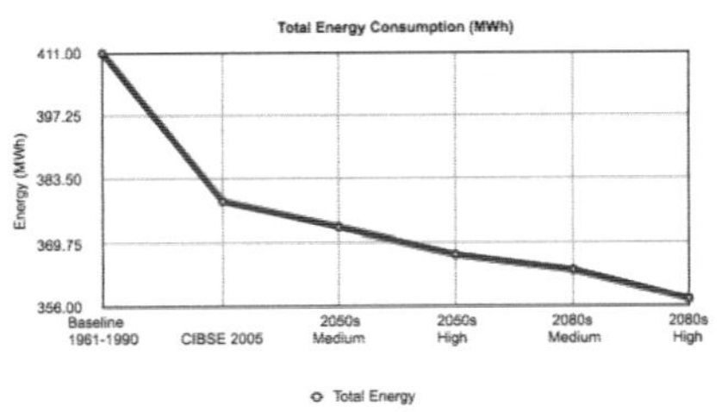

Figure 9: Adapted case, normalised energy breakdown and total energy per scenario

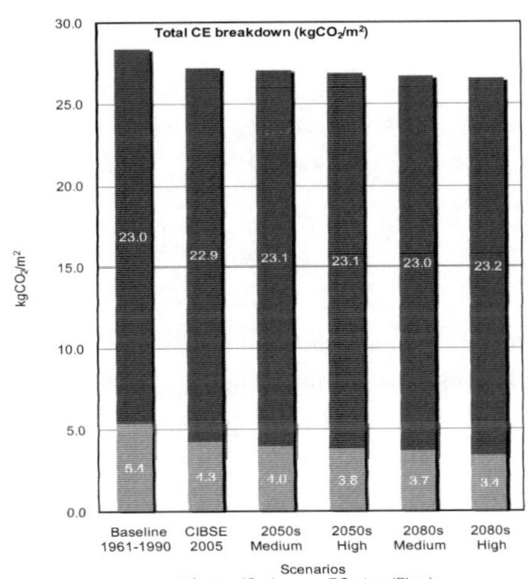

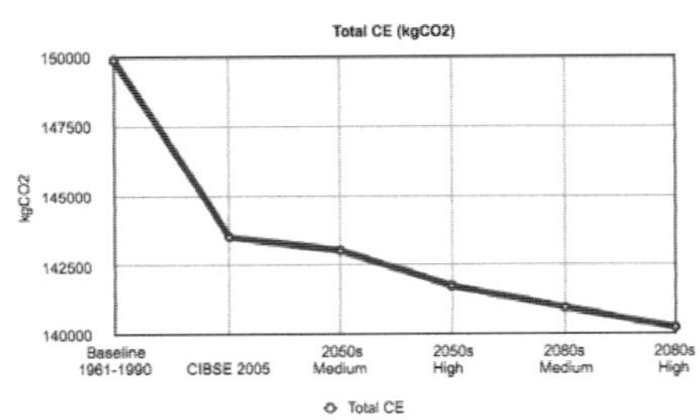

Figure 10: Adapted case, normalised carbon emissions breakdown and total carbon emissions per scenario

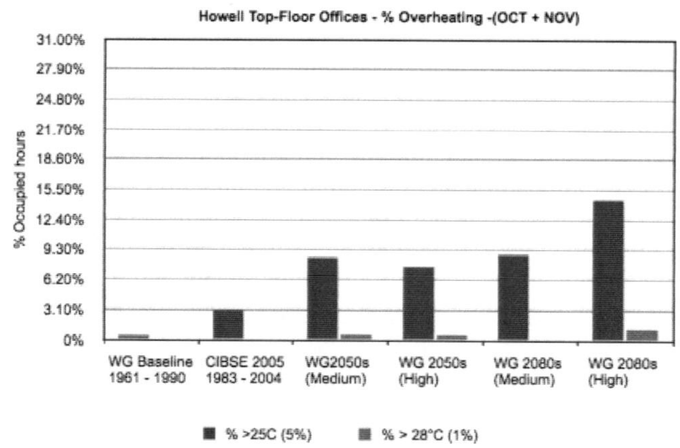

Figure 11: Adapted Case, Top-Floor Offices overheating hours minus the Oct & Nov hours.

In terms of energy consumption, improvements were seen in the winter months, and the analysis of the boiler plant consumption showed that there were reductions for the adapted case, similarly, the peak heating plant load was reduced in each weather scenario.

The adapted case (relative to the base case) was shown to reduce the specific energy consumption and carbon emissions by the quantities shown in Tables 2 and 3.

ΔkWh/m^2	Total Gas	Total Elec.	Total kWh/m^2	%
Baseline, 1961-1990	-263.5	-89	-353.0	82
CIBSE 2005	-236.8	-89	-326.1	82
2050s, Medium	-224.3	-89	-313.6	82
2050s, High	-224.1	-89	-313.5	82
2080s, Medium	-212.3	-89	-301.7	81
2080s, High	-206.2	-89	-295.4	81

Table 2. Difference between Adapted & Base cases ΔkWh/m^2 and % reduction.

$\Delta kgCO_2/m^2$	Total Gas	Total Elec.	Total kWh/m^2	%
Baseline, 1961-1990	-50.1	-41.6	-91.7	76
CIBSE 2005	-45.0	-41.5	-86.5	76
2050s, Medium	-42.6	-41.5	-84.2	76
2050s, High	-42.6	-41.5	-84.2	76
2080s, Medium	-40.4	-41.6	-82.0	75
2080s, High	-39.2	-41.4	-80.7	75

Table 3. Difference between Adapted & Base cases $\Delta kgCO_2/m^2$ and % reduction

CONCLUSION

The adapted case reductions have shown that the adaptation strategy has provided the required level of thermal comfort, (in accordance with the defined thermal comfort criteria), while, simultaneously reducing energy consumption and carbon emissions. For this particular case study, there is good evidence that, with the right approach, engineers are able to mitigate the effects of climate change with passive design measures which consists of improved insulation and air-tighness, solar and internal heat gain reduction and utilising day and night natural ventilation for cooling during the summer months., However, there is need to be mindful of the fact that the weather data are probabilistic in nature and the degree of adaptation required for a actual refurbishment project may preclude certain changes due to cost factors.

It appears, that, there are greater reductions to be achieved now, than to wait until the future and there were greater reductions that could have been achieved in the past baseline period. This point of view is one of an economic (return of investment) view, where government incentives based on carbon reductions would provide a greater return of investment, if one took advantage of such a scheme to reduce their operational emissions now, rather than later. This is because a passive refurbishment will alleviate the need to active cooling and climate change predicts higher external temperatures thus reducing the need for heating.

ACKNOWLEDGEMENTS

Work for this paper was carried out as part of the USE-Efficiency project funded by EACI (no IEE/08/526/S12.528395); website: www.useeffciency.eu

REFERENCES

[1] IPCC, (2007). The IPCC Assessment Reports. Available at: http://www.ipcc.ch/

[2] UKCIP, (2010). UKCP09: UK Climate Projections. Available at: http://www.ukcip.org.uk/ukcp09/

[3] Lu N, Taylor T, Jiang W, Jin C, Correia J, Leung L R, and Wong P C (2010). Climate Change Impacts on Residential and Commercial Loads in the Western U.S. Grid, IEEE Transactions on power systems, Vol. 25, No. 1.

[4] Crawley DB and Drury B, (2008). Estimating the impacts of climate change and urbanization on building performance, Journal of Building Performance Simulation, 1: 2, 91 — 115

[5] Mansura E T, Mendelsohnb R, Morrisonc W, (2008). Climate change adaptation: A study of fuel choice and consumption in the US energy sector, Journal of Environmental Economics and Management 55, 175–193

[6] Holmes M J, Hacker J N, (2007). Climate change, thermal comfort and energy: Meeting the design challenges of the 21st century, Energy and Buildings 39, 802–814

[7] Lam T N T, Wan K.W, Wong S.L., Lam J C, (2010). Impact of climate change on commercial sector air conditioning energy consumption in subtropical Hong Kong, Applied Energy 87 2321–2327

[8] Wang X, Chen D, Ren Z, (2010). Assessment of climate change impact on residential building heating and cooling energy requirement in Australia, Building and Environment 45 1663–1682

[9] Jenkins D, Liu Y and Peacock A D. (2008). Climatic and internal factors affecting future UK office heating and cooling energy consumptions, Energy and Buildings, Vol 40 Issue 5, 2008, pp 874-881

[10] De Wilde P and Tian W, (2010). Predicting the performance of an office under climate change: A study of metrics, sensitivity and zonal resolution, Energy and Buildings, Vol 42, Issue 10, pp 1674-1684

[11] Belcher SE, Hacker JN and Powell DS, (2005). Constructing design weather data for future climates, Building Services Engineering Research and Technology. 2005; 26: 49-61.

[12] Jentsch MF, Bajaj AS and James P A B (2008). Climate change future proofing of buildings—Generation and assessment of building simulation weather files, Energy and Buildings, Vol 40, Issue 12, 2008, pp 2148-2168

[13] Eames, M., Kershaw, T. & Coley, D., (2011). On the creation of future probabilistic design weather years from UKCP09. Building Services Engineering Research and Technology, Vol. 32 Issue 2, p127-142

[14] Kolokotroni M., Ren X., Davies M., Mavrogianni A (submitted, July 2011), London's urban heat island: impact on current and future energy consumption for heating and cooling. Energy and Buildings.

[15] Demanuele C, Mavrogianni A, Davies M, Kolokotroni M, Rajapaksha I, (to appear). *London's urban heat island: impact on overheating in naturally ventilated offices*. Building Services Research and Technology

[16] IPCC SRES (2000), http://www.ipcc.ch/pdf/special-reports/spm/sres-en.pdf

[17] Murphy, J.M., Sexton, D.M.H., Jenkins, G.J., Booth, B.B.B., Brown, C.C., Clark, R.T., Collins, M., Harris, G.R., Kendon, E.J., Betts, R.A., Brown, S.J., Humphrey, K.A., McCarthy, M.P., McDonald, R.E., Stephens, A., Wallace, C., Warren, R., Wilby, R. and Wood, R.A. (2009) *UK Climate Projections Science Report: Climate Change Projections*, Exeter, UK, Meteorological Office Hadley Centre.

[18] Eames, M., Kershaw, T. & Coley, D. (2011) On the creation of future probabilistic design weather years from UKCP09. Building Services Engineering Research and Technology, 32, 127.

[19] http://centres.exeter.ac.uk/cee/prometheus/

[20] Hacker, J.H., M. Belcher, S. Davies, G. (2005a) CIBSE TM36: Climate change and the indoor environment: Impacts and adaptation. Chartered Institute of Building Services Engineers, London,

[21] Hacker, J.B., SE, Connell, RK (2005b) Beating the Heat: keeping UK buildings cool in a warming climate. UKCIP Briefing Report. UKCIP.

VENTILATION SOLUTIONS IN NET ZERO ENERGY BUILDINGS, THE ELITHIS TOWER CASE STUDY

Oscar HERNANDEZ[1], Francis ALLARD[2]

1. Elithis Groupe *1C Boulevard de Champagne* *21000 Dijon* *France*	*2. Université de La Rochelle* *Avenue Albert Einstein* *17000 La Rochelle* *France*

ABSTRACT

This paper focuses on ventilation solutions of a net zero energy building. We present the monitoring results after two years as part of the PHD programme research "Design, simulation and control of hybrid ventilation systems in high energy performances buildings" founded by Elithis Groupe in parternship with the LEPTIAB of La Rochelle.

The objective of this paper is to study the ventilation solution of a net zero energy building, the thermal comfort and the indoor air quality in terms of CO_2 [1]. We present in the first part, a building description and ventilation principle, in second part we present the energy monitoring results of the Elithis Tower and ventilation consumption system during 2009 and 2010. Finally we show the ventilation performances in terms of energy, CO_2 concentrations and thermal comfort.

Results show an office building with an acceptable thermal comfort with a high energy performance but with a relatively low indoor air quality.

KEYWORDS

HVAC, Triple flux, Ventilation, Energy consumption, positive energy.

INTRODUCTION

The buildings account for more than 40% of energy demand in Europe and more than a third of greenhouse gas emissions, a major effort to improve the energy efficiency of buildings associated with a drastic reduction of their gas emissions of the greenhouse is now necessary. In this sense, changes in regulations, both European and national, converge on large-scale development of new buildings or renovated with very low energy demand. In addition, we consider today that more than a third of the total energy consumed on earth is for the air-conditionning of interior spaces. In air-conditioning strategies of interior spaces, optimal management of ventilation occupies the first position. In addition, ventilation is indeed the oldest and most widely used strategy for controlling the indoor environmental quality.

This paper presents the building together with ventilation - and their related ventilation and control strategies. It also describes the monitoring programme, which includes two years, for heating, cooling, fans (ventilators), pumps, lighting, elevators and plugs and loads equipments. The monitoring also includes one day per month of thermal comfort and average CO_2 concentrations of one level.

BUILDING DESCRIPTION

The Elithis tower (figure 1 and 2) is the first postive energy office building according to the French regulation[1], it is located in Dijon, France. This building is composed of 9 levels and 1 technical level (HVAC systems). It is 33.5 meters high. 4 levels are occupied by Elithis Engineering Company, and the others by the Ademe (Departmental Agency of Energy Management), X-rays medical services, a restaurant and other civil engineering companies.

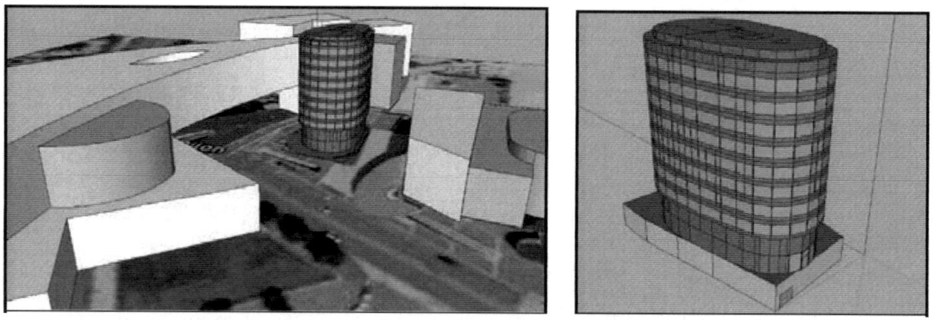

Figure 1. Elithis Tower Energy Plus model

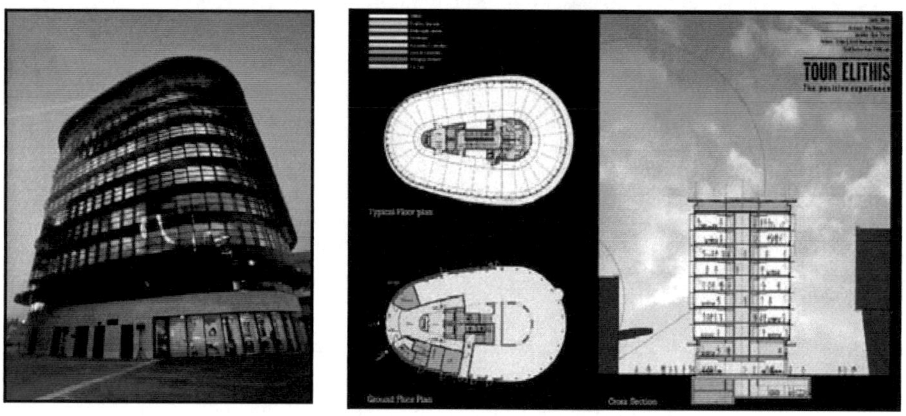

Figure 2. Elithis Tower - Govin SOREL - ARTE CHARPENTIER

The aim of this project was to combine aesthetics, urban integration, thermal comfort, energy, environmental performance and price. The Elithis Tower is a net zero energy building. For this purpose, 7 millions of Euros only were required to construct it.This represents 1400 € per m², which is the usual cost for a standard building in France.

The use of passive means and natural resources was a priority so as to reduce the environnemental impact. To achieve thermal and visual comfort in the building, sun and wood were selected as the main energy sources. Thermal comfort was retained because the sun heats the building naturally and provides light all the day long. Visual comfort because the surface fenestration is about 75% of the facades.

[1.] BEPOS (Positve energie building) Only the heating, cooling, ventilation, lighting and hot water are taken into account. (www.effinergie.org)

The thermal mass of the building can be considered as medium because only the "central core" (figure 3) is in concrete. The facades are made of wood and recyclable insulation (cellulose wadding) and 75% of surface fenestration.

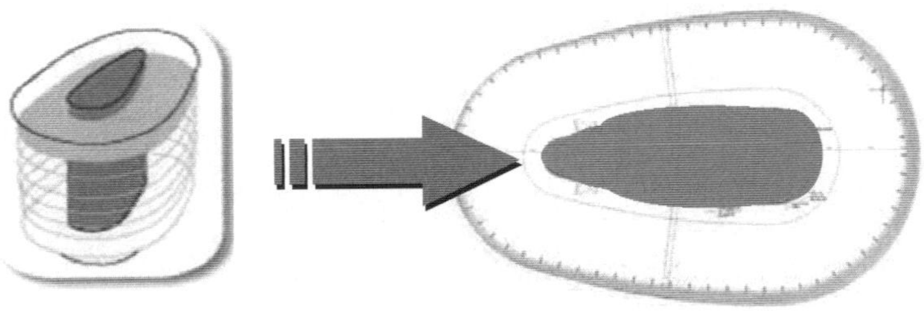

Figure 3. Central core

The design criteria of the building are show in table 1.

Window U	1.1 W/(m² K)
Window g	0.4
Exterior wall U	0.32 W/(m² K)
Base floor U	0.39 W/(m² K)
Roof U	0.22 W/(m² K)
Mean occupant density	15 m²/person (overall average)
Occupied hours	2450 h
Design outdoor temperature for heating	-11°C
Design outdoor temperature and RH for cooling	32°C / 38%
Heating degree days (include base temperature)	2 650 Degree days (base 18°C)

Tableau 1. Enveloppe values

In order to improve a thermal and visual comfort all the levels of the Elithis Tower were designed in an open plan distribution (figure 4.). This solution gives the possibility to improve the air contact with the thermal mass, and increases the efficiency of the natural lighting.

Figure 4.Tipycal level - Elithis Tower

VENTILATION PRINCIPLE
The ventilation principle is mechanical and the air distribution principle is mixing. The building is ventilated by façade slits or by cold beams (figure 5) depends on season.

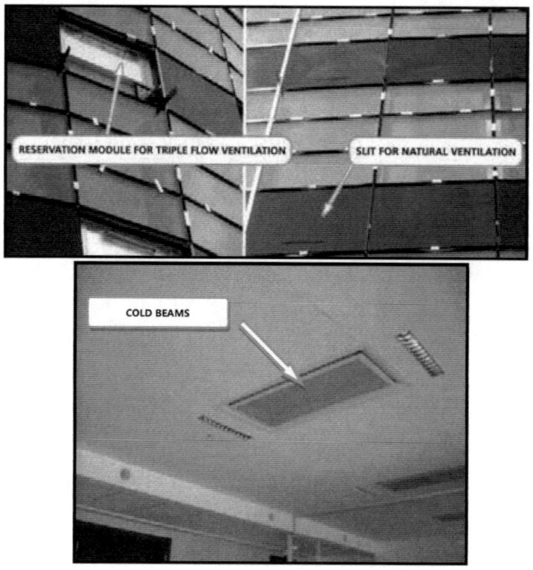

Figure 5. Air distribution principle

The system is controlled by BEMS-system in order to comply with the French ventilation standard code [6] (25 m^3/h per person-Office). Three differents ventilation modes were implemented in the Elithis Tower, winter mode, middle season and summer mode (triple flux).

Winter mode
The first one is used for the average cold periods (below or close 0°C). (Figure 6). The enveloppe (1) insulates the building from the exterior temperatures, the solar shield (2) protects the building from solar radiation in summer but gives the possibilty to warm up the building in winter. This combined with the indoor activity (human metabolism and computers) permits gives recovery via a heat exchanger (3). 90% of the indoor energy air can be retrieved to the outside air. This enables to blow air between 16 and 18°C in winter.

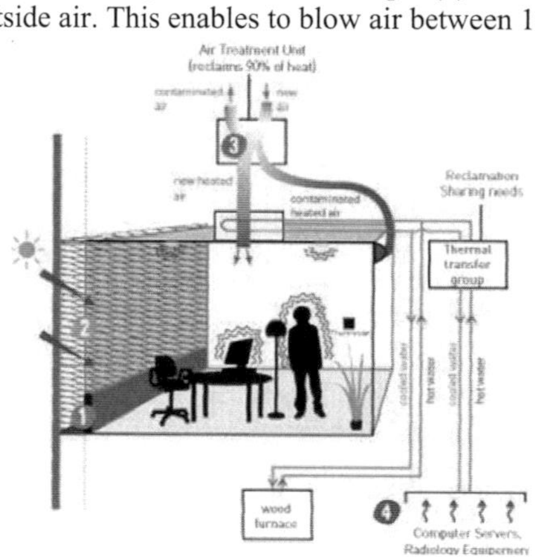

Figure 6. Winter mode

If the outside air temperature are very low a wood furnace produces hot water to the water network of the cold beams. In order to keep the inside air temperature to 22°C (figure 7).

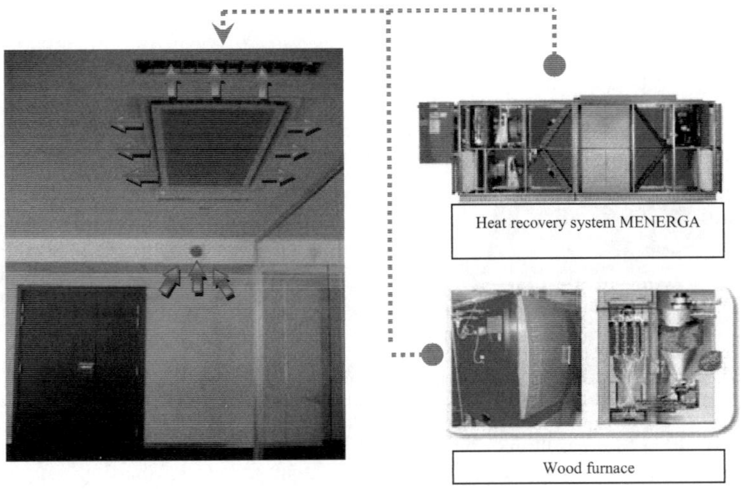

Figure 7. Air Distribution - cold beams

Mid seasons mode (spring and fall) :
In spring and fall when the outside air temperature increases (T°C > 10). The outdoor air conditions are interesting to cooling the building. At this moment (mid-seasons) the winter mode is turned-off, the BEMS-system opens the facade slits (1) and the air extraction is insured by a lower pression fan (2) (Figure 8). This is called "Triple flux®".

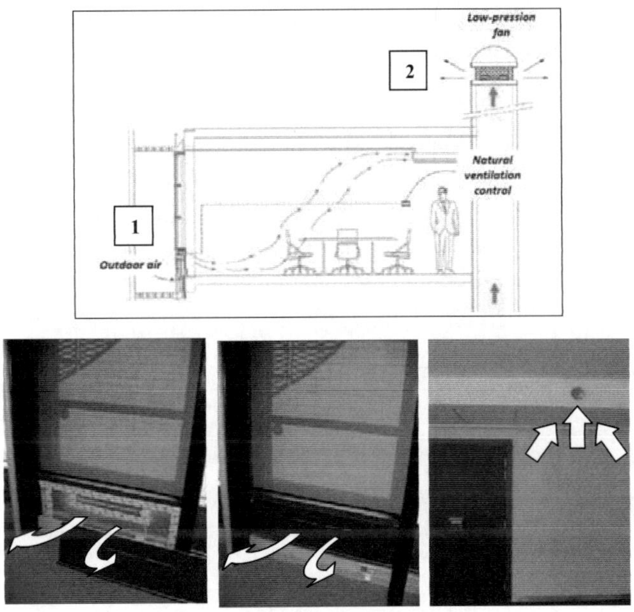

Figure 8. Mid seasons system

As is shown in figure 9, the solar shading blocks solar radiations (1) but provides a perfect natural lighting, the enveloppe (3) and the central core combined to the system, allow low temperatures to be maintained throughout the building.

Figure 9. Solar shading protection

Night Ventilation:
This system also uses night ventilation. When the building is empty (nights and weekends) and when the outside air temperatures are lower than indoors, the building can be overventilated by 3 air changes per hour, 30000 m^3/h.

Summer mode
Figure 10 describes the principles of the ventilation system implemented during the summer period. When the outdoor air temperature is below 26°C, the mid seasons system is turned-off, free-cooling at this moment isn't enough due to the outside air conditions.

When this happens (T out > 26°C) an adiabatic evaporative system treats the air in order to reduce the temperature. On the first stage the outside air is sprayed with fresh water so as to reduce its temperature (adiabatic) and to blow fresh air into the building. The second stage applies only when the outside temperature increases (T out > 30°C). A heat pump produces "cold" water for the cold beams. Each cold-beam is equiped with a water network, so it is possible to drive along "cold water" or "hot water". This combination allows the indoor air temperature to be maintained below 26°C whatever outdoor conditions are in summer period. This is illustrated in figure 10.

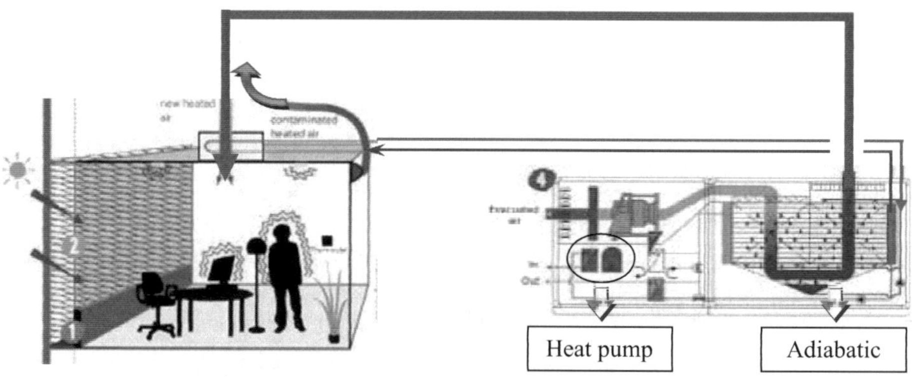

Figure 10.Cooling system

VENTILATION CONTROL STRATEGY

We saw the ventilation solutions of the Elithis Tower, now we are going to focus us on the functional analysis. The control of the ventilation strategy depends on temperature, humidity levels and occupancy.

288 temperature sensors and 9 humidity sensors control the inside air temperature and humidity (32 T°C sensors per level – 1 H% sensor per level). A weather station gives the outdoor conditions (T°C, H%, wind direction, wind speed). As regards occupancy, no control is performed, since a schedule was "a priori" defined (7:30 – 18:30). The ventilation system is controlled as presented on Figure 11.

When the indoor air temperatures are below to 21 or higher to 24, winter and summer respectively, the air central unit (MENERGA) treats the air in order to comply the requirements (heating or cooling). When the outdoor air temperatures are lower than indoor but the indoor air temperature is between 21-24°C or higher to 22°C (Only winter) the BEMS-system gives priority to the "triple flux®" system. For night ventilation a previous schedule is defined and the BEMS-system control this strategy.

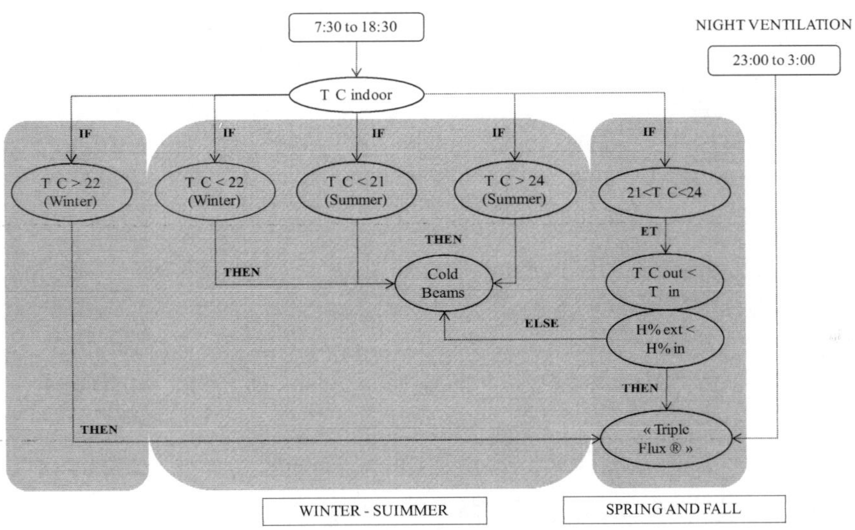

Figure 11. BEMS-System - Elithis Tower

MONITORING PROGRAMME

The monitoring programme was divided in three parts: energy consumption, thermal comfort survey and CO_2 concentrations survey. As regards energy consumption, the building was monitored all over 2009 and 2010. For thermal comfort survey, the monitoring period was started in June 2010 and for the CO_2 concentrations survey only the 7th level was monitored during the spring season 2011.

The BEMS-system is composed of 1600 sensors and provides access to the total energy consumption of the building. It is also possible to survey other parameters such as inside temperatures, humidity, illumination, occupancy. An overview of the measured parameters in each level is showed in the next figure.

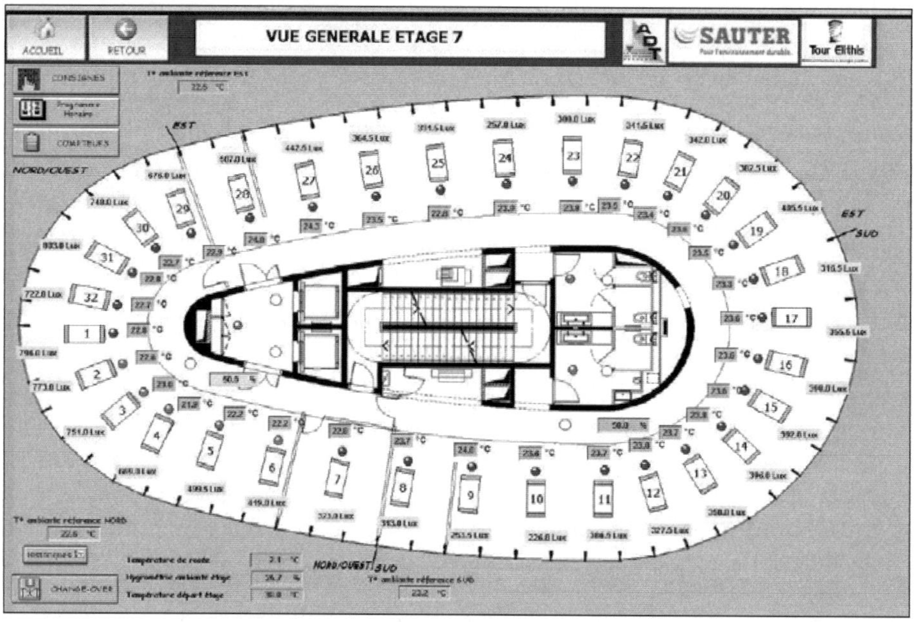

Figure 12. Inside parametres measured - SAUTER

An additional sensor was installed on the 7th level in order to fulfil the requirement of the programme (CO_2). In general, date was recorded every 10 seconds, which is very precise.

For thermal comfort survey (ISO 7730 [2], ASHRAE 55 [3]), a questionnaire was created into a site web. This gave us the possibility to manage the questionnaire by internet, no paper was required therefore we could handle the results more efficiently (figure 13). More than 600 questionnaires were sended to the Elithis Tower users. Only 400 answers were collected during this period. An answer peak was recorded one day per month.

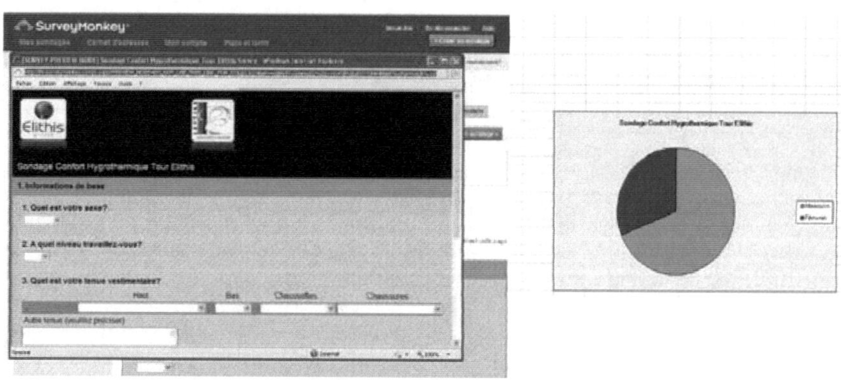

Figure 13. Survey comfort questionnaire

MEASUREMENT RESULTS

The monitoring period for energy consumptions has been calculated from the 1st April 2009 to the 31 March 2011. Two years were measured and the balance is showed in figure 14. On the left side (blue) we have the total energy used by the Elithis tower, heating, cooling, ventilation, lighting, pumps, elevators, plugs and PV production are included. For the French standards only the energy used by elevators and plugs aren't take in count.

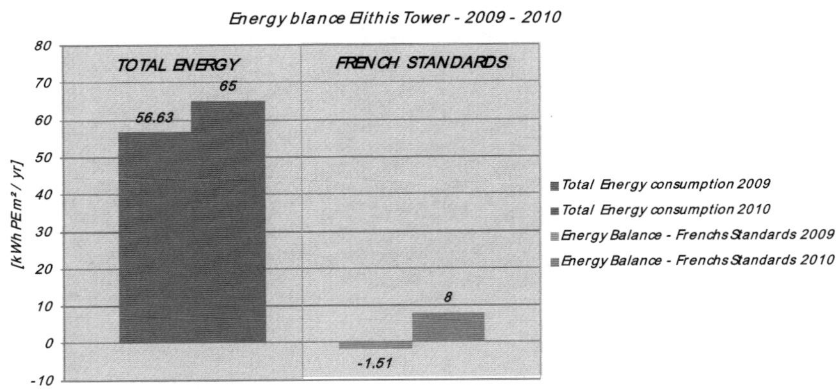

Figure 14. Energy Balance Elithis Tower

If we compare 2009 and 2010 values a difference can be noted. Many reasons could account for this situation. One of the most significant involved the heating consumption. Comfort was achieved during the winter season. At the beginning 2009, the building was heated at 19°C. This temperature wasn't high enough to insure the thermal comfort of the occupants. So the heating temperature in 2010 was increased to 22°C. As a consequence, 2°C in the heating temperature requires twice more energy. (Figure 15). Furthermore, the total energy production by the photovoltaics pannels. 2 kWh PE m² / yr less produced in 2010. This could be explain the difference between 2009 and 2010.

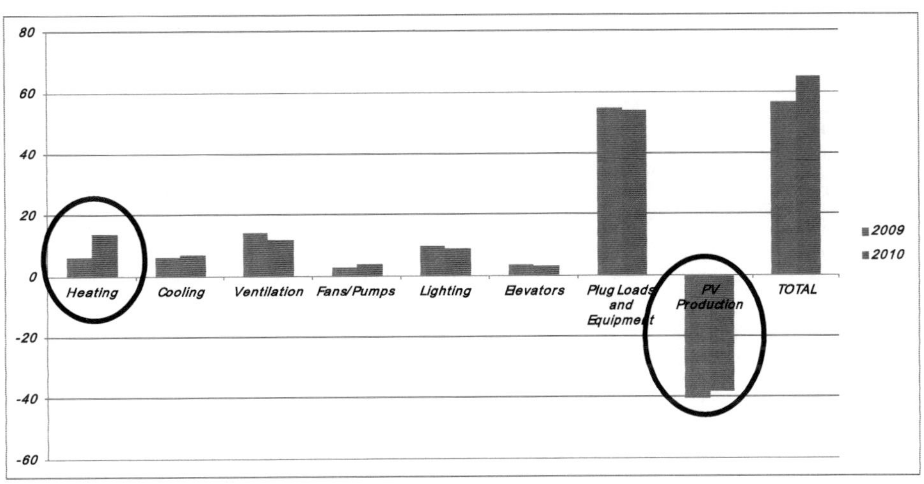

Figure 15. Total Energy Balance share

Now if we take a look at the ventilation system energy consumption the difference between 2009 and 2010 was significant. In 2009 the part of the ventilation system represented 15% of the total energy whereas in 2010 it amounted only to 12%. (Figure 16)

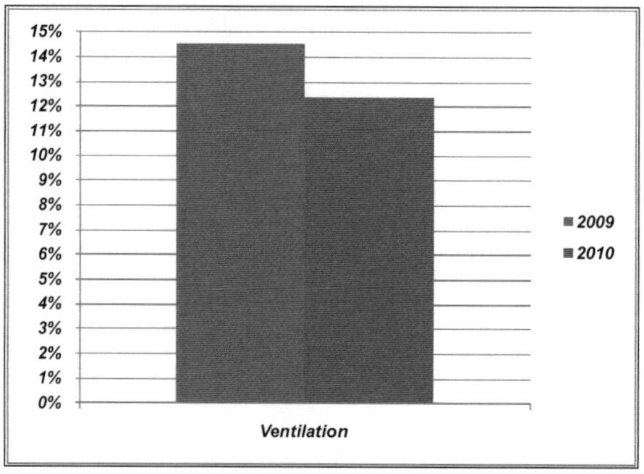

Figure 16. Ventilation energy consumption

The first year (2009) was an optimisation period. During this year, many modifications were implemented. First of all, the "triple flux" system was performed, the regulation band was modified in order to give priority to this system. Second, a high loss pressure was measured in 2009 due to an important number of acoustic pannels. Many acoustic pannel were removed but this solution didn't modify the acoustic comfort of the building. No complaints were registered during the comfort survey. And third, leakages in the air ventilation network [4,5] were detected and the periodicity of the maintenance was increased. This modifications enabled a final gain of 3% of the total energy, which represents 2.1 kWh PE m² / yr.

As stated above, the total energy used by the building in 2010 was 103 kWh PE m² / yr. The part of the ventilation was 12 kWh PE m² / yr. Now we are going to focus on the energy consumption of the different ventilation strategies.

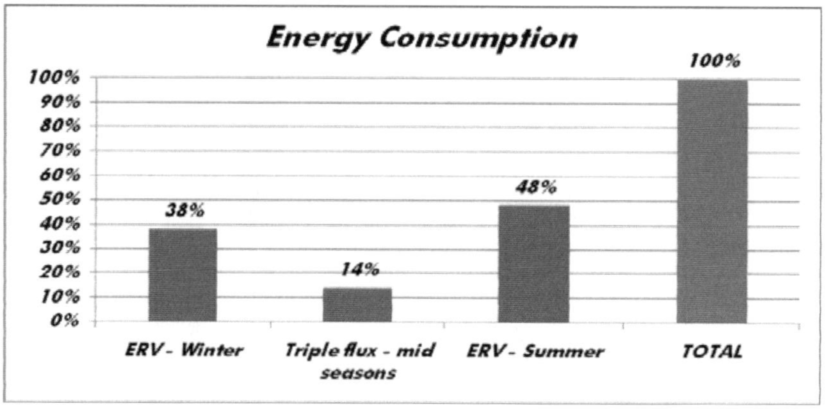

Figure 17. Energy consumption - Elithis Tower ventilation strategies

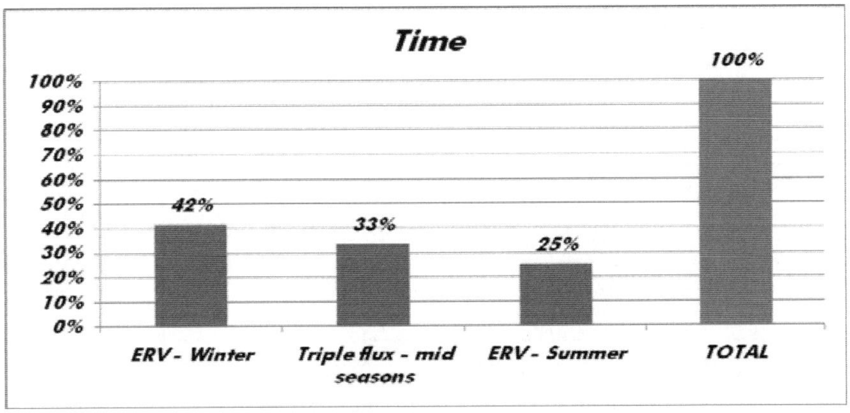

Figure 18. Time consumption - Elithis Tower ventilation strategies

The figures 17 and 18 show the energy and time consumption of each ventilation strategy. If we compare these graphs, we can see that the energy used by the mid seasons system represents only 14% of the energy required for ventilation while time consumption represents 33%. Only 1.68 kWh PE m² / yr was used by this strategy (mid-season), regarding time only 808 hours were consumed. That represents 0.00207 kW per hour. That is very low compared with the other strategies (winter and summer).

The regulation of this system remains very delicate. The weather plays an important role because, any change in humidity (Ex: storm) will force the system to turn-off. As a consequence, the energy balance may be different. That problem is related to the cold-beams. Now we focus on thermal comfort and CO_2 concentrations.

Thermal confort

The objective of this survey was to characterize the thermal comfort inside the building and the air quality. The results are given below according to a level of satisfaction.

Parameter	Satisfaction level (in %)
Temperature in the building	71.3%
Air quality	61%

Table 2. Sastisfaction percentage - Elithis Tower

The temperature level of the atmosphere in the building seems slightly comfortable. According to ASHRAE 55, indoor environnement can be considered comfortable when the satisfaction rate is above 80%. It is to be noted that the atmosphere of the Tower Elithis is not as comfortable as we had hoped.

But since 2009 several improvements have been made allowing an increase in the occupants' satisfaction rate. If we take a close look at the level of comfort throughout the campaign between 2010 and 2011, we notice in several instances, a satisfaction rate above 80% (Table 3).

Lower values were also recorded but this could be explained by the fact that the daily average temperature exceeded the monthly values:

Month	Day			Day Average T°C	Month average T°C	Satisfaction [%]
January	2011	27	Jeudi	4	-0.4	89
Febrary	2011	3	Jeudi	2.5	2.5	75
March	2011	14	Lundi	11	6.3	75
April	2011	11	Lundi	15.5	13.8	45
May	2011	12	Jeudi	18	16	60
June	2011	22	Mercredi	21	18.1	63
July	2011	8	Vendredi	19.5	21.4	80
August	2011	29	Lundi	20	18.8	75
September	2011	Not finished				
October	2010	27	Vendredi	9.5	10.9	87
November	2010	29	Lundi	0	6.7	65
Décember	2010	20	Lundi	4	0.3	74
					Average	71.6

Tableau 3. Thermal comfort - Survey monitoring

Unfortunately, explaining and better characterizing the level of comfort in the building didn't allow the collection of data several days a month, as some answers are not representative in this kind of questionnaires.

Indoor air quality

The only parameter monitored during this campaign was CO2. Several values were recorded, the maximum value was 2000 ppm in a meeting room, but in the local office "open space R 7", the maximum concentration recorded was about 1000 ppm (figure 19).

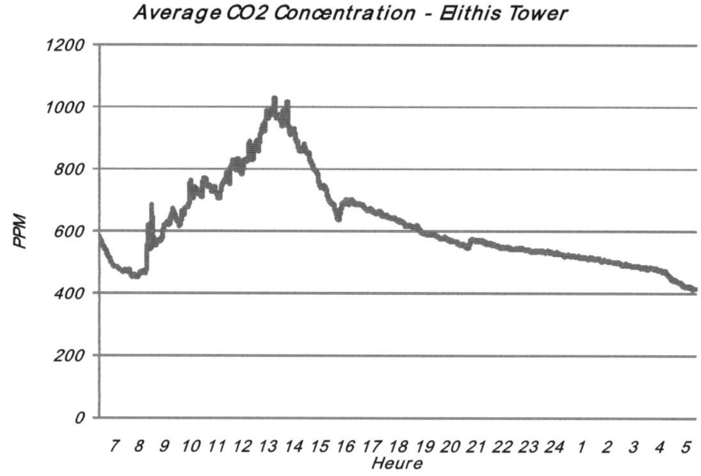

Figure 19. CO2 Concentrations

CO2 is not a gas that can assess the air quality but the level of confinnement; As consequence it hasn't been possible to determine the cause of discomfort by almost 40%. Air analysis (VOC and aldehydes) are underway to get a clearer picture of the problem.

CONCLUSIONS

The results of the monitoring of the Tower Elithis were very interesting. Compared with French standards, the Elithis Tower in 2009 ranked as the first office building with positive energy. And compared with other buildings the Elithis Tower used less energy than a classical office building. The environmental impact is reduced and so the Elithis Group goals have been reached.

We saw that the part of the ventilation energy consumption is about 15% compared with the total energy used by the building. This value could be different depending on the building. Controlled heat loss and low values air tightness modify the energy used by ventilation. As we saw in this paper, in the Elithis Tower the heating energy used in 2010 was higher. This difference alters the part of the ventilation system. It's therefore important to take care of the energy used individually and not globally. It's not possible to have this generalized.

The ventilation system "triple flow®" could produce significant savings, but its performances are linked to outside conditions and internal gains. Ventilation systems themselves aren't effective. The combination with other parameters as envelope, solar protection, or employee's behavior could guarantee the energy performances.

Thermal comfort is a parameter that was closely monitored. In the thermal comfort analysis many parameters will modify the thermal perception. The results show a thermal satisfaction percentage of 70%. However this was analyzed during a short term monitoring (1 day per month), consequently we don't know if in a long term analysis the result would be the same.

High windows surface and "open space" requires tighter ventilation control. In temperature conditions above average, thermal comfort is modified. The cause could be radiant temperature. The users of the Elithis Tower are more exposed to outside conditions. In the next papers, this parameter will be studied.

A building with a high efficiency level is quite feasible but when we take indoor air quality into account, thermal comfort and energy performance become much more complicated.

The control system is an important point. Innovative systems are necessary but they should be regulated and anticipate the changing conditions of temperature. Poor control can lead to overconsumption.

The air quality is a parameter that could not be monitored in a fairly precise way. We disclosed problems in air quality which, however, are not linked to CO2 levels. We expect therefore to do indoor air analysis (COV and aldehydes).

PERSPECTIVES

Other more accurate monitoring programmes will be implemented shortly. The size of database was a difficulty. A significant number of data are recorded every 10 seconds, as a consequence the management is complicated.

We found 40% of dissatisfaction and we couldn't find a relation. As part of the PHD these funded by Elithis Group and supported by Francis ALLARD of the LEPTIAB of La Rochelle the relation between energy performance, indoor air quality and thermal comfort will be studied.

New surveillance surveys regarding comfort will be soon implemented. A long term monitoring will be conducted. A software tool coupled with the GTC will also be implemented in order to recover the necessary information (temperature, humidity).

Other modes of regulation will be developed to allow the building to be more efficient. In addition, they will be modeled and tested as part of the PHD these.

ACKNOWLEDGEMENTS

This work was supported by the Elithis Groupe and the LEPTIAB of La Rochelle as part of the PHD work thesis funded by Elithis Groupe and the National Agency of Research and Technique (ANRT)

I would also like to thank the Elithis Groupe and his employees, the employees of the ADEME, Cabinet radiologue – Tour Elithis and Bourgogne solaire for their participation in the thermal comfort survey.

REFERENCES

[1] CO_2 as indicator for the indoor air quality. W.F. de Gids, TNO, The Netherlands. P Wouters, BBRI, Belgium.

[2] ISO 7730. Ergonomics of the thermal environment -- Analytical determination and interpretation of thermal comfort using calculation of the PMV and PPD indices and local thermal comfort criteria.

[3] ANSI/ASHRAE Standard 55-2010. Thermal Environmental Conditions for Human Occupancy

[4] Duct System Air Leakage — How Scandinavia tackled the problem. ASIEPI. Peter G. Schild – Jorma Raiilio

[5] Etat des pratiques européennes concernant l'étanchéité à l'air de l'enveloppe et des réseaux. ASIEPI – CETE Lyon. Août 2008

[6] Règlement sanitaire départemental type – France

LOW-ENERGY BUILDINGS WITH NIGHT VENTILATION AND AIR-TO-AIR HEAT EXCHANGERS - CASE STUDIES AND ANALYSIS

Jens Pfafferott[1*], Martin Fischer[2], Doreen E. Kalz[2]

1 Hochschule Offenburg
University of Applied Sciences
Badstraße 24
77652 Offenburg, Germany

2 Fraunhofer Institute
for Solar Energy Systems
Heidenhofstraße 2
79110 Freiburg, Germany

Corresponding author: jens.pfafferott@hs-offenburg.de

ABSTRACT

This study presents some results from a monitoring project with night ventilation and earth-to-air heat exchanger. Both techniques refer to air-based low-energy cooling. As these technologies are limited to specific boundary conditions (e.g. moderate summer climate, low temperatures during night, or low ground temperatures, respectively), water-based low-energy cooling may be preferred in many projects. A comparison of the night-ventilated building with a ground-cooled building shows major differences in both concepts.

KEYWORDS

low-energy cooling, night ventilation, earth-to-air heat exchangers, ground cooling, thermo-active building systems, energy performance, monitoring, data evaluation and analysis

INTRODUCTION

Low-energy cooling is strongly recommended by EPBD §25 [1] in non-residential buildings and has been widely applied to residential and non-residential buildings all over Europe. In recent years many research projects supported the understanding of passive and low-energy cooling techniques. Many reports, guidebooks and tools may be found in Ref. [2], [3] and [4], respectively. This study is strictly based on monitoring data only.

A clear definition of low-energy cooling is still missing. In the context of this study,

- low-energy cooling refers to cooling technologies which mainly use ambient energy heat sinks such as the cool night air or the ground for cooling (in combination with reversible heat pumps).
- These concepts base on passive cooling technologies, e.g. free ventilation, solar shading, reduced internal heat gains and high thermal inertia.
- Mechanical cooling is reduced to a minimum.

Figure 1 gives an overview on different cooling concepts.

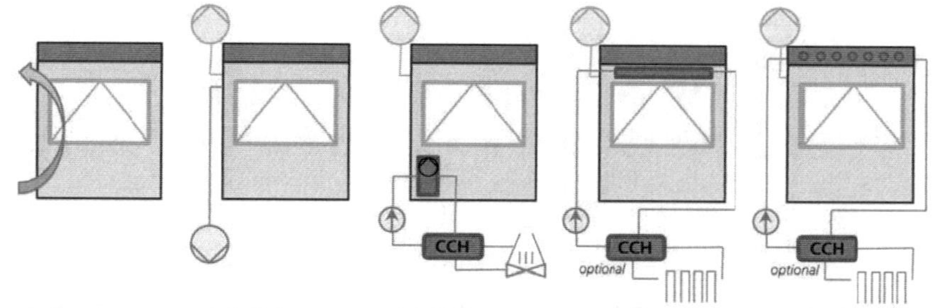

vent.	free day vent.	hybrid day vent.	hybrid day vent.	hybrid day vent.	hybrid day vent.
cooling concept	free night ventilation	mechanical night ventilation	hp active cooling w/ fan coil	radiant cooling w/ ceil. panel	radiant cooling w/ TABS
heat sink	ambient air	ambient air	cooling tower	BHEx [8-18]	BHEx [22-6]

Figure 1. Low-energy cooling concepts.

The ThermCo project [5] investigated the potential of different ventilation and cooling strategies with regard to energy efficiency and thermal comfort in different European climates. The results demonstrate a high potential for night ventilation strategies in North-European climate with low ambient air temperatures. In the Mid-European climate, water based low-energy cooling technologies based on radiant cooling make use of the cool ground in summer. Active cooling provides good thermal comfort in South-European climate with high and fluctuating cooling loads. *Figure 2* shows which cooling concept for a typical low-energy office building (according to EPBD requirements) may be favourably applied in different climate zones. Additionally, local specific conditions have to be considered such as microclimate, building design and location.

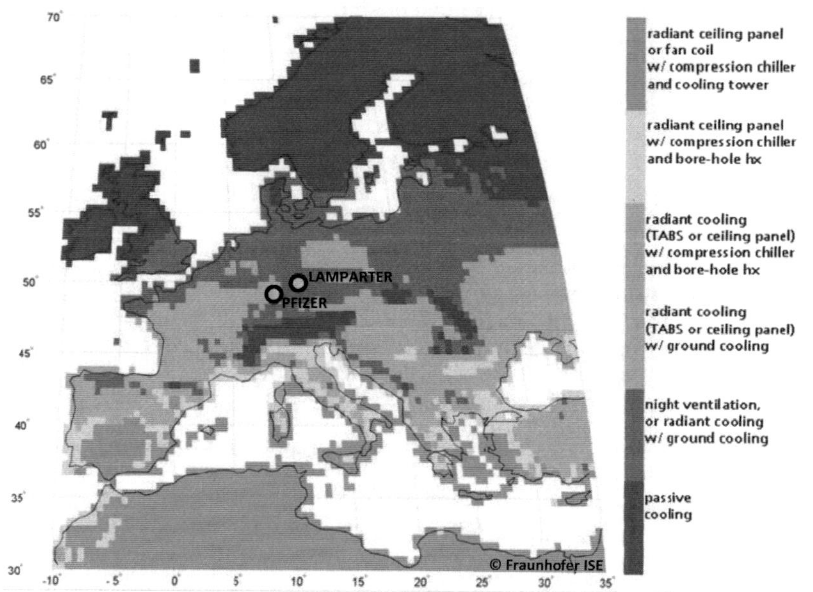

Figure 2. Cooling concepts (from *Figure 1*) in different European climate zones.

METHODOLOGY: ENERGY EFFICIENCY AND THERMAL COMFORT

Energy efficiency is calculated for the overall systems based on monitored data. *Figure 3* shows the adopted methodology for the concept based on night-ventilation [6] and earth-to-air heat excanger [7] and for the water-based low-energy cooling concept [8].

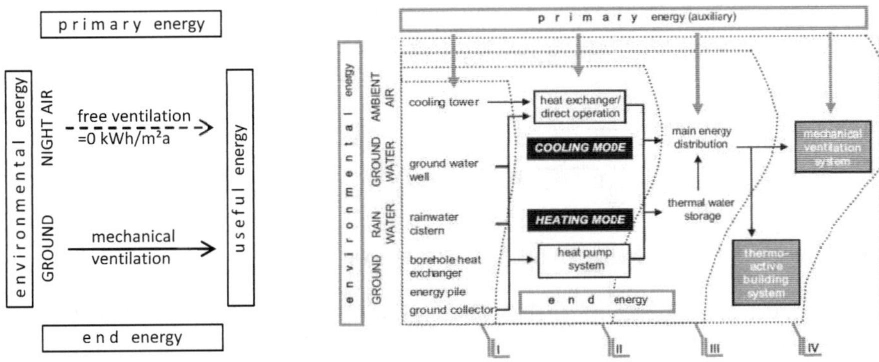

Figure 3. Methodology for the energy and efficiency evaluation.
Left: For low-energy cooling concepts based on night ventilation and earth-to-air heat exchanger.
Right: For water-based low-energy cooling according to four balance boundaries I–IV.

Figure 4 shows a building signature which combines four evaluation criteria (anti-clockwise):

- The energy efficiency is calculated for the useful energy [kWh_{cool}] and the primary energy demand for the whole system [kWh_{prim}] as shown in *Figure 3*. As most low-energy cooling concepts are driven by electricity (e.g. mechanical night ventilation, pumps for ground cooling, or reversible heat pumps), the target value is calculated for best practice systems considering a primary energy factor of 3.0 kWh_{prim}/kWh_{el}. Realised best practice buildings and theoretical calculations show, that low-energy cooling may reach seasonal performance factors (SPF) for the overall system (balance boundary IV) of 20 $kWh_{cool}/kWh_{el,free,IV}$ for free cooling and 5 $kWh_{cool}/kWh_{el,mech,IV}$ for mechanical cooling. As in moderate summer climates the free cooling potential is high, this may result in an overall SPF of 10 $kWh_{cool}/kWh_{el,IV}$ or 3.3 kWh_{cool}/kWh_{prim}, respectively..
- The cooling energy demand [$kWh_{cool}/m^2_{net}a$] characterises the quality of the building envelope and the passive cooling concept. The cooling energy demand should not exceed 20 $kWh/m^2_{net}a$ in moderate or warm summer climate.
- The overall primary energy demand for heating, cooling, ventilation and lighting [$kWh_{prim}/m^2_{net}a$] characterises the building performance, incl. the use of renewables or high-performance HVAC&R components. The primary energy demand should not exceed 100 $kWh/m^2_{net}a$.
- The thermal comfort in passively cooled buildings with free-night ventilation and earth-to-air heat exchanger (LAMPARTER) is evaluated according to the adaptive and in the air-conditioned reference building (PFIZER) according to the static approach in standard EN 15251 [9].

Different building and energy concepts can be comprehensively evaluated by these benchmark numbers.

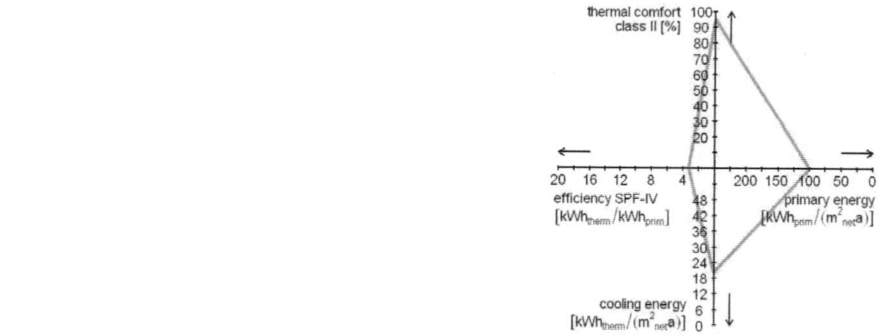

Figure 4. Building signature with target values referring to [8].

NIGHT VENTILATION AND EARTH-TO-AIR HEAT EXCHANGER

Next to the prevailing weather conditions, there are many other boundary conditions with regard to a successful implementation of night ventilation concepts. Textbooks [10], [11]and guidelines [12] give practical information. With regard to this study, microclimate and building concept are the most important specific parameters.

The LAMPARTER building, *Figure 5*, represents a prototypical night-ventilated building in a moderate summer zone (*Figure 2*) The passive-house building is located in a rural area with fresh summer winds from the high plains of the Swabian Alb. The small office building allows for free cross-ventilation. The members of staff (±20 persons) are well-informed about the building concept and, hence, adapt their user behaviour (window opening and external Venetian blinds) according to the indoor and outdoor climate [13]. Furthermore, the night cooling is enhanced by an earth-to-air heat exchanger which was built in the existing excavation (for economic reasons).

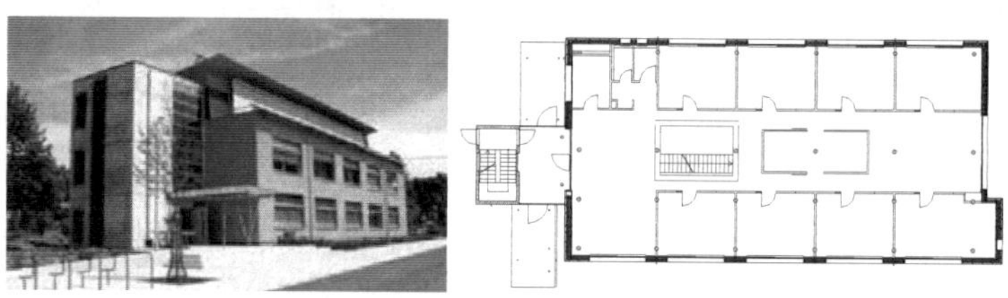

Figure 5. The LAMPARTER Building, Weilheim, Germany. © Hans Lamparter GbR.

Monitoring data and building description are given in Ref. [14]. Figure 6 shows the summer energy concept and the monitored air-change rates.

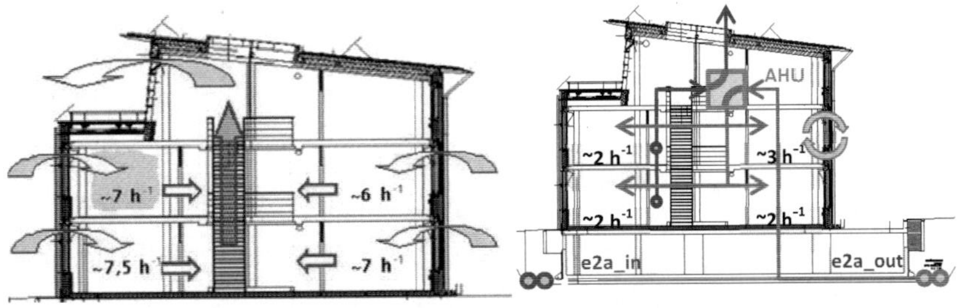

Figure 6. Free night-ventilation and hybrid day-ventilation with earth-to-air heat exchanger.

Thermal Comfort

As the LAMPARTER building is passively cooled, the thermal comfort is evaluated according to the adaptive approach. Noteworthy, the room temperatures exceed more often the lower than the upper limit of comfort class B. Even during the European heatwave in summer 2003, the thermal comfort complies with comfort class B.

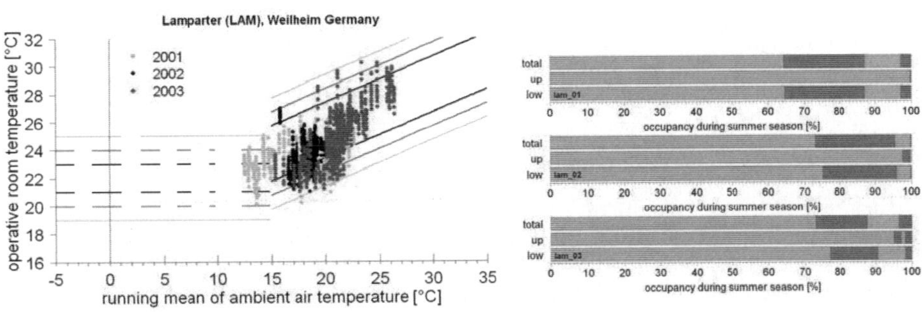

Figure 7. LAMPARTER: Thermal comfort and footprints for three years, summer only.

Energy Efficiency and Building Signature

The cooling energy from free night-ventilation is a function of the temperature difference between inside and outside and cannot be measured. Furthermore, the cooling energy is an implicit function of thermal comfort. Thus, the cooling energy is set to 0 $kWh/m^2_{net}a$ for free-cooling as no mechanical system is needed. The energy efficiency of free night-ventilation is infinite by definition.

Accordingly, the cooling energy is only the cooling energy from the earth-to-air heat exchanger which provides 3.4 $kWh_{cool}/m^2_{net}a$. As the whole mechanical ventilation runs during the daytime to provide the hygienic air-change rate, the additional electricity demand for the operation of the earth-to-air heat exchanger is calculated only for the (very low) pressure drop in the buried pipes. This results in a seasonal performance factor of 53.2 kWh_{cool}/kWh_{el} or 17.7 kWh_{cool}/kWh_{prim}, respectively.

Due to the lean building concept, high performance HVAC components, a daylight concept with controlled artificial lighting and solar power, the primary energy consumption is 50 $kWh/m^2_{net}a$ only.

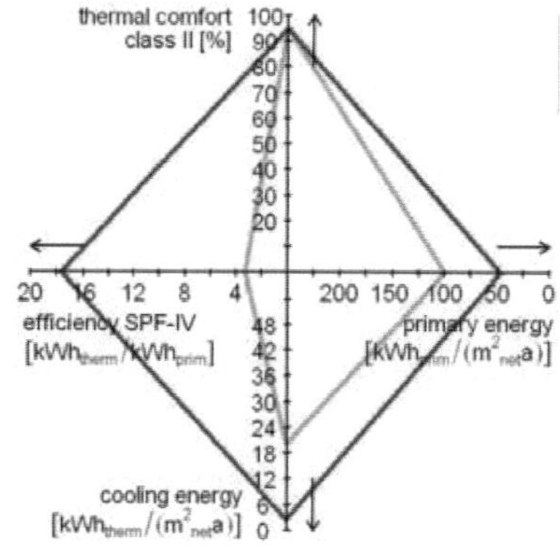

Figure 8. LAMPARTER: Building signature.

GROUND COOLING AND THERMO-ACTIVE BUILDING SYSTEMS

There are many boundary conditions which avoid a successful implementation of a night-ventilation concept. However, a low-energy cooling concept based on ground cooling reduces the primary energy demand significantly compared to a mechanically cooled building. Guidelines [15] and manufacturer information give practical information on the implementation of these concepts.

The PFIZER building, *Figure 9*, represents a typical office and laboratory building in a warm summer zone (*Figure 2*) The retrofit concept is based on passive-house technologies and a low-energy cooling concept. The air-conditioned building is located in an industrial area with poor air change in the Rhine valley. The big building with a sealed façade (labs) does not allow for cross-ventilation and is fully air-conditioned for safety-at-work reasons.

Figure 9. The PFIZER Building, Freiburg, Germany. © Fototeam Vollmer.

Monitoring data and building description are given in Ref. [16]. *Figure 10* shows the heating and cooling concept with the main components in summer operation mode.

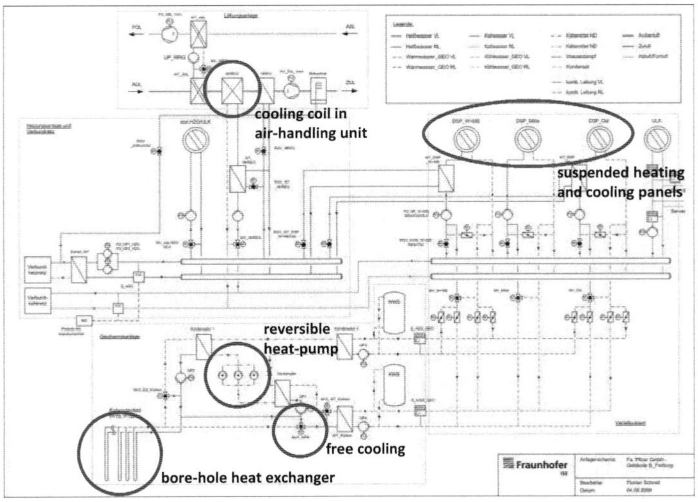

Figure 10. Energy concept with ground cooling, reversible heat pump and thermo-active building systems.

Thermal Comfort

As the PFIZER building is air-conditioned, the thermal comfort is evaluated according to the static approach. Noteworthy, the room temperatures exceed more often the lower than the upper limit of comfort class B, since no summer / winter comfort control was implemented in 2008. (Note: After optimisation, the summer room temperatures are higher today.)

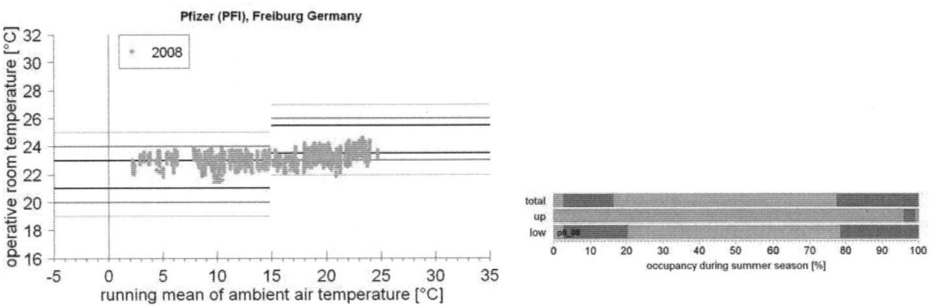

Figure 11. PFIZER: Thermal comfort and footprints for 2008.

Energy Efficiency and Building Signature

The cooling energy demand is 50 $kWh/m^2_{net}a$ for dehumidification, convective and radiant cooling.

The energy efficiency for balance boundary II (electricity for primary pump and compressor according to the definition of SPF in DIN V 18599 [17]) is 6.9 $kWh_{cool}/kWh_{el,free,II}$ in free cooling mode and 5.6 $kWh_{cool}/kWh_{el,free,II}$ in mechanical cooling mode. The comparatively small SPF in free cooling mode is due to the low temperatures and the accordingly high volume flow rate which results in a comparatively high energy consumption of the primary pump. The comparatively high SPF in mechanical mode is due to low ground temperatures and high cooling temperatures for the radiant cooling systems. Considering the percentage of operation time in free and mechanical cooling mode and the whole distribution system (secondary pumps and energy losses), the overall SPF_{IV} is reduced to 3.9 kWh_{cool}/kWh_{el} or 1.3 kWh_{cool}/kWh_{prim}, respectively.

This shows clearly the practical limitations of cooling concepts. In spite of high-performance components and a well-designed HVAC&R concept, the overall energy efficiency is considerably lower than the potential SPF_{IV} of 10 kWh_{cool}/kWh_{el}. Nevertheless, this concept is much more energy efficient than a conventional one: A compression chiller / cooling tower with a COP of 3 kW_{cool}/kW_{el} may result in an overall SPF_{IV} of 1.8 kWh_{cool}/kWh_{el} or 0.6 kWh_{cool}/kWh_{prim}, respectively.

The primary energy is calculated for the local energy consumption and production only. Due to high-performance HVAC components, ground cooling with low-temperature thermo-active building systems and heat production from a wood fired boiler, the primary energy consumption is 300 $kWh/m^2_{net}a$. Though this is considerably higher than the target value of 100 $kWh/m^2_{net}a$, it is much lower than the benchmark for laboratories with 800 $kWh/m^2_{net}a$.

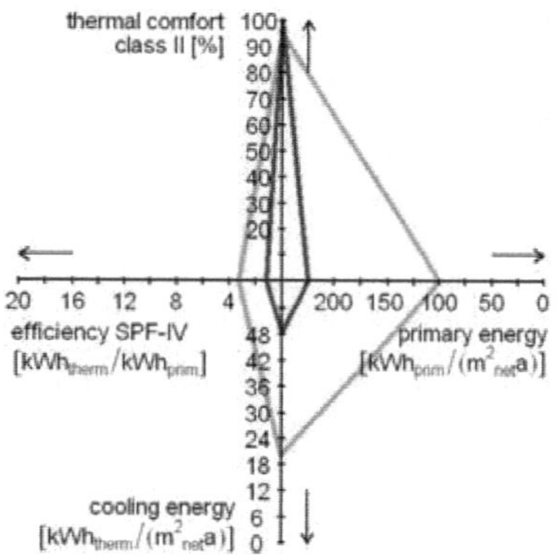

Figure 12. PFIZER: Building signature.

CONCLUSION

Passive cooling with free night-ventilation and earth-to-air heat exchanger provides good thermal comfort with high energy efficiency. However, the practical application is strictly limited to specific boundary conditions such as moderate summer climate, favourably microclimate, or small buildings with high thermal inertia. The whole building design and operation fosters the passive cooling concept.

Though mechanical night-ventilation enhances the heat dissipation compared to free night-ventilation, often a water-based cooling concept may be more reliable and similarly energy efficient and will provide better thermal comfort.

Ground cooling in combination with a reversible heat pump is often a better solution for warm summer climate, locations with inappropriate microclimate or other practical limitations. If thermo-active building systems are used for heat transmission to the room, these systems can be operated with high energy efficiency.

ACKNOWLEDGEMENTS

The project LowEx:MONITOR „Exergetisches Monitoring für Gebäude mit Erdwärmenutzung“ is funded by the German Ministry of Economics and Labour BMWi under the program “Energy Optimized Building” (0327466B).

REFERENCES

[1] DIRECTIVE 2010/31/EU OF THE EUROPEAN PARLIAMENT AND OF THE COUNCIL of 19 May 2010 on the energy performance of buildings (recast)

[2] http://www.buildup.eu/ [Sep. 2011]

[3] http://www.epbd-ca.eu/ [Sep. 2011]

[4] http://ec.europa.eu/energy/intelligent/index_en.html [Sep. 2011]

[5] http://www.thermco.org/ [Sep. 2011]

[6] J. Pfafferott. *Enhancing the Design and Operation of Passive Cooling Concepts.* Fraunhofer IRB Verlag, 2004.

[7] J. Pfafferott. *Evaluation of three earth-to-air heat exchangers with a standardised method to calculate energy efficiency criteria.* Energy and Buildings 35 (2003)

[8] D. E. Kalz, J. Pfafferott, S. Herkel. *Building signatures: A holistic approach to the evaluation of heating and cooling concepts.* Building and Environment 45 (2010)

[9] EN 15251:2007-08. *Indoor environmental input parameters for design and assessment of energy performance of buildings addressing indoor air quality, thermal environment, lighting and acoustics.* European committee for standardization CEN(2007).

[10] M. Santamouris, D. Asimakopoulos (eds.). *Passive Cooling of Buildings.* James&James, 1996.

[11] J. Pfafferott. *Erdwärmetauscher und Nachtlüftung.*in: *Bürogebäude mit Zukunft*, eds. K. Voss, S. Herkel, G. Löhnert, A. Wagner, M. Wambsganß.. TÜV-Verlag, Köln, 2005.

[12] J. Pfafferott. *Passive Kühlung mit Nachtlüftung.* BINE themeninfo I/03 (2003).

[13] S. Herkel, U. Knapp, J. Pfafferott. *Towards a model of user behaviour regarding the manual control of windows in office buildings.* Building and Environment 43 (2008)

[14] http://www.enob.info/de/neubau/projekt/details/passiv-buerohaus-lamparter/ [Sep. 2011]

[15] D. Kalz, J. Pfafferott. *Thermoaktive Bauteilsysteme in Nichtwohngebäuden.* BINE themeninfo I/07 (2007).

[16] http://lowexmonitor.ise.fraunhofer.de/demonstrationsgebaeude/buro-und-laborgebaude-der-firma-pfizer-freiburg [Sep. 2011]

[17] DIN V 18599 :2007-02. *Energetische Bewertung von Gebäuden - Berechnung des Nutz-, End- und Primärenergiebedarfs für Heizung, Kühlung, Lüftung, Trinkwarmwasser und Beleuchtung.* Deutsches Institiut für Normung (2007).

PRELIMINARY ANALYSIS OF U.S. RESIDENTIAL AIR LEAKAGE DATABASE v.2011

Wanyu R. Chan* and Max H. Sherman

Lawrence Berkeley National Laboratory
1 Cyclotron Road, MS 90R3058
Berkeley, CA 94720, USA
**Corresponding author: wrchan@lbl.gov*

ABSTRACT

Air leakage and other diagnostic measurements are being added to LBNL's Residential Diagnostics Database (ResDB). We describe the sources of data that amount to more than 80,000 blower door measurements. We present summary statistics of selected parameters, such as floor area and year built. We compare the house characteristics of new additions to ResDB with prior data. Distributions of normalized leakage are computed for income-qualified homes that were weatherized, homes that were participants of various residential energy efficiency programs, and new constructions built between 2006 and 2011. Further work is underway to relate air leakage to house characteristics utilizing the full database for predictive modelling of air infiltration and to support studies of energy efficiency. Current status of ResDB can be found at http://resdb.lbl.gov/.

KEYWORDS

Air leakage, blower door, fan pressurization measurements, infiltration

INTRODUCTION

The needs of residential energy efficiency and weatherization programs have led to many measurements of air leakage being made in existing homes and new constructions in recent years. We gathered this data to characterize the air leakage distribution of homes in the US. This effort is necessary to evaluate the energy implications of uncontrolled airflow through the building envelope. It also allows predictions of the expected improvements from various energy efficiency measures, given the baseline of current building stock.

Previous versions of LBNL's Residential Diagnostic Database [1] (ResDB) were dominated by homes from an income-qualified weatherization assistant program in Ohio, and also by homes that were built for the extreme weather in Alaska. Our latest data collection effort not only increase data counts, but also improve spatial representation of the dataset. Data collection is nearly completed at the time when we prepare this conference paper. Because data analysis is on-going, summary statistics and prelminary analyses presented here are likely to be revised when the final report is released.

DATABASE DESCRIPTION

Data Sources

We collected blower door data on over 81,000 homes, of which over 70% are single-family detached homes. Mobile home and mulit-family dwellings made up approximately 20% and 10% of the remaining data. Income-qualified weatherization assistant programs (WAPs)

remain the major sources of data, accounting for almost 60% of the blower door measurements. The database contains WAPs data from 15 states. Over 95% of the WAPs homes were tested at least twice, once before and once after weatherization. WAPs are adminstrated by state. As a result, there are many differences in how the work was performed and the data was collected; see [2] for an overview of national evaluation of WAPs. Some data were provided to us by agencies responsible for the programs in the form of a database. Others are contributed by contractors who performed the work. In future analysis, we plan to compare state-by-state or regional differences among WAPs, if any, in reducing air leakage.

Residential energy efficiency programs are another major sources of data. For example, the Home Performance with Energy Star program is implemented in over 30 states in US [3]. Many utility sponsored programs also offer incentives for energy efficiency upgrades. The majority of the energy efficiency programs data were contributed by energy auditors who performed the work. Some enegy efficiency programs that contributed data provided pre- and post-retrofit blower door measurements, which are available in about 60% of the data.

We defined new constructions as those built 2006 and later, i.e. after the release date of the prior ResDB. New constuctions account for approximately 20% of the data in the current database. Many of the new constructions were tested for air leakage in order to obtain an energy certification. These data were contributed mostly by energy auditors who performed the tests or by verification organizations. In addition, there are also a few research programs that collected data on new homes, such as Department of Energy's Building America Program [4]. California, North Carolina, Neveda, Texas, and Washington are the states with the most new constructions data available in the current ResDB.

Data Summary

New data being addded to ResDB are mostly single-point blower door measurements at 50 Pa. These measurements were converted to normalized leakage (*NL*) assuming a power law flow exponent of 0.65 as follows:

$$NL = 1000\left(\frac{ELA_{4\text{ Pa}}}{Area}\right)\left(\frac{H}{2.5\text{ m}}\right)^{0.3} \quad \text{where} \quad ELA_{4\text{ Pa}} = \sqrt{\frac{\rho}{2 \times 4\text{ Pa}}}\left(Q_{50\text{ Pa}}\right)\left(\frac{4\text{ Pa}}{50\text{ Pa}}\right)^{0.65} \qquad (1)$$

$ELA_{4\text{ Pa}}$ (m^2) is the effective leakage area at 4 Pa, *Area* (m^2) is the dwelling floor area, H (m) is the dwelling height, $\rho = 1.2$ kg/m^3, and $Q_{50\text{ Pa}}$ (m^3/s) is the airflow rate at 50 Pa measured by the blower door. In some cases, data contributors reported other measures of air leakage, such as specific leakage area, which is *ELA* normalized by floor area. All data were converted to normalized leakage for the analysis.

Dwelling locations (state, county, city, or zip code) are known for all data but at varying levels of detail. Figure 1 shows the spatial distribution of single-family detached homes roughly mapped to 19 cimate zones [3] by state-line. The number of data shown, N, exclude entries that are known to be mobile homes and multi-family homes (note that as we continue to check ResDB for errors and missing data, some entries might be reclassified). Figure 1 shows that most of the climate zones are represented in ResDB v2011, including populous areas in US, such as the Northeast states (NY-PA-NJ), Florida (FL), Texas (TX), and California (CA). Kentucky and the South remain two areas that we lack data.

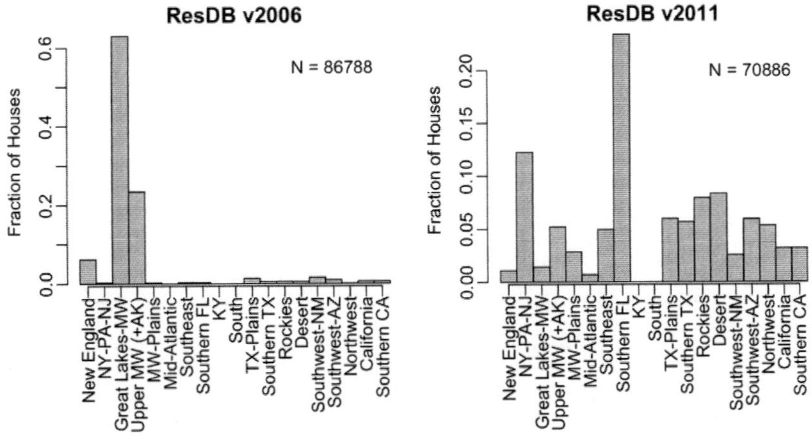

Figure 1. Single-family homes in prior (v2006) and current (v2011) ResDB organized by climate zones. Many climate zones cross state-line [3], but for simplicity this is not reflected here because the data is segregated by state-line. This is acceptable for most climate zones except for CA and TX. For these two states, the data counts are split equally between two zones: TX-Plains and Southern TX, California and Southern-CA, respectively.

Other parameters available in ResDB include, in descending order of availability, dwelling size (floor area, height, number of story, volume), foundation type, duct system, number of bedrooms and occupants. Comapre to prior version, the current ResDB contains more information about the heating, cooling, and ventilation system. There are approximately 30,000 duct blaster data entered into ResDB that were measured using different test configurations, e.g. leakage to outside, supply air leakage to outside, total leakage, etc. Combustion test data, indoor air contaminants measurements, and energy usage or projected savings are avaliable in a few datasets. We use a SQL database management system for data entry, storage, and retrieval. This software program allows us to better organize the multi-dimensional datasets for analysis.

EXPLORATORY ANALSYSIS

Houses Characteristics

Figure 3 compares the basic house characteristics represented in the prior (v2006) and current (v2011) version of ResDB. Data that falls outside of the acceptable ranges are excluded from the comparison. Acceptable ranges are defined as floor area between 30 and 1000 m^2, number of stories between 1 to 3, year built between 1800 and 2011, and house age at the time of testing is non-negative. About 2600 data points were excluded based on these criteria.

New additions to the current ResDB tend to be larger in floor area, which is reflective of the trend of US homes being built [5]. In ResDB v2006, houses with floor area <92 m^2 (1000 ft^2) were assumed to be single-story, and larger homes were assumed to be one and a half story. For the purpose of computing *NL* (Equation (1)), this is not a large source of uncertainty [6] even though two-story dwellings are more common in the US than split-level, as shown by v2011 data in Figure 2. There are comparable proportion of one- and two-story homes in both verisons of ResDB, assuming that many of the previously classified one and a half story homes are more likely to be two-story homes.

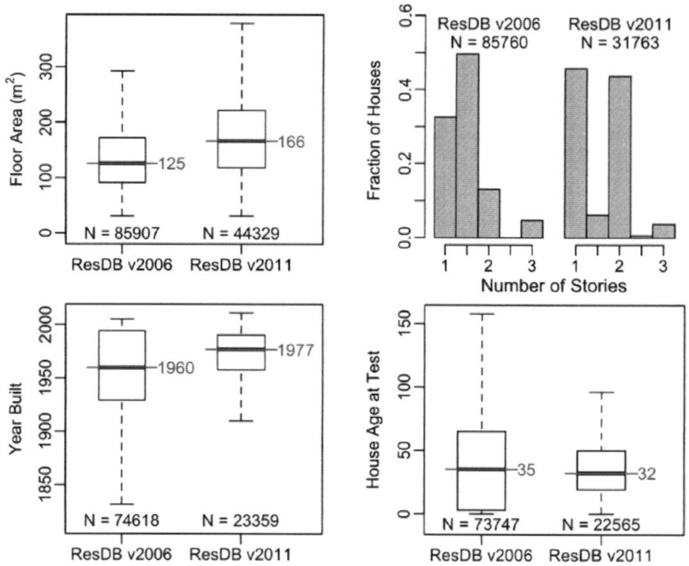

Figure 2. Comparison of house characteristics of new data added to v2011 ResDB with data in v2006. Boxplot shows the range where the central 50% of the data are between the 1st and 3rd quartiles. Extreme values beyond extent of the whiskers are not shown. Median floor area, year built, and house age at test are highlighted in red.

There are many data with missing year built in the new data that are added to ResDB v2011. Since year built is a predictor for air leakage [1], we are in the process of seeking this information if available from data contributors or estimating the values via other means, such as real estate listings and records. Overall, the homes that are added to ResDB v2011 are about the same age as those in v2006 at the time when the blower door tests were performed (Figure 2). The boxplot also shows that homes added to ResDB v2011 are more similar in age as a group, as indicated by a narrower distribution, relative to data in v2006.

Despite the similarity in house age, there is a difference in the year built of homes. First, v2006 does not contain the new constructions built between 2006 and 2011. ResDB v2006 is also the cumulative result of many years of data compilation since 1990, whereas most of the new additions to v2011 were blower door tests performed in the past few years. The combined dataset will allow us to evaluate if new constructions built today more airtight than homes that were built in the 1980's and 1990's when those homes were tested as new. We also plan to separately assess the relationship of air leakage with year built and house age, if possible, with the combined dataset. For example, new constructions today may become more leaky with age at a different rate than homes that were built in the 1980's and 1990's.

Normalized Leakage

Normalized leakage was computed using all valid blower door measurements. We erred on the inclusive side to estimate floor area and house height, the two parameters necessary for computing $ELA_{4\ Pa}$, using all available means, such as by inferring from house volume and number of story. We also included blower door measurements that were performed at pressure differential other than 50 Pa, typically between 25 and 50 Pa. Some of these are multi-point blower door tests that measured airflow at different pressure differentials, but there are also single-point measurements where the target pressure of 50 Pa was not reached. If provided, we used the reported flow exponent instead of the 0.65 default value. Approximately 1000 flow exponent estimates are available in ResDB v2011. Most values (90%) fall between 0.58 and 0.78.

Figure 3 shows the resulted *NL* having a distribution that is roughly lognormal with a geometric mean of 0.51 and a geometric standard deviation of 1.98. This distribution includes all valid *NL* estimates, so some homes were represented more than once if multiple blower door measurements were performed. This distribution also has not been adjusted for the house characteristics, so it should not be considered as representative but rather, a preview of the air leakage measurements in ResDB v2011.

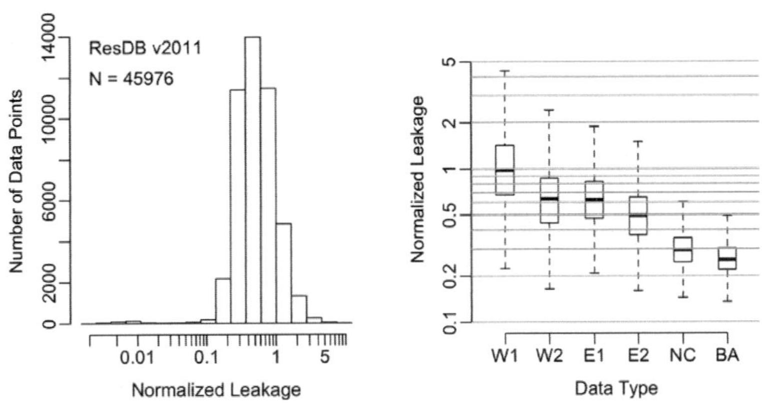

Figure 3. Unadjusted normalized leakage distribution computed from blower door measurements in ResDB v2011 (left panel). The right panel shows differences between the unadjusted *NL* distributions when estimates of *NL* are categorized into six types according to the source of data (see Table 1 for data type descriptions).

	Data Type	Normalized Leakage	Number of Data	Floor area (m^2)	Year Built	Age
	Income-Qualified WAPs					
W1	Pre-weatherization	0.98 (0.68 – 1.43)	6576	118	1962	34
W2	Post-weatherization	0.63 (0.44 – 0.87)		(91 – 166)	(1928 – 1977)	(50 – 83)
	Energy efficiency program					
E1	Pre-retrofit	0.63 (0.47 – 0.83)	8225	190	1980	30
E2	Post-retrofit	0.49 (0.37 – 0.65)		(146 – 248)	(1960 – 2008)	(9 – 50)
NC	New constructions	0.29 (0.25 – 0.35)	9745	199 (162 – 260)	2008 (2007 – 2008)	1 (1 – 2)
BA	Building America Program	0.25 (0.22 – 0.30)	724	178 (141 – 223)	2008 (2008 – 2008)	1 (1 – 2)

Table 1. Summary statistics of the unadjusted normalized leakage and corresponding house characteristics. Median values are shown with the 1st and 3rd quartile values in parentheses.

As previously observed in ResDB v2006, income-qualified WAPs homes tend to be the most leaky. This is partially because they tend to be older and smaller in size compare to other homes, but also potentially because of disrepair of the building structure. The difference in median *NL* between pre- and post-weatherization is about 35%. This reduction is slightly more substantial than the apparent change from retrofits performed on homes that participated in the various residential energy efficiency programs, where the change in median *NL* is about 22%. There are many possible explanations for this, some of which can be investigated by comparing the before and after *NL* by first adjusting for parameters that have a known effect on air leakage. For example, a comparison can be made between WAPs and energy efficiency programs by selecting only homes that have similar characteristics and from the same state.

New constructions built in 2006 and after have a median *NL* of 0.3. This value approaches the median *NL* from the Building America research program, which aim is to accelerate the development and adoption of building energy technologies. In addition, the range of *NL* values are similar among the new constructions and Building America data. The factor of three differences as shown by the extent of the whiskers from *NL* of about 0.15 to 0.5 is likely a reasonable estimate of the inherent differences among new homes.

CONCLUSION

Large number of blower door and other dignostic measurements have been added to LBNL's ResDB. We preformed exploratory analyses to look for relationships between normalized leakage and house characteristics such as floor area, year built, and if the house tested is part of an energy efficiency or income-qualified weatherization program. Findings are compared with previous published analyses of ResDB. Once the current data has been checked for quality it will be combined with v2006, and the full ResDB v2011 will be analyzed to characertized the stock of US housing. Such analyses will support studies of energy efficiency and related concerns such as indoor air quality.

ACKNOWLEDGEMENTS

We greatly appreciate organizations and individuals who shared their blower door and other diagnostic data with us. Funding was provided by the U.S. Dept. of Energy Building Technologies Program, Office of Energy Efficiency and Renewable Energy under DOE Contract No. DE-AC02-05CH11231 and by the California Energy Commission through Contract 500-08-061. This task is also supported by the U.S. Dept. of Housing and Urban Development Office of Healthy Homes and Lead Hazard Control through Interagency Agreement I-PHI-01070, by the U.S. Environmental Protection Agency Office of Air and Radiation through Interagency Agreement DW-89-92322201.

REFERENCES

[1] McWilliams, J., and Jung, E. 2006. *Development of a Mathematical Air-Leakage Model from Measured Data*, LBNL-59041, Lawernce Berkeley National Laboratory, Berkeley, CA.

[2] Eisenberg, J.F. 2010. *Weatherization Assistance Program Technical Memorandum Background Data and Statistics*, ORNL/TM-2010/66, Oak Ridge National Laboratory, Oak Ridge, TN.

[3] *Home Energy Saver: Engineering Documentation. The Home Energy Scoring Tool for Asset Rating.* https://sites.google.com/a/lbl.gov/hes-public/home-energy-scoring-tool.

[4] *Building America – Resources for Energy Efficient Homes.* http://www1.eere.energy.gov/buildings/building_america/.

[5] US Census Bureau. 2011. *Median and Average Square Feet of Floor Area in New Single-Family Houses Completed by Location.* http://www.census.gov/const/C25Ann/sfforsalemedavgsqft.pdf.

[6] Chan, W.R., Nazaroff, W.W., Price, P.N., Sohn, M.D., and Gadgil, A.J. 2005. *Analyzing a Database of Residential Air Leakage in the United States.* Atmos. Environ. 39, 3445–3455.

U.S. COMMERCIAL BUILDING AIRTIGHTNESS REQUIREMENTS AND MEASUREMENTS

Steven J Emmerich[*1], Andrew K Persily[1]

[1] *National Institute of Standards and Technology*
100 Bureau Drive, MS 8633
Gaithersburg, MD USA

ABSTRACT

In 1998, Persily published a review of commercial and institutional building airtightness data that found significant levels of air leakage and debunked the myth of the airtight commercial building. Since that time, the U.S. National Institute of Standards and Technology (NIST) has maintained a database of measured airtightness levels of U.S. commercial building leakages, in part to support the development and technical evaluation of airtightness requirements for national and state codes, standards and programs. This paper presents the airtightness data from the NIST database and a summary of recent developments in U.S. codes and standards. The average airtightness of the 228 buildings in the database is 25 $m^3/h{\cdot}m^2$ at 75 Pa, which is a little over 10 % tighter than the average reported by Persily in 1998. This average commercial and institutional building airtightness is tighter than the average of all U.S. houses but leakier than conventional new houses based on a large database of U.S. residential building airtightness. The data are examined for trends related to size, year of construction, type of construction and climate. The paper also discusses recent code and standards developments including the adoption of a continuous air barrier requirement in ASHRAE Standard 90.1-2010 *Energy Standard for Buildings Except Low-Rise Residential Buildings*, various U.S. state building codes, and the proposed International Green Construction Code.

KEYWORDS
Airtightness, air barrier, commercial buildings, infiltration

INTRODUCTION

Past NIST efforts have demonstrated that, despite assumptions to the contrary, typical modern U.S. commercial building envelopes are not particularly airtight (1, 2), building envelope leakiness results in a significant energy cost (3), and substantial energy savings would result through the requirement of an effective air barrier for new commercial buildings (4). This work has led to the consideration and adoption of prescriptive air barrier requirements in a number of building standards and codes, e.g., ASHRAE Standard 90.1, the U.S. Army Corps of Engineers (USACE), and several states in the U.S.. This paper presents the currently available airtightness data from the NIST database and summarizes recent developments in U.S. codes and standards.

The airtightness of building envelopes is measured using a fan pressurization (blower door) test in which a fan is used to create a series of pressure differences across the building envelope between the building interior and the outdoors. ASTM Standard E779 (5) is a test method that describes the fan pressurization test procedure in detail, including the specifications of the test equipment and the analysis of the test data. In conducting a fan pressurization test in a large building, the building's own air-handling equipment sometimes can be employed to induce the test pressures. A Canadian General Standards Board test method, CGSB 149.15, describes the use of the air-handling equipment in a building to conduct such a test (6). Typically, the test results are reported in terms of the airflow rate at some reference pressure difference divided by the building volume, floor area or envelope surface area.

The airtightness data presented here are collected from a number of different studies that use a variety of units and reference pressure differences. The results are presented here as airflow rates at an indoor-outdoor pressure difference of 75 Pa normalized by the above-grade surface area of the building envelope. When necessary, this conversion is based on an assumed pressure exponent value of 0.65. The values of envelope airtightness are given in units of $m^3/h{\cdot}m^2$, which can be converted to cfm/ft^2 by multiplying by 0.055.

DATABASE AND ANALYSIS

Table 1 contains a summary of the air leakage data for the 228 U.S. commercial and institutional buildings included in the NIST database. Sources of these data included 9 buildings tested by NIST (7, 8, 9), 89 buildings tested by the Florida Solar Energy Center (10, 11), 41 buildings tested by Terry Brennan (13, 14, 15), and 88 buildings tested by the U.S. Army Corps of Engineers (USACE) (unpublished data including some partial school buildings), and three other buildings (12, 16). The buildings were tested for a variety of purposes and were not randomly selected to constitute a representative sample of U.S. commercial buildings. None of the Table 1 buildings are known to have been constructed to meet a specified air leakage criterion, which has been identified as a key to achieving tight building envelopes.

		Air Leakage at 75 Pa ($m^3/h{\cdot}m^2$)			
Dataset	Qty	Mean	Standard Deviation	Min	Max
NIST	9	15.1	11.5	3.9	43.3
FSEC[1]	89	33.8	23.2	4.0	124
Brennan	39	19.8	18.2	2.7	80.7
USACE	88	19.3	10.3	3.4	63.4
Other	3	8.7	2.0	6.4	10.1
All buildings	**228**	**24.8**	**19.1**	**2.7**	**124**

[1]FSEC values differ from earlier publication due to corrected envelope surface area.

Table 1. Summary of Building Airtightness Data

As seen in Table 1, the average air leakage at 75 Pa for the 228 buildings is 24.8 $m^3/h{\cdot}m^2$, which is 12 % tighter than the average of 28.7 $m^3/h{\cdot}m^2$ for the U.S. buildings included in the earlier analysis by Persily (1). This average airtightness is tighter than the average of all U.S. houses, but leakier than conventional new houses based on a database of residential building airtightness (17).

The airtightness data were also analyzed to assess the impact of a number of factors on envelope airtightness including number of stories, year of construction, and climate. It is important to note that the lack of random sampling and sample size limits the strength of any conclusions concerning the impacts of these factors. Also, not all of these parameters were available for all buildings in the database. Figure 1 is a plot of the air leakage at 75 Pa vs. the reported number of stories of the building and shows a tendency toward more consistent tightness for taller buildings. The shorter buildings display a wide range of building leakage. This result is consistent with past analyses [1][2].

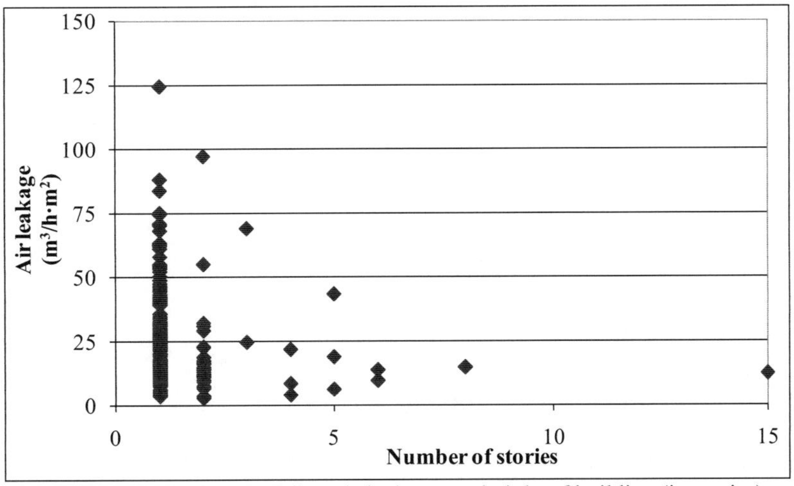

Figure 1: Normalized building air leakage vs. height of building (in stories)

Figure 2 is a plot of the air leakage at 75 Pa vs. the year of construction of the building for buildings built more recently than 1955. While common expectation is that newer commercial buildings would be tighter than older ones, the data indicate no significant trend.

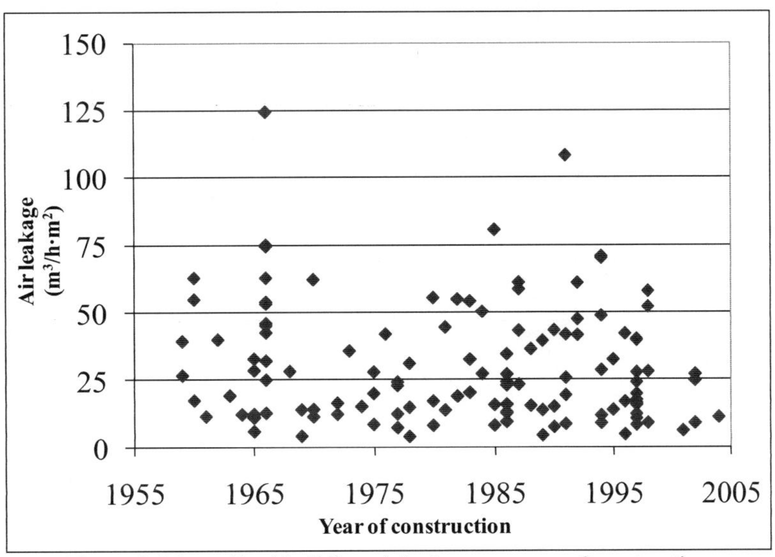

Figure 2: Normalized building air leakage vs. year of construction

Figure 3 is a plot of the air leakage at 75 Pa vs. wall construction type for 200 of the buildings from the database. While the data suggests that buildings with frame and frame/masonry wall types are somewhat leakier than the other types, the large standard deviations for the individual categories do not support any firm conclusions. Additionally, data interpretation is complicated by a lack of clear definition of construction types and and because the use of different terms for wall construction may not be consistent among those reporting the leakage data.

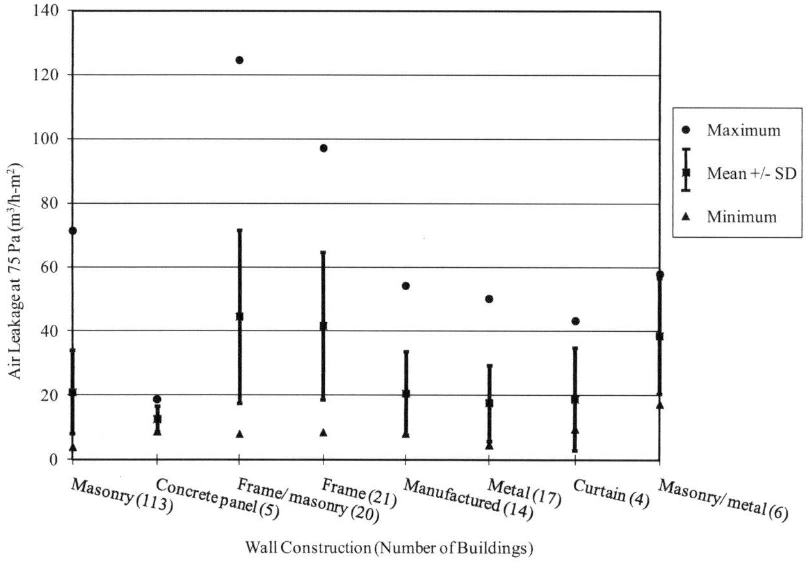

Figure 3: Normalized building air leakage vs. Wall type

The USACE requires new army buildings to meet a maximum whole building airtighness specification of 4.5 $m^3/h\ m^2$ at 75 Pa based on the entire building enclosure area including the slab and any below grade walls (USACE 2009). New buildings are tested and improvements to airtightness are made if they fail to meet the standard. The tightness of the 'typical' buildings in the NIST database (after conversion to a basis including the below-grade walls and slab) are compared to the tightness of the new USACE buildings (Zhivov 2010) in Figure 4. The average USACE building is about 84 % tighter than the average 'typical' building, conclusively demonstrating improvements in tightness can be achieved through a rigorous specification and testing program in new buildings. Importantly, the variation in tightness of the USACE buildings is also significantly reduced, with a standard deviation that is only 37 % of the average compared to a standard deviation of 77 % of the average for the NIST database. By specifying and testing to a maximum tightness limit, the increased certainty enables better prediction of infiltration, building energy use, and design and operation of a building's HVAC system.

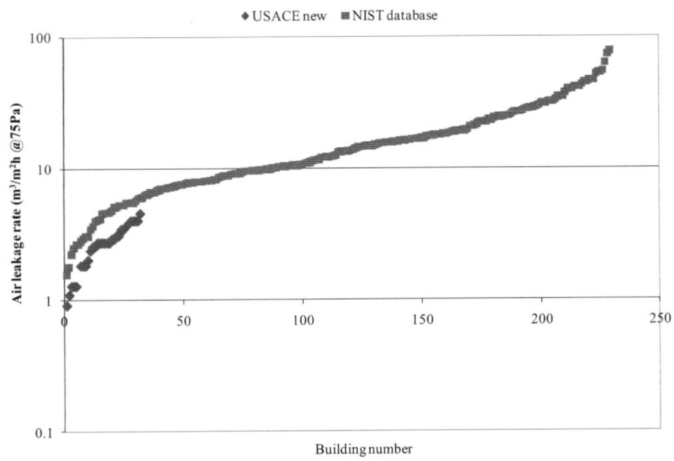

Figure 4. Comparison of commercial building airtightness for NIST database (typical buildings) and USACE data (designed and tested to meet 4.5 $m^3/h\ m^2$ at 75 Pa).

U.S. STANDARDS AND REGULATIONS

	Air Leakage at 75 Pa ($L/s{\cdot}m^2$)[1]		
Standard or code	**Material**	**Assembly**	**Whole building[2]**
ASHRAE 90.1	0.02	0.2	-
ASHRAE 189.1	0.02	0.2	2.0
IECC	0.02	0.2	2.0
IGCC	-	-	1.3 (required)
USACE	0.02	-	1.3 (required)
GSA	0.02	0.2	1.3 (required)
GA	0.02	-	-
MA	0.02	-	-
MN	0.02	-	-
NH	0.02	-	-
OR	0.02	-	-
RI	0.02	-	-
WA	0.02	-	-
Seattle, WA	-	-	1.3 (required)

[1] 1 $L/s{\cdot}m^2 \approx 0.2\ cfm/ft^2$

[2] Whole building limits are based on enclosure including slab and below-grade walls (not specified in 189.1). Test optional unless indicated.

Table 2. Summary of Building Airtightness Data

In the U.S., commercial building construction practices are addressed by various standards, codes, and green building program requirements, and Table 2 summarizes some of the relevant air leakage limits from these requirements. ASHRAE requires continous air barriers (CAB) for most commercial buildings in both Standard 90.1 *Energy Standard for Buildings Except Low-Rise Residential Buildings* (23) and Standard 189.1 *Standard for the Design of High-Performance Green Buildings* (22). Since 2010, Standard 90.1 requires the CAB to meet either a material tightness limit (0.02 $L/s{\bullet}m^2$ under a pressure differential of 75 Pa) or an assembly tightness limit (0.2 $L/s{\bullet}m^2$ under a pressure differential of 75 Pa), but does not include a whole building tightness limit nor a requirement for whole building pressurization testing. When using the prescriptive option for energy efficiency, Standard 189.1 includes an option of a whole building test demonstrating the building meets a tightness limit of 2.0 L/s•m^2 under a pressure differential of 75 Pa although the normalization area is not specified.

The 2012 International Energy Conservation Code (IECC, 24) has similar requirements to Standard 189.1 with options for a CAB with material or assembly tightness or a whole building test with the same limit as 189.1. The Internationl Green Construction Code (IgCC) Public Version 2.0 (25) includes a requirement for a whole building test with a leakage limit of 4.57 m^3/hr/m^2 at 75 Pa in its prescriptive compliance option, which is equivalent to 1.3 L/s•m^2. Note that Standard 189.1 is an alternative compliance path within the IgCC. Several U.S. state building codes (including Georgia, Massachusetts, Minnesota, New Hampshire, Oregon, Rhode Island, and Washington) include requirements for air barriers or whole building pressurization tests (18).

Since 2009, the USACE has required that conditioned buildings be built or retrofitted to include a continuous air barrier to control air leakage through the building envelope (19). The specification requires whole building testing with a maximum leakage of 1.3 L/s•m^2 at 75 Pa based on the building enclosure area including the slab and subgrade walls. The average tightness for a set of 31 new USACE buildings was reported to be 0.8 L/s•m^2. Zhivov estimates the first cost for new construction to be \$5.40/m^2 of floor area with a simple payback in energy savings of 2 yr to 10 yr (20). Similarly, the U.S. General Services Administration (21) now requires all new U.S. federal buildings for the Public Buildings Service to include an air barrier with the whole building having an air leakage rate of not more than 1.3 L/s•m^2 at 75 Pa.

CONCLUSION

Past NIST efforts have demonstrated that, despite assumptions to the contrary, typical modern U.S. commercial building envelopes are not particularly airtight, building envelope leakiness results in a significant energy cost, and substantial energy savings would result through the requirement of an effective air barrier for new commercial buildings. The average airtightness of the 228 buildings currently available in the NIST database is 25 m^3/h·m^2 at 75 Pa, which is a little over 10 % tighter than the average reported by Persily in 1998. The data show no significant trends related to size, year of construction, or type of construction, but when compared to recent USACE tested buildings, do demonstrate that typical commercial buildings are much leakier than buildings designed and tested to meet a whole building tightness specification. Important recent developments in U.S. codes and standards include the adoption of a continuous air barrier requirement in ASHRAE Standard 90.1-2010 *Energy Standard for Buildings Except Low-Rise Residential Buildings*, the requirement for a whole building tightness limit with testing by the USACE, and inclusion of air barrier or whole building tightness limits in various U.S. state building codes and the proposed International Green Construction Code.

REFERENCES

[1] Persily, A.K. 1998. Airtightness of Commercial and Institutional Buildings: Blowing Holes in the Myth of Tight Buildings. Proceedings of Thermal Performance of the Exterior Envelopes of Buildings VII.

[2] Emmerich, S.J., Persily, A.K. 2005. *Airtightness of Commercial Buildings in the U.S.* AIVC Conference.

[3] Emmerich, S.J., Persily, A.K. 2005. Impact of Infiltration on Heating and Cooling Loads in U.S. Office Buildings. AIVC Conference.
[4] Emmerich, S.J., McDowell, T., Anis, W., Investigation of the Impact of Commercial Building Envelope Airtightness on HVAC Energy Use (2005) NISTIR 7238.
[5] ASTM. (2003). E779-03, Standard Test Method for Determining Air Leakage Rate by Fan Pressurization. West Conshohocken, PA: American Society for Testing and Materials.
[6] CGSB. (1996). *CAN/CGSB-149.15-96, Determination of The Overall Envelope Airtightness of Buildings by the Fan Pressurization Method Using the Building's Air-Handling Systems*. Canadian General Standards Board.
[7] Persily, A.K. and Grot, R.A. (1986). "Pressurization testing of federal buildings." Measured Air Leakage of Buildings, ASTM STP 904, H.R. Treschsel and P.L. Lagus, Eds., American Society for Testing and Materials, pp. 184-200.
[8] Persily, A.K., Dols, W.S., Nabinger, S.J. and Kirchner, S. (1991). Preliminary Results of the Environmental Evaluation of the Federal Records Center in Overland Missouri. NISTIR 4634.
[9] Musser, A. and Persily, A. (2002). Multizone Modeling Approaches to Contaminant-Based Design. Proceedings of ASHRAE Transactions Vol. 108.2.
[10] Cummings, J.B., Withers, C.R., Moyer, N., Fairey, P. and B. McKendry. (1996). Uncontrolled Airflow in Non-residential Buildings. Florida Solar Energy Center, FSEC-CR-878-96.
[11] Cummings, J.B., Shirey, D.B., Withers, C., Raustad, R. and Moyer, N. (2000). Evaluating the Impacts of Uncontrolled Air Flow and HVAC Performance Problems on Florida's Commercial and Institutional Buildings, Final Report, FSEC-CR-1210-00.
[12] Bahnfleth, W.P., Yuill, G.K. and Lee, B.W. (1999). "Protocol for Field Testing of Tall Buildings to Determine Envelope Air Leakage Rate," ASHRAE Transactions, Vol. 105 (2).
[13] Brennan, T., Turner, W., Fisher, G., Thompson, B., and B. Ligman. (1992). "Fan pressurization of school buildings." Proceedings of Thermal Performance of the Exterior Envelopes of Buildings V, American Society of Heating, Refrigerating and Air-Conditioning Engineers, Inc., pp. 643–45.
[14] Brennan, T., Persily, A.K., Clarkin, M., and S.J. Emmerich. *Measuring Airtightness at ASHRAE Headquarters*. ASHRAE Journal 2007.
[15] Henderson, H.I., Cummings, J.B., Zhang, J. and T. Brennan. 2007. MITIGATING THE IMPACTS OF UNCONTROLLED AIR FLOW ON INDOOR ENVIRONMENTAL QUALITY AND ENERGY DEMAND IN NON-RESIDENTIAL BUILDINGS. Report to New York State Energy Research and Development Authority on Agreement #6770.
[16] Crovella, P.L. 2010. A Comparison of Techniques to Measure Commercial Building Infiltration Rates. IAQVEC 2010. Syracuse, NY.
[17] Sherman, M.H. and Matson, N.E. (2002). Airtightness of New U.S. Homes: A Preliminary Report. LBNL-48671, Lawrence Berkeley National Laboratory.
[18] Spinu, M., and B. Erickson. 2011. Impact of Air Leakage on the Building Envelope: Myths and Facts About Airtightness. The Construction Specifier Feb (2011) pp 30-41.
[19] USACE (2009) US Army Corps of Engineers Engineering and Construciton Bulletin No. 2009-29 *Building Air Tightness Requirements*.
[20] Zhivov, A. 2011. New Requirements to the United States Army Buildings Air Tightness. AIVC Workshop, Brussels, Belgium.
[21] GSA. 2011. *Facilities Standards for the Public Buildings Service*. U.S. General Services Administration.

[22] ASHRAE (2010). ANSI/ASHRAE/USGBC/IES Standard 189.1-2010, Standard for the Design of High-Performance Green Buildings, American Society of Heating, Refrigerating and Air-Conditioning Engineers, Inc., Atlanta, GA.
[23] ASHRAE (2010). ANSI/ASHRAE/USGBC/IES Standard 90.1-2010, Energy Standard for Buildings Except Low-Rise Residential Buildings, American Society of Heating, Refrigerating and Air-Conditioning Engineers, Inc., Atlanta, GA.
[24] ICC. 2010. International Green Construction Code, Public Version 2.0. International Code Council, Inc.
[25] ICC. 2011. International Energy Conservation Code. International Code Council, Inc.

Thursday 13 October 2011

09.00 – 10.30 Parallel Session 4A with short oral presentations and posters - Ventilation in high performance buildings - Special applications

- Measured public benefits from energy-efficient homes (Jonathan Coulter, USA)
- Measurement of pollutant emissions in two similar very low energy houses with cast concrete and timber frame (Franck Alessi, France)
- Low pressure drop air transfer between rooms in buildings with balanced ventilation - A commonly ignored issue (Peter Schild, Norway)
- Impact of the filtration system on the indoor-outdoor particles concentration relationships in an air conditioned office building (Alain Ginestet, France)
- Measured performance of three types of energy program houses in two US cities (Jonathan Coulter, USA)
- Experimental evaluation of Supply-Only ventilation effectiveness (Mireille Rahmeh, France)
- The Applicability of Glazing System with Dynamic Insulation for Residential Buildings (Shinsuke Kato, Japan)
- Integrated Approach of CFD and SIR Epidemiological Model for Infectious Transmission Analysis in Hospital (Hiroaki Asanuma, Japan)
- Shelter-in-place effectiveness in the event of toxic gas releases: French and Catalan assessment

approach (Montoya Maria Isabel, Spain)

09.00 – 10.30 Parallel Session 4B - Quality frameworks for airtightness assessment

Rewarding or imposing good airtightness in a regulation directly calls into question the reliability and accuracy of the measurements that are performed in practice. Several schemes will be presented to increase or assess the quality of the measurements and execution. This session is part of an AIVC-TightVent project.

- Quality system for airtightness measurement of buildings (Oliver Solcher, Germany)
- The quality framework for Air-tightness measurers in France: assessment after 3 years of operation (Valérie Leprince, France)
- Interlaboratory tests for the for the determination of repeatability and reproducibility of airtightness measurements (Christophe Delmotte, Belgium)
- Pressure distribution inside large buildings during airtightness tests (Stefanie Rolfsmeier, Germany)

10.30 – 11.00 Room Change and coffee break

MEASURED PUBLIC BENEFITS FROM ENERGY-EFFICIENT HOMES

Jonathan Coulter[1]*, Benjamin Hannas[1], Colby Swanson[1], Michael Blasnik[2]*, and Eric Calhoun[1]

[1]Advanced Energy
909 Capability Drive Suite 2100
Raleigh, North Carolina, USA
*Email: jcoulter@advancedenergy.org

[2]M. Blasnik & Associates
150 Poplar Street
Roslindale, Massachusetts, USA
*Email: michael.blasnik@verizon.net

ABSTRACT

The objective of this paper is to summarize results of two studies that compared Baseline, ENERGY STAR® and Guaranteed Performance homes co-located in Phoenix, Arizona, USA and determine if homes in these three groups could be distinguished from each other in terms of actual summer/cooling energy usage or homeowner satisfaction related to the HVAC systems. The summer/cooling energy use study surveyed 7,141 houses, of which 3,336 were Baseline homes, 2,979 were ENERGY STAR homes and 826 were Guaranteed Performance homes. Statistically valid energy data shows that ENERGY STAR homes saved on average 16% in summer/cooling energy use (kWh/m^2) as compared to the typical Baseline homes. The Guaranteed Performance homes saved on average 33% in summer/cooling energy use over the Baseline homes and saved 20% compared to ENERGY STAR homes. During the spring and summer of 2005, the homeowner satisfaction study was administered to 708 houses from the same 7,141 house sample set. The second study found that 49% of the Guaranteed Performance homeowners said they were completely satisfied with their home's "ability to keep them comfortable year round" compared to 35% of ENERGY STAR homeowners and only 27% of Baseline homeowners. In fact, this survey found that Guaranteed Performance homeowners were more satisfied with every aspect of their home's HVAC performance – year round comfort, the freshness of air inside the house, evenness of temperatures from room to room, reliability and cooling cost compared to Baseline and ENERGY STAR houses. Combining the results from these two studies shows that Guaranteed Performance homes consume less energy than comparable ENERGY STAR or code-built homes while simultaneously improving homeowner satisfaction.

KEYWORDS

Energy efficiency programs, energy consumption, homeowner satisfaction, guaranteed performance

INTRODUCTION

For more than 30 years in the United States, a variety of approaches have been tried to improve the energy efficiency of newly-constructed homes. Before 2004, millions of homes had been constructed to local building codes, about 400,000 were ENERGY STAR compliant and over 60,000 qualified for Guaranteed Performance recognition. For all the attention on projected and deemed energy savings these programs were claiming, there was not enough data being analyzed to determine the actual energy reduction impact or homeowner satisfaction these programs were having post-occupancy.

In 2004, the United States with 4.6% of the world's population accounted for 24.9% of the world's primary energy consumption. The housing sector accounted for 36% of all U.S. electrical demand with a predicted 39% growth in this sector alone between 2000 and 2010 [1].

Over the past several decades, rising energy prices have driven a demand for more energy-efficient homes. Builders initially responded with simple energy saving remedies such as increased insulation, double-paned glass, tighter door seals, window awnings and other measures. Recent field applications of building physics advancements (such as high-efficiency HVAC equipment, improved duct sealing, building infiltration barriers, low-e glass and compact fluorescent lighting) have continued to offer more sophisticated and effective methods of providing predicted energy savings. Each of these measures reduces overall home energy bills in computer modeling, but little is known about how these changes effect the homeowners as they live in these homes, or whether overall homeowner satisfaction is being influenced by these building changes. This report summarizes results from two studies to compare local code-built or Baseline, ENERGY STAR and Guaranteed Performance homes and determines if homes in these three groups have different summer cooling energy usage or homeowner satisfaction.

ENERGY STAR Homes Background

In 1995, the U.S. Environmental Protection Agency (EPA) launched its ENERGY STAR Homes program, which established guidelines for reducing home energy use and promoted partnerships with homebuilders to construct homes to be more energy efficient than code-built homes. A thesis study of 291 homes in Phoenix, Arizona in 2000 compared ENERGY STAR homes to code-built homes and concluded that ENERGY STAR homes used 2.3% less energy per square foot than the code-built homes [2]. At the time the energy use and homeowner satisfaction studies were started in 2004, the program standards based on computer modeling proposed savings of 30% for home heating, cooling and water heating as compared to homes built to the requirements of the 1993 Model Energy Code (MEC) or 15% more efficient than state energy code, whichever was more rigorous. These savings were based on guidelines of reduced building envelope infiltration and HVAC duct leakage, improved wall, ceiling and floor insulation levels and more efficient windows, doors and HVAC system. The adherence of these guidelines was field-verified in houses on a random basis. ENERGY STAR certified homes, it was reasoned, would offer homeowners dependable savings on their monthly energy bills while collectively reducing the overall energy consumption and impact of the residential sector nationwide. However, factors such as homeowners' lifestyles (with respect to energy use), effective product installations, operation and maintenance of HVAC systems, house sizes and others, made it difficult to assess the actual impact that these energy conservation intentions had on lowering home energy. Computer-modeled and deemed energy savings were available for the 400,000 plus homes built by 2004, but little was known about the as-built energy performance or homeowner comfort of these houses post-occupancy. Additionally, by 2011 over 1 million homes have now been built in this program.

Guaranteed Performance Homes Background

More recently, several organizations have created and promoted an ENERGY STAR "Plus" program for the new construction market. Called Guaranteed Performance homes, these homes are designed and built to go beyond the EPA ENERGY STAR program by:

- Requiring the use of more energy-efficient building components.
- Requiring 100% of the houses to have field quality assurance checks after framing, wall insulation and completed construction to make sure that the specifications are met at every stage.
- Providing a two-year heating and cooling energy use guarantee to the homeowners (typical house average of $1 U.S. or 12.5 kWh used per day).
- Providing a two-year comfort guarantee to the homeowners (defined as a temperature differential of no greater than plus or minus two degrees C (three degrees F) from the thermostat location to the center of any conditioned room within the zone) to ensure that the house is performing as designed after the homeowners have moved in.

Before this study was conducted in 2004, there were more than 60,000 houses nationwide built and certified to the Guaranteed Performance standards. To date in 2011, there are more than 150,000 homes built in this program.

Studies Objectives

The energy efficiency study was structured to compare the actual summer cooling energy usages of Baseline, ENERGY STAR and Guaranteed Performance homes, while taking into consideration a large number of variables in home design. That study looked at real data and real energy performance of occupied houses – not computer-model data. The results of the study could then be used to answer several fundamental questions about the effectiveness of these efficiency programs:

- How much energy did the Baseline, ENERGY STAR and Guaranteed Performance homes actually consume for summer space cooling?
- How much cooling energy savings are actually realized by ENERGY STAR and Guaranteed Performance homes, compared to similar Baseline homes?

The homeowner satisfaction study followed the energy efficiency study to determine homeowner satisfaction with right-sized HVAC systems. Right-sizing is typically applied in conjunction with other energy-reducing materials and construction techniques, such as low-E windows, higher Seasonal Energy Efficiency Rating (SEER) levels, tighter duct leakage standards and proper installation. As a result, overall homeowner satisfaction depends on more than just a right-sized HVAC system. The approach for this survey was to use a quantitative questionnaire to compare attitudes of homeowners of the three major categories of new homes in the Phoenix, Arizona market: Baseline, ENERGY STAR and Guaranteed Performance.

METHODS

Energy Efficiency

Because the Phoenix, Arizona market was an early adopter of both the ENERGY STAR and Guaranteed Performance programs, there is a high concentration of homes with many years worth of energy use data that provided an excellent opportunity to verify energy consumption data on the three home types under real-world conditions.

For the 7,141 houses included in the energy study, data was compiled and analyzed based on the following three categories: Baseline homes (3,336 homes not built as part of any energy efficiency program, but resembled other homes in the study), ENERGY STAR homes (2,979 homes built per U.S. EPA ENERGY STAR program standards) and Guaranteed Performance homes (826 ENERGY STAR homes plus additional energy efficiency improvements, as well as a comfort and heating/cooling use guarantee). Once assigned to a category above, the homes were then segregated by builder, year built, square footage, presence of swimming pool, solar orientation, HVAC type and zip code. These groupings helped identify patterns in the data that can point to factors with the greatest effect on home efficiency within the boundaries of the study.

Direct comparisons of energy use between the three home categories – Baseline, ENERGY STAR and Guaranteed Performance – are difficult at best, given the vast number of variables that can affect both home performance and total energy use. Swimming pools, in particular, add significantly to the overall energy use of a home. Even seasonal differences in the costs of operating electric water heaters versus gas heaters can alter energy use profiles by as much as 900 kWh/year, invalidating certain study results. To reduce the chance for such variables to skew the results of this study, homes were only compared within certain definite data sets. Since no all-electric Guaranteed Performance homes were available for this study, the cleanest comparison is to look at gas-heated homes with no pool, within different size ranges.

Electric use data provided by the utility for years 2003 and 2004 were analyzed using a variable-base Heating Degree Day (HDD) regression analysis. In addition, periods with unusually low usage were also excluded, which were defined as use of less than 150 kWh/month or less than 400 kWh, and either less than 25% of the median use of the month or less than 40% of the 25th percentile of use for that home. These data screens excluded about 11% of the total 400,027 meter readings from 1998 through 2004, but only about 6% of the 85,963 meter readings in 2004. The use analysis results were considered reliable if they were based on at least nine meter readings that spanned at least half of the heating degree days (base 65°F, HDD65) and cooling degree days (base 75°F, CDD75) of a typical year, included at least one period of true baseload use (very few CDD or HDD) and resulted in an estimated baseload use of at least 2000 kWh/yr (to eliminate likely unoccupied homes not caught by the meter reading screens). This screening eliminated 596 homes (8%), primarily due to the requirement for nine meter readings, leaving 6,545 houses with apparently reliable records of electric use for the study.

The primary method for electric-use data weather normalization was a CDD and HDD adjustment. This approach classified and summed meter reading data and degree days for summer, winter and baseload categories based on HDD65 and CDD75. The resulting three equations were solved to estimate baseload use per day, summer cooling use per CDD75 and winter/heating use per HDD65, assuming a linear relation between use, CDD and HDD. This analysis approach allows for heating and cooling occurring within all seasons and appears to provide more reliable results in many cases than using a regression model. This analysis was run separately for each home during each calendar year.

A regression analysis of summer cooling use (kWh/m^2) was also performed as a function of living area. These regression models attempt to control for factors such as house size, orientation and baseload to avoid having to perform comparisons on increasingly smaller groups of buildings. The study team used the regression modeling to estimate the energy use for a 167 m^2, gas-heated home without a pool across the three home categories.

Homeowner Satisfaction Survey

The homeowner satisfaction survey was conducted in Phoenix during the spring and summer of 2005. It was conducted in two phases: 1) qualitative research among homeowners, builders and mechanical contractors and 2) a quantitative survey of 708 homeowners.

1. Qualitative Research

   The first steps began with qualitative research among homeowners, builders and contractors to understand what drives homeowner satisfaction and to ensure that the survey document was comprehensive and written in the homeowner's language.

2. Quantitative Research

   A written, four-page survey was mailed to 7,000 homeowners during July and August 2005. In all, 708 homeowners responded with completed surveys. A market research firm designed the survey and tabulated the returned surveys with an overall sampling margin of error of 3.7 percentage points at a 95% confidence level.

Homeowners identified that several factors were important when deciding whether to buy a particular new home. The principal factors were location, design and functionality. However, once the decision was made and the homeowner experienced living in the house, energy cost and performance of the HVAC system became very important. Qualitative research among homeowners and builders in Phoenix identified a set of drivers that determine how satisfied a homeowner is with their house after they have had at least a year's experience living in it. HVAC systems must satisfy four basic needs:

1. **Comfort**. Comfort is a basic need. Clearly, it is hard to be satisfied in a house that does not deliver basic thermal comfort in all seasons of the year. To homeowners, comfort is a function of several things: air freshness, evenness of temperatures from room to room, the ability to regulate temperatures and responsiveness, or the HVAC system's ability to heat or cool the house quickly.

2. **Energy Efficiency**. Homeowners need a system that delivers comfort in all seasons at a reasonable cost.

3. **Reliable Performance**. Along with cost, homeowners also need a system that is easy to operate, reliable and does not require constant repair. Furthermore, they want a system that is not noisy.

4. **Healthiness**. There is emerging awareness of the link between the home and family health. Homeowners are beginning to see how their home can be a source of health threats

from mold, mildew and airborne allergens. The link between health and home systems makes common sense to consumers.

RESULTS

Figure 1 shows the concentration of house type per year built.

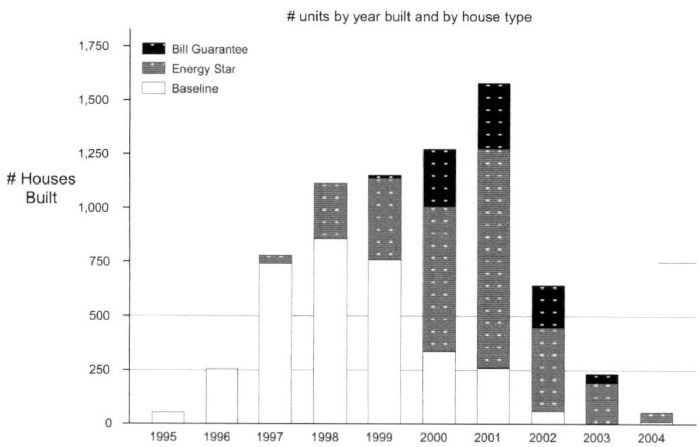

Figure 1: Study Homes by Year Built and Category

Energy Intensity (kWh/m²)

Table 1 outlines the summer/cooling energy use of gas-heated Baseline, ENERGY STAR and Guaranteed Performance homes with no swimming pools. Results are separated by house size into small (< 148 m^2), medium (149-223 m^2) and large (> 223 m^2) homes.

	Baseline	ENERGY STAR	Guaranteed Performance
Homes < 148 m^2			
# Homes	282	326	141
kWh/m^2/yr	0.41	0.38	0.33
Homes 149-223 m^2			
# Homes	37	660	282
kWh/m^2/yr	n/a	0.33	0.25
Homes > 223 m^2			
# Homes	20	208	136
kWh/m^2/yr	n/a	0.28	0.21

Table 1. Summer/Cooling Comparison

Energy Savings

After applying regression analysis, the annual summer/cooling intensities were estimated to be 16% lower for ENERGY STAR homes compared to Baseline homes (0.325 kWh/m2 versus 0.386 kWh/m2). Guaranteed Performance homes realized an energy savings of 33% over Baseline homes (0.260 kWh/m2 versus 0.386 kWh/ft2) or roughly 1800 kWh/year. As shown in Table 2, there were 708 responses to the homeowner satisfaction survey.

CATEGORY	CODE	# OF SURVEYS	% OF TOTAL
Baseline Homeowners	B	205	29%
ENERGY STAR Homeowners	ES	255	36%
Guaranteed Performance Homeowners	GP	235	33%
Unknown		13	2%
Grand Total		708	100%

Table 2: Composition of Survey Respondents

There are differences in the demographics of the three groups of homeowners:

- Baseline homeowners were the least affluent and live in the smallest houses. They were also between Guaranteed Performance and ENERGY STAR homeowners in age and the presence of occupants under the age of 18.
- ENERGY STAR homeowners had the biggest households with the most children. They were young and fairly affluent.
- Guaranteed Performance homeowners were older, more affluent and lived in smaller households. Their homes were large and presumably more expensive.

All the homes in the survey were built between 1994 and 2004, with the majority built since 2000. Figure 2 shows the number of homes constructed by year and by category.

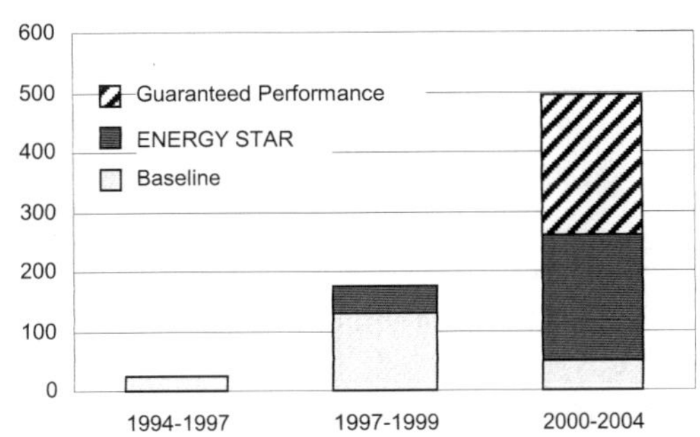

Figure 2: Houses Built By Year and Category

The characteristics of the houses in each of the three categories are depicted in Table 3. Guaranteed Performance homes tend to be larger, one story dwellings. The presence of a swimming pool is not dependent on the category of house, nor is SEER rating of the HVAC system.

CHARACTERISTIC	BASELINE	ENERGY STAR	Guaranteed Performance
Average square meters (Feet^2)	157 (1,685)	174 (1,877)	197 (2,125)
Percent of homes w/gas heat	26%	47%	73%
Percent of homes w/2 or more floors	20%	23%	15%
Percent of homes w/swimming pool	19%	20%	18%
Average SEER rating	11.6	11.9	11.7
Average square meter/ton of cooling (feet^2)	39 (423)	39 (418)	44 (474)

Table 3: Characteristics of Homes That Responded to the Survey

One of the key findings in this survey shows that Guaranteed Performance homeowners are more satisfied than ENERGY STAR or Baseline homeowners on almost every one of the following influences or drivers of satisfaction: comfort, energy efficiency, reliable performance, healthiness. Table 4 summarizes the percent of homeowners that are completely satisfied with each driver of satisfaction as well as the statistical relevance of each measure. Each measurement was statistically significant to 99%.

Category of Home	Need	Driver of Satisfaction	Guaranteed Performance	ENERGY STAR	Baseline
YEAR ROUND DRIVERS FOR ALL HOMEOWNERS	COMFORT	The ability of your home to keep you comfortable year round	49%	35%	27%
		The freshness of the air inside your home during those times you keep doors and windows shut	40%	24%	22%
	HEALTHI-NESS	Ability of heating and cooling systems to reduce allergies and other airborne ailments in your home	32%	19%	15%
SUMMER PERFORMANCE DRIVERS FOR ALL HOMEOWNERS	COMFORT	Your ability to regulate temperatures during the summer	41%	28%	23%
		The ability of your home to keep you comfortable during the summer	37%	22%	20%
		The ability of your air conditioner to cool your home down quickly	36%	25%	19%
		The evenness of temperatures from room to room during the summer	25%	14%	12%
	ENERGY EFFICIENCY	The cost of cooling your home	22%	14%	12%
	RELIABLE PERFORM-ANCE	The reliability of your cooling system (i.e., repair frequency)	53%	41%	32%
		The noise of your cooling system when it's running	29%	19%	15%

Table 4: Homeowner Satisfaction with Drivers of Satisfaction

DISCUSSION

At the beginning of this project, there was skepticism around the validity of conducting a study to compare homes across the three categories we selected (Baseline, ENERGY STAR and Guaranteed Performance). The concern was that the amount of variability due to factors that have nothing to do with the programs we were studying or that cannot be controlled, would mask any noticeable differences. However, after having done this study and confirming there is tremendous variability, the sample size was still large enough to see statistically significant differences among the three categories. While we have accumulated a body of evidence which indicates that the programs are a driver of these savings in one specific geographic region, the data should not be viewed as proof for all regions of the U.S. We recognize there are issues and we cannot prove the exact amount of savings nationwide, but we now have a jumping point for further investigation and benchmarking in other locations. Bottom line – this kind of study can produce valid results and those results will be strengthened with additional data. That being said, the important findings from this study include:

- The number of homeowners surveyed was large enough to give statistical reliability, but due to variations in climate, construction practices, etc., these results are not extendable to other markets outside Phoenix.
- Compared to Baseline homes, ENERGY STAR homes used 16% less energy and Guaranteed Performance homes used 33% less energy for total summer/cooling, saving about 1800 kWh/year.
- Larger homes had lower cooling intensity.
- Two-story homes had higher cooling intensity than one-story homes even after accounting for living area.
- Baseload electric use constituted about 0.15 kWh of additional summer/cooling load for each 1 kWh of annual baseload electric used. This baseload electric impact was substantial – equal to about 1300 kWh of total summer/cooling for the average baseload use of 8571 kWh in the analysis sample, equal to about 20% of the entire summer/cooling load.
- Homes facing northeast had significantly lower summer/cooling energy use than homes facing east (the default category), but no other orientations showed statistically significant differences.
- This survey demonstrates that right-size HVAC systems, along with other energy efficient features, results in greater homeowner satisfaction.
- The researchers believe that there is a latent demand for higher performance, or better building science, on the part of the homeowner. Unfortunately, this demand seems to be overshadowed by other factors at the time of purchase.

CONCLUSION

Implementation of the ENERGY STAR and Guaranteed Performance programs can yield improvements in the overall energy efficiency of new homes, as compared to homes built to code. The quantitative actual energy use results and homeowner satisfaction results from three levels of home construction (Baseline, ENERGY STAR and Guaranteed Performance) demonstrate that when the energy-efficient building requirements of the ENERGY STAR homes program were followed, the houses used 16% less summer/cooling energy and were 35% more comfortable than Baseline homes. But even more energy savings of 33%, as well as increased overall homeowner satisfaction of 49%, were possible when additional energy efficient requirements, 100% field quality assurance checks and a two-year comfort/heating/cooling use guarantee were added, meeting the Guaranteed Performance homes program standards.

These win-win findings will help the managers of the EPA ENERGY STAR, Guaranteed Performance and other energy-efficient new building programs adjust their respective program guidelines to ensure that the most cost-effective, energy-saving measures are identified and implemented into new home construction. For code officials, this may provide ideas for future changes. For homebuilders, contractors and other industry professionals, these measured results will provide evidence to support their claims of increased energy savings, validate the benefits of commissioning and help expand the market share of energy-efficient homes. Utility services may also benefit from this study by using the data to help identify future trends in the housing market and predict patterns of energy use.

One concern that came out of the data analysis is the appearance of the increase in average home size which is a trend the entire U.S. is experiencing (and the subsequent increasing energy use). This tendency offsets, to a large degree, the savings achieved through improvements in home energy efficiency. To realize actual overall energy reductions along with the environmental, economic and health benefits associated with those reductions, the trend to build larger homes may be useful to address.

The performance and satisfaction "bar" for new energy-efficient home construction has been raised in the Phoenix area as a result of the ENERGY STAR and Guaranteed Performance home programs. These programs have been instrumental in the education and training of consumers, builders and contractors about the benefits and construction differences of higher energy performing homes and homes with higher homeowner satisfaction. This study was initiated to provide a model for ongoing efforts to illuminate impact, as well as a feedback mechanism to support continuous improvement of energy-efficient programs for new home construction in the rest of the United States, as well as the world.

ACKNOWLEDGEMENT

U.S. Environmental Protection Agency, U.S. Department of Energy, Arizona Public Service and Southwest Gas should be especially recognized for their support of this study.

REFERENCES

[1] 2004 Buildings Energy Databook, U.S. Department of Energy.

[2] Energy Consumption Comparison of ENERGY STAR® Homes in Phoenix, Arizona, Bradley Jay Bashford, Arizona State University, May 2000.

Swanson, C, Blasnik, M, and Calhoun, E. 2005. Measuring Public Benefit From Energy Efficient Homes. Report submitted to U.S. Environmental Protection Agency by Advanced Energy, Raleigh, NC.

Dixon, T, Swanson, C, and Heineman, D. 2006. Phoenix Homeowner Satisfaction Survey. Report submitted to U. S. Department of Energy by Advanced Energy, Raleigh, NC.

MEASUREMENT OF POLLUTANT EMISSIONS IN TWO SIMILAR VERY LOW ENERGY HOUSES WITH CAST CONCRETE AND TIMBER FRAME

Franck Alessi*[1], Michael Sollaris[2]

1 CEA INES
50 avenue du lac Léman
73370 le Bourget du Lac, France
**Corresponding author: franck.alessi@cea.fr*

2 CEA INES
50 avenue du lac Léman
73370 le Bourget du Lac, France

ABSTRACT

This article is devoted to Indoor Air Quality (IAQ) in two low energy houses, with different frames (cast concrete (I-BB) and timber frame (I-OB)) built in platform INCAS of INES (National Institute of Solar Energy - in french: Institut National de l'Energie Solaire). In order to quantify pollutant emissions due to building materials and products, an experimental protocol consisted in stopping ventilation systems -"balanced ventilation"- of each house (a little before and during the measurement campaign), closing doors and windows, and not allowing occupant. Measurements started quickly after the end of construction: 70 days for I-BB and 224 days for I-OB. To measure IAQ, an experimental protocol was developed using continuous recorders (TSI Q-TRAK 7565 for CO_2, multi-gas PID monitor Graywolf with TG502 VOC Probe for TVOC, NO_2, O_3; KIMO KH200 KISTOCK for t_a and RH) and passive samplers (radial diffusive samplers Radiello® for specific VOCs - hydrocarbons and glycol ethers- and aldehydes). The duration of each measurement campaign was 7 days (from 02/04/2010 to 09/04/2010 for I-BB and from 23/05/2011 to 30/05/2011 for I-OB). Logging intervals for continuous recorders were 10 minutes. So that to estimate emissions due to building materials and eventual associated health effects, we focused on Volatile Organic Compounds (VOC: including specific VOCs and aldehydes) both with passive samplers Radiello® and continuous recorder Graywolf. Results reveal that in these low energy houses, building materials and products emit a great quantity of VOC with an indoor concentration 7 times higher than outdoor levels (both for I-OB and I-BB). Each frame has a main pollutant: Toluene for I-BB with a concentration almost twice higher than WHO guideline value (260µg/m^3, 7days), and hexanal for I-OB : averaged value around 570 µg/m^3 against $\approx$ 140 µg/m^3 for I-BB. This study confirms clearly the importance of ventilation systems in these new low energy buildings and in general rules. If well dimensioned they should improve energy efficiency in order to increase IAQ, comfort and human well being.

KEYWORDS

Indoor Air Quality; Chemical pollutants; Pollution; Outdoor air; TVOC; VOCs; Aldehydes; Concentration; Measurement campaign; Continuous recorder; Passive sampler; Platform INCAS; Frame; Cast concrete; Timber frame; Building materials, Products; Health effects.

INTRODUCTION

The Indoor Air Quality in low energy buildings should be a preoccupation in order to minimize health risks for occupants. For several decades the air exchange rate has been reduced and the leakage of dwellings has also decreased [1]. Moreover, building materials hold several chemical products unknown in term of health effects. Consequently, ventilation systems are very important to dilute indoor air pollutants in order to minimize health risks for occupants. Unfortunately in France, ventilation is sometimes inadequate for many reasons: a complex regulation, too many workers (both in construction of installation), a poor or nonexistent maintenance, and an impact of renovation and rehabilitation unappreciated [1].

A Recent article of OQAI ("Observatoire de la Qualité d'Air Intérieur") has shown that the air exchange rate in French houses is less than the prescribed rate for 56% of them [2] and it doesn't depend on ventilation systems (mechanical ventilation, natural ventilation,...). [2]
Despite an extensive literature search, it appears that a similar approach has never been followed until now. For other studies conducted so far, the VOC concentrations measured are the result of the presence of multiple sources of pollution (materials, equipments, furniture, household products, human activities, external environment...). This study is going to determine with more certainty the impact of construction materials and products on indoor air pollution.
Volatile Organic Compounds (VOC): Organic compounds with boiling points ranging from a lower limit between 50 °C and 100 °C, and an upper limit between 240 °C and 260 °C, where the upper limits represent mostly polar compounds. NOTE: in this article VOC includes VOCs (glycol ethers and hydrocarbons) and aldehydes (formaldehyde, acetaldehyde...) (Table 2).
The study was conducted with an experimental protocol different from already made national and international studies [3,4,5,6,7,8,9]. Having two experimental homes with ventilation systems stopped, closed windows and doors, identical geometries, but different frames allowed us to quantify VOCs and aldehydes emissions emitted by products and building materials.
The study focuses on the houses of the platform INCAS from INES (Institut National de l'Energie Solaire – National Institute of Solar Energy) in Le Bourget du Lac (Savoie - 73).
There are currently 3 un-inhabited 110m² houses (an additional one is under planning), with the same internal geometry, same architecture, same level of insulation (Figure 3), and located under the same climate, therefore with the same solar inputs. The three houses are respectively made of concrete blocks (house I-DM), cast concrete (house I-BB) and timber frame (house I-OB), with different insulation technical and materials in order to comply with the "Passivhaus" energy standard (less than 15 kWh/m² heating needs per year). Note that measurements on Indoor Air Quality (IAQ) were only done in I-OB and I-BB houses.
The experimental protocol is based on the work of the OQAI for the selection of pollutants to target [3].Chosen Pollutants are the same as in this investigation in order to have a magnitude of expected concentrations. In 2010, a campaign measurement was realised in I-BB house (from 02/04/2010 to 09/04/2010) and a second (I-OB) was done from 23/05/2011 to 30/05/2011.

Figure 1. Houses INCAS I-BB, I-DM and I-OB.

MATERIALS AND METHODS

In this article we are focusing on two un-inhabited 110m² houses of the platform INCAS: I-BB and I-OB. The goal : quantifying the level of indoor pollutants of I-BB (made of cast concrete) and I-OB (made of timber frame) due to building material emissions without ventilation, occupant, and with low leakage paths (permeability values for Passivhaus label : N_{50} = 0.54 Vol/h for I-OB and 0.26 Vol/h for I-BB). All measurements were realised during a week (7 days). The main logging interval used for the recorders was 10 minutes.

Building description

I-BB and I-OB houses were equipped with a balanced ventilation (BV). This system is a mechanical equipment which integrates a heat exchanger combined with a ventilation system. To characterize precisely pollution from building materials and products, this BV system was stopped. Consequently, indoor pollutant levels increased.
So that to describe I-BB and I-OB building materials and to identify their different pollutant sources (Table 1), we realised an investigation based on a French standard [10].
I-BB house was built with cast concrete walls and coated with a 20 cm extruded polystyrene exterior insulation. There was a reinforced concrete slab on the basement insulated with polyurethane and a concrete slab on the underside of the first floor ceiling insulated with glass wool as well as the gable walls. The linings were tiles on

the ground floor and PVC on the second floor. On the walls, gypsum board covered with a paint solvent ensured the finishing touches. For the ceiling of each floor, gypsum blocks were placed.
For I-OB house, walls had well spread wood wool insulation. A reinforced concrete slab was placed on the basement insulated from the underside by polyurethane. The floor of the first floor and the ceiling were made of wooden beams and wood chipboard. The same finishing touches were done for the house I-BB : tiled floor on the ground floor, covering vinyl PVC -Polyvinyl chloride- on the first floor (except in the bathroom : tiles), ceiling with gypsum blocks and on the walls gypsum board plus paint solvent.

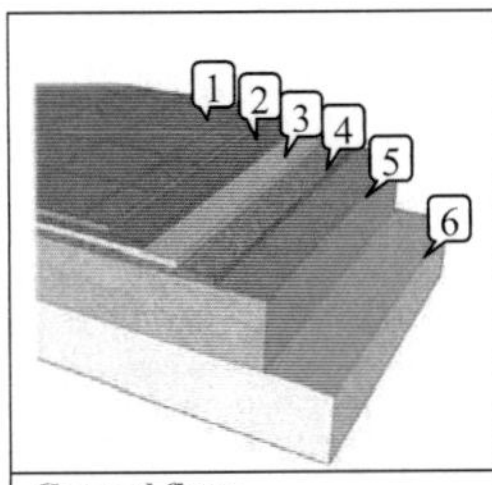

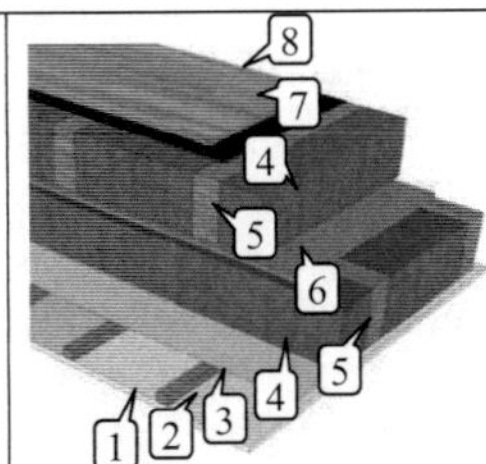

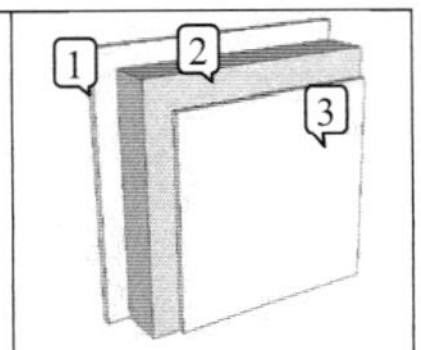

Ground floor	Ceiling of first floor	Walls	Interior Walls
1: Tiled floor + glue (1cm) 2: Gifafloor knauf® (2,5cm) 3: Phenolic foam (3,0cm) 4: Insulation pellets (2,4cm) 5: Reinforced concrete(18cm) 6: Polyurethane (14cm)	1: Gypsum blocks (1cm) 2: Wooden slats (6,4cm) 3: Vapor barrier 4: Wood wool (2 x 24cm) 5: Wooden beam (2 x 24cm) 6: Wooden chipboard (1,9cm) 7: Rain barrier 8: Wood flooring (2cm)	1: Roughcast (0,6cm) 2: Polystyrene (3cm) 3: Wooden chipboard (2,5cm) + Cement 4: Rain barrier 5: Wood wool(22cm) 6: Vapor barrier 7: Wood wool(16cm) 8: Wooden chipboard(1,6cm) 9: Gypsum board (1cm)	1: Gypsum board (1cm) 2: Wood wool (5cm) 3: Gypsum board (1cm)

Table 1. Description of building materials use in I-OB (timber frame).

This study is only considering the pollutant emissions in indoor air due to the first layers of materials and paints (interior faces).

Measurement materials

Carbon dioxide (CO_2) was measured thanks to TSI Q-track 7565 with a probe "IAQ Probe Model 980". It is a continuous recorder. CO_2 is an index of confinement in office buildings. It permits to check the ventilation system efficiency. In this study the protocol didn't allow human presence during the measurement campaigns. Moreover, no sources of burning were present in buildings, the balanced ventilation was stopped during the week of the campaign. The monitoring of carbon dioxide got information about the behaviour of the I-BB and I-OB houses when ventilation system was off, doors and windows closed, no occupant.
Temperature (t_a) and relative humidity (RH) in the air were continuously measured (KIMO KH200 KISTOCK) at the centre of each house room, at a height of 1.2 m directly above the floor close to radial diffusive samplers Radiello®. This information is very important because Total Volatile Organic Compounds (TVOC), Volatile Organic Compounds (VOCs) and aldehydes emissions increase when the temperature increases [11].
So that, to characterise building material emissions, a continuous recorder (multi-gas PID monitor Graywolf with TG502 VOC Probe) was used to measure the global level of TVOC, Nitrogen dioxide (NO_2) and Ozone (O_3). Considering the surface and the number of rooms of each house (I-BB and I-OB), it was impossible to position a TVOC continuous recorder in each room. These measurements were made at the centre of the biggest area of I-BB and I-OB: "kitchen/living room" at a height of 1.2 m above the floor. The PID monitor was positioned on an anodized aluminium desk to avoid chemical emissions due to this table.
In addition, radial diffusive samplers Radiello® were used to measure specific chosen VOCs and aldehydes in each house room (Cf. Table 2). Moreover, radial diffusive samplers were placed on air inlet of the air system supplier so that to estimate the difference of concentration between indoor and outdoor air. Measurements on Indoor Air Quality (IAQ) for specific VOCs and aldehydes in I-OB and I-BB houses have been based on OQAI study "Campaign National Homes: State of Indoor Air Quality in French dwellers"[3]. Pollutants chosen were the same as in this investigation in order to have a magnitude of expected concentrations.
With these specific measurements, we have attempted to identify different building material sources and associated health effects. The concentrations of VOCs and aldehydes were calculated on a sampling period of a

week (the duration of measurement campaigns). All chemical analysis were realised by the Fondazione Salvatore Maugeri. Aldehydes were analysed by High Performance Liquid Chromatography (HPLC) and VOCs by a Thermal Desorption system attached to a Gas Chromatograph with Flame Ionisation Detector (TD/GCFID).

VOCs				Aldehydes	
Hydrocarbons	*SOURCES*	*Glycol Ethers*	*SOURCES*		*SOURCES*
Benzene	glue, interior paint, plasticizer, maintenance products	1-methoxy-2-propanal	interior paint, varnish, fungicide, herbicide, wood treatment, silicone clog	Formaldehyde	wooden panel, fiberboard, crude wood panel, solvent paint, tobacco smoke
Toluene	interior paint, glue, ink, carpet, silicone clog	2-butoxyethyl acetate		Acetaldehyde	tobacco smoke, crude wood panel, wooden panel
Ethylbenzene	fuel, solvent, pesticide	1-Methoxy-2-propyl acetate		Acrolein	burning, tobacco smoke, drippings
m/p-xylene	interior paint, varnish, insecticide	2-butoxyethanol		Propanal*	solvent paint
Styrene	plastic material, heat insulation, fuel, tobacco smoke			Butanal*	wooden panel, parquet treatment
o-xylene	interior paint, varnish, insecticide			Benzaldehyde*	solvent paint, parquet treatment
n-decane	white spirit, floor glue, wax, wood varnish, cleaner, carpet			Isopentanal*	parquet treatment, wooden panel
1,2,4 trimethylbenzene	oil solvent, fuel, tar, varnish			Pentanal*	solvent paint, wooden panel
n-undecane	white spirit, floor glue, wax, wood varnish, cleaner, carpet			Hexanal	wooden panel, solvent paint, wood working, crude wood panel

Table 2. Choice of specific VOCs and aldehydes and theirs relative sources in order to quantify emissions due to building materials (*not taken into account in campaign national homes [3])

Thanks to the investigation based on a French standard in I-BB and I-OB, we noted that there are more potential sources of pollutants into the house I-OB. Because the I-OB house is made of wood, wood wool, wooden chipboard, ..., (Table 1) which are organic materials and therefore VOC emitters. Whereas I-BB has an exterior insulation which doesn't emit pollutant inside the home and its frame is made of cast concrete which is an very low VOC emitters.

RESULTS

Results of carbon dioxide concentration confirm that air inlets are very small. For I-BB and I-OB, the measurement campaigns show a higher CO_2 concentration of 200ppm between the outside and inside air. Moreover, indoor variation of carbon dioxide is not correlated with outdoor concentrations (Figure 2). Furthermore, Figure 2 shows a CO_2 level inside the houses below the background noise led us to put forward two hypothesises. Firstly, there is a CO_2 sink due to indoor materials, chemical reactions,... [12]. Secondly, there is a sedimentation/stratification of the CO_2 concentration because of its gas density (d=1,53v [13]). We expected a gradient concentration between the floor and the ceiling with a higher concentration close to the ground. (Note that a measurement campaign was realised to check this phenomenon but the uncertainty of measurements due to continuous recorders used, was too big to validate this hypothesis).

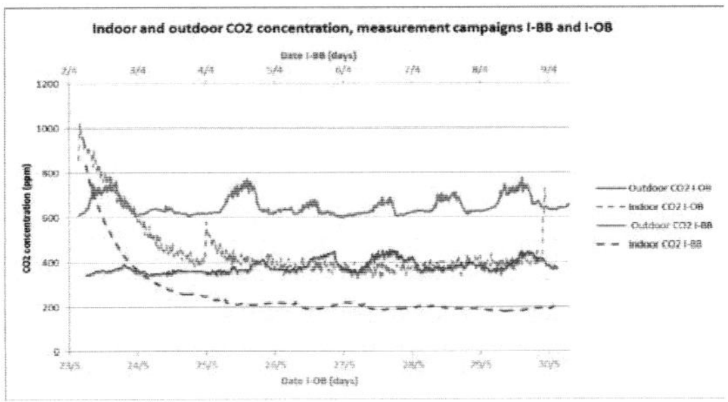

Figure 2 : Variation on a week of indoor and outdoor CO_2 concentration both for I-BB and I-OB.

During measurement campaigns, ozone (O_3) was measured at concentrations below the Detection Limit (DL) of the Graywolf device (DL=40µg/m^3). Outdoor averaged values measured by AASQA[1] 'Air APS' were higher than DL (60.2µg/m^3 for I-BB and 81.9µg/m^3 for I–OB). As a result, low indoor level of O_3 was certainly due to reactions with VOCs [2], building materials and products. Furthermore, impact of photochemical pollution on indoor air quality was studied during the summers of 2003 and 2004 in MARIA CSTB experimental house [14]. About 80% to 95% of the ozone was removed inside the room, showing the presence of major ozone sinks. [14]
Nitrogen dioxide (NO_2) was also measured below the detection value of the Graywolf material. As ozone gas, it is involved in chemical reactions with VOC [12].
In order to have continuous information about the TVOC concentration, a Graywolf PID monitor was used. Its goal was to obtain on the one hand, an historical every ten minutes to show the trends of these pollutant concentrations, and on the other hand, to quantify a global level of TVOC in real time. It was a further approach in regards with radial diffusive samplers Radiello®. This passive method gives an averaged concentration value on the duration of the measurement campaign for each compound targeted. As a result it is impossible to notice different variations of level during nights and days unlike PID Monitor. However the PID monitor doesn't permit to identify an emitting source produced by building materials for a given level of TVOC. This device takes into account a sum of several hundred volatile pollutants (including VOCs and Aldehydes, Table 2).
That is why we used these two different methods to get a global approach in order to identify different sources and associated health effects (when possible).

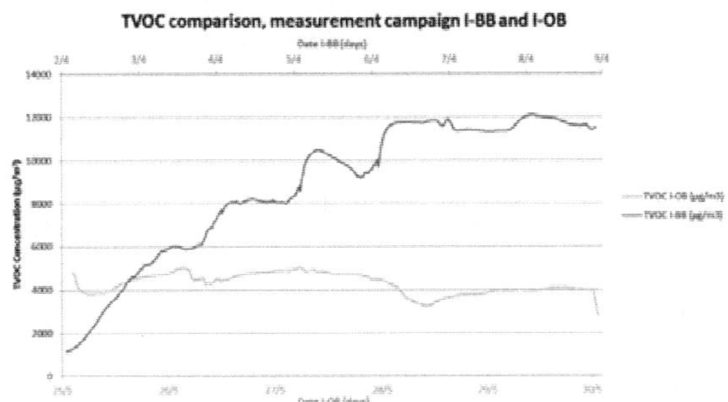

Figure 3. Variation on a week of TVOC in I-OB and I-BB houses.

Figure 3 shows the trends of the emissions of TVOC in I-BB and I-OB houses. The average TVOC concentration equals 4,3mg/m^3 for I-OB against 8.8mg/m^3 for I-BB. These values are very high compared to guideline values found in articles (e.g.: 0.3mg/m^3 in Deutschland [15]).
Even if I-BB TVOC concentration is multiplied by 2 compared to I-OB, the difference between these 2 concentrations might come from several causes: firstly the period of time between the end of the construction and the measurement campaigns is three times longer for I-OB than I-BB (respectively 224 days to70 days), secondly ventilation systems weren't exactly stopped at the same time before the I-OB and I-BB campaigns (2 days before for I-OB and at the beginning of the measurements for I-BB). Consequently, pollutants were more diluted. Unfortunately, it is difficult to conclude about the origin of these high concentrations and difference

[1] AASQA : Association Agréée de Surveillance de la Qualité de l'Air

between these 2 frames at this step of the study because of the average temperature difference (Figure 4) between I-OB and I-BB which equals $\approx 10°C$ ($\overline{t_{aI-OB}} \approx \overline{t_{aI-BB}} + 10°C$). Note that an increase of temperature has impacts on VOC emissions [11].

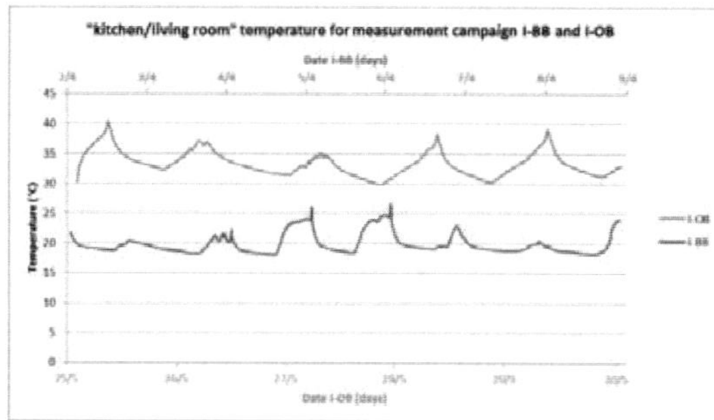

Figure 4. Variation of air temperature (t_a in °C) in I-OB and I-BB houses during the measurement campaigns (a week).

About specific VOCs and Aldehydes, in a first time we were compared the sum of VOCs concentrations (Table 2 : list of VOCs and chosen aldehydes) obtained with passive samplers in each room of I-BB and I-OB. Then the same protocol was realised for Aldehydes (Figure 6).

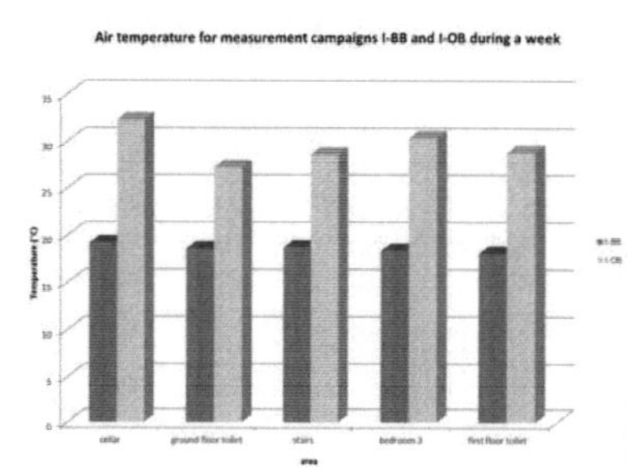

Figure 5. Examples of air temperature averaged on a week (only rooms equipped with air temperature KIMO KH 200).

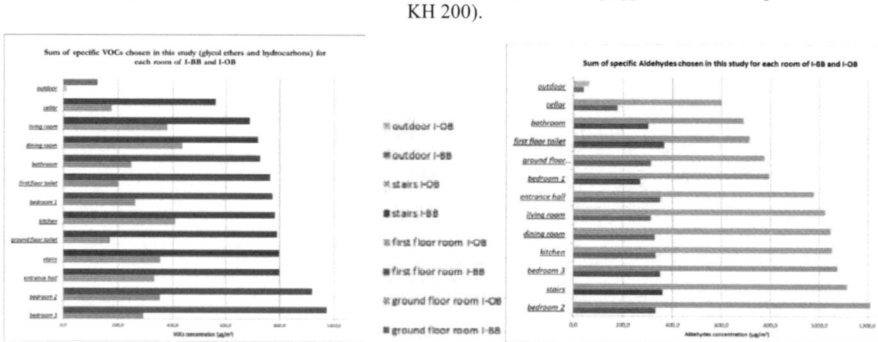

Figure 6. On the left : Sum of specific VOCs targeted, values obtained and averaged on a week, and for each room (I-OB and I-BB) ; On the right: sum of specific aldehydes targeted, values obtained and averaged on a week, and for each room (I-OB and I-BB).

The difference between the sum of the indoor VOCs and outdoor is significant and is multiplied by 7: $[VOCs]_{indoor} \approx 7\ [VOCs]_{outdoor}$. Moreover, the VOCs concentration in I-BB is approximately homogeneous [16] in each

room but it is not the case in I-OB (Figure 6). Furthermore, in I-OB the average concentration is two times as small as I-BB. To finish, except the stairs and the entrance hall in I-OB, we observe that each room with south exposure has a higher concentration in VOCs certainly because of a higher air temperature in these areas [11].
Figure 6 shows again a difference between indoor and outdoor aldehydes concentrations: $[\text{Aldehydes}]_{\text{indoor}} \approx 7\,[\text{Aldehydes}]_{\text{outdoor}}$. However and contrary to the sum of VOCs concentration, the level of aldehydes in I-OB is higher than I-BB.
In a second time, we were detailed the value of measurements of each VOCs and aldehydes during the campaign, in each room of I-BB and I-OB in order to obtain their specific levels and their associated health effects with their associated current guideline values (when they exist).

Results for specific VOCs:

Each concentration of chemical pollutants measured by radial diffusive samplers Radiello® is shown on Figure 7 below. On these histograms it is possible to visualize the variations of specific VOCs in different rooms of I-BB and I-OB houses. Note that the value of concentration is averaged on 7 days (duration of measurement campaigns).

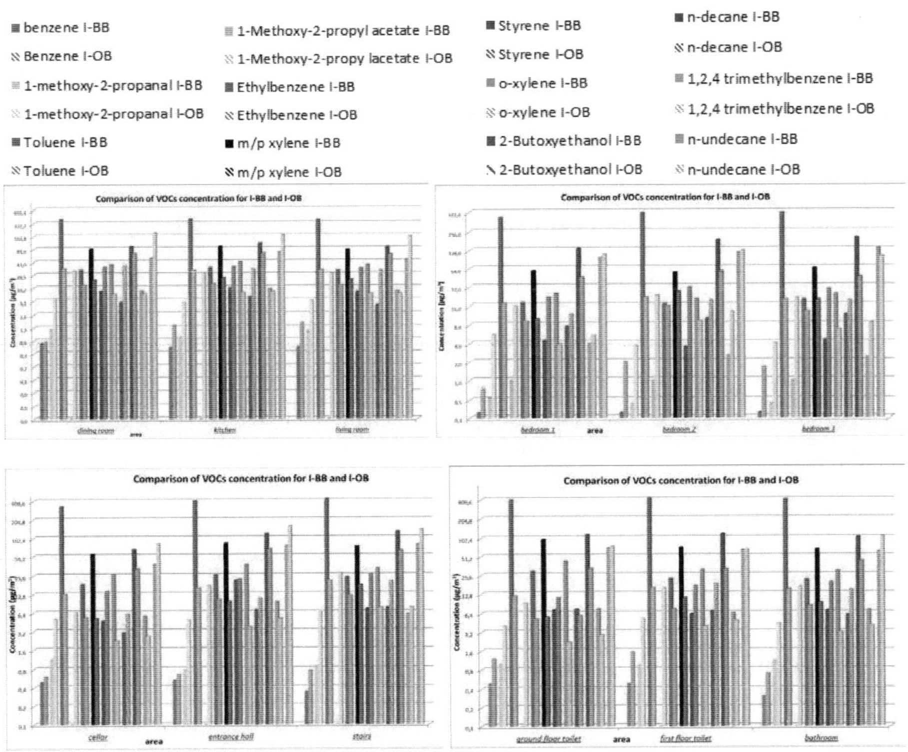

Figure 7. Comparison in each room of I-OB and I-BB houses of specific VOCs concentrations thanks to passive samplers RADIELLO® (averaged values on a week per pollutant).

Glycol ethers (Table 2) had a higher concentration in I-OB than I-BB probably due to woodworking products and fungicides. Moreover, outdoor concentrations about glycol ethers were negligible (between 0 and $0.2\mu g/m^3$). Consequently, we may estimate that emissions were linked to building materials.

	Measurement campaign		WHO	OQAI Values (µg/m³)			Health Effects
	Maximum value **I-BB** (µg/m³)	Maximum value **I-OB** (µg/m³)	Guideline values	Median	90th percentile	maximum	
1-methoxy-2-propanal	1,2	6,5	*No value*	1,9	10,8	170,1	Eyes and breath soreness
1-methoxy-2-propylacétate	<DL (²)	**28,3**	*No value*	<DL	<QL (³)	39,5	Eyes and breath soreness
2-butoxyethanol	**11,4**	**34,2**	*No value*	1,6	5,5	60,6	Hemolitic anaemia
2-butoxyethyl-acétate	<DL	**2,1**	*No value*	<DL	<DL	12,2	Eyes and skin soreness

Table 3. Averaged values of glycol ethers and associated health effects (from measurement campaigns I-OB and I-BB, and OQAI campaigns plus WHO guideline values [17]).

For three Glycol ethers (1 methoxy-2-propanol, 1-methoxy-2-propyl acetate and 2-butoxyethanol), measured concentrations in I-OB were higher than the 90th percentile of the OQAI campaign national homes [3], this means that 90% of measured values in French homes were below the levels found in I-OB. Products of wood processing and fungicides used in building materials (frame, particle boards, raw wood panels, wood wool, ...) were certainly the sources responsible for these high levels.

	Measurement campaign		WHO	OQAI Values (µg/m³)			Health effects
	Maximum Value **I-BB** (µg/m³)	Maximum Value **I-OB** (µg/m³)	Guideline values (µg/m³)	Median	90th percentile	maximum	
Benzene	0,9	2	30(7 days)	2,1	5,7	22,8	Bone marrow affected,
Trichloroethylene	0,1	0,1	20 (chronic exposition)	1	3,3	4087,2	Psycho organic syndrome, skin soreness
Toluene	**504,1**	29,5	260 (7days)	12,2	46,9	414,2	Central nervous system effect, somnolence, headache
Tetrachloroethylene	0,1	0,6	250 (chronic exposition)	1,4	5,2	684,3	Soreness, neurological disease, renal effects, hepatic effects
Ethylbenzene	**31**	**16,2**	*No value*	2,3	7,5	85,3	Lung and liver cancer
(m+p) xylene	**96,4**	**27,4**	4800 (24h)	5,6	22,0	232,8	Headache, giddiness
o-xylene	**43**	**9,2**	4800 (24h)	2,3	8,1	112,3	Headache, giddiness
Styrene	**21,3**	**32,5**	260 (7 days)	1	2	35,1	Eyes and lung soreness
n-decane	**202,1**	**68,3**	*No value*	5,3	29,1	1774,1	No estimate
1,2,4-trimethylbenzene	9,6	13	*No value*	4,1	13,7	111,7	Breath soreness
1,4 dichlorobenzene	<DL	0,1	*No value*	4,2	68,5	4809,8	Lung soreness and cancer
n-undecane	**139**	**199,2**	*No value*	6,2	33,6	502,1	No estimate

Table 4. Averaged values of hydrocarbons and health effects obtained by passive samplers (results of measurement campaigns in I-BB and I-OB, and OQAI campaigns plus WHO guideline values [17]).

Toluene was the higher value of hydrocarbons measured in I-BB house. We can't explain it. This high level was found everywhere in the I-BB house. The average value on 7 days was 504µg/m³ compared to 29.5µg/m³ for I-OB. Currently the WHO guideline value on 7 days is 260µg/m³. Paints, glues and varnishes were certainly the products responsible for these high concentrations. In I-BB and I-OB, five other compounds had high concentrations compared to OQAI study: ethylbenzene, (m + p) xylene, o-xylene, styrene, n-decane and n-undecane.

(1) Detection Limit
(2) Quantification Limit

Results for specific aldehydes:

Each concentration of chemical pollutants measured by radial diffusive samplers Radiello® is shown on Figure 8 below. On these histograms it is possible to visualize the variations of specific aldehydes in different rooms of I-BB and I-OB houses. Note that the value of concentration is averaged on 7 days (the duration of measurement campaigns).

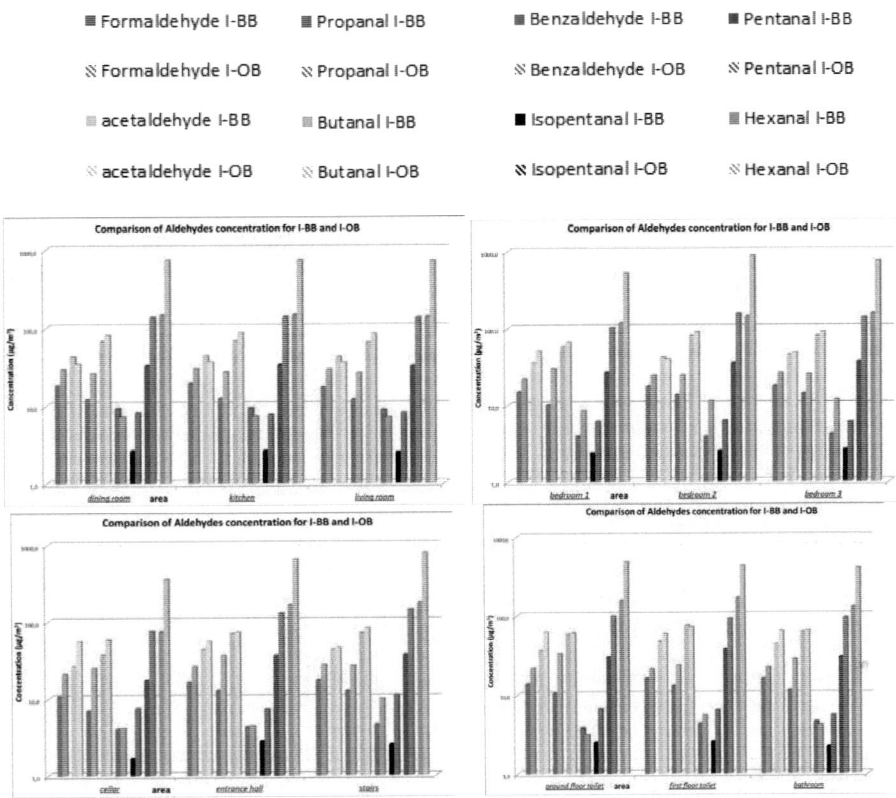

Figure 8. Comparison in each room of I-OB and I-BB houses of specific aldehydes concentrations thanks to passive samplers RADIELLO® (averaged values on a week per pollutant).

The concentration of aldehydes was higher in I-OB than I-BB for all pollutants (excepted acrolein). Note that the aldehydes come from mostly wood materials (raw wood panels, particle boards...), so the higher concentration in the house I-OB makes sense. The hexanal was the pollutant found in greater quantities at each site with an average of 140µg/m^3 for I-BB and 570.7µg/m^3 for I-OB. Bedroom 2 of I-OB had the maximum value. It was twice the maximum measured in French study "Campaign National Homes" by OQAI [3]. It was a very high value. A study[18] revealed that Hexanal represented 21% of indoor air pollutant emissions whereas in I-OB campaign, hexanal represented 60% of the total concentration of pollutant. Currently health risks due to this pollutant are poorly understood but skin irritation, eye and respiratory symptoms are related to hexanal [19]. A recent article [2] confirmed a high presence of hexanal in low energy buildings made of timber frame. No guideline value exist for this compound, however, the national campaign of OQAI [3] indicated a median value of 13.6 µg/m^3, 10 times less than in I-BB and 40 times less than I-OB.

Main sources of hexanal come from particle boards, woodworkings and paint solvents. That is why the wooden staircase of the house I-OB had the highest concentration of hexanal (863.6 µg/m^3).

	Measurement campaign		WHO	OQAI Values (µg/m³)			Health effects
	Maximum Value **I-BB** **(µg/m³)**	Maximum Value **I-OB** **(µg/m³)**	Guideline value (µg/m³)	Median	90th percentile	maximum	
Formaldehyde	**19,1**	**29,3**	50 (2 h) 10 (1year)	19,6	39,9	86,3	Eyes nose and lung soreness, asthma
Acetaldehyde	**46,9**	**63,1**	*No value*	11,6	24,3	94,6	Eyes nose and lung soreness, asthma
Acrolein	<0,1	<0,1	*No value*	1,1	2,6	12,9	soreness
Propanal	14,1	35,9	*No value*	-	-	-	Ocular lesion
Butanal	78	88,7	*No value*	-	-	-	soreness
Benzaldehyde	9,4	11,6	*No value*	-	-	-	Eyes nose and lung soreness, asthma
Isopentanal	2,8	11,0	*No value*	-	-	-	Eyes nose and lung soreness, asthma
Pentanal	37,8	154,9	*No value*	-	-	-	Eyes nose and lung soreness, asthma
Hexanal	**174,6**	**863,6**	*No value*	13,6	35,6	368,5	Headache, eyes nose and lung soreness, asthma

Table 5. Averaged values of aldehydes and associated health effects obtained by passive samplers (results of measurement campaigns in I-BB and I-OB, and OQAI campaigns plus WHO guideline values [17])

CONCLUSION

The protocol of this study was developed to quantify the pollutant emissions due to building materials and products with no balanced ventilation, with closed doors and windows and without any occupant. These two low energy buildings with two different frames (cast concrete and timber frame) were studied in order to determine emission sources, level of chemical pollutants of each dwelling, and associated health effects (when feasible).
Carbone dioxide measurements reveal "strange" values below the CO_2 outdoor background noise. We think that this phenomenon is due to either a sink effect or a stratification of CO_2 between floor and ceiling.
Ozone and nitrogen dioxide are below the detection limit. Perhaps, there is a sink effect due to chemical reactions with VOC gases and building materials for O_3.
The sum of glycol ethers, of hydrocarbons, and of aldehydes has highlighted that concentration of these groups of chemical pollutants is homogeneous enough in each room of I-BB and a bit less than in I-OB. In general rules, we determined that emissions of chemical volatile compounds were higher in Southern rooms of I-OB certainly due to different weather conditions and the average of indoor air temperature ($\overline{t_{al-OB}} \approx \overline{t_{al-BB}} + 10°C$).
Furthermore, both for I-BB and I-OB houses, the indoor VOCs emissions were multiplied by 7 compared to outdoor pollution ($[VOCs]_{indoor} / [VOCs]_{outdoor} \approx 7$).
The studies of two houses show that I-OB had more sources of pollutants and the concentrations of aldehydes and glycol ethers were higher.
As well, we noted a very high concentration value in toluene (504.1 µg/m^3) for I-BB. This pollutant level is almost twice greater than guideline value recommanded by WHO (260µg/m^3 of 7 days). These high concentrations are certainly due to paints, glues and varnishes. This rate might (260µg/m^3) cause headaches, somnolence.... Similar effects have been perceived in I-OB house by engineer but we think that hexanal may have caused these symptoms (toluene average concentration in I-OB was low and below its guideline value). Note that hexanal pollutant was found in each room at a very high level.
However, it is very difficult to associate pollutants emitted by sources to health effects in this article. In fact, a health effect may have several sources and a building material source may have several health effects.
Moreover, current studies realized in I-BB and I-OB show that the ventilation systems when operating drastically remove pollutants (hydrocarbons and aldehydes concentrations were removed up to 75% and 70% respectively for I-BB and I-OB).
Despite the same protocol used for I-OB and I-BB campaigns, it is very difficult to compare emissions arising from these different frames because air temperature is higher in I-OB than I-BB ($\approx 10°C$) and measurements started on the one hand 70 days after the end of construction for I-BB and on the other hand 224 days for I-OB.
To conclude, in these low energy consumption buildings with very little leakage paths, it is very important to use ventilation systems well dimensioned, with good maintenance, in order to ensure the recommended air exchange rate. Currently, energy efficiency is not the only parameter to watch. That is why, to get a global approach relating both occupant comfort and energy efficiency, indexes of Indoor Environment Quality (indoor

air quality, visual comfort, acoustic comfort, thermal comfort) must be taken into account, monitored, regulated, as soon as possible when designing low energy buidings.

REFERENCES

[1] J. P Lucas, O. Ramalho, S. Kirchner, J. Ribéron, June 2009, *Etat de la ventilation dans le parc de logement français,* OQAI. DESE/SB – 2009-037

[2] S. Kirchner et al., 2011, *Qualité d'air intérieur, qualité de vie : 10ans de recherche pour mieux respirer,* OQAI, p°71

[3] S. Kirchner, J. F. Arenes, C. Cochet et al., 2006. *Campagne nationale logements : Etat de la qualité de l'air dans les logements Français, Rapport final,* OQAI

[4] L. Mosqueron, V. Nedellec, 2004, *Revue des enquêtes sur la qualité de l'air intérieur dans les logements en Europe et aux Etats Unis,* OQAI

[5] Dimosthenis, A. Sarigiannis, Spyros, et al. 2010. *Exposure to major volatile organic compounds and carbonyls in European indoor environments and associated health risk,* Environment international

[6] J. M. Logue, T. E. McKone, M. H. Sherman, B. C. Singer, 2010, *Hazard assessment of chemical air contaminants measured in residences,* INDOOR AIR

[7] J. S. Park, K. Ikeda, 2006, *Variation of formaldehyde and VOC levels during 3 years in new and older homes,* INDOOR AIR

[8] M. Mueller, 2009, *Ventilation and indoor air quality in new homes,* California Energy Commission

[9] A. T. Hodgson and H. Levin, 2003, *Volatile Organic Compounds in Indoor Air: A Review of Concentrations Measured in North America since 1990,* LBNL-51715

[10] Normalisation française, 1999, *Audit de la qualité de l'air dans les locaux non industriels,* AFNOR XP X43-403

[11] J. Xiong, Y. Zhang. *Impact of temperature on the initial emittable concentration of formaldehyde in buildings materials: experimental observation.* 16 March 2010, Department of Building Science, Tsinghua University, Beijing, China

[12] *L'impact des COV,* 2005, ADEME

[13] N. Bonnard, M.-T. Brondeau, M. Falcy, et al., 2005. *Fiche toxicologique Dioxyde de carbone,* INRS. FT238.

[14] M. Nicolas, 2006, , *Ozone et qualité de l'air intérieur : Interactions avec les produits de construction et de décoration,* CSTB

[15] L. Schriver-Mazzuoli. 2009, *La Pollution de l'Air INTERIEUR. DUNOD,* P°104. (Environnement et Santé)

[16] L. Mosqueron, V. Nedellec, October 2004, *Revue des enquêtes sur la QAI en Europe et aux Etats-Unis,* Observatoire de la Qualité de l'Air Intérieur, p°20.

[17]K. Charles, R.J. Magee, D. Won, E. Lusztyk, 2005, *Indoor Air Quality Guidelines and Standards,* National Research Council Canada.

[18] A. Mahevas, J. Grall, M. Corron, September 2009, *Etude de la qualité de l'air dans les locaux du conseil général des Côtes-d'Armor.* Air Breizh

[19] *Air Intérieur Impact sanitaire, see website,* AIRAQ Atmo Aquitaine.

Impact of the filtration system on the indoor-outdoor particles concentration relationships in an air conditioned office building

Eva Rendek[1*], Alain Ginestet[1], Dominique Pugnet[1]

1 CETIAT
BP52042,
69603VILLEURBANNE cedex
FRANCE
**Corresponding author: eva.rendek@cetiat.fr*

ABSTRACT

The objective of this study is to measure the impact of various classes of filters in the HVAC system on the IAQ in an office building. It aims at understanding outdoor-indoor concentrations and filtration relashionships in order to guide filtration system designs.
Various filters classified G4, F7 and F9 (according to the standard EN 779), new and used, were placed in the HVAC system of an office building ventilated with 100 % outdoor air and located in the suburb of Lyon (southeast France). Particle size distributions were measured outdoors, indoors and in the return duct of the ventilation system (balanced ventilation system).
These data were used to investigate filtration efficiency and relationships between indoor and outdoor particle concentrations.
Results show a dramatic reduction of fine particle using high efficiency filters (F7, F9). As for larger particles (> 1 µm) no significant improvement of IAQ is measured, indoor air can even present higher concentrations of large particles due to the presence of various indoor sources.
A high efficiency filtration system can then be used sucessfully to reduce indoor exposures to fine particles but its impact on total particles concentration is limited. This solution has to be taken into account and integrated in a global approach for a better IAQ.

KEYWORDS
Filters, particles, indoor air quality

INTRODUCTION
Indoor air quality and particles exposures are today of major concern in building systems designs.
Concentrations of particles indoors depend upon the fraction of outdoor particles that penetrate through the building shell or are transported via the air handling system, the generation of particles by indoor sources, and the loss mechanisms that occurs indoors such as filtration or deposition [1-6].
At the interface between indoor and outdoor air, filters play a significant role. As for the regulation concerning filtration systems, they are today requirements for filtration levels on fresh and recycled air (French regulation : filtration of fresh air with filters classified G4 at least and air recycled with filters classified F5). However regarding to indoor air quality no specific particles concentrations or specified or required.
This study aims at investigating filtration efficiency and relationships between indoor and outdoor particle concentrations in order to guide filtration system designs.

MATERIEL AND METHODS

Building

The office building (figure 1) is located in an industrial area, close to a beltway. Only the first floor is instrumented. The area is 450 m2 and includes an open space, various offices, a kitchen and lavatories.

Figure 1 : Outside view of the office building

Figure 2 : Air supply device

HVAC system

The air handling unit is equipped with an energy recovery system (balanced ventilation system). It operates 24h/24h with an air flow of 1000 m^3/h. There is no recycled air inside the building.

Ventilation fans are equipped with electronic variable speed transmission but there is no automatic regulation based on air pressure drop.

Fresh air is supplied through ceiling mounted fan coils equipped with G3 filters. They are used for air conditioning and heating and operate with 100 % recycled air (figure 2).

Filters

Filters classified G4, F7 and F9 (according to the europena standard EN 779) are alternatively tested on the air handling unit. Front filters dimensions are 305 mm x 500 mm x 48 mm.

The filter is set on the air handling unit and the system operates one full night to renew the indoor air. Measurements are carried out during the day after.

Measurements

Initial filter efficiencies are first measured on a test rig accordind to the EN 779 standard. The measurments are then carried out on site. Table 1 summarizes all the measurements carried out and the location of the measurement devices.

Measurement	Location
Air flow rate	Supply air duct Air outlet duct
Particle size distribution and concentrations	inlet air before filter inlet air after filter indoor (office) outlet air

Table 1: locations of measurement devices

Particles concentrations are measured five times a day at 8 a.m, 10 a.m, 12 a.m, 2 p.m. and 4 p.m. At each hour 3 sets of measurements are averaged to obtain the result.

RESULTS

Air flow rate

Table 2 presents the results of the air flow measurements in the ventilation ducts. Results show that despite the different pressure drops due to the filters, the air flowrate is monitored to remain almost constant .

filter	G4		F7		F9	
	air inlet	air outlet	air inlet	air outlet	air inlet	air outlet
air flow rate (m^3/h)	1075	114	1057	1152	1039	1124

Table 2 : Air flowrates in the ventilation system

Filters efficiency

Table 3 show the efficiency of the different filters measured in laboratory (accordind to the EN 779 standard test) and on site, in the supply air duct.

	G4	F7	F9	G4	F7	F9
particle diameter (µm)	efficiency (%) (EN 779 test)			efficiency (%) (on site)		
0.2-0.3	2	67	84	6	54	69
0.3-0.5	2	74	89	10	64	77
0.5-0.7	4	80	92	15	73	84
0.7-1	6	86	95	21	80	87
1.2	11	90	97	39	86	89
2.3	38	95	98	64	90	90
3.5	74	98	99	74	94	91

Table 3 : Filters efficiency in the supply air duct, flowrate 1000 m^3/h

The effiency obtained on site in the suplly air duct is lower then the efficiency measured on test rig for with the ashrae dust for the F7 and F9 filters and higher for the G4 filter. This is partly due to the difference between the ashrae dust and the atmospheric aerosol. Results match however the different efficiency classes.

Indoor/Outdoor particles concentration ratio

Figure 3 show the indoor/outdoor particle concentration ratio obtained for each filter.
On site results show a dramatic reduction of fine particle (<1µm) using fine filters.
The indoor/outdoor ratio is below 60 % with a G4 filter, 42 % with a F7 filter and 35 % with a F9 filter. The indoor/outdoor concentration ratio decreases as the class of the filter is better.
As for larger particles (> 1 µm) no significant improvement of IAQ is measured. Indoor air can even present higher concentrations of large particles due to the presence of various indoor sources.

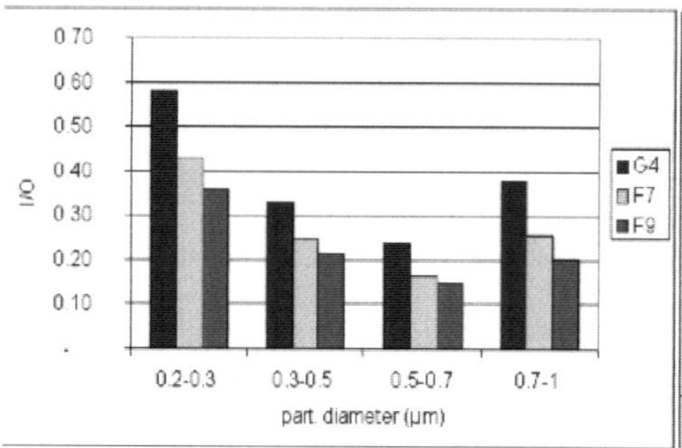

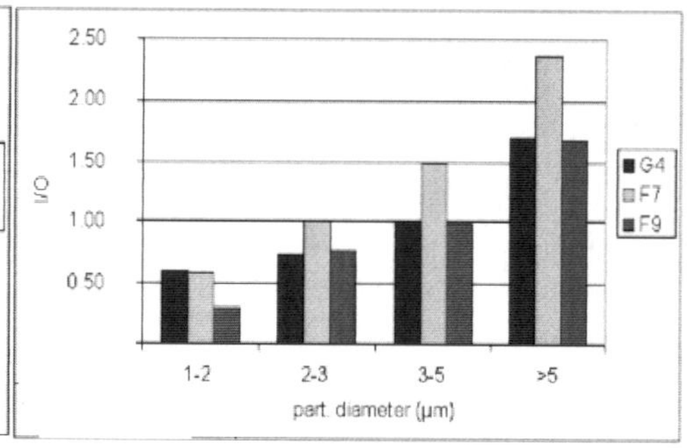

Figure 3 : indoor/outdoor particle concentration ratio

CONCLUSION

Results show that high efficiency filtration systems can then be used successfully to reduce indoor exposures to fine particles but its impact on larger particles concentration is limited. This solution has to be integrated in a more global approach for a better IAQ where indoor sources are also taken into account.

REFERENCES

[1] Howard-Reed, C. et al. 2003. *Effect of ventilation systems and air filters on decay rates of particles produced by indoor sources in an occupied townhouse.* Atmospheric Environment 37, 5295-5306

[2] Thatcher, T.L et al. 2001. *Factors affecting the concentration of outdoor particles indoors (COPI) : Identification of data needs and existing data.* Lawrence Berkeley National Laboratory Report n° 49321

[3] Eileen Abt et al. 2000. *Relative contribution of outdoor and indoor particle sources to indoor concentrations.* Environmental Science and Technology, 34, 3579-3587

[4] Partti-Pellinen K. et al 2000. *Penetration of nitrogen oxides and particles from outdoor into indoor air and removal of the pollutants through filtration of incoming air.* Indoor air 10 126-132.

[5] Fisk W.J, Faulkner D. 2003. *Particle air filtration in HVAC supply-air streams – HPAC Engineering* 24-36.

[6] Croxford C. et al. 2000. *A Study of local electrostatic filtration and main pre-filtration on airborne and surface dust levels in air-conditioned office premises.* Indoor Air, 10, 170-177.

MEASURED PERFORMANCE OF THREE TYPES OF ENERGY PROGRAM HOUSES IN TWO U.S. CITIES

Shaun Hassel[1], Jonathan Coulter[1]*, Ben Hannas[1]

[1]Advanced Energy
909 Capability Drive, Suite 2100
Raleigh, North Carolina 27606, USA
*Email: jcoulter@advancedenergy.org

ABSTRACT

This paper examines two similar studies in Phoenix, Arizona and Houston, Texas that compared the actual energy performance of three classes of homes – ENERGY STAR®, Guaranteed Performance homes and Baseline homes (homes built using standard construction practices and local energy codes at a minimum). In addition to the specific questions of individual residential energy performance, the results illuminate the impact that energy efficiency programs have on the overall building marketplace. The Phoenix study surveyed 7,141 houses, of which 3,336 were Baseline homes, 2,979 were ENERGY STAR homes and 826 were Guaranteed Performance homes. Statistically valid energy data show that ENERGY STAR homes are up to 12% more efficient (kWh/m^2) as compared to the typical Baseline homes. The Guaranteed Performance homes were up to 23% more efficient than Baseline homes and 12% more efficient than ENERGY STAR homes. The Houston analysis draws from over 158,698 houses, of which 70,828 were Baseline homes, 81,755 were ENERGY STAR homes and 6,115 were Guaranteed Performance homes. Key findings from this study include:

- Average Baseline homes consumed less energy than the reference homes defined by the modeling program.
- All homes in Houston have become more energy efficient over time - 16% from 2002 to 2007.
- Energy usage differences between the three groups of homes were small.
- Modeling predictions of the energy usage of ENERGY STAR homes are reasonably accurate.
- Regression modeling provided some more detailed results on code influences and construction practices.

An important difference between the Houston study and many others is the robust analysis of real-world data and billing histories rather than models. The data set used here is large enough to have adequate statistical power for high confidence in the results. Finally, evaluating home programs with real-world data allows us to best identify construction techniques and products that deliver truly energy-efficient buildings.

KEYWORDS

Predicted energy usage, actual energy usage, market transformation, energy efficiency programs

INTRODUCTION

For more than 40 years in the United States, a variety of approaches have been tried to improve the energy efficiency of newly-constructed homes. Before 2004, millions of homes had been constructed to local building codes, about 400 000 were ENERGY STAR compliant and over 60,000 qualified for Guaranteed Performance recognition. For all the attention on projected and deemed energy savings these programs were claiming, there was not enough data being analyzed to determine the actual energy reduction impact these programs were having post-occupancy.

In 2004, the United States had 4.6% of the world's population accounting for 24.9% of the world's primary energy consumption. The housing sector accounted for 36% of all U.S. electrical demand with a predicted 39% growth in this sector alone between 2000 and 2010 [1].

Over the past several decades, rising energy prices have driven a demand for more energy-efficient homes. Builders initially responded with simple energy-saving remedies such as increased insulation, double-paned glass, tighter door seals, window awnings and other measures. Recent field applications of building physics advancements (such as high-efficiency HVAC equipment, improved duct sealing, building infiltration barriers, Low-E glass and compact fluorescent lighting) have continued to offer more sophisticated and effective methods of providing predicted energy savings. Each of these measures reduce overall home energy bills in computer modeling, but little is known how these changes effect the homeowners as they live in these homes?

This report documents the methodology and findings of two similar studies: Phoenix Measuring Public Benefits From Energy Efficient Homes study and Houston Home Energy Efficiency study. Both studies were managed and conducted by Advanced Energy with data analysis performed by Michael Blasnik and Associates. The purpose of the studies was to assess and compare energy consumption patterns of homes built to three different energy efficiency standards – Baseline homes, ENERGY STAR homes and Guaranteed Performance homes [2] [3].

In an effort to promote energy-efficient new homes and reduce the emissions associated with home energy use, the EPA launched the ENERGY STAR qualified new homes program in 1995. The program established guidelines for building energy-efficient buildings and developed partnerships with homebuilders to construct energy-efficient homes. To qualify for labeling as an ENERGY STAR home, construction plans and building components must meet specific criteria for energy performance and be certified by a qualified third-party Home Energy Rater. Two methods can be used to assess predicted energy consumption: computer energy simulation modeling or prescriptive construction standards approved by the EPA[1]. In Houston, nearly 100 percent of the ENERGY STAR homes built are modeled with software to demonstrate they will meet the ENERGY STAR guidelines. This computer modeling produces a HERS (Home Energy Rating System) score that indicates the predicted energy performance of the home as compared to a reference home built to the appropriate version of the IECC or local energy code, whichever is more stringent.

[1] http://www.energystar.gov

More recently, several organizations – Masco Corporation with their Environments for Living® program[2], General Electric with their homes inspired by ecomagination℠ program[3], Tucson Electric Power with their Guarantee Home program[4] and Advanced Energy with their *System**Vision***™ program[5] – have been promoting the construction of guaranteed performance homes. These homes are designed to go a step beyond the ENERGY STAR program, using advanced building science materials and techniques to lower home energy use even further. For guaranteed performance homes, the standards and testing protocol are more stringent than ENERGY STAR to ensure predictable energy performance.

To offset the slightly higher cost of these guaranteed performance homes and enhance their marketability, the builders or program administrators guarantee that the annual energy usage for heating and cooling the home will not exceed a modeled annual amount. Any excess costs for heating and cooling energy use are reimbursed to the homeowners. The programs also include a comfort guarantee that compliments the heating and cooling usage guarantee, stating that any room in the home will be within three degrees of the thermostat set point. To date, more than 130,000 houses nationwide have been built and certified to the guaranteed performance standards (Masco, GE, *System**Vision*** and Tucson Electric Power).

Historically, billing data for Baseline, ENERGY STAR and Guaranteed Performance homes has not been collected and analyzed en masse to determine how the homes have performed while occupied under real-world conditions. A handful of studies in Wisconsin [4], New York [5] and Arizona [1] [6] have analyzed actual energy bills in an effort to evaluate the performance of various new home energy standards. The analysis of these studies showed an interesting trend: smaller savings than anticipated between ENERGY STAR and Baseline homes, primarily driven by inaccurate assumptions about the reference homes that ENERGY STAR homes are compared to when predicting energy savings.

While past studies have shown that varying amounts of energy savings were being achieved across distinct programs with different standards, there have always been a number of issues relating to study size, composition of the construction market and the difficulty of acquiring a truly representative data set in terms of types of homes and scope of data. These issues have revealed an ongoing need to conduct additional studies utilizing real-world data that meet the following criteria: develop a data set representative of all of the homes built to different standards in a given market and randomly select an unbiased and statistically significant number of homes in each subset of homes. The Phoenix Measuring Public Benefits From Energy Efficient Homes study was small, but the Houston Energy Efficiency Study was able to develop a much more robust data set for looking at energy efficiency influences.

Study Design and Methodology

Because the Phoenix, Arizona market was an early adopter of both the ENERGY STAR and Guaranteed Performance programs, there is a high concentration of homes with many years worth of energy use data that provided an excellent opportunity to verify energy consumption data on the three home types under real-world conditions.

---

[2] http://www.environmentsforliving.com/
[3] http://ge.ecomagination.com/index.html
[4] http://www.tucsonelectric.com/Green/GuaranteeHome/index.asp
[5] http://www.systemvision.org/

Metropolitan Houston has become one of the largest markets in the country for new housing construction, with more than 350,000 new home starts since 2000. Houston was also an early adopter of the ENERGY STAR label and currently has one of the highest market shares in the nation with over 50 percent of new homes in 2008 certified as ENERGY STAR. As the market share of ENERGY STAR homes grew in Houston, some builders began to look for new ways of differentiating their homes. The Guaranteed Performance labels from various organizations provided them with the opportunity to take a step beyond ENERGY STAR in terms of energy performance, but without having to sacrifice many of the benefits they received from their participation in the ENERGY STAR program.

Study Objective

The Phoenix and Houston studies were structured to compare the actual energy efficiency of Baseline homes, ENERGY STAR qualified homes and Guaranteed Performance program homes, while taking into consideration a large number of variables in home design. These studies looked at real data and real energy performance of thousands of occupied houses, not computer models. The results could then be used to answer several fundamental questions about the effectiveness of these efficiency programs:

- How much energy did the Baseline, ENERGY STAR and Guaranteed Performance homes actually consume?
- How much energy savings are realized by ENERGY STAR and Guaranteed Performance homes, compared to similar Baseline homes?
- Has the implementation of energy efficiency programs in new home construction resulted in a reduction of total energy consumption?
- Did the energy-efficient homes perform as predicted by the models and program sponsors?
- What construction techniques, components and strategies delivered efficiency?

Survey Design

For the 7,141 houses included in the Phoenix energy study, data was compiled and analyzed based on the following three categories: Baseline homes (3,336 homes not built as part of any energy efficiency program, but resembled other homes in the study), ENERGY STAR homes (2,979 homes built per U.S. EPA ENERGY STAR program standards) and Guaranteed Performance Homes (826 ENERGY STAR homes with additional energy efficiency improvements, as well as a comfort and heating/cooling use guarantee). Once assigned to a category above, the homes were then segregated by builder, year built, square footage, presence of swimming pool, solar orientation, HVAC type and zip code. These groupings helped identify patterns in the data that can point to factors with the greatest effect on home efficiency within the boundaries of the study.

More than 226,000 homes built by dozens of different production building companies in Houston from 2002 to 2007 were included in the Houston Home Energy Efficiency Study: 114,000 Baseline homes, 106,000 ENERGY STAR homes and 6,600 Guaranteed Performance homes. Details on the physical design and construction of the various homes (such as HVAC ratings, window size, types and volume) were obtained from home builders, utilities, contractors and testing companies. Energy use histories for the homes were provided by the local utility provider over the period of 2002 through 2008. One significant factor not addressed by this study is the impact of the energy consumption habits (lifestyle) of the home occupants on overall energy use. To account for this variability, the authors used a statistically large sample of homes to diffuse the impact of the lifestyle variable on the results of the study. It is assumed that the range of homeowner behavior is equally represented across all three categories of homes.

ANALYSIS AND RESULTS

Electric Use Survey Results

Of more than 7,000 original Phoenix study homes, acceptable electric-use records were only available for 6,545 homes. Similarly, 86 attached townhouses and three houses without information on living area were eliminated, leaving 6,480 houses with electric data suitable for the analysis. The Electric Use Summary table displays electric use by home category. The table also provides a breakout by heating fuel type for the total heating and cooling usage per square foot. All-electric homes use heat pumps and electric hot water, while gas heated homes also use gas hot water.

The Baseline homes, as predicted, have the highest overall energy use intensity (kWh/ft²). However, the ENERGY STAR homes actually showed slightly higher summer/cooling load intensity than the average Baseline homes. The Guaranteed Performance homes averaged roughly 10% lower energy use than the Baseline homes. These simple comparisons are useful in a "big picture" sense. The presumably more efficient homes use about the same total amount of electricity as the Baseline homes because they are larger. These comparisons do not represent a fair assessment of the energy performance of the different homes, as many other factors besides square footage may differ between the home groups and have an effect on energy use, particularly the choice of heating/hot water fuel [2].

The regression modeling yielded a number of conclusions in the Phoenix study:

- Guaranteed Performance homes have significantly lower summer/cooling intensity than Baseline homes. When adjusted to a standardized 1,800 ft² home, the estimated savings equals about 1,000 kWh/year or about 16% of the load.
- ENERGY STAR homes have about the same summer/cooling intensity as Baseline homes – the small difference ranging from -1% to 2% is not statistically significant.
- Larger homes have lower cooling intensity (as we found previously).
- Two-story homes have higher cooling intensity than one-story homes even after accounting for living area.
- Baseload electric usage constitutes about 0.15 kWh of additional summer/cooling load for each 1 kWh of annual baseload electric usage. This baseload electric impact is substantial – equal to about 1,300 kWh of summer/cooling for the average baseload usage of 8,571 kWh in the analysis sample – about 20% of the entire summer/cooling load.
- Homes facing northeast have significantly lower summer/cooling intensity than homes facing east (the default category), but no other orientations show statistically significant differences.
- Annual summer/cooling loads estimated to average 6,413 kWh for Baseline homes, 6,493 kWh for ENERGY STAR® (1% savings) and 5,409 kWh for Guaranteed Performance homes (16% savings).

Perhaps the most surprising outcome of the analysis of the Houston Energy Study is the fact that electricity consumption in new homes in Houston dropped dramatically for all three groups built from 2002 to 2007. Figure 1 below shows the 2008 total energy use and baseload energy use for homes built in different years. Both values trend down by construction, as can be seen by the average trend line. The total energy use decreases on average by 16 percent from homes built in 2002 to homes built in 2007.

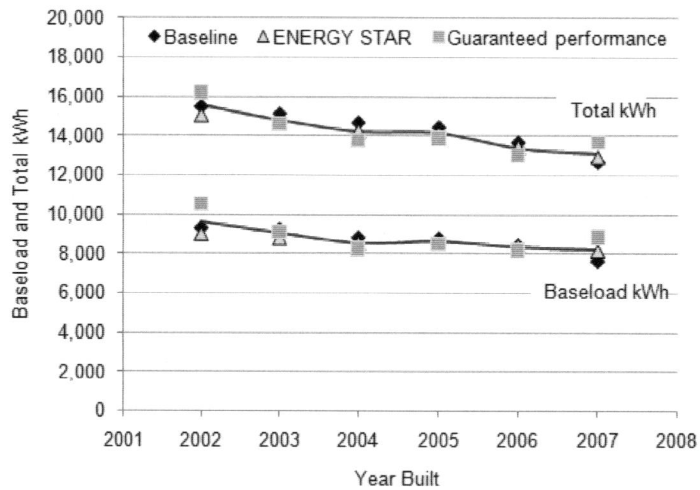

Figure 1: 2008 total and baseload kWh usage by year of construction

This 16 percent drop in total electricity consumption appears to be explained by three factors:

- Establishment of a statewide residential energy code in 2001
- Change in federal standards from SEER 10 to SEER 13 in 2006
- Influential affects of high-performance home programs and initiatives adopted throughout the Houston market, including but not limited to:
    - Programs and incentives
    - Training and technical support
    - Home energy rater infrastructure
    - Consumer marketing
    - Support from product manufacturers

The entire residential new construction marketplace cooperated towards a common goal of reducing energy use across ALL homes. The change in federal SEER standards appears to have accounted for approximately half of the reduction in cooling usage, while the degree of impact from other code changes and spillover effects from the ENERGY STAR program cannot be determined.

These data reveal that all homes in Houston experienced this drop in electricity consumption, and that differences in overall usage and summer/cooling usage (the best usage for comparisons in a cooling dominated climate like Houston) across different groups of homes was small. The summer/cooling load of Baseline homes declined by 18 percent over the period – from 6,194 kWh for homes built in 2002 to 5,068 kWh for homes built in 2007. Over the same period, ENERGY STAR homes dropped by 21 percent and Guaranteed Performance homes dropped by 14 percent. Figure 2 below shows the 2008 summer/cooling energy use for homes built in different years.

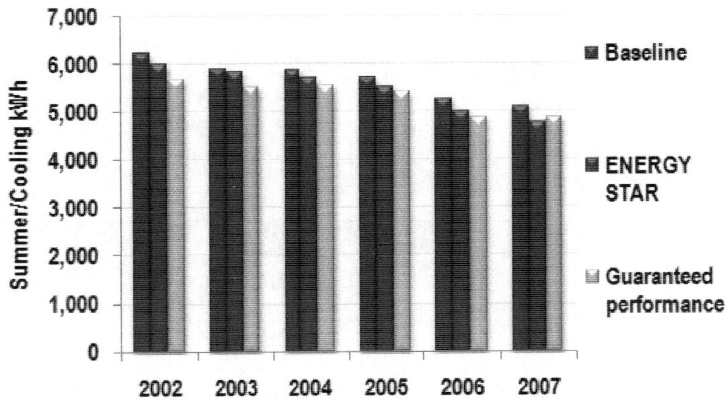

Figure 2: Trends in 2008 summer/cooling usage by year of construction

Although consumption differences across groups of homes are smaller than advertised, ENERGY STAR homes perform very close to the predictions of the models as a whole, while Baseline homes perform better than the reference homes defined by the HERS standard.

Market Impacts

The ENERGY STAR program brought duct leakage testing and building envelope leakage testing into widespread use in the new construction market in Houston. This testing is likely to have contributed toward the common use of better duct installation and building framing practices so that homes could pass the ENERGY STAR testing requirements. Contractors then applied these same approaches to all new homes. This phenomenon is referred to as market transformation or "spillover." The impact of these changes is that Baseline home performance improved, narrowing any observed difference in energy usage between the ENERGY STAR homes and the Baseline homes. Therefore, market transformation effects can make a program appear to have less impact, when in reality it may be having a bigger impact.

Although we are unable to measure the impact of market transformation on the findings, it is clear that, in Houston, typical construction practices are much better than the reference homes as defined by the ENERGY STAR standards. The reference home is defined as having minimum local code specifications combined with the least efficient cooling, heating and water heating equipment available, a building envelope with high infiltration and a leaky duct system. Typical new construction clearly exceeded this level of performance even before the code change, as higher SEER air conditioners were common.

Accuracy of Modeling Predictions

The data collected in this project allowed the study team to examine the relationship between actual and projected energy usage. In this application, the primary quantity of interest is the projected summer/cooling load since baseload usage depends strongly on post-occupancy appliance acquisitions and behaviors. Also, heating loads are small in this market and quite sensitive to behavioral preferences.

Utilizing REM/Rate cooling load projections from 10,258 homes with electric usage results, the study team found that the REM/Rate projected average cooling load of 5,506 kWh/yr was three percent higher than the billing analysis average cooling load of 5,677 kWh/yr. REM/Rate also estimated the average heating usage of program homes fairly well – only four percent lower than the measured loads.

Cooling Load Projections and Usage	
Average Load (kWh/yr)	
REM/Rate estimate	5,506
Billing data	5,677
Difference	171 (3%)
Absolute Error	
Mean	1,235 (21.8%)
Median	992 (17.5%)
% Homes where REM/Rate within	
10% of billing data	28%
25%	64%
50%	91%
Correlations with billing data	
REM/Rate	0.62
Floor area, envelope area	0.67

Table 1: Statistical comparison of the REM/Rate projects and billing analysis results

Although the analysis found no systematic bias in the REM/rate cooling projections, there was a large amount of variability in the data. It was found that the correlation was higher between house size and cooling load than between the REM/Rate projected cooling load and actual usage. However, the study team feels confident in stating that when using current modeling software with energy efficient new homes, there is a strong and fairly consistent relationship between actual and projected performance using REM/Rate for both heating and cooling.

Further Analysis – Regression Modeling

Simple comparisons of energy usage between groups of homes can be informative, but more sophisticated analyses are needed to disentangle the impacts of multiple factors operating at once.

Regression modeling is used to assess the differences in energy usage over time and between groups. A regression modeling approach of homes with REM/Rate file data is also used to explore some technical performance issues. Some of the following results could prove useful for designing programs and determining priorities in terms of technical standards:

- Savings from higher SEER air conditioners are generally consistent with simple projections based on the SEER ratings, although perhaps declining a little for SEER 15 units – this can be seen in Figure 3 below.
- About two-thirds of the reduction in summer/cooling loads from 2005 to 2007 can be accounted for by changes in SEER ratings.
- Building shell leakage appears to increase summer/cooling loads by about 0.4 kWh per CFM50 of leakage, accounting for about 14 percent of summer/cooling loads in ENERGY STAR homes.
- Baseload electric usage is strongly related to summer/cooling loads - at 0.13 kWh cooling per annual kWh consumed, about 1150 kWh (20 percent of cooling load) is removing baseload heat.

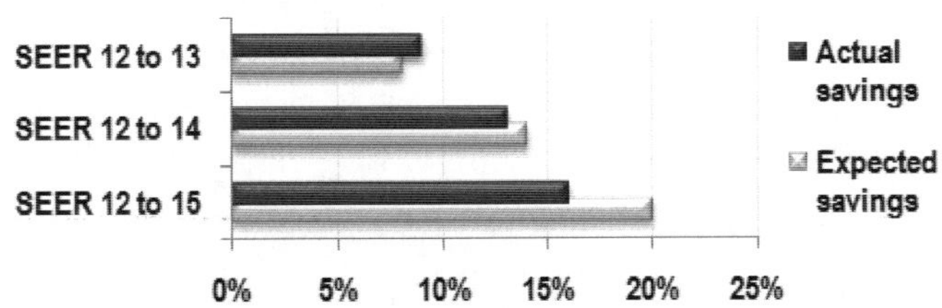

Figure 3: Actual versus predicted SEER savings

CONCLUSION

Market Transformation Impacts

A number of different factors have helped to transform the Phoenix area housing market in terms of energy performance. When the ENERGY STAR program entered the Phoenix market area and gathered a number of avid supporters, it raised the energy performance bar for *all* housing in the market, since few people wanted to purchase an energy *inefficient* home, given the alternatives. Consumers are requesting energy efficient features because programs such as ENERGY STAR and Environments For Living® have educated them. Another major market transformation occurred when the HVAC contracting company that installs more than 70% of all HVAC systems in Phoenix required duct sealing training for all of its installers. In doing so, they initiated a significant change in the entire market which resulted in a steady decline in duct leakage performance numbers. Similarly, training in other energy performance enhancements brought testers, builders and contractors up to speed quickly.

As Low-E glass and higher efficiency HVAC units were incorporated into early ENERGY STAR homes, the price for these products dropped noticeably, due mostly to increased market penetration and competition for supply. Lowered prices equate to more frequent requests by homeowners for efficient glass.

ENERGY STAR and Guaranteed Performance programs brought the concept of "right-sizing" into the market, leading HVAC system designers to use more sophisticated software, such as Manual J, to size HVAC units for homes. This has resulted in smaller units being installed, saving the homeowner money over purchasing a larger unit, while preventing short cycling and allowing the HVAC units to reach their maximum SEER efficiency. Typical HVAC sizing in Phoenix previously was 400 sq. ft. of livable floor space per ton. This figure has increased, as more attention is being given to improved thermal performance.

From the Houston study, savings between program homes and Baseline homes are small. The data indicates that ENERGY STAR does deliver savings in Houston – but the amount of savings is small. When looked at over a five-year period the Houston ENERGY STAR homes used approximately five percent less energy for summer/cooling than a similar group of baseline homes. Guaranteed Performance homes used six percent less summer/cooling energy than Baseline homes in Houston.

It is important to clarify that these results do not mean ENERGY STAR homes are using more energy than predicted. On average, ENERGY STAR homes perform very close to the predictions of the models, but Baseline homes perform much better than the reference homes defined by the HERS standard. The better-than-code construction practices of Baseline homes reduced the difference between ENERGY STAR and Baseline homes substantially.

ENERGY STAR has played an important role in positively impacting standard construction practices and energy savings in residential buildings. For example, the ENERGY STAR program brought duct leakage testing and building envelope leakage testing into widespread use in the new construction market in Houston. This testing is likely to have contributed toward the common use of better duct installation and building framing practices so ENERGY STAR homes would pass the testing requirements. Contractors then applied these same approaches to all new homes. This phenomenon is referred to as market transformation or "spillover." Although we are unable to measure the impact of market transformation on the findings, it is clear that market transformation has taken place in Houston and resulted in very positive benefits to consumers and also delivered electricity savings.

The Phoenix Study Produced Statistically Valid Results

At the beginning of this project, there was much skepticism around the validity of conducting a study to compare homes across the three categories we selected (Baseline, ENERGY STAR and Guaranteed Performance). The concern was that the amount of variability due to factors that have nothing to do with the programs we were studying or that could not be controlled would mask any noticeable differences. However, after having completed this study, and confirming that there is tremendous variability, we still see statistically significant differences among the three categories.

While we have accumulated a body of evidence which indicates that the programs are a driver of these savings, the data should not be viewed as proof. We recognize that there are many issues, such as the exact amount of savings cannot be proved, but we now have a jumping point for further investigation and benchmarking. **The bottom line is that this kind of study can produce valid results, and those results will be strengthened with additional data.**

Statistically valid energy savings were found for both the ENERGY STAR and Guaranteed Performance homes, when compared to Baseline homes. One surprise was how well the Baseline houses performed, but it is our belief that this is partly due to the impact made on the marketplace by the ENERGY STAR and Guaranteed Performance programs. Obviously, savings are directly related to how far a builder/contractor pushes specifications (toward energy efficiency) and improves installation. A major catalyst for this push (market transformation) is programs such as ENERGY STAR and Guaranteed Performance, and all the activities that go into supporting them.

Homes in Houston are More Efficient

The usage data indicated that new homes in Houston have become considerably more efficient in terms of summer/cooling loads for homes built over the period 2002 through 2007. Total electricity usage declined by 16 percent on average while cooling loads dropped by 18 percent. This drop in electricity consumption appears to be explained by three factors: the establishment of a statewide residential energy code in Texas in 2001, the change in federal SEER standards from SEER 10 to SEER 13 in 2006 and market transformation effects resulting from the ENERGY STAR program.

During the early years of the study period, the decline in electricity usage was most likely due to the implementation of a new building code in late-2001 and increased code compliance over time. While high-SEER equipment was already prevalent in most new homes, the move to low-solar gain windows and more efficient distribution systems resulted in clear drops in energy use in Baseline homes.

About half of the decline in electricity use occurred from 2005 to 2007 – most likely related to the increase in the federal air conditioner efficiency standard from SEER 10 to SEER 13. While this decline is substantial, it is only half of what the models would have predicted in moving from SEER 10 to SEER 13 (comparing 2005 to 2007 cooling use). Because of HVAC trade-offs allowed by the code in Houston, the average SEER most likely changed from about 11.5 or 12 to about 13.5 or 14, resulting in a drop in cooling usage closer to the 11 percent observed in the data.

Modeling Software and Energy Use Predictions

On average, REM/Rate accurately predicts heating and cooling loads. The relationship between REM/Rate cooling load projections and actual electric usage was examined graphically and statistically for 10,258 homes with sufficient data. REM/Rate projected an average cooling load of 5,506 kWh/yr while the billing analysis estimated average cooling loads at 5,677 kWh/yr, about three percent higher – excellent overall agreement. Although the data found no systematic bias in the REM/rate cooling projections, there was a large amount of variability and the correlation was higher between house size and cooling load than between REM/Rate-projected cooling load and actual usage.

RECOMMENDATIONS

Baseloads are Large and Need to be Addressed

As there are continued efforts to reduce overall energy usage in Phoenix, the sole focus of residential buildings should not be on space cooling/heating and water heating. While the savings are positive, the larger context of these savings is not as impressive. Space cooling/heating and water heating are the largest individual energy users in a home, although they represent roughly 40% of the home's overall energy usage. This means that even a 10% reduction in cooling/heating and water heating costs – a significant reduction – only equates to a 4% savings on the home's total energy bill. Obviously, all areas of energy use within residential buildings must be investigated to discover the maximum energy savings potential.

Need for Increasing the ENERGY STAR® Standard

The narrowing of energy savings between Baseline and ENERGY STAR homes justifies an increase in the ENERGY STAR standard. In both studies above, the ENERGY STAR specifications are no longer stringent enough. Version three of ENERGY STAR, slated for 2012, intends to answer this need. More clear and stringent standards need to be developed while providing the necessary support to raters so they can push the new construction market and truly differentiate committed builders.

Real-World Data Matters

Billing analysis provides the most accurate measurement of program results and clarifies what specifications provide energy savings in new construction programs such as ENERGY STAR. While modeling and projected savings provide an excellent starting point, there is always a need for ongoing evaluation and feedback loops involving real-world data. Doing so will help clarify our models and develop more accurate assumptions. Likewise, there is a need for conducting studies such as this in less mature markets. Perhaps in a city or region with less market share for energy-efficient homes programs, differences between baseline homes and program homes would be larger. Spillover from programs may have less impact on standard practice in these markets.

ACKNOWLEDGEMENTS

U.S. Environmental Protection Agency, Arizona Public Service, Beazer Homes, Chas Roberts Air Conditioning, Continental Homes, D.R. Wastchak, LLC, Hacienda Builders, Pulte Homes, Royce Homes, Sonoran Air, Inc., Southwest Gas

REFERENCES

[1] 2004 Buildings Energy Databook, U.S. Department of Energy.

[2] Advanced Energy. (2005). *Phoenix Home Energy Efficiency Study.* Retrieved from http://www.advancedenergy.org/buildings/services/docs/FINAL%20PAPER%20EPA%20Phx%20Study_v2.pdf

[3] Hassel, S., Hannas, B. (2009). *ENERGY EFFICIENT HOMES: Predictions, Performance and Real-World Results -What can we learn from 160,000 houses in Houston?.* Advanced Energy.

[4] Pigg, S. (2002). *Energy Savings from the Wisconsin ENERGY STAR Homes Program.* Madison: Energy Center of Wisconsin.

[5] Pedrick, G., Harris, J., Blasnik, M. (2007) *Reference design guide for highly energy efficient residential construction.* NYSERDA Report 07-13. July 2007.

[6] Bashford, B. (May 2000). *Energy consumption comparison of ENERGY STAR homes in Phoenix, Arizona.* Arizona State University.

EXPERIMENTAL EVALUATION OF SUPPLY-ONLY VENTILATION EFFECTIVENESS

Mireille Rahmeh[*1, 2], Claude Algourdin [1], Michèle Potard [2] and Karim Limam [1]

[1] *LEPTIAB, University of La Rochelle, Avenue Michel Crépeau 17042 La Rochelle Cedex 1, France*
[2] *VENTILAIRSEC, Zone des Moulinets, 16 Bd Charles de Gaulle, F-44800 Saint Herblain, France*
**mrahme01@univ-lr.fr*

ABSTRACT

Nowadays, indoor air quality has become a major concern. Regarding the fact that people spend most of their time indoors, it is necessary to study the performance of the ventilation system in order to limit the risks on occupants' health. This study evaluates the ventilation effectiveness of supply-only ventilation (SOV) and extract-only ventilation (EOV) in terms of air exchange efficiency and contaminants removal effectiveness. These indicators are measured as function of air change rate and inlet/outlet devices positions using the gas tracer technic. The results show that air change rate has no influence on the air change efficiency but on room mean age of air. Also it shows that the effectiveness of SOV could be better than EOV.

KEYWORDS
Supply-only ventilation, contaminant removal effectiveness, Air change efficiency.

INTRODUCTION

Several decades ago, dwelling's ventilation wasn't an important issue for builders. Indeed, houses were leaky enough to provide adequate air exchange. Yet, air leaks were at the same time energy leaks representing an important problem and especially after the crisis in 1973. And since the thermal discomfort and high energetic bills are more noticed by occupants than poor indoor air quality (IAQ), building's airtightness took more designer's attention and became essential in construction design. However, more dwellings are made airtight, greater the impact of internal pollutant sources on IAQ and occupants health unless ventilation is effective. This emphasizes the need for good ventilation design. Today, beside the natural ventilation, there are three basic types of residential ventilation system: exhaust-only, supply-only and balanced.

In this context, several studies have been dedicated to the improvement and characterization of different ventilation systems [1][2][3]. However, there is a critical lack of information and knowledge regarding the supply only ventilation (SOV).

This study is about evaluating the SOV effectiveness as a function of air exchange rate and diffuser location. Also, it shows the major differences that impose exhaust only ventilation positioned in the same conditions. The experiments were carried out in a full scale laboratory cell test where the gas tracer technic is used to assess the ventilation effectiveness.

VENTILATION EFFECTIVENESS

As awareness and concern about efficient ventilation increase, so does the research for developing different methods and IAQ indicators to evaluate the ventilation process. For example, we find in the literature [9]:

- Number of air changes (N_{AC}) which gives information about the intensity but not the quality of ventilation system
- Ventilation effectiveness (Ev) as a function of recirculation rate (r) and a fraction of the supply air (s) that bypasses directly to the exhaust. Still, there's no information provided by standards for the calculation of (s).
- Also, we can find large number of new IAQ indicators that have been developed by Japanese researchers, who concentrated on airflow characteristics such as SVE1 and SVE2.

Nevertheless, the most common used indicators are those defined by Sandberg and Sjoberg in 1983 which are based on the air age concept [5][6][7]: the air change efficiency ε^a and the contaminant removal effectiveness ε^c. These indices appear to be the most suitable for use in design and standards because they are wide and almost all other indicators are an extension of both of them [9]. In addition, considering average or local values, ε^a and ε^c can provide information about the overall room ventilation quality or in a particular point in the room or maybe larger such as breathing zone.

Furthermore, Skaaret [6] has defined 3 basic rules for effective ventilation. At the design stage, the ventilation should be designed to give an (1) ε^a above 50%, (2) a local age in the breathing zone lower than the room average, i.e. ε_p^a>1 and (3) ε^c greater than 1 and a lower concentrations of contaminants in the zone of occupation than the room average i.e. $\varepsilon_p^c > \varepsilon^c$.

Air change efficiency ε^a

Air change efficiency characterises the mixing behaviour of incoming air with the air already present in the room [8] and it is independent of the distribution and emission characteristics of pollutants.

ε^a is defined as the ratio between the shortest possible time needed for replacing the air in the room $\tau_n/2$ and the actual air change time $<\tau>$

$$\varepsilon^a = \frac{\tau_n}{\bar{\tau}_r} = \frac{\tau_n}{2 < \bar{\tau} >} \tag{1}$$

Where $<\tau>$ is calculated by measuring the tracer gas concentration at the exhaust:

$$< \tau >= \frac{\int_0^\infty t.C_e(t)dt}{\int_0^\infty C_e\,(t)dt} \tag{2}$$

Local air change efficiency (ε_p^a) is defined as follows:

$$\varepsilon_p^a = \frac{\tau_n}{\bar{\tau}_p} \tag{3}$$

Where $\bar{\tau}_p$ is the local mean age of air at point P.

Contaminant removal effectiveness ε^c

Contaminant removal effectiveness characterizes the ability of the ventilation air to dilute and remove pollutant from the room. It depends on the source intensity and location. This index is defined as the ratio between the nominal time constant of the ventilation air and the turnover time for the contaminant (τ_t^c)

$$\varepsilon^c = \frac{\tau_n}{\tau_t^c} \tag{4}$$

Also, it can be expressed in terms of steady state pollutant concentrations:

$$\varepsilon^c = \frac{C_e - C_s}{< C > - C_s} \tag{5}$$

Where: C_e, C_s *and* $< C > C_e, C_s$ *and* C_p are the pollutant concentrations in the exhaust, supply and the average room concentration.

The local air quality index can be obtained by substituting $< C >$ in the equation (5) by C_p, the pollutant concentration in the point of interest p:

$$\varepsilon_p^c = \frac{C_e - C_s}{C_p - C_s} \tag{6}$$

METHOD

Experimental set-up

Measurements were carried out in a 35m^3 test cell (4.28 × 3.14 × 2.6) composed of steel panels (Figure 1). The airtightness of the room was determined by a BlowerDoor test and with a pressure difference of 50 Pascal, (n50) is about 1.4 vol/h. Air temperature (type K thermocouple), air velocity (TSI 8475 omnidirectional transducer) and carbon dioxide concentrations (Vaisala GMM 222) were acquired at three different heights: 0.1m (ankle), 1.1m (head of a sitting person) and 1.7m (head of a standing person) and at nine different positions in the room, resulting in 27 sampling positions inside the room. Additionally, temperature and carbon dioxide concentrations were measured in the inlet and exhaust ducts, and the temperature of the walls were estimated with five type K thermocouples per wall to ensure that the tests were performed in isothermal conditions. The inlet flow was determined with a Höntzsch ZS25 vane anemometer. The measurements were performed every 30 seconds and acquired on two Campbell Scientific CR1000 data loggers. Occupancy was modelled by two cylindrical manikins, with carbon dioxide being supplied at a height of 1.1m for each manikin, through a diffusing device.

Two types of ventilation strategies were assessed: supply-only and extract-only, and three ventilation configurations were taken into account (Figure 2):

- Configuration A: upper wall supply/extract and lower-opposite air output/input,
- Configuration B: upper wall supply/extract and lower-side air output/input,
- Configuration C: Ceiling supply/extract and lower air output/input.

For each configuration, four different air change rates were considered (0.5, 1, 1.5, 2.5 ACH). Depending on the configuration, the air terminal device employed varied. For A and B, part of the air was supplied toward the ceiling at an angle of 270° and the other part was supplied axially. For C, 10% of the air was supplied axially and the remaining part was supplied toward the ceiling at an angle of 360°.

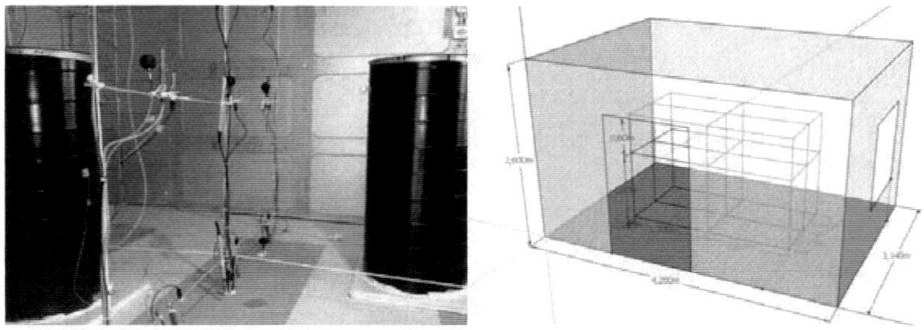

Figure 1-Experimental chamber

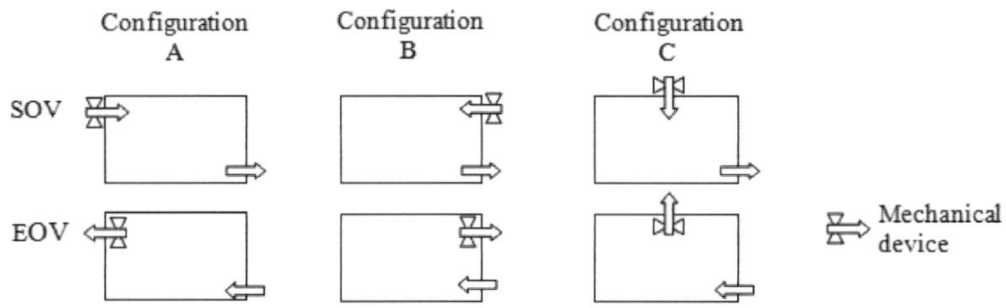

Figure 2- Three different configurations in combination of two types of ventilation

Test procedure

The tracer gas concentration decay technic (step-down method) was used to measure the ability to exchange air in the room. The mixing fan was turned on and the CO2 released rapidly. After obtaining a suitable level and homogenous concentration in the room, the gas supply and the mixing fan was switched off along with switching on the ventilation system. Then the concentration decay at the 29 sampling points is recorded every minute.

For measuring the ability of the ventilation system to remove pollutant, the CO2 is released continuously by both occupants at a rate of 18 L/h/occupant while the ventilation system was kept on. When steady-state conditions are reached, the concentration in the 27 sampling points and in the exhaust are measured for local $\varepsilon_p^c \varepsilon^c$ calculations. The CO2 sources were then stopped and the decay concentration at the exhaust is recorded for the calculation of the turnover time of contaminant τ_t^c meanwhile of global ε^c. Each test was repeated in order to verify the repeatability of the measurements.

ANALYSIS OF AIR CHANGE EFFICIENCY

Influence of air change rate N_{AC}

The configuration A was chosen to study the influence of the air change rate on the ventilation effectiveness in term of replacing the existing air in the room by new fresh air. Figure 3 (a) and (b) present the results for the room mean age of air $<\tau>$ and the air change efficiency ε^a respectively for SOV and EOV systems.

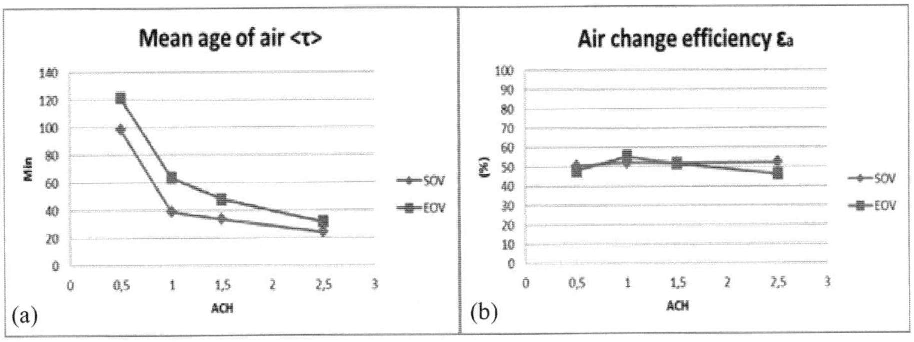

Figure 3-(a) Mean age of air and (b) air change efficiency of configuration A as a function of air change rate

For the EOV system, the room mean age decreases with increasing the N_{AC}, which returns to the fact that the air velocity is proportional to the air flow rate. Therefore by increasing the N_{AC} the new air reaches the entire room rapidly. In addition, it's so remarkable that for 0.5 Ach, an air change rate typically used in residential ventilation, <τ> is much higher than for the other N_{AC}. The difference between 0.5 and 1 Ach (~58 min) is almost 2 times greater than between 1 and 2.5 Ach (32 min). As for the air change efficiency ε^a, it varies slightly around 50 % which correspond to the usual value of perfect mixing ventilation.

On the other hand, the SOV system presents the same decay of room mean age as for EOV but with lower values. The difference is ranged between 10 min for 2.5 Ach and 24 min for 0.5 Ach. Also, the ε^a is almost constant and equal to 50%, which means that, for both systems, the air change rate (N_{AC}) has no influence on the air change efficiency.

Finally, the two systems present the same efficiency, though the measurement of the room mean age shows that SOV replaces the born air faster than EOV. It's believed that the evaluated τ_n of SOV is lower than the real value. Indeed, the SOV maintains a positive pressure in the room, so the air leakages contribute in the evacuation of the CO2 resulting in lower concentration at the exhaust.

Influence of air terminal devices position

The ventilation effectiveness depends on the air flow pattern which in its turn is related to the inlet and outlet devices positions. Therefore a comparison between three different and realistic configurations (A, B and C) was performed. An N_{AC} of 1 Ach was chosen, since it implies a lower mean age of air than for 0.5 Ach and above it, the decay of <τ> is small

For the EOV system, Figure 4 (a) shows that configuration A presents the lowest room mean age (~62 min) followed by B (~78 min) then C (~90 min). For the SOV system, A also presents the lowest <τ> (~40 min) while it is the same for B and C (~50Min). Whereas comparing the two mechanisms (supply/extraction), it's found that for EOV <τ> is globally higher than for SOV.

Figure 4(b) is a representation of the air change efficiency of the different configurations. The results displayed by configurations A and B, for both systems, are typical of perfect mixing ventilation (ε_a~50%). For configuration C, ε_a is about 40% for EOV system which indicates a short circuiting of the air while for the SOV, ε_a is about 62% (>50%) which means that the flow pattern tends to a unidirectional flow [9].

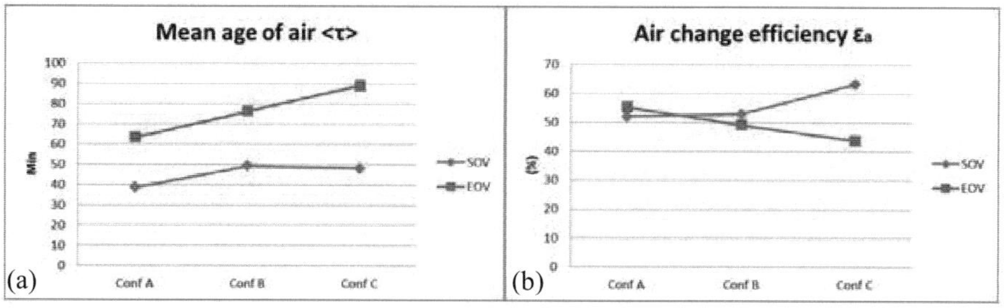

Figure 4-(a) Mean age of air and (b) air change efficiency of the three configurations

ANALYSIS OF CONTAMINANT REMOVAL EFFECTIVENESS

Influence of air change rate N_{AC}

These following results provide information about the ability of the studied ventilation systems to remove the pollutant emitted by a specific source (2 occupants releasing 18 L/h of CO_2).

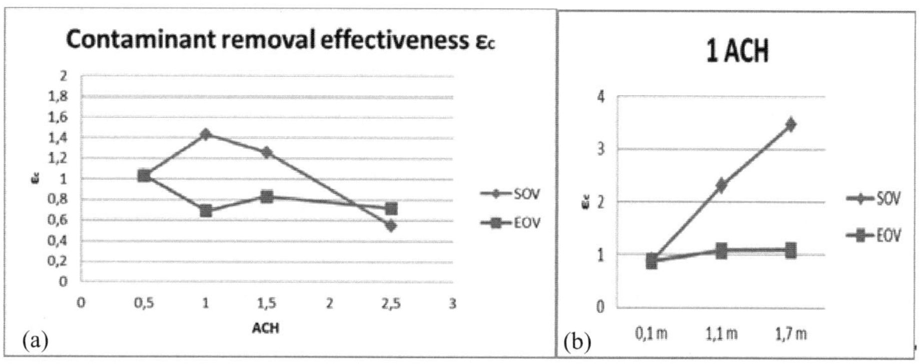

Figure 5-The results of contaminants removal effectiveness for configuration A

In the Figure 5 (a), the contaminant removal effectiveness is plotted as a function of the air change rate. It's noticed that for 0.5 ACH the two systems present an effectiveness of 1 which indicates that the mixing is complete and instantaneous and the pollutant source position does not have an influence [9]. In contrast, for the other N_{AC} values, their behaviour is different: as it is illustrated, for the SOV system, ε^c increases till a maximum value (1.4) for 1 ACH then decreases by increasing the air change rate till 0.6. It's believed that by increasing the velocity of the air flow, the air flow is short-circuited and the pollutant source is in the recirculation area, which is completely separated from the bypass area. As for the EOV, ε^c decreases till a minimum value (0.68) for 1 ACH then increase slightly with N_{AC}.

This different behaviour of the two systems can be referred to different reasons: position, form, and type of the air terminal devices. The SOV has a mechanical circular inlet positioned in the top of the wall and free rectangular outlet representing the bottom of the door while it's the opposite for the EOV.

In the other hand, Figure 5 (b) presents the contaminant removal effectiveness at different levels (0.1, 1.1 and 1.7m) for 1 ACH. The measurements show that for EOV system, ε^c is equal to 1 in the breathing zones of a standing and sitting person while for the SOV, ε^c is greater than 1 (~2.5 at 1.1m and 3.5 at 1.7m height). The latter meets the rules of Skaaret for effective ventilation [6]. These results can be explained by the fact that the air inlet for SOV is in the upper part of the wall, so the new air push the CO2 down to the outlet device. While for the EOV the inlet is in the lower part of the wall, so the pollutant is pushed up to the breathing zones before reaches the outlet.

Influence of air terminal devices position

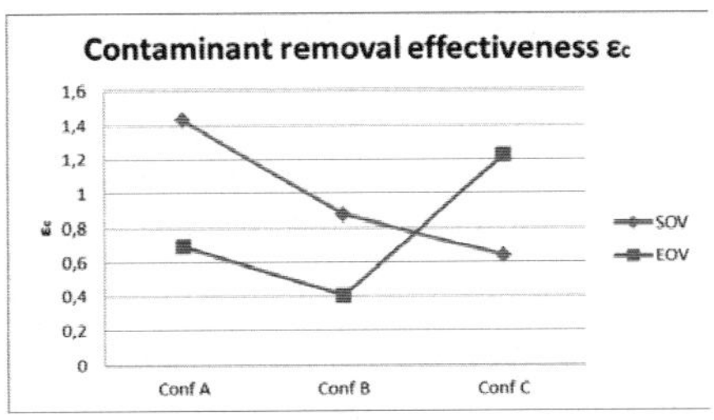

Figure 6- Contaminant removal effectiveness for the three configurations

For 1 ACH, Figure 6 illustrates the contaminant removal effectiveness for the three studied configurations. The results show that for configurations A and B, the SOV present the highest effectiveness while for the configuration C, it's the EOV. In fact, for SOV in configuration C the ceiling supply is radial and it's directly above the occupants, thus the new air can't reach the pollutant source rapidly. Finally, we can conclude that configuration A using SOV system presents the highest ε^c (1.4).

CONCLUSION

In this study, the ventilation effectiveness of supply-only ventilation (SOV) and extract-only ventilation (EOV) systems has been investigated by assessing two indicators: air change efficiency and contaminant removal effectiveness. The results show that 1 ACH can be a suitable air change rate for residential ventilation. In addition, the combination SOV-configuration C seems to be the most efficient in terms of replacing the air in the room while the combination SOV-configuration A is more adequate for removing contaminants released by occupants. The reason that the two indicators did not give the same tendency is that the behaviour of air and pollutant are different, especially when the pollutant is not uniformly distributed.

In the other hand, it's easier to measure ε^a than ε^c which depends not only on the airflow pattern but also on the intensity, area, and positions of contaminant sources relatively to this airflow pattern. Besides, the use of CO2 as a tracer gas has caused difficulties in evaluating the two indicators since it exists in the supply air.

More studies will be done to evaluate the SOV effectiveness in a passive house with real conditions (non-isothermal environment, effect of wind, more types and sources of contaminants).

ACKNOWLEDGEMENTS

The authors would like to thank Christian Inard and Axel Cablé, in charge of the test cell used for our experiments.

REFERENCES

[1] Koffi, J. F., Allard, F. Akoua, J. J., 2011. *Numerical Assessement of the performance of Ventilation Strategies in a Single-Family Building*. The international journal of Ventilation ISSN 1473-3315, vol. 9, No 4 march 2011, 337-350.

[2] Pavlovas, V. 2004. *Demanded controlled ventilation: A case study for existing Swedish multifamily buildings*. Energy and Building. 36. 1029-1034

[3] Olesen, B.W. et al., 2011. *Experimental Study of Air Distribution and Ventilation Effectiveness in a Room with Combination of Different Mechanical Ventilation and Heating/Cooling Systems*. International Journal of Ventilation ISSN 1473-3315, vol.9, No 4 march 2011, 371-383.

[4] Roberson, J.A. et al., *Recommended ventilation strategies for energy-efficient production homes.* 1998. Lawrence Berkeley National Laboratory LBNL-40378

[5] Sandberg, M. and Sjoeberg, M. 1983. *The use of moments for assessing air quality in ventilated rooms*, Build. Environ. 18, 181-97.

[6] Skaaret, E. 1986. *Contaminant removal performance in terms of ventilation effectiveness. Environment International*. 12, 419-427.

[7] Etheridge, D., and Sandeberg, M. 1996. *Building ventilation: theory and measurement.* John Wiley & Sons. 607-626.

[8] Liddament M.W. 1996. *A guide to energy efficient ventilation*. Air Infiltration and Ventilation Centre AIVC.

[9] Srebric J. 2003. *Comparison of air exchange efficiency and contaminant removal effectiveness as IAQ indices*. ASHRAE transactions, vol.109, part 2, 339-349.

THE APPLICABILITY OF GLAZING SYSTEM WITH DYNAMIC INSULATION FOR RESIDENTIAL BUILDINGS

Shinsuke Kato[1], Sihwan Lee[*2]

1 Institute of Industrial Science, The University of Tokyo
4-6-1-Cw403, Komaba, Meguro-ku, Tokyo
The University of Tokyo, Japan

2 Department of Architecture, The University of Tokyo
4-6-1-Cw403, Komaba, Meguro-ku, Tokyo
The University of Tokyo, Japan
**Corresponding author:shany@iis.u-tokyo.ac.jp*

ABSTRACT

It is essential to reduce the inordinate amount of energy used for climate control in buildings. To reduce heat loss in residential buildings, it is necessary to insulate building envelopes more airtight. Many air tightness and insulation methods have been proposed and successfully applied to the building envelope, including areas such as walls, windows and the others. However, if it concentrates only air tightness and insulation to save energy consumption in the buildings, that'll make a problem to maintain indoor air quality within acceptable levels, such as sick building syndrome.

To solve this problem, this paper proposed a new dynamic insulation system applied to the glazing and frame of the windows. Dynamic insulation refers to the use of porous insulation material through which ventilation air enters a building, thereby reducing the conductive heat loss through the material to very low level. Moreover, the proposed system is composed of three parts by installing two additional parts to control indoor/outdoor pressure difference and to reduce ventilation loads: a double pane airflow window system with window frame made of a porous material, a mechanical ventilation system, and a heat-recovery heat pump system.

The aim of this paper is to evaluate the thermal insulation efficiency and probability of moisture condensation in the proposed glazing system in order to confirm its feasibility and applicability. First, a double pane airflow window system was designed to ventilate through the window frame and the air space of a double pane window. Then, to verify its thermal insulation efficiency, the temperature distribution of the window system was evaluated using computer fluid dynamics with different coupled conditions, such as the indoor/outdoor pressure difference and outdoor temperature, after confirming calculation accuracy using glazing model. In addition, to verify the probability of moisture condensation, the relative humidity in the window system was calculated based on the various conditions.

The calculated results show the thermal load was proportional to the outdoor temperature and inversely proportional to the indoor/outdoor pressure difference. Moisture condensation depends on the outdoor temperature and humidity ratio and it does not occurred when outdoor temperature is more than 6.0 ºC in the proposed system. Therefore, the proposed system is technically feasible to reduce the home energy consumption by installed residential buildings.

KEYWORDS

dynamic insulation, airflow window, thermal insulation efficiency, moisture condensation

INTRODUCTION

Solution offerings to the inordinate amount of energy used for climate control in buildings are of paramount importance. To insulate buildings more efficiently, many insulation methods have been proposed and successfully applied to the building envelope. One technical solution involves dynamic insulation that blocks heat transport by making the incoming airflow pass through a porous material, same principle as airflow window system. Because dynamic insulation not only reduces heat loss but also helps maintain indoor air quality, several different dynamic insulation systems have been proposed to improve in the performance of specific building elements and successfully applied to the building envelope. Furthermore, because the Building Standard Law of Japan has been in force since July 1st, 2003 in Japan requiring minimum indoor ventilation rates of 0.5 air changes per hour for the entire 24 hours, many studies have shown that it is possible to use dynamic insulation efficiently in residential buildings. In spite of their efforts, it still remains a heat loss at glass and frame of window because they have relatively poor insulating qualities and usually contribute the greatest heat loss by heat conduction. Moreover, they have also a high risk of moisture condensation occurs because they have low surface temperature than the other building envelope. Although many studies have shown that it is possible to use low-emissivity glazing, gas-filled glazing, or vacuum glazing to solve this problem, it has also a demerit such as high cost [1], [2], [3], [4], [5], [6]. This paper deals with a new insulation system to insulate glass and frame of the window efficiently by using dynamic insulation and airflow window system.

OBJECTIVE

In this paper, a new insulation system is proposed to insulate glass and frame of the window efficiently in residential buildings, because they usually exhibit the greatest heat loss. The aim of this paper is to evaluate the thermal insulation efficiency of the insulation material in the proposed insulation system in order to confirm its applicability. We also evaluate whether it produces excessive moisture condensation, depending on the outdoor temperature, the indoor/outdoor pressure difference.

METHODS

1. The new system proposed to increase thermal insulation of window

Figure 1 shows the concept of the proposed system to increase thermal insulation and air-tightness of glass and frame of the window in residential buildings. This system is composed of three parts: a double pane airflow window system with window frames made of a porous material, a mechanical ventilation system, and a heat-recovery heat pump system.

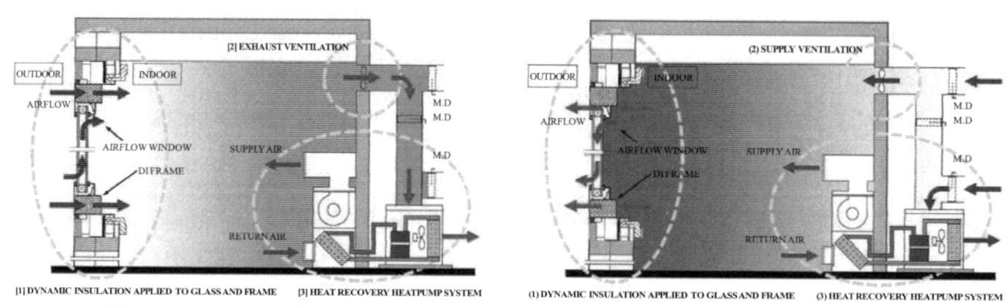

(a) Winter mode [negative indoor pressure] (b) Summer mode [positive indoor pressure]

Figure 1. Mechanisms of the proposed system.

1.1 Dynamic insulation incorporated in glass and frame of the window

1.1.1 Dynamic insulation applied to frame of the window

In this paper, in order to apply dynamic insulation to window frames, porous insulation material such as a packed bed of particulate material (i.e. glass wool, mineral wool, aluminium particles, etc.) is installed around the window so that fresh air, water vapour and heat can pass through. Dynamic insulation refers to the use of porous insulation material through which ventilation air can enter a building, thereby reducing the conductive heat loss through the material to a very low level. Assuming uniform air flow, heat transfer in the dynamic insulation can be described using a 1-D steady-state model, as given by Equation 1 [7], [8], [9], [10], [11], [12], [13], [14], [15].

$$k\frac{d^2T(x)}{dx^2} - u\rho_a C_P \frac{dT(x)}{dx} = 0 \qquad (1)$$

where k [W/(m·K)] is the thermal conductivity of the insulation material, T [K] is the temperature, ρ_a [kg/m^3] and C_P [J/(kg·K)] are the density and the heat capacity of air, and u [m/s] is the air velocity.

Assuming a constant temperature boundary condition for the indoor and outdoor environments, the following dynamic U-value can be derived to represent the overall heat loss, as given by Equation 2. The dynamic U-value, U_{dyn} is the U-value of the wall modified by the air velocity across it and the insulation thickness.

$$U_{dyn} = \frac{u\rho_a C_P}{e^{\frac{u\rho_a C_P L}{k}} - 1} \qquad (2)$$

where L [m] is the insulation thickness. It is clear that the dynamic U-value is a function of the air velocity, or, more exactly, the Pe number, as given by Equation 3. This is defined to be the ratio of the rate of advection of a physical quantity by flow to the rate of diffusion of the same quantity driven by an appropriate gradient.

$$Pe = \frac{u\rho_a C_p L}{k} \qquad (3)$$

1.1.2 Airflow window system applied to glass of the window

An airflow window system is used to reduce thermal loss from glass of the window in dwelling houses. Outdoor air can be directed through the window from either bottom to top. It has useful function to reduce the energy transfer of indoor/outdoor because it has minimal temperature difference of outdoor and air space between glass panels of the window.

1.2 Mechanical ventilation system (exhaust / supply ventilation)

A mechanical ventilation system is used to maintain an adequate supply of fresh air and to maintain the thermal insulation efficiency of the dynamic insulation. This means that indoor air leaves the room through the ventilation system, while fresh air enters the room through the porous material in the window frames. A mechanical ventilation system works as exhaust ventilation functionally in winter to prevent moisture condensation by entering low humidity air from outdoor at building envelops as shown in Fig. 1(a). On the contrary, it works as supply ventilation functionally in summer to prevent high humidity air entered at building envelops as shown in Fig. 1(b).

1.3 Heat-recovery heat pump system

The heat pump system recovers exhaust heat, enabling the creation of a zero-energy house. However, the heat recovery function of the heat pump system is used only in winter, because a positive pressure condition must be maintained indoors to prevent moisture condensation in summer.

2. Numerical evaluation of thermal insulation efficiency

Computer fluid dynamics (CFD) was used to model fluid, heat flow, and moisture condensation in the porous material of the proposed system. CFD has been widely used to simulate air movement, heat transfer, mass transfer, and the interaction between indoor and outdoor environments.

2.1 Calculation model

To evaluate the thermal insulation efficiency of the proposed system, the temperature contributions of frame and glass of the window were calculated using 3-D steady-state CFD simulation includes detailed ray-tracing-based radiation modelling as well as direct modelling of convective heat transfer under different coupled conditions, i.e. the indoor/outdoor pressure difference or outdoor temperature etc. Figure 2(a) shows the plan of the building model used for the calculation, which is proposed as a standard dwelling house model in Japan by the Institute for Building Environment and Energy Conservation. In this study, a children's room of 2nd floor (2.40 m (H) × 2.90 m (W) × 3.50 m (D) = 24.36 m^3) is only used for calculations as shown in Fig. 2(b). Figure 2(c) shows the detail of window (1.95 m (H) × 1.65 m (W) × 0.10 m (D)) installed the calculation model with Fig. 2(d) and Fig. 2(e). The model included an indoor zone, an outdoor zone, a window, and a window frame. The insulation material applied to the window frame acts as an air supply opening and an airflow window system applied to the window glass allowing fresh air into the room. The porosity of the porous material at window frame is 0.5 [-] and the frame area is about 10% of the whole window area. Opening size (0.001 m (H) × 0.0765 m (W) : 2EA) of airflow window system was determined to satisfy about 90% of a room required ventilation rates at the indoor/outdoor pressure difference of 10 Pa. In Japan, indoor ventilation rates need to provide at least one half air changes per hour (0.5 ACH) for the entire 24 hours. Unfortunately, the effect of the heat-recovery heat pump system is not considered, in order to keep the calculation model simple in this study.

2.2 Flow model and boundary conditions

A Abe-Kondoh-Nagano low-Reynolds number k-epsilon turbulence model [16], [17] was used for computing the turbulent viscosity and diffusivity. For modelling of fluid, heat flow (detailed ray-tracing-based radiation model, direct model of convective heat transfer), and moisture condensation (species transfer model), the boundary conditions required for the simulation were those associated with heat transfer through air and porous material, and properties such as porosity and diameter. The material properties for the simulation are summarized in Table 1. This calculation used 4 materials, such as air, water vapour, porous material and glass pane.

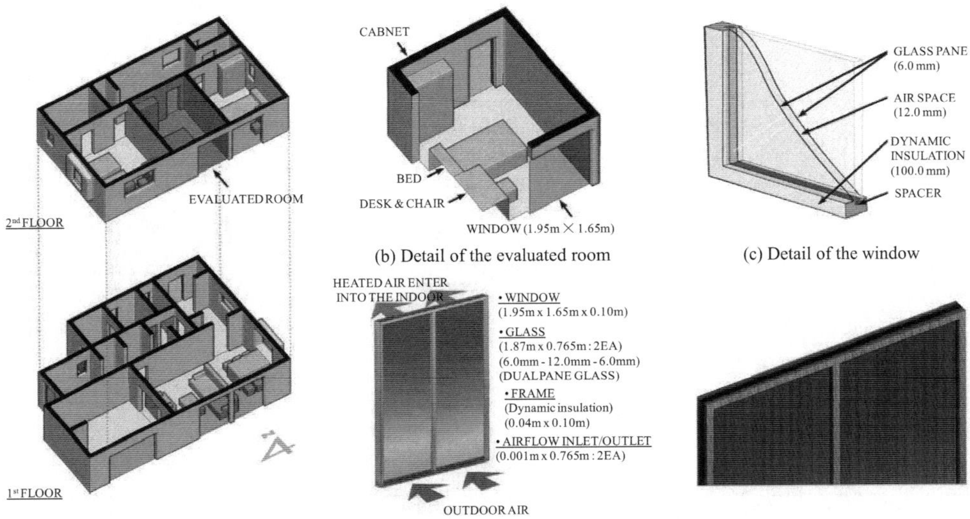

(a) Applied dwelling house (b) Detail of the evaluated room (c) Detail of the window (d) Detail of the calculation model (e) Calculation mesh

Figure 2. Principle of dynamic insulation and calculation model

Item	Air	Water vapour	Porous material	Glass pane
Specific heat (C_P)	1006.43 [J/(kg·K)]	2014 [J/(kg·K)]	2100 [J/(kg·K)]	753 [J/(kg·K)]
Conductivity (λ)	0.0242 [W/(m·K)]	0.0261 [W/(m·K)]	0.0558 [W/(m·K)]	0.65 [W/(m·K)]
Viscosity (μ)	1.79×10^{-5} [kg/(m·s)]	1.34×10^{-5} [kg/(m·s)]	-	-
Moler weight (M)	28.97 [kg/kgmol]	18.02 [kg/kgmol]	-	-
Thickness (l)	-	-	-	6.0-12.0-6.0 [mm]
Emissivity (ε)	-	-	-	0.90 [-]

Table 1. Material properties.

2.2.1 Flow model for the porous media at window frame

The steady state flow model for simulation of the porous material is represented by a mass conservation equation as follows.

$$\frac{\partial(\rho U_i)}{\partial x_i} = 0 \tag{4}$$

where U_i is the mean superficial air velocity in the x_i direction. The superficial air velocity (velocity based on volumetric flow rate) is related to the physical velocity V_i (true velocity through pores of the medium for instance) by the porosity, $U_i = \varepsilon V_i$.

The permeability and inertial resistance factor of the porous medium were derived from the Ergun equation defined by the Turkish chemical engineer Sabri Ergun in 1952, as given by Equation 5 [18]. The Ergun equation is applied here for the pressure drop, dp, over the length, dL, determined by the diameter of the porous material [19], [20], assuming that the U-value is 0 W/m^2 when the pressure difference is 10 Pa.

$$\frac{dp}{dL} = \frac{150\mu(1-\varepsilon)^2}{\varepsilon^3 \phi_C{}^2 D_P^2} U + \frac{1.75\rho(1-\varepsilon)}{\varepsilon^3 \phi_C D_P} U^2 \tag{5}$$

where φ_C [-] is the shape factor, U [m/s] is the superficial fluid velocity through the porous medium, ρ [kg/m^3] is the fluid density, D_P [m] is the diameter of the solid part on porous medium, μ [kg/(m·s)] is the molecular viscosity, and ε [-] is the porosity of the porous medium.

2.2.2 Air density

The density ρ of a mixture of dry air molecules and water vapour molecules can be expressed by Equation 6. To determine the density of the air, it is necessary to know the actual air pressure, the water vapour pressure, and the air temperature.

$$\rho = \frac{P_{dry}}{R_{dry} \times T} + \frac{P_{vapour}}{R_{vapour} \times T} \tag{6}$$

where P_{dry} [Pa] is the pressure of dry air in pascals, P_{vapour} [Pa] is the pressure of water vapour in pascals, R_{dry} [287.05 J/(kg·K)] is the gas constant for dry air, and R_{vapour} [461.49 J/(kg·K)] is the gas constant for water vapour.

2.2.3 Thermal conductivity

The effective thermal conductivity of the porous material, k_{eff}, can be expressed as the volume average of the fluid conductivity and the solid conductivity, as shown in Equation 7.

$$k_{eff} = \varepsilon \times k_{fluid} + (1-\varepsilon) \times k_{solid} \tag{7}$$

where k_{solid} [W/(m·K)] is the solid medium thermal conductivity, and k_{fluid} [W/(m·K)] is the fluid phase thermal conductivity.

2.2.4 Mass diffusivity

The diffusion of moisture through building materials is a natural phenomenon, and it has significant effects on the comfort of the indoor environment. The effective diffusion coefficient in porous media, D_e, describes diffusion through the pore space of porous media, as shown in Equation 8 [21], [22].

$$D_e = \frac{D\varepsilon_t \delta}{\tau} \tag{8}$$

where D [m^2/s] is the diffusion coefficient of the gas or liquid filling the pores, ε_t [-] is the porosity available for transport, δ [-] is constrictivity, and τ [-] is tortuosity. In this study, we calculate the diffusion coefficient, D, using the following formula.

$$D = 2.31 \times 10^{-5} \left(\frac{T}{273.16}\right)^{1.81} \left(\frac{P_a}{P_a + P_v}\right) \tag{9}$$

where P_a [Pa] is the air pressure, in this case the atmospheric pressure, and P_v [Pa] is the vapour pressure.

2.3 Calculation cases

Table 2 shows the calculated cases by indoor/outdoor pressure difference (case 1, 2) and Table 3 shows the calculated cases by outdoor temperature (case 3, 4) in winter/summer operating mode for the CFD simulation. These cases were used temperature and relative humidity of indoor based on design basis conditions to build dwelling houses generally in Japan. A total heat transfer coefficient of window surface is assumed 23.25 W/(m^2·K) at outdoor surface and 9.09 W/(m^2·K) at indoor surface for calculation.

Case 1 (Winter operating mode)			Case 2 (Summer operating mode)		
Indoor	Temperature	22 [°C]	Indoor	Temperature	26 [°C]
	Relative humidity	50 [%RH]		Relative humidity	50 [%RH]
Outdoor	Temperature	0 [°C]	Outdoor	Temperature	36 [°C]
	Relative humidity	30 [%RH]		Relative humidity	80 [%RH]
Indoor/outdoor ΔP		0, -1, ... , -9, -10 [Pa]	Indoor/outdoor ΔP		0, 1, ... , 9, 10 [Pa]

Table 2. Calculation cases by indoor/outdoor pressure difference.

Case 3 (Winter operating mode)			Case 4 (Summer operating mode)		
Indoor	Temperature	22 [°C]	Indoor	Temperature	26 [°C]
	Relative humidity	50 [%RH]		Relative humidity	50 [%RH]
Outdoor	Temperature	-20, -18, ... , -2, 0 [°C]	Outdoor	Temperature	28, 30, ... , 34, 36 [°C]
	Relative humidity	30 [%RH]		Relative humidity	80 [%RH]
Indoor/outdoor ΔP		-10 [Pa]	Indoor/outdoor ΔP		10 [Pa]

Table 3. Calculation cases by outdoor temperature.

RESULTS AND DISCUSSIONS

1 Effect of indoor/outdoor pressure difference

Figure 3 is showed the calculated results by indoor/outdoor pressure difference from 1 Pa to 10 Pa in summer and winter. Figure 3(a) is airflow rates for indoor ventilation from glass opening and frame made by porous material. It is decrease by indoor/outdoor pressure difference. The heat loss/gain in the room was reduced with increases in the indoor/outdoor pressure difference, as shown Fig. 3(b). Figure 3(c) and 3(d) show the surface temperature of frame and glass at indoor. Moisture condensation is not occurred in surface of indoor frame but it is occurred in surface of indoor glass because dew point temperature is 11.1 °C when indoor air condition is 22.0 °C and 50 %RH.

Figure 3(e) shows the temperature distribution versus the indoor/outdoor pressure difference in air space of glass. In winter, the air temperature rises slowly as the outdoor air passes through the insulation materials and air space of glass. On the contrary, it is cooled slowly in summer. However, it is possible to increase ventilation loads by increasing the indoor/outdoor pressure difference. Therefore, this is the best way to use the heat pump system to recover heat from the exhaust air to realize zero-energy residential buildings.

Figure 3(f) shows relative humidity distribution versus the indoor/outdoor pressure difference in air space of glass. It is not occurred in porous material and air space of glass in any calculated case because supply/exhaust air flow make reduce the moisture condensation probability.

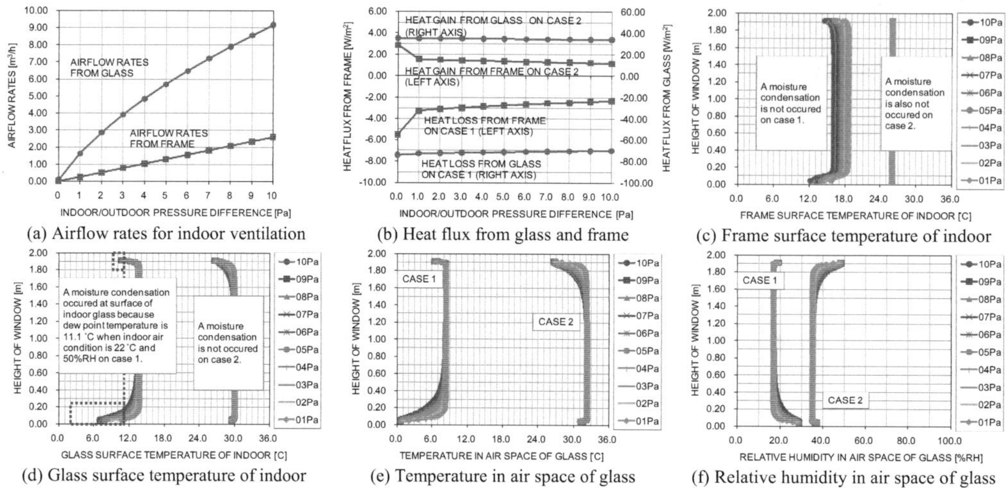

(a) Airflow rates for indoor ventilation (b) Heat flux from glass and frame (c) Frame surface temperature of indoor

(d) Glass surface temperature of indoor (e) Temperature in air space of glass (f) Relative humidity in air space of glass

Figure 3. Simulation results by indoor/outdoor pressure difference (case 1, case 2)

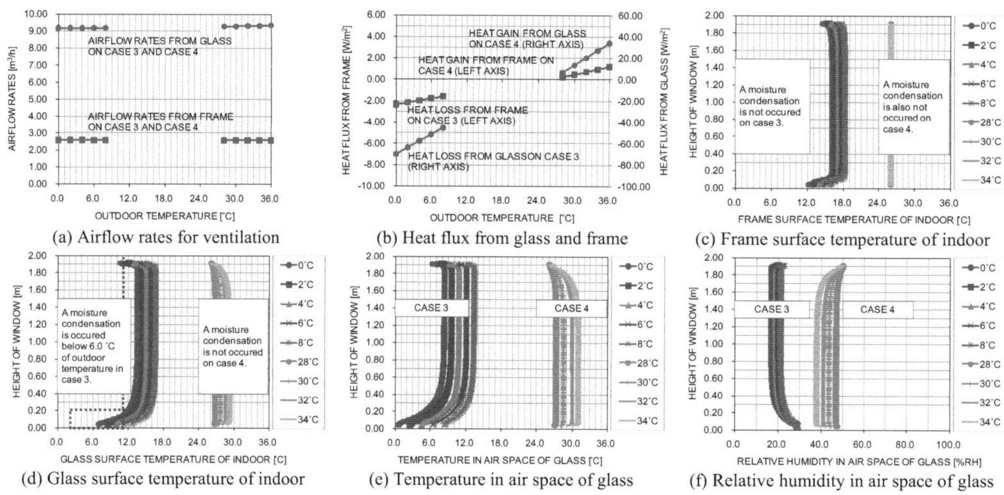

(a) Airflow rates for ventilation (b) Heat flux from glass and frame (c) Frame surface temperature of indoor

(d) Glass surface temperature of indoor (e) Temperature in air space of glass (f) Relative humidity in air space of glass

Figure 4. Simulation results by outdoor temperature (case 3, case 4)

Effect of outdoor temperature

Figure 4 shows the calculated results by the outdoor temperature at the fixed indoor/outdoor pressure difference of 10 Pa. Figure 4(a) shows constant airflow rates for indoor ventilation from glass opening and frame made by porous material. The heat loss/gain in the room was smaller with a smaller indoor/outdoor temperature difference, as shown Fig. 4(b). Furthermore, the variation of the indoor wall surface temperature increased with a decrease in the indoor/outdoor temperature difference at a fixed pressure difference.

Figure 4(c) and 4(d) show the surface temperature of frame and glass at indoor. Moisture condensation is occurred when outdoor temperature is below than 6.0 ºC.

Figure 4(e) and 4(f) show the temperature and relative humidity distribution versus the indoor/outdoor pressure difference in air space of glass. It is not occurred in porous material and air space of glass in any calculated case. In winter, low-temperature outdoor air comes into the indoor through the porous material and the dual glass pane. The relationship of the temperature difference between indoor air/outdoor air and supply air/outdoor air is calculated and leads to Equation 10 and 11 as a linear approximation:

$$T_{s\_DIframe} = -0.3989 \times (T_{indoor} - T_{outdoor}) + T_{outdoor} \quad (10)$$

$$T_{s\_DIframe} = -0.3989 \times (T_{indoor} - T_{outdoor}) + T_{outdoor} \quad (11)$$

where $T_{s\_airflowwindow}$ [°C] is the supply air temperature at opening boundary of window glass, $T_{s\_DIframe}$ [°C] is the supply air temperature at opening boundary of window frame made by dynamic insulation material, $T_{outdoor}$ [°C] is outdoor air temperature, T_{indoor} [°C] is the indoor air temperature.

CONCLUSION

This paper proposed a new double pane airflow window system with window frames made of a porous material applied dynamic insulation technology to reduce energy consumption in residential buildings, and conducted an applicability study on its effectiveness using CFD simulation. In this paper, we found that the thermal insulation efficiency of the proposed system increased with increasing indoor/outdoor pressure difference, with decreasing indoor/outdoor temperature difference. A summary of the general findings of this study is as follows.

- The heat loss/gain in the room was reduced with increasing indoor/outdoor pressure difference, depending upon the estimated pressure drop across the porous material. Furthermore, it was also reduced with decreasing indoor/outdoor temperature difference.

- Moisture condensation occurred in the surface of indoor glass because dew point temperature is 11.1 °C when indoor condition is 22.0 °C and 55.0 %RH. However, it is not occurred when outdoor temperature is more than 6.0 °C. It is also not occurred in porous material and air space of glass in any calculated case because supply/exhaust air flow make reduce the moisture condensation probability.

- Next study, it need to evaluate various design model such as double dual pane, triple glass, low-e glass, drain materials etc., because it is important to prevent moisture condensation probability in indoor glass surface. Moreover, thermal comfort should be calculated for a realizable room model by examining the effects of any cold drafts and the energy-saving effects of the heat pumps included in the proposed system, and this will be evaluated in future investigations.

ACKNOWLEDGEMENTS

This study was made possible by the financial support received from the Ministry of the Environment Government of Japan (MOEJ, http://www.env.go.jp/en/). Moreover, we would like to thank the staff at Sankyo Tateyama Aluminium Inc., J Architecture System Inc., and Tsuchiya Hometopia Inc. for their generous cooperation and support.

REFERENCES

[1] Guohui Gan. 2001. *Thermal transmittance of multiple glazing: computational fluid dynamic prediction*, Applied Thermal Engineering Vol.21, pp.1583-1592

[2] Carli, Inc. 2006. *Mathematical models for calculation of thermal performance of glazing systems with or without shading devices*, Technical Report

[3] ISO 10292. 1994. *Glass in building - Calculation of steady-state U value (thermal transmittance) of multiple glazing*

[4] EN 673/A2. 1998. *Glass in building - Determination of thermal transmittance (U value) - Calculation method*

[5] JIS R 3107. 1998. *Evaluation on thermal resistance of flat glasses and thermal transmittance of glazing*

[6] JIS R 3209. 1998. *Sealed insulating glass*

[7] A. Dimoudi, A. Androutsopoulos, S. Lykoudis. 2004. *Experimental work on a linked, dynamic and ventilated, wall component*, Energy and Buildings 36, 443-453

[8] A. Dalehaug, A. Fukushima, Y. Honma. 1993. *Dynamic Insulation in a wall*, Report Collection of Architectural Institute of Japan, No. 66

[9] M. S. Imbabi. 2006. *Modular breathing panels for energy efficient, healthy building construction*, Renewable Energy 31, pp.729-738

[10] J.M. Wong, F.P. Glasser, M.S. Imbabi. 2007. *Evaluation of thermal conductivity in air permeable concrete for dynamic breathing wall construction*, Cement & Concrete Composites 29, pp.647-655

[11] Guohui Gan. 2000. *Numerical evaluation of thermal comfort in rooms with dynamic insulation*, Building and Environment 35, pp.445-453

[12] L. Jensen. 1993. *Energy impact of ventilation and dynamic insulation*, Proceeding of the 14th AIVC conference, Copenhagen, Denmark, pp.251-260

[13] Annex 44 A3, *Dynamic Insulation Systems*, Annex 44 State-of-the-Art Working Report

[14] Sihwan Lee, Miho Tanaka, and Shinsuke Kato. 2010. *The evaluation of the thermal insulation efficiency at the proposed window frames - Dynamic insulation system applied to window frames (Part 1)*, Proceeding of the 31st AIVC Conference, Seoul, Korea

[15] Miho Tanaka, Sihwan Lee, and Shinsuke Kato. 2010. *Energy saving effects of the proposed system on the residential buildings - Dynamic insulation system applied to window frames (Part 2)*, Proceeding of the 31st AIVC Conference, Seoul, Korea

[16] K.ABE, T.Kondoh, and Y.Nagano. 1994. *A new turbulence for predicting fluid flow and heat transfer in separating and reattaching flows – I. Flow field calculations*, Heat Mass Transfer Vol.37, pp.139-151

[17] K.ABE, T.Kondoh, and Y.Nagano. 1995. *A new turbulence for predicting fluid flow and heat transfer in separating and reattaching flows – II. Thermal field calculations*, Heat Mass Transfer Vol.38, pp.1467-1481

[18] Sabri Ergun. 1952. *Fluid Flow Through Packed Columns*, Vol. 48, No. 2, Chemical Engineering Progress, pp.89-94

[19] M. Mayerhofer, J. Govaerts, N. Parmentier, H. Jeanmart, and L. Helsen. 2011. *Experimental investigation of pressure drop in packed beds of irregular shaped wood particles*, Powder Technology, Volume 205, Issues 1-3, pp.30-35

[20] P. Grathwohl. 1998. *Diffusion in natural porous media, Contaminant transport, sorption / desorption and dissolution kinetics*, Kluwer Academic Publishers

[21] Schirmer, R. 1938. *The diffusion coefficient water vapour-air mixtures and the rate of evaporation*, VDI Beihaft Verfahrestechnik 6

[22] W.P. Boynton, W.H. Brattain. 1929. *Interdiffusion of gases and vapors (1st edition), International Critical Tables of Numerical Data Physics*, Chemistry, and Technology Vol. 1, McGraw-Hill, New York, pp.62-63

Integrated Approach of CFD and SIR Epidemiological Model for Infectious Transmission Analysis in Hospital

Hiroaki Asanuma and Kazuhide Ito*

IGSES, Kyushu University
6-1 Kasuga-koen, Kasuga
Fukuoka, Japan

**Corresponding author: ito@kyudai.jp*

ABSTRACT

The indoor environment can play a significant role in the transmission of and exposure to various contaminants. In the case of some emerging aerial infections, such as those caused by influenza virus and tuberculosis virus, the airborne route of transmission is considered to be important for evaluating the health risk associated with exposure to contaminants. In this paper, we first present an analytical procedure for coupling the computational fluid dynamics (CFD)-based prediction of unsteady contaminant concentration distribution with a basic epidemiological model (here, SIR model) and show how this procedure can be used to predict exposure risk of people in hospital space.

In this study, we focus on the coupled simulation of unsteady and non-uniform distribution of infectious particle concentration and infectious risk, which directly indicates the changes in the population densities of *Susceptible* (*S*) and *Infective* (*I*) in a hospital space. The numerical analysis involved changes in the type of contaminant and infection probability and was performed for a hospital waiting space with a complicated geometry. The results showed a non-uniform distribution of (*S*) and (*I*) in such a space. Moreover, these results indicated the dependence on unsteady and inhomogeneous contaminant distribution.

KEYWORDS

Epidemiological model, SIR model, CFD, Hospital, Infectious transmission

INTRODUCTION

A majority of people in the developed world spend much of their lives in indoor spaces and hence the quality of indoor environments has a strong impact on their health. In indoor environments, huge numbers of contaminants exist and influence human health by long-term and short-term exposures. Concerning the sustainability of healthy and safe indoor environments, the control of airborne transmission by infectious bio-aerosols is one of the most important factors, especially in designing indoor spaces that many people use. This is especially true in hospitals, as the 2003 worldwide outbreak of severe acute respiratory syndrome (SARS) and a super-spreading event in a hospital ward in Hong Kong provided new understanding of the significance of indoor environmental control, especially ventilation in infection control.

The objective of this study is to develop an integrated numerical approach of computational fluid dynamics (CFD) and the classic SIR-type epidemiological model proposed by Kermack and McKendrich to analyze unsteady infectious transmission in hospital. Toward this end, first, we introduce an analytical procedure of coupled analysis of CFD-based prediction of unsteady contaminant concentration distribution and the classic SIR model to predict time-dependent exposure risk of people in enclosed spaces. Second, we demonstrate a coupled

CFD-SIR model simulation targeting a university hospital and investigate its effectiveness as a predictive tool for indoor air quality management.

OUTLINE OF EPIDEMIOLOGICAL MODEL

Various epidemiological models have been proposed to reproduce infection diffusion/propagation theoretically and mathematically in an enclosed space, and the SIR model proposed by Kermack and McKendrick is one of the basic epidemiological models that represent propagation of transmission. The classic SIR model consists of three differential equations coupling the changes in the population density of susceptibles (*S*), the population density of infectors (*I*), and the population density of those who recover to an immune state (*R*). The basic equations of the SIR model are indicated in Table 1 (equations (1) – (3)). The original SIR model defined by Kermack and McKendrick consists of only a reaction term, and equations (1) – (3) in Table 1 constitute an extended model including the diffusion term. β is the contact rate between susceptibles and infectors [1/person/s], and γ is the recovery rate [1/s]. The basic reproductive ratio R_o is defined from β and γ as shown in equation (5), and has been used widely as a parameter that describes the average number of new cases that an infector produces in a particular population. $R_o>1$ denotes that the infection rate is larger than the recovery or removal rate, which may lead to an epidemic and the spread of infection. $R_o<1$ indicates that the infection rate is smaller than the recovery rate, which may lead to an endemic situation. Here, the equation $S+I+R=N$ ($=S_0$) is used.

Table 1 Governing Equations of Epidemiological Model and Scalar Transport

[1] SIR-type epidemiological model (enhancing expression that added a diffusion term)

$$\frac{dS}{dt} = -\beta SI + \nabla(D_S \nabla S) \qquad (1) \qquad\qquad \frac{dI}{dt} = -\gamma I + \beta SI + \nabla(D_I \nabla I) \qquad (2)$$

$$\frac{dR}{dt} = \gamma I + \nabla(D_R \nabla R) \qquad (3) \qquad\qquad \beta = p \cdot a\overline{C} \cdot \overline{\Delta} \qquad (4)$$

[2] Basic reproduction number

$$R_o = \frac{\beta}{\gamma} S_0 \qquad (5)$$

[3] Scalar transport equation for contaminant (Eulerian approach)

$$\frac{\partial \overline{C}}{\partial t} + \overline{u_i} \frac{\partial \overline{C}}{\partial x_i} - V_g \frac{\partial \overline{C}}{\partial x_3} = \frac{\partial}{\partial x_i}\left(\left(D_p + \frac{\nu_t}{\sigma_c}\right)\frac{\partial \overline{C}}{\partial x_i}\right) + S_0 \qquad (6)$$

[4] Stokes' law of aerosol settling velocity

$$V_g = 2.46\left[\left(\frac{\rho_p}{\rho} - 1\right) d_p g\right]^{1/2} \qquad (7)$$

PROCEDURE OF INTEGRATION ANALYSIS OF UNSTEADY CFD AND SIR MODEL

It was possible to predict unsteady and non-uniform concentration distribution by solving the scalar (aerial infectious contaminant) transport equation shown in equation (6) in Table 1 on the basis of flow field analysis by CFD. The time change of contact rate β distributions through unsteady concentration simulation in enclosed spaces and then equations (1)-(3) were solved using this time-dependent β (equation (4)). In the case that the infectors (*I*) emit infectious contaminant, namely, new contaminant sources, the source term S_0 that denotes the formation of scalar (new infectious contaminant) in equation (6) must be considered. In this study, S_0 and feedback from SIR analysis to scalar transport equation were disregarded. The SIR model in Table 1 is an enhancing model that considers diffusion terms in (*S*), (*I*), and (*R*), and the movement of susceptibles or infectious could be reproduced approximately by giving the diffusion coefficient. In particular, when the grid size of CFD is set to larger than human scale, the numerical results of diffusion and propagation of infectors become reasonable analyses that reproduce an actual phenomenon. In this case, SIR parameters indicate population density, that is, [population/m^2], and the correction that corresponds to the width of the grid scale $\overline{\Delta}$ was built into β to ensure the correspondence of the dimension. In this study, unsteady and non-uniform concentration distribution was analyzed in three dimensions by CFD and governing equations of the SIR model were analyzed in two dimensions at a height of 1.6 m (y) from floor level (breathing zone level).

OUTLINE OF NUMERICAL ANALYSIS

Waiting Space in University Hospital

In this study, a waiting lounge of a university hospital is analyzed. Figure 1 shows an outline of the hospital space and a photo of its exterior.

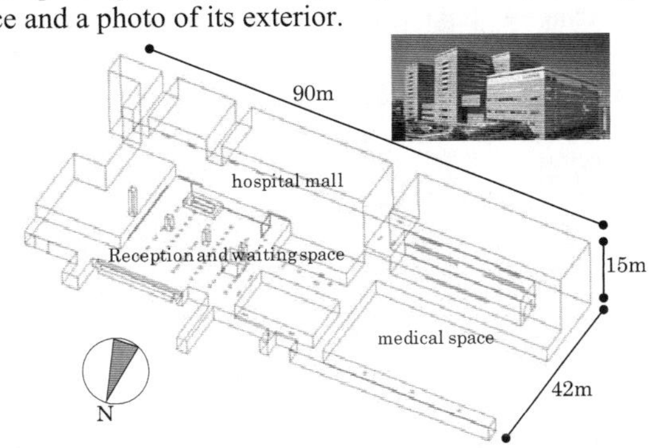

Figure 1 Perspective view of university hospital

The plan of this analytical space is 42 m × 90 m with a total floor area of 2020 m^2. The first floor of this hospital consists of three zones: (i) reception and waiting space on the north side, (ii) hospital mall on the south side, and (iii) medical space on the west side. The height of zones (i) and (iii) is 3.0 m and that of zone (ii) is 15 m. The heating, ventilation, and air-conditioning (HVAC) system was designed and constructed in accordance with these three zones. At (iii) medical space on the west side of the hospital, because the doors of each room were always closed and the HVAC system was also independent, zone (iii) medical space was excluded for the numerical analytical domain. Accordingly, the flow fields and contaminant distribution in (i) reception and waiting space on the north side and (ii) hospital mall on the south side were analyzed.

Multiple fan coil units (FCU) are arranged in zones (i) and (ii) in order to control indoor temperature. A total of 79 supply inlet openings of the air-conditioning system are installed on the ceiling and air is exhausted through lavatories in three places. The geometries of hospital space, furniture, and the supply inlet and exhaust outlet opening were simplistically modeled in order to avoid major effects on the prediction accuracy.

Numerical and Boundary Conditions and Cases Analyzed

Numerical and boundary conditions of CFD simulation and analyses of SIR model are shown in Table 2 and Table 3, which summarize the cases analyzed.

Flow fields were analyzed by the RNG k-ε model in the steady state condition. The SIMPLE algorithm was used with the QUICK scheme for the convective terms, and a second-order center difference scheme was used for the others. After steady flow field analysis under isothermal condition, unsteady aerosol contaminant concentration distributions were analyzed by solving ensemble-averaged scalar transport equation (6) in Table 1 based on a Eulerian approach. Three types of contaminant were assumed: passive contaminant as gas phase, and aerosols of 10 μm and 50 μm in diameter. Concerning the aerosol dynamic equation, convection, diffusion, and gravitational settling were considered. Gradient zero of scalar was adopted as wall surface boundary condition. The contaminant was assumed to be an infectious bio-aerosol generated through 16 points of supple inlet opening of the air-conditioning system in zone (i) reception and waiting space constantly for two hours, and subsequently changed to zero concentration at the supply inlet positions. The contaminant concentrations of aerosol in this analysis were normalized by using supply inlet concentration of the air-conditioning system.

The total number of meshes was set to approximately 600,000 and unstructured mesh was used for the analysis. The analysis was carried out in three dimensions. The air inlet velocity from the air-conditioning system and the turbulent intensity were set to U_{in} = 0.5 m/s and 10%, respectively, in accordance with the drawing and specifications of the target hospital.

Table 2 Numerical and Boundary Conditions

Turbulent model	RNG k-ε model (3-dimensional Cal.)
Difference scheme	Convection Term: QUICK
Inflow boundary	U_{in}=0.5 [m/s] for n=4.36[h^{-1}], $k_{in}=3/2\times(U_{in}\times0.1)^2$, $\varepsilon_{in}=C_\mu^{3/4}\times k_{in}^{3/2}/l_{in}$, l_{in} =0.03[m]
Outflow boundary	U_{out}= free slip, k_{out}= free slip, ε_{out}= free slip
Wall treatment	Velocity, generalized log low
Contaminant	Passive Contaminant Gravitational settling velocity: v_p= 3.0×10^{-3} m/s (D_p=10 μm), 7.5×10^{-2} m/s (D_p=50 μm) D_ϕ= 2.4× 10^{-12} [m^2/s] (D_p=10 μm), D_ϕ= 4.7× 10^{-13} [m^2/s] (D_p=50 μm),
SIR model	S_0=0.15[population/ m^2], I_0=5.0^{-4}[population/ m^2], p=1.67×10^{-4}[m^3/s], $\overline{\Delta}$ = 0.25[m^2], μ=2.8×10^{-4}[1/s], D_S=D_I=2.8×10^{-4} [m^2/s]

Table 3 Cases Analyzed

Analysis case	Temp. condition	Aerosol Source	Aerosol diameter	Basic reproduction number R_0 [-]	Infectivity titer a [-]	Diffusion coeffi. D_S, D_I [m^2/s]
Case 1	Isothermal	HVAC system of zone (i)	Passive	15	670	0
Case 2				6	270	
Case 3				15	670	2.8×10^{-4}, 2.8×10^{-4}
Case 4				6	270	
Case 5			10 μm	15	670	0
Case 6				6	270	
Case 7			50 μm	15	670	
Case 8				6	270	

The transport equations of (S), (I), and (R) were analyzed in conjunction with solving the unsteady contaminant concentration distribution in hospital space. In the Wells-Riley model, which can evaluate airborne infection risk and predict the number of susceptibles (S) as a function of infectious contaminant concentration and exposure time by respiration, contaminant concentration is expressed as 'quanta' intensity. In modeling 'quanta' separately for simple scalar C transport and the infectivity titer a, independent analyses of scalar transport and infectious risk become possible, as shown in equation (4) in Table 1. In this analysis, infectivity titer "a" was set at two levels: 670 [-] and 270 [-], and these values agree with the quanta concentrations of the Wells-Riley model of 670 [quanta/m^3] and 270 [quanta/m^3], respectively.

In this study, unsteady and non-uniform concentration distribution was analyzed in three dimensions by CFD and governing equations of the SIR model were analyzed in two dimensions at the height of the breathing zone (1.6 m height).

CFD simulation is preferable to understand a non-uniform distribution in a comparatively short time scale. On the other hand, SIR-type epidemiology model is often used for infection transmission prediction for a long time scale, such as weeks or months. In this analysis, the condition for highly infectious contaminant generated in the target space was assumed, and unsteady analyses were carried out for 6 hours in consideration of the time usually spent in the hospital lobby.

In this analysis, the population of infectors did not recover (γ=0) and they were assumed to die at a constant rate after infection. γ in equation (2) was replaced with death rate (μ) in this analysis. The initial population of susceptibles S_0 was set at 0.15 [population/m^2], which corresponds to a total of 300 people in hospital. I_0 was assumed to be 5×10^{-4} [population/m^2], which corresponds to 1 infected person in the hospital. The contact rate (β) was calculated by infectivity titer (a) and prediction results of contaminant concentration in the breathing zone (C at y=1.6 m). The basic reproductive ratio (R_o) was set at two levels: R_0=15, which corresponds to measles, and R_0=6, which corresponds to rubella, and are extremely high values.

The movement of susceptibles or infectors could be reproduced approximately by giving the diffusion coefficient. This diffusion coefficient was estimated by using representative length scale in hospital space, nominal time constant (time scale estimated by space volume and ventilation rate), and diffusion speed.

RESULTS

Here, the results of flow fluid in hospital waiting space, concentration distribution of infectious contaminant, and population density of susceptibles (S) and infectors (I) by coupled simulation of unsteady CFD and SIR-type epidemiological model are shown in series.

Flow field

Figure 2(1) shows the result of flow fluid analysis of x-z plane at y=1.6 m height and representative x-y section with isothermal condition. The flow field in hospital waiting space was almost stagnant and air velocity was below 0.1 [m/s] except for the vicinity of jet flow from a supply inlet (U_{in}=0.5 [m/s]). A non-uniform distribution of flow field was confirmed.

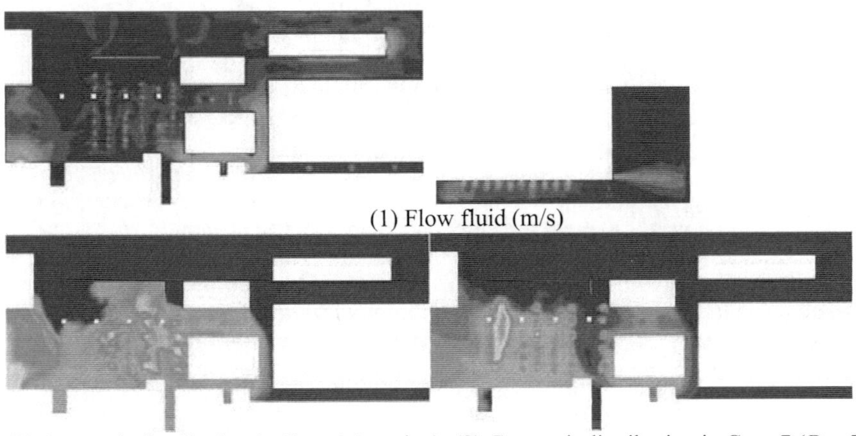

(1) Flow fluid (m/s)

(2) Contami. distribution in Case 1 (passive) (3) Contami. distribution in Case 7 (Dp=50 μm)

Figure 2 Results of flow fluid and contaminant distribution in hospital (y=1.6 m, t=2 hours)

Contaminant concentration distribution

Figures 2(2) and (3) show the results of contaminant concentration distribution at two hours after the contaminant generation start in two representative cases: Case 1 for passive contaminant and Case 7 for aerosol particle (Dp=50 μm). The results of Case 5 (Dp=10 μm) were almost the same distribution as Case 1 and are omitted in this paper. Infectious contaminant was generated through the supply inlet opening of the air-conditioning system in zone (i). The concentrations in Figure 2(2) and (3) were normalized by supply inlet concentration (C_{in}).

Non-uniform concentration distributions were formed in the space for both bases. In the case of passive contaminant (Case 1 in Figure 2(2)), the contaminant stayed and widely diffused within zone (1) of the hospital waiting space. On the other hand, in the case of Dp=50 μm size aerosol (Case 7 in Figure 2(3)), non-uniformity of concentration distribution became clear and A remaining in the high-concentration zone was confirmed in the center of the zone (i) compared with that in Case 1. This was because of the effect of gravitational sedimentation of Dp=50 μm size aerosol.

Transmission analysis and distributions of (S) and (I)

Figure 3 shows the population density of susceptible (S) and infectious (I) distributions in the x-z plane at y=1.6 m height at 2 hours from the start of analysis for two representative cases: Case 1 and Case 7.

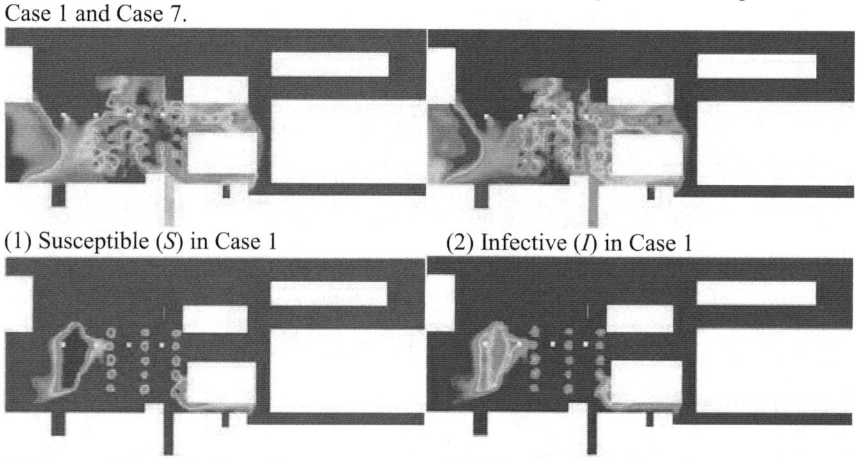

(1) Susceptible (S) in Case 1 (2) Infective (I) in Case 1

(3) Susceptible (S) in Case 7 (4) Infective (I) in Case 7

Figure 3 Distributions of population density of Susceptible (S) and Infective (I) (y=1.6 m, t=2 hours)

In Case 1 as shown in Figures 3(1) and (2), the distributions of (S) and (I) were similar, which was because (S) and (I) were assumed to be motionless and converted from (S) to (I) at the same points. (S) and (I) exhibited inhomogeneous distributions caused by non-uniform distribution of contaminant. In other words, there were inhomogeneous distributions of infection risk in this hospital waiting space. In Case 7, for which the dispersion of Dp=50 μm size aerosol was assumed, the distributions of (S) and (I) were not spread and were caused only in the high concentration region at the center of zone (i) corresponding to Figure 2(3).

Figure 4 shows the time series of average values of (S) and (I) for each case. For Case 2, Case 4, Case 6, and Case 8, for which the infectivity titer "a" was set at a higher value (R_0=6 and a=270), contact rate β became small in a short time scale and the time series of (S) and (I) showed constant profile. On the other hand, for Case 1, Case 3, Case 5, and Case 7, for which the infectivity titer "a" was set at a higher value (R_0=15 and a=670), space-averaged (S) and (I) greatly changed and, especially the time profile of (I), had a clear local maximum value and then decreased gradually. Considering the diffusion of (S) and (I) in Case 3 and Case 4, the number of (S) was decreased and that of (I) was increased compared with those for Case 1 and Case 2. This is because the mixture of (S) and (I) was accelerated and the reaction/generation term (βSI) of the transport equation of (I) was increased.

Compared with Case 1 and Case 7, the time profiles of (S) and (I) were completely different in these two cases and the characteristics of unsteady and non-uniform concentration distribution of infectious contaminant were confirmed to impact on infectious risk in hospital space.

The peak value of (I) in Case 1 was 0.032 [population/m^2], which corresponded to 64 people infected.

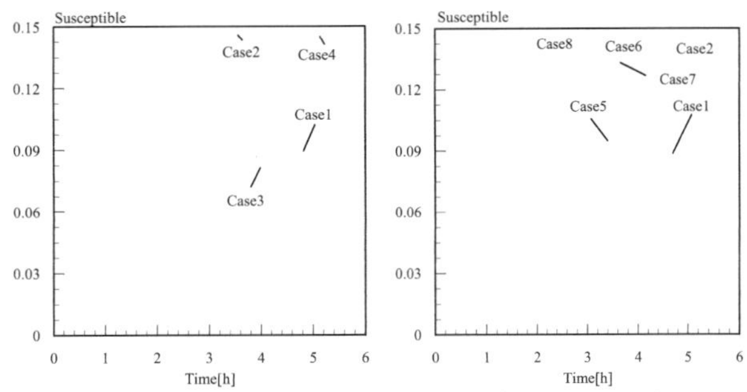

(1) S in Cases 1 - 4 (2) S in Cases 5 - 8

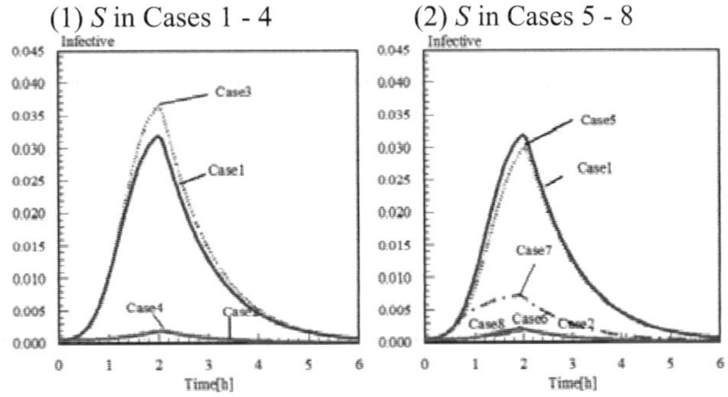

(3) I in Cases 1-4 (4) I in Cases 5-8

Figure 4 Time series of population density of (S) and (I) for each case
(Vertical axis indicates space-averaged value of population density)

DISCUSSION

In this study, numerical analyses of unsteady and non-uniform concentration distribution by three-dimensional CFD and the procedures of coupled simulation between CFD and SIR-type epidemiological models were carried out. On the basis of these proposed procedures, the time dependence and distribution of infectious risk in indoor environments could be predicted quantitatively. Contaminant concentration in the breathing zone was strongly influenced by the setting of the ventilation condition and position of contaminant source, and therefore contact rate (β) was estimated as unsteady and inhomogeneous; then, space distributions of (S) and (I) were greatly changed in accordance with the boundary condition.

As shown in equations (1)-(3), constant (linear) and isotropic diffusion coefficients were adopted to reproduce movement of residents in hospital space. In order to reproduce human behavior more accurately, it is necessary to adopt a non-isotropic diffusion model and analysis by cellular automata. This will also be investigated in future.

CONCLUSION

In this paper, an integrated analytical procedure of unsteady CFD simulation and SIR-type epidemiological model for infectious transmission in hospital space was proposed and the results of numerical simulation were also demonstrated.

Under the assumption of a constant linear rate of pulmonary ventilation rate and contaminant concentration in the breathing zone, contact rate β can be directly estimated. Through the analysis of infectious contaminant concentration level and non-uniform distribution in a large hospital space with CFD, the prediction of the number changes of (S), (I), and (R) becomes possible through the analysis of β.

As a result of this analysis, the contaminant concentration in the breathing zone was found to be strongly dependent on the characteristic of target contaminant, that is, the aerosol diameter, and then the distributions of (S) and (I) were greatly changed in accordance with unsteady contaminant conditions. We believe that the prediction procedures of coupled analysis of CFD and SIR model are effective for infectious risk assessment in indoor environments with reasonable accuracy.

ACKNOWLEDGEMENTS

This research was partly supported by a Grant-in-Aid for Scientific Research (JSPS KAKENHI for Young Scientists (S), 21676005). The authors would like to express special thanks to the funding source.

REFERENCES

[1] Li, Y., Huang, X., Yu, I. T. S., Wong, T. W., Qian, H., Role of air distribution in SARS transmission during the largest nosocomial outbreak in Hong Kong, ***Indoor Air***, 15 (2) pp83-95, 2004

[2] Kermack, WO. and McKendrick, AG., A contribution to the mathematical theory of epidemics, ***Proceedings of the Royal Society of London Series*** *A*, *115*, pp 700-721, 1927

[3] Fine,P.E.M., Herd immunity: history ,theory, practice, ***Epidemiological Review*** *15(2)*: pp265-302, 1993

[4] Ito, K., Numerical prediction of Infection Process by Coupled CFD and SIR model in Indoor Space, *Indoor Air 2011*, The 12th International Conference on Indoor Air Quality and Climate, Austin, Texas, June 5-10, 2011, Manuscript number (a258-2)

[5] Fukuoka, T., Ito, K., Exposure Risk Assessment by Coupled Analysis of CFD and SIR model in Enclosed Space, *AIVC 2010*, Low energy and sustainable ventilation, Technologies for green buildings, Seoul, Korea, Manuscript number (1A-3)

[6] Yakhot, V. and Orszag, S.A., Renormalization group analysis of turbulence, I. Basic theory. ***J. Sci. Comput.***, ***1***: 3-51, 1986

SHELTER-IN-PLACE EFFECTIVENESS IN THE EVENT OF TOXIC GAS RELEASES: FRENCH AND CATALAN ASSESSMENT APPROACH

M.I. Montoya[1], G. Guyot[2,*], E. Planas[1]

1 Centre d'Estudis de Risc Tecnològic (CERTEC), Universitat Politècnica de Catalunya Av. Diagonal 647, CP-08028 Barcelona, Spain

2 Centre d'Etudes Techniques de l'Equipement (CETE) de Lyon - 46 rue Saint Théobald. F – 38081 L'Isle d'Abeau Cedex, France
**Corresponding author:*
Gaelle.Guyot@developpement-durable.gouv.fr

ABSTRACT

Origins of toxic gas clouds may be diverse, including accidental releases due to industry or to hazardous materials transportation, or biological or chemical attacks. A protection to such a phenomenon consists in taking advantage of the protection offered by buildings against airborne pollutants. In this event, people can shelter in a building and wait until the toxic plume has gone. European directive concerning major accidents hazards control, known as SEVESO II, requires from Member State's local authorities, to ensure that appropriate measures are taken to guarantee people's protection when located in areas close to Seveso facilities.

In France and in Catalonia, two different political approaches are set to assess the effectiveness of shelter in place in the event of toxic gas releases. This paper exposes and discusses both approaches, including: 1- a brief presentation of the political, legal and technical strategy developed in both countries to face up the imminent risk due to industrial premises, storing and manipulating hazardous materials; 2- a description of the methodology developed in both countries to assess shelter in place effectiveness in buildings; 3- a study case illustrating both approaches in order to assess their consistency.

KEYWORDS

Air infiltration, leakage, shelter-in-place, airflow calculation, vulnerability, toxic releases, outdoor dispersion, dwellings, airtightness

INTRODUCTION

During the last decades, the number of accidents in chemical industries and during transportation of hazardous substances has significantly increased, most of them occurring in inhabited areas. This fact is a huge challenge for local administrations who must minimize risks around industrial facilities where major accidents can occur, and must provide a safe community emergency response in case of accident.

In Europe, following the Seveso accident in Italy, the European Council adopted the SEVESO Directive (EU 85/501/EEC) in 1982, which deal with major accident hazards of certain industrial activities. Later, in 1996, this directive was replaced with the Council Directive 96/82/EC on the control of major accident hazards, known as the Seveso II Directive. Then, in 2003, it was enlarged with the Directive 2003/105/EC. Since 1999, the Directive obligation have been compulsory for industry as well as for the Member States public authorities that are responsible for the implementation and enforcement of the Directive in every country [1]. The legislation considers as potentially dangerous any activities where specified hazardous substances are present in certain quantities. In fact, this Directive classifies industrial facilities that manipulate these hazardous (called Seveso facilities) in two categories, high and low level, depending on the quantity of on-site classified substances.

Major hazards faced by industrial facilities comprise fires, explosions and toxic releases. Fires (i.e pool fire, flash fire, jet fire) are most common, but explosions (i.e BLEVE, confined and unconfined explosion) are more significant in terms of potential damage. Nevertheless, along the history, toxic clouds are recognized for their greatest potential to kill, or injure people and pollute zones for weeks or months. They can affect wider areas than fires or explosions, and may be extremely dangerous in case of released highly toxic substance: the methyl isocyanate Bhopal catastrophe (1984), entailed more than 2500 deaths and 10000 injured people [2][3]. Toxic clouds may originate in direct releases, as a consequence of domino effect or in formation of hazardous compounds as a result of combustion, runaway reactions or unwanted reactions. Article 12 of the Seveso II requires from local authorities of every Member State, to ensure that appropriate measures are taken to guarantee the protection of people living in areas close to Seveso facilities. Hence, the study and assessment of shelter in place effectiveness is of great importance for local administrations: in addition to organize protection of the community in the event of such an accident, they must deal with different urban and land use planning situations (authorization for the implantation of new Seveso facilities, authorization of any substantial change and urban growth in the vicinity of present Seveso facilities) near Seveso facilities [4].

The shelter-in-place (SIP) strategy consists in taking advantage of the protection offered by buildings against airborne pollutants. This protection is based on the fact that the building acts as a barrier that slows down the toxic substance entrance. Therefore, the substance inside concentration would be lower than outside, as well as the toxic load (TL) to which people are exposed. The simplest way of sheltering in place consists in closing all external openings, such as doors and windows, turning off all mechanical ventilation systems and closing openings to reduce outdoor air entrance. So, the only way for pollutant entrance is air infiltration. Then, protection can be improved by seeking refuge in a room with limited air infiltration: 1- seal doors and windows with tape; 2- implement structure modifications to improve the building or an internal room envelope airtightness.

ASSESSMENT OF SIP EFFECTIVENESS: CATALAN APPROACH

Catalunya is a high-industrialized region located in the north-eastern part of Spain, and concerned with numerous high risk level companies according to the Seveso II Directive. Current Catalan legislation regarding the control and planning of severe accidents involving hazardous materials [4][5], requires these facilities to present a risk analysis where different risk zones must be established. In case of toxic gas dispersion, 4 zones are determined: the alert (based on the threshold AEGL-1, [6]), the intervention (based on the AEGL-2), the lethal area of 1% and the fourth zone corresponding to a 0.1% lethal dose inside buildings, when taking into account the implementation of shelter in place. The fourth zone is used to delimit the evacuation area (the evacuation radius). It requires the estimation of indoor concentration (C_i) and indoor toxic load (TL_i). The methodology currently used by the Catalan Government considers several factors such as the involved substance, the site conditions, the cloud duration, and the air infiltration exchange rate (*ACH*), considered as constant. However, the use of a fix *ACH*, does not take into account neither, buildings' airtightness distribution in the affected area, nor the *ACH* distribution under prevailing and worst meteorological conditions, which can lead to under or overestimate the real evacuation radius [7].

In this paper, we used a simplified methodology (see method 1 in Figure 1), proposed by Montoya [8]. It includes the estimation of airtightness distribution in the affected area depending on buildings' features and meteorological conditions, as described below:

1. Define source term conditions: meteorology, data related to substance properties.

2. Estimate outdoor gas dispersion and establish the affected zone.
3. Create a grid over this zone and identify the census tracts involved.
4. Calculate the airtightness (c') distribution by census tract using the UPC-CETE airtightness model [8].
5. Compute the *ACH* by census tract, using the AIM-2 ventilation model and the 100th percentile of c' distribution.
6. For each grid cell, compute C_i and TL_i profiles using the *ACH* of the cell's census tract.
7. Estimate the casualty probability for each grid cell, through the Probit analysis, and establish the evacuation radius, i.e the largest distance between the source and the last downwind cell with a casualty probability of 0.1%.

In addition to closing all ventilation openings and turning off the mechanical ventilation systems, people can increase the protection level by seeking shelter in an indoor room. In that case, Montoya et al. [9] found that the indoor shelter *ACH* when sealed; consists approximately in a 35% reduction of the dwelling *ACH*. However no structural works on shelter are performed, in order to improve building or indoor room airtightness.

ASSESSMENT OF SIP EFFECTIVENESS: FRENCH APPROACH

In France a prevention strategy based on mandatory works in an inner room is being implemented. People should seek refugee in such a room and be protected during 2 hours against irreversible effects while the toxic cloud passes away. It applies to existing and future buildings located near Seveso II facilities. Local land-use plans, named technological risk prevention plan (PPRT) [10], specify airtightness requirements for such shelter inside dwellings. However, buildings, when very close to the toxic site, when the risk is very high, could be expropriated.

This second methodology (see method 2 of Figure 1) is based on abacus we elaborated with CONFINE [11], assuming that every dwelling can be modeled as a standard 3-zones dwelling (Figure 2) with a default envelope airtightness level (Q_{4Pa_surf}, the airtightness indicator in French Thermal regulation [12]) estimated from the CETE airtightness database[1] (the 95th percentile).

The method is described here below:

1. Define the PPRT impacted area, based on the safety report supplied by the operator (list of all the possible dangerous phenomena, their probability and the forecast intensity of their effects).
2. Define different zones along with the intensity of the aggression (irreversible, lethal 1%, or lethal 5% effects) and on the types of pollutants. For each zone, a conventional toxic cloud (60 min duration) is also defined.
3. Select the abacus corresponding to the meteorological conditions used for precedent zoning. Extract the shelter airtightness requirement (n_{50},[12]), corresponding to this conventional toxic cloud, and to the limit indoor concentration (French threshold is close to AEGL-2 120 min). Abacus are drawn for down and upwind shelters.
4. For each zone, inscribe those shelter airtightness requirements into the final PPRT-plan.

Contrarily to non-residential buildings, there is no need to use a modeling software such as CONFINE to define the shelter airtightness level of dwellings.

[1] 402 dwellings in 2007

PERFORMANCE OF BOTH APPROACHES: A STUDY CASE

In order to assess the consistency of both scientific approaches, we realized a comparative study described in Figure 1. It consists in determining the evacuation radius with method 1. Then, according to method 2, for a standard single family dwelling located at different points inside and outside the evacuation radius, we estimate through CONFINE the required airtightness for a limit indoor concentration equal to AEGL-3_30 min. If the required airtightness value is lower than the representative airtightness value taken from the SIP-CETE database[2] (e.g. the 90th or 50th percentile of the distribution, which would represent the probable airtightness level of the shelters in the zone), the conclusion is that sealing works are needed in order to achieve a safe shelter. On the contrary, if the required airtightness value is higher than the representative airtightness, the dwelling provides a safe shelter. Our study case comprises a 60 t chlorine release from a storage tank. For the calculations, we considered an industrial area site with a ground roughness (z_o) of 0.3 m, 15°C of temperature, a 4 m/s wind speed and neutral stability (4D).

Results obtained

Following the above mentioned steps, the application of method 1 gave an evacuation radius of 2700 m. Outside, shelter-in-place is implemented. Therefore we selected 5 points downwind the release source: 1200, 1800, 2400, 2700 and 3200 m. For each location, we estimated the airtightness level needed to provide a safe environment for a shelter in a standard dwelling. The required airtightness value was calculated using CONFINE for two different shelter situations: upwind and downwind. As can be seen in Table 2, we can firstly observe that downwind shelters are more protective. Then, for the 50th percentile of the SIP-CETE airtightness database, shelters in both situations, downwind and upwind, are protective for a distance over 2700 m, the evacuation radius estimated with method 1. On the contrary, when the representative airtightness level was assumed to be the 90th percentile of the SIP-CETE airtightness database, only downwind shelters provided a protected environment for 2700 and 3200 m. In this case, upwind shelters with this same airtightness level would require structural sealing works. From these results, carried on only one case study, we can say that equivalent results between both methods were obtained when assuming the representative airtightness level as the 50th percentile. If assuming the 90th percentile as the representative airtightness value method 2 is more conservative than method 1.

CONCLUSION

This paper gave a brief description of both French and Catalan strategies to comply with the Seveso II Directive in their territory. We have assessed the consistency of two scientific approaches, one used in France and the other proposed for Catalonia, to face up the risk due to a toxic cloud, considering shelter in place as a protection measure. We found a complete coherence between both approaches when the 50th percentile of the SIP-CETE airtightness database was used as the representative airtightness level for shelters in the zone. However, the small size of the SIP-CETE database and the differences in construction techniques may affect the extreme values of the distribution. Therefore more work is needed, both in France and Catalonia: 1- to obtain, experimental data on building airtightness and *ACH* measurements, especially on internal rooms; 2- to carry many more case studies in order to specify the most convenient airtightness level percentile.

---

[2] 39 shelters in 2011, this database has just been started in July 2011

ACKNOWLEDGEMENTS

The authors are very grateful to the French ministry for ecology, sustainable development, transport and housing (MEDDTL) and to the Autonomous Government of Catalonia. The sole responsibility for the content of this publication lies with the authors. It does not necessarily reflect the opinion of the Ministries.

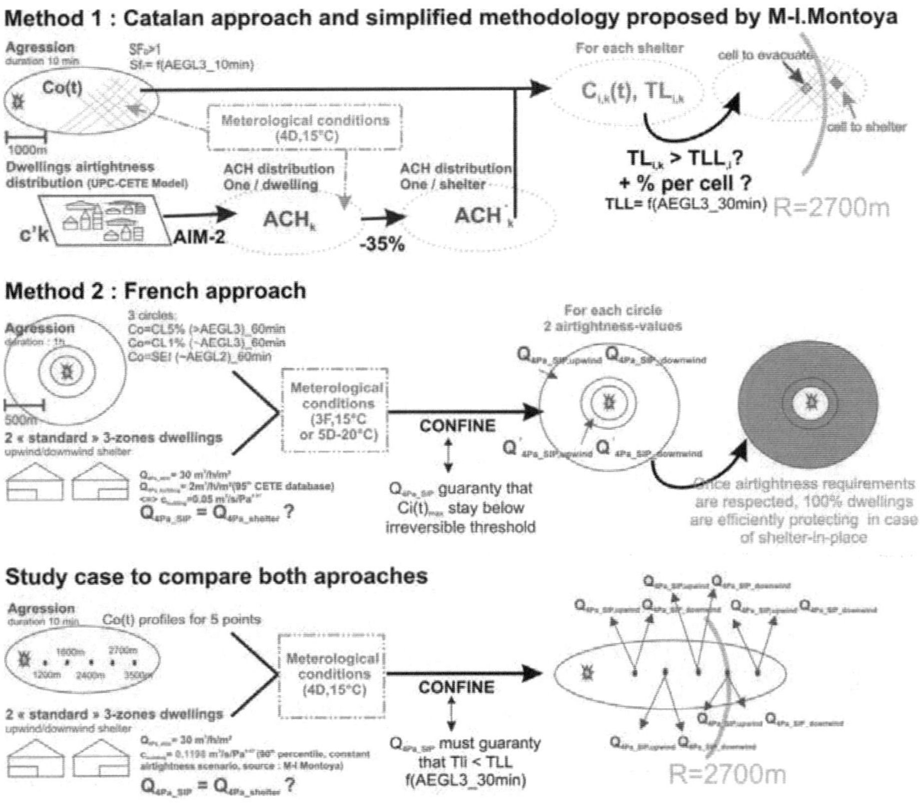

Figure 1 : Description of the different approaches to assess shelter in place effectiveness

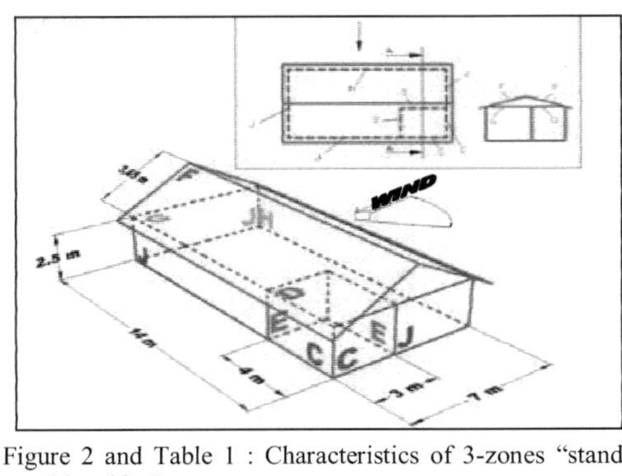

$V_{shelter}$ (m^3)	27
V_{attic} (m^3)	98
$V_{rest\ of\ the\ building}$ (m^3)	251
$H_{building}$ (m)	4.2
Slope of the roof (°)	25
$Q_{4Pa\_Surf,attic}$ ($m^3/h/m^2$)	30
$Q_{4Pa\_Surf,building}$ ($m^3/h/m^2$)	2

Figure 2 and Table 1 : Characteristics of 3-zones "standard single-family dwelling", with downwind shelter considered in France

Location (m)	Conclusion method 1	Standard dwelling	Shelter airtightness required n_{50} (h^{-1})	Conclusion based on the n_{50} (90th)*: 13,3 h^{-1}	Conclusion based on the n_{50} (50th)*: 8,3 h^{-1}
1200	Evacuation	Downwind	9,0	sealing works	sealing works
		Upwind	2,0	sealing works	sealing works
1800	Evacuation	Downwind	14,8	protected	Protected
		Upwind	3,2	sealing works	sealing works
2400	Evacuation	Downwind	34,0	protected	Protected
		Upwind	6,7	sealing works	sealing works
2700	Shelter in place	Downwind	46,0	protected	Protected
		Upwind	8,6	sealing works	Protected
3200	Shelter in place	Downwind	84,2	protected	Protected
		Upwind	12,9	sealing works	Protected

* The n_{50} correspond to the 90th and 50th percentile of the SIP-CETE airtightness database, respectively.

Table 2. Results obtained from comparative study

REFERENCES

[1] European Comission : Environment. *Chemical Accidents (Seveso II) - Prevention, Preparedness and Response.* http://ec.europa.eu/environment/seveso/index.htm.

[2] Khan F.I., Abbasi S.A. 1999. *Major accidents in process industries and an analysis of causes and consequences.* J Loss Prevent Proc, 12, 361–378.

[3] Vílchez J.A., Sevilla S., Montiel H., Casal J. 1995. Historical analysis of accidents in chemical plants and in the transportation of hazardous materials. J Loss Prevent Proc, 8 (2), 87-96.

[4] Tost S. 2008. *El panorama de la seguretat industrial a Catalunya. El cas de planificació del territori als voltants dels establiments amb risc d'accident greu.* Revista Catalana de Seguretat Pública.

[5] Llei 12/2008 de 31 de julliol, de Seguretat Industrial. Parlament de Catalunya.

[6] EPA. 2009. U.S Environmental Protection Agency. Website available at: www.epa.gov/oppt/aegl/

[7] Montoya M.I. Pastor E. Carrié F.R. Guyot G. Planas E. 2010. *Air leakage in Catalan dwellings: Developing an airtightness model and leakage airflow predictions.* Building and Environment 45 (6), 1458-1469.

[8] Montoya M.I. 2010. Contribution to the assessment of shelter in place effectiveness as a community protection measure in the event of a toxic gas release. Ph.D Dissertation, Universitat Politècnica de Catalunya, Spain.

[9] Montoya M.I., Pastor E., Planas E. 2011. *Air infiltration in Catalan dwellings and sealed rooms: An experimental study.* Building and Environment 46(10), 2006-2011.

[10] French ministry of Ecology and Sustainable Development, *Technological risk prevention plan (PPRT)-Acting together to control risks*, 3p., 2007.

[11] Guyot, G., Gentilhomme, O., Carrié, F-R., *Shelter-in-place strategy : CONFINE, an airtightness level calculation tool to protect people against accidental toxic releases*, 6p., 32nd AIVC Conference, Brussels, October 2011.

[12] EN 13829: February 2001 – Thermal performance of buildings - Determination of air permeability of buildings - Fan pressurization method (ISO 9972:1996, modified)

QUALITY SYSTEM FOR AIRTIGHTNESS MEASUREMENT OF BUILDINGS

Oliver Solcher

Fachverband Luftdichtheit im Bauwesen e.V.,
Wissenschafts- und Technologiepark, Berlin Adlershof, Kekuléstraße 2-4, 12489 Berlin, Deutschland,
Tel: +49 (0) 30 / 6392 - 5394, Fax: +49 (0) 30 / 6392 - 5396, E-Mail: info@flib.de, www.flib.de

ABSTRACT

In 2002 the Association for Air Tightness in Buildings (FLiB e. V.) established a certification procedure for airtightness testers in Germany. As airtightness tests are part of national implementation of EPBD but no qualification requirements for testers are defined FLiB e. V. took the lead and defined a procedure.

Part of the certification procedure is the proof of knowledge of airtightness testing. The tester shall be engineer, technician or a master craftsman. Furthermore the ability of testing must be proven be sending 5 test protocols acc. EN 13829 or by attending a FLiB approved education program.

The FLIB certification is recognised of several funding organisations in Germany that make it mandatory to have the airtightness test been performed by a certified tester as e.g. a tester with FLiB certificate.

KEYWORDS

Certification, airtightness measurement, round robin test, leakage tester

INTRODUCTION

With the EnEV 2002 the national implementation of EPBD airtightness of new buildings should not exceed certain values. The test shall be performed acc. EN 13289 without any qualification requirement for the tester.
The Association for Air Tightness in Buildings (FLiB e.V.) had worked out the qualification program before so with start of EnEV 2002 FLiB e.V. started its certification program for airtightness testers.

NECESSITY OF CERTIFICATION

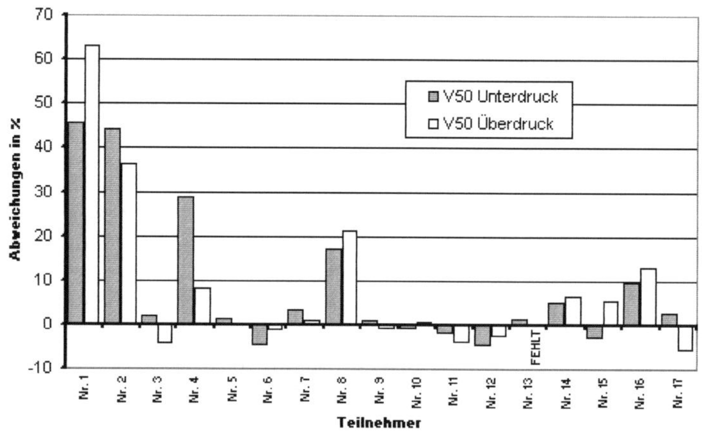

Figure 2. Variation of Leakage airflow during a round robin test [1]

The goal of certification of airtightness testers is to have unified test results that are comparable and reproducible. As different preparation of the building, calculation of the volume or handling of the test units produces different test results on the same building it is necessary to perform the test the same/right way in order to have reliable test results.
Especially for heating and energy demand calculations correct figures are needed, as a wrong n_{50}-value might lead to a false classification of the building or a wrong dimension of the heating system.

FLiB is working for several years to establish a checklist for building preparation as basis for leakage tests acc. EN 13829.

EDUCATION OF TESTERS

As tests shall be performed according to the standards the testers must be able to understand how a building must be prepared and under which conditions the test must be performed to produce the needed figures.

To be able to pass the certification process of FLiB, leakage testers shall have a technical education as engineer, technician or master craftsman. Although there is a way for those to be certified who took a side entry to airtightness testing via a reasoned request to the examination board.

Testers shall proof their testing ability. There are two ways to do so. The easiest is to send 5 test reports acc. EN 13829 to the examination board. If those reports meet the requirements of the standard the tester is allowed to participate at the certification procedure.
The second way to proof their testing ability is to attend a FLiB approved education program.

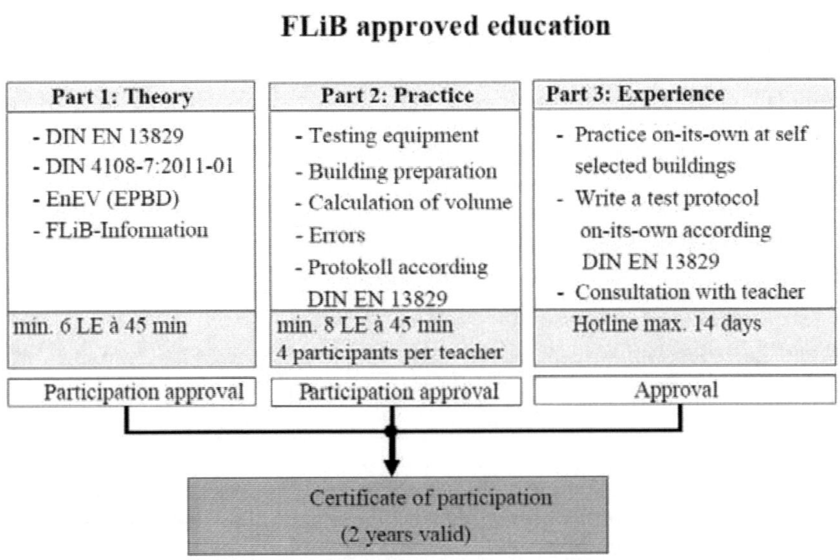

Figure 2. FLiB approved education

In Germany we have three training facilities that have the FLiB approval: Energie und Umweltzentrum (euz) in Springe (since 2003), Zentrum für umweltbewusstes Bauen (ZUB) in Kassel (since 2003) and Kerschsteiner Schule in Reutlingen (since 2011).

CERTIFICATION

Certification procedure contains of two parts:

- Theoretical test
- Practical test

Certificate cam only be issued if both parts were past.

Theoretical test

Test contains around 30 questions, partly multiple choices. The questions are about the following themes:

- Part of the building that needs to be tested
- Convenient time of testing
- Building preparation
- Requirements on weather conditions
- Test procedure
- Test equipment
- Function and inspection of test equipment
- Limitation and error possibility of the used test equipment
- Layer and sections of air tightness
- What must be checked - what could go wrong during test?
- Necessity of building inspection
- Evaluation of leakages
- Test report

- Calculation of building volume
- Specifications n_{50}, q_{50}, w_{50},
- Standards
- FLiB-Information to EN 13289
- Knowledge of typical errors and its sources
- How to avoid systemic errors
- Knowledge of its own limitation
- Basic knowledge of building physics to humidity, dew point, etc.

Test is performed usually direct before practical test under the survey of the commissioner and one attendant. Testers must have 60 % of the available points to pass the test. The test is checked from the commissioner and one member of the examination board. In critical cases a third check is performed.

Practical test

The practical test is performed at a building or part of a building one tester at the time. The tester received the buildings drawings in advance to prepare the volume calculation. The tester shall bring its own equipment; test can be performed with any available and usable measuring device. The commissioner interviews the tester during he/she performs the test. The tester shall show that he is able to work with his/her test equipment, to prepare the building all right and to find the leakage by performing a leakage test.
The commissioner judges the test quality and the testing ability of the tester.

Recertification

The certificate is valid for 3 years. It can be extended for 3 more years each time. The request must be send to the examination board with a proof of their ability. This can be:

- Attendance at a seminar
    - Advanced education
    - Training
    - Symposium
- Seending in 5 test protocols according EN 13829

CERTIFICATION IN THE FUTURE

Some other certification procedures have been established in Germany after the FLiB but FLiB certification is one with a very high reputation. .So it's recognised of several funding organisations in Germany that make it mandatory to have an airtightness test been performed by a certified tester such as a tester with FLiB certificate.
Other certificates are available: Some are given out after a short education from some manufactures. Some others are more sophisticated and follow the rules that have been established by the FLiB.
To give an overview of certification procedures that produce well educated leakage testers FLiB e.V. decided to check out other systems. To have an independent classification of the different certificates, other certification institutes can have their certification procedure been checked by the FLiB e.V.. In case the procedure meets the standards set by FLiB testers are able to recertify at the FLiB.

FIGURES

Since 2002 150 FLiB certificates have been passed out. The figures varied over the years and in 2010/11 there is a strong increase in requests. Certification procedure costs are about 1.900,- EUR including the education program at the training facilities; certification alone is 770,- EUR.

CONCLUSION

As neither the leakage test nor a certified leakage tester is mandatory in Germany but test results still vary a lot (see above) FLiB feels there is still a lot work to do in achieving a good testing ability of leakage testers. FLiB-certification is a good way in securing well educated leakage testers.

Certified testers are often requested as legal experts as testing building leakage, giving a dependable report on the test and to identify the quality of the building needs more then the ability to run a test program.

In certifying and recertifying process testers are checked over the years and give a very positive response when discussing the given test protocols with the examination board. A lot of knowledge is lost or buried over the years of constant testing and to have somebody else looking its work over is appreciated a lot.

REFERENCES

[1] S. Rolfsmeier, K. Vogel, T. Bolender, Ringversuche zu Luftdurchlässigkeitsmessungen vom Fachverband Luftdichtheit im Bauwesen e.V. - Round robin tests to leakage tests in buildings of Association for Air Tightness in Buildings (FLiB e. V.)

The quality framework for Air-tightness measurers in France: assessment after 3 years of operation.

Valerie Leprince*[1], Rémi Carrié [2] and Myriam Olivier [1]

1 Centre d'Etudes Techniques de l'Equipement de Lyon.
46 rue Saint Théobald. F – 38081 L'Isle d'Abeau Cedex, France,
**Corresponding author*
valerie.leprince@developpement-durable.gouv.fr

2 INIVE,
Lozenberg 7
BE-1932 Sint-Stevens-Woluwe, Belgium

ABSTRACT

The 2012 French thermal regulation will include a minimum requirement for residential buildings envelope air-tightness, with two options to justify its treatment: a) measurement at commissioning or b) adoption of an approved quality management approach. This paper describes the qualification process for air-tightness measurement authorized technicians when their results are to be used in the EP-calculation method. Our analyses underline the importance of the qualification process to ensure homogeneous measurement practice among technicians. It also shows the limits of the controls that can be implemented if the process is not appropriately sized to absorb a large number of applications. This process highlight nevertheless the great interest of collecting trusted data through a controlled measurers network .

KEYWORDS

air infiltration, envelope, ductwork, leakage, energy use, low-energy buildings, licensed measurer

INTRODUCTION

The objective of the French energy performance regulation revision is to generalize low-energy buildings. The threshold is set to 50 kWh of primary energy per m^2 per year for residences, with some modulation depending on climate zone and altitude; this value includes the primary energy use for heating, cooling, domestic hot water, auxiliary equipment, and lighting. Given the severe impact of envelope leakage in such low energy buildings, a minimum requirement is set for the envelope air-tightness of residential buildings in this 2012 energy performance regulation. Two options are proposed to comply with this minimum requirement : a) measurement at commissioning or b) adoption of an approved quality management approach. The first option raises the question of the measurement quality to ensure homogeneous practice among technicians, and therefore avoid perverse effects that could entail discredit to the whole framework and distort competition. This paper discusses the measurer qualification process, the difficulties to set such a process and the advantages linked to data capitalization.

This paper is based on a former presentation of the process [1].

QUALITY FRAMEWORK FOR MEASURERS

Although there exists a European and an International Standard (EN 13829 and ISO 9972) on the measurement of envelope air-tightness, there remains unanswered questions when the result is to be used as an input in an energy performance calculation method.

One key problem lies in the building preparation, i.e., the openings that have to be sealed or closed during the measurement. One often quoted example is the case of a biomass boiler with a combustion air intake in the living space, for which it could be interpreted that method A requires that the intake remains in its normal condition of use, although it may make sense to seal it if the calculation method appropriately takes into account the energy impact of the additional air drawn into the house.
This is the reason why the BBC-Effinergie label specifies in more detail the openings that can and cannot be sealed or closed during the test [1]. The basic rule is that the building envelope must represent the conditions prevailing during the season when heating or cooling systems are used; however, openings whose contributions are taken into account in the energy calculation method are sealed. This rule is now integrated in an application guide for EN 13829 that was released in February 2010 (GA P 50-784).

A second issue is the estimation of the measured volume. In order to avoid this uncertainty, the French indicator is $Q_{4Pasurf}$, that is to say the flow-rate at 4Pa divided by loss surfaces area (excluding basement floor). This area is automatically calculated in the thermal study, thus a measurer does not evaluate it himself, which avoids any cheating : if the area is over-estimated (to decrease the $Q_{4Pasurf}$), it is also penalizing as it automatically increases conductive losses.

Another major issue lies in the measurer's competences. As it is clearly bringing to light in Rolfsmeier et al.'s paper [6], there can be serious misinterpretations of the measurement protocol and analysis that can of course be intentional but that can also be done in good faith, simply because of a lack of basic knowledge on the energy performance regulation, HVAC systems, or airflow and pressure measurement techniques. This can entail serious errors in the estimations of derived quantities that are used in the calculation method. This is why the ministry decided to require that the measurers be authorized to perform such measurements. The authorization is delivered according to a 3-step approach :
1. The candidate must attend a training programme, based on a state referential, and approved by the ministry. Five trainings bodies are now offering this kind of training.
2. The candidate must pass a theoretical and a practical examination. The theoretical examination is based on a multiple-choice questionnaire. The practical part is divide in two parts:
  - Examination of a test performing process (including building preparation, etc.)
  - Examination of a test report by the training body.

  The training body is given a grid to help in this evaluation process.

3.The candidate must then submit 5 reports to a commission in charge of advising the ministry to deliver authorizations. This commission is constituted of about 20 anonymous experts in the field. The experts can ask for complements if mistakes are not fundamental or if clarification is needed. In this case, the agreement is discussed at the following meeting based on the new materials provided by the candidate. An application cannot be evaluated more than 3 times; over 3 trials, the candidate must submit a new application.

In addition, a "Frequently Asked Question" internet site has been set to deal with questions that are not answered in standards. When a new technical question appears, it is discussed in an expert working group (that meet five times a year) and the consensual answer is published.

On September 2011, more than 230 persons have been authorized (they were merely 50 one year and half ago). Applicants usually have to go through 2 commission meetings to obtain their authorization. The mean elapsed time between the application and the authorization is about 5.5 months. Figure 1 show the number of authorisation per month and the total from mid 2008 (beginning of the commission), till the end of 2010.

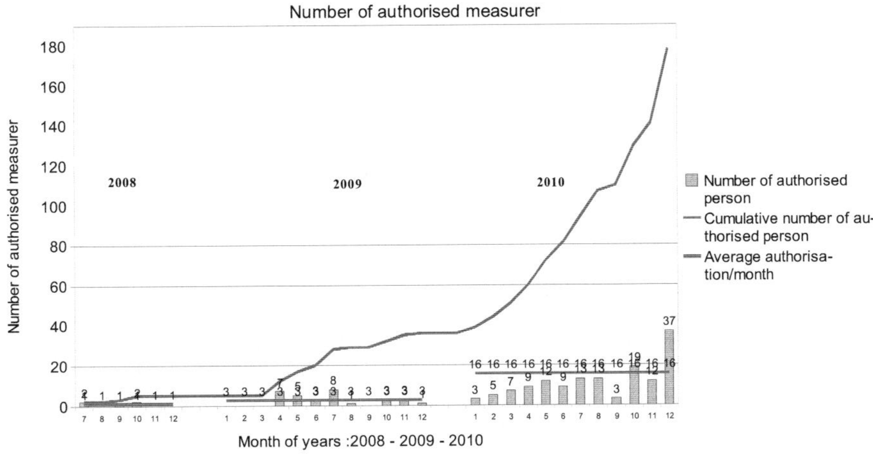

Figure 1: Evolution of number of authorised measurers

There is mostly three profiles for authorised measurer : either they belong to a design office, or to a thermal diagnostician company, or they switch to this new full time measurement activity.

Till December 31th, 2010, the commission was only carried by the sole ministry, and all examinations were performed by unpaid experts. But, by the end of 2010, the number of new applications received per month was over 15. This number should even increase as the 2012 regulation will require this authorization for all measurements when used as proof of compliance to the EP-regulation.

So, in order to prepare the 2012 regulation, a new organisation was set to be able to deal with over 500 applications a year.

DELEGATION TO QUALIBAT

The basic idea was to delegate the process to a private body who will organize the reports examination by pool of experts approved by the ministry.

The "Qualibat" association has been chosen to be in charge of the commission. Qualibat is a qualification and certification body for construction companies. It is a well known organisation in France that has today qualified more than 30 000 companies in the whole construction field. A referential for this new committee has been set in collaboration between the ministry and Qualibat to ensure that former requirements will be maintained and that the authorisation will concern a person authorisation and not a company authorisation. An agreement has been signed between the ministry and Qualibat. Thus, starting January 1st 2011, the ministerial authorisation is delivered through a Qualibat qualification. Qualibat manages the commission and delegates the expertise to ministry approved experts .

Training experts

In January 2011 the function of the ministry was then changed, it is no longer in charge to directly evaluate measurers' competencies, but to "evaluate those who evaluate measurers". Thus, a two days training is organise twice a year for people willing to become expert for the commission.

On the first half day, the standard analysis grid is detailed. The grid summarizes the NF EN 13-829 and GA P 50-784 key elements, the training consists in explaining what should be consider as reserve, as unconformity or just as recommendation. Then a fictive case is evaluated by trainer and corrected in group. Finally, the trainer is examined on his evaluation of a second fictive case.
Candidates for expertise are experimented measurers, trainers from an approved training center and experts recommended by other members of the commission. It's necessary to dispose of expert from all over France, as an expert should never examine a candidate from his own region. To ensure this dissemination the training is organised along a web conference. At the end of the first training session, 13 candidates out of 15 validated the test and were added to the seven former experts. A new session with ten or so candidates will soon be organised.

During this new commission' s first months, the number of measurer candidates decreased compared to the end of 2010, with an average of only seven qualifications per month, compared to 16 in 2010. In fact, on one hand, in the beginning, the delegation lead to some setting difficulties, and on the other hand, as getting the qualification was no longer free of charge (now around 500€ per candidate), it discouraged some of potential candidates.
Nevertheless, Qualibat has involved itself in promotion and advertisement of the measurer qualification which should arise new applications.

RECOVERING MEASURMENT DATA

A framework has been specified to monitor the authorized measurers' activities. To this end, the measurers must file a standard database for each measurement they perform. This record is a simple spreadsheet table , with pre-defined fields. Each measurer must send this record along with his first application; he must also update it every year to have his authorization extended.
Thus the ministry dispose of a national database with measurements done by all authorised measurers. Mid-2011, when we began to make use of this database, it already gathered almost 2500 measurements, from 25 authorised measurers. This number will exponentially increase as the 2010 authorised measurers will soon send their updated file for annual prolongation.
This database gives us a good view of buildings air-tightness in France, and allowed us to produce interesting statistics on buildings air-tightness according to their localisation, kind of buildings, year of construction, type of construction, etc. [4,5]. Nevertheless, it's necessary to keep in mind that the database is biased towards low-energy buildings: for example 47% of tested detached houses were involved in a BBC-Effinergie certification process, whereas this certification has a market share of only 7% of all new constructions. Since the certification imposes a maximum air-tightness level for houses of $Q_{4Pasurf}$=0.6m^3/h.m^2, it is foreseen that the average of the database is better than the national average.
Figure 1 gives example of statistic that are done with the database [4,5].

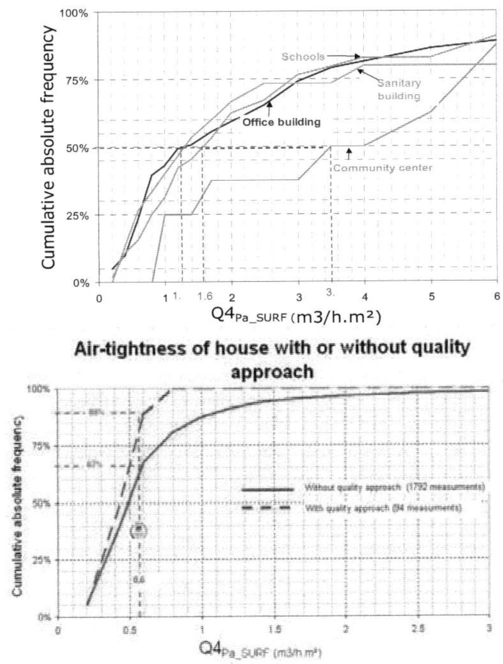

Figure 2: Example of use of the database [4,5]

CONCLUSION

Even if the overall framework for the commission is still under test period, our feeling is that it has greatly helped the measurers' community to keep a positive image. The potential discredit that could fall on a new profession has been avoided so far, even though this subject remains controversial among building professionals.

Moreover, as all measurers are identified, it's easier to disseminate new requirements or adaptations, or to promote potential developments such as ventilation network measurement for example.

Finally the measurer network produce a valuable database for an air-tightness observatory of in France.

ACKNOWLEDGEMENTS

The contribution of CETE de Lyon to the implementation of these quality frameworks is funded by the ministry of ecology, sustainable development, transport and housing (MEDDTL).

The sole responsibility for the content of this publication lies with the authors. It does not necessarily reflect the opinion of the Ministry.

REFERENCES

[1] Carrie, F.R., Charrier, S. and Leprince, V. 2010. *Implementation of measurement and quality frameworks in the French regulation for achieving airtight envelopes*, 31th AIVC Conference, 26-28 October 2010, Seoul Korea

[2]*Arrêté du 26 octobre 2010 relatif aux caractéristiques thermiques et aux exigences de performance énergétique des bâtiments nouveaux et des parties nouvelles de bâtiments*, JO 27 octobre 2010
[3]*Arrêté du 24 mai 2006 relatif aux caractéristiques thermiques des bâtiments nouveaux et des parties nouvelles de bâtiments*, JO 25 mai 2006
[4] Leprince, V., Bailly, A., Carrie, F.R., and Olivier, M. 2011. *State of Art of Non-Residential Buildings Air-tightness and Impact on the Energy Consumption*, 32th AIVC Conference, 12-13 October 2011, Bruxelles, Belgium
[5] Leprince, V., Biaunier, J., Carrie, F.R., and Olivier, M. 2011. *Quality Management Approach to Assess Buildings Air-tightness, Requirements and Checking*, 32th AIVC Conference, 12-13 October 2011, Bruxelles, Belgium
[6] Rolfsmeier, S., Vogel, K., Bolender, T., 2010. *Ringversuche zu Luftdurchlässigkeitsmessungen vom Fachverband Luftdichtheit im Bauwesen e.V.*, 5 th International Symposium on Building and Ductwork Air-tightness, October 21-22, 2010, Copenhagen/Lyngby, Denmark

INTERLABORATORY TESTS FOR THE DETERMINATION OF REPEATABILITY AND REPRODUCIBILITY OF BUILDINGS AIRTIGHTNESS MEASUREMENTS

Christophe Delmotte*[1], Jelle Laverge[2]

1 Belgian Building Research Institute
Rue du Lombard, 42
1000 Brussels, Belgium
**Corresponding author: cde@bbri.be*

2 Department of Architecture and Urban Planning
Faculty of Engineering and Architecture
Ghent University
Jozef Plateaustraat 22
9000 Gent, Belgium

ABSTRACT
The issue of the uncertainty of building airtightness measurements has built up a greater importance since this topic was introduced in many regulations regarding the energy performance of buildings. Different studies have contributed to the evaluation of the uncertainty but the question is still incompletely solved in practice.

To contribute to the determination of the repeatability and reproducibility of these measurements in practice, the Belgian Building Research Institute organized interlaboratory tests with 10 other laboratories.

This paper presents the details of the study together with the repeatability and reproducibility standard deviation calculated at different pressure differences. The issue of unweighted vs. weighted least square regression is also discussed.

The reproducibility standard deviation calculated at 50 Pa was 2.4%. It was below 3% between 30 and 100 Pa but was noticeably higher at 4 and 10 Pa. However the application of a weighted least square regression showed a possibility to reduce the standard deviation of the results calculated at low pressure difference.

KEYWORDS
Airtightness, blower door, uncertainty, round robin

INTRODUCTION
In European countries, increasing importance has been given to airtightness of buildings since the first publication of the directive on the energy performance of buildings in 2002. In some countries there are even requirements or considerable financial incentives linked with the airtightness level. It is therefore more and more important to pay attention to the uncertainty of airtightness measurements.

The issue of uncertainty of airtightness measurements has already been dealt with in various publications (e.g. [8]) but is still incompletely solved in practice. This is also a point of discussion in the current revision of the related ISO standard [6].

Beside the traditional mathematical analysis of the problem, a contribution to a better knowledge of uncertainty of airtightness measurements can be made through the determination of their repeatability and reproducibility in practice. Such study was organized during the summer 2011 by the Belgian Building Research Institute (BBRI) in collaboration with 10 other laboratories.

This paper presents the details of the study together with the repeatability and reproducibility standard deviation calculated at different pressure differences. The issue of unweighted vs. weighted least square regression is also discussed.

OPERATING MODE

According to ISO 5725-1 [3], repeatability is the precision under conditions where independent test results are obtained with the same method on identical test items in the same laboratory by the same operator using the same equipment within short intervals of time. While reproducibility is the precision under conditions where test results are obtained with the same method on identical test items in different laboratories with different operators using different equipment. Thus repeatability and reproducibility are the two extremes of precision, the first describing the minimum and the second the maximum variability in results.

Taking into account the time required for a full measurement (about 2 hours), the limited financial resources and the availability of the test building, the BBRI made 10 replicate tests (depressurisation and pressurisation) under repeatability conditions and 10 other laboratories made 1 test each under reproducibility conditions. One of them however made 2 replicate tests on one's own initiative. All tests took place between 14/06/2011 and 15/07/2011. The preparation of the building was the same for all tests.

During the tests, external temperature and wind speed on the measurement site were monitored by the BBRI. The maximum wind speed varied from 1.3 to 5.1 m/s and the mean temperature varied from 11.7 to 28.1°C with a maximum of 4.2°C variation during a test.

The tests were made according to EN 13829:2001 (ISO9972:1996 modified) [2]. The measurements were taken in the range of 10 to 100 Pa indoor-outdoor pressure difference in increments of about 10 Pa.

10 laboratories used a Minneapolis Blower Door Model 4 while only one laboratory used a Retotec 2200 blower door. All of them used electronic manometers connected to dedicated software that automatically manages the tests and stores the data (Tectite Express plus Blower Door Excel sheet or Fan Testic). One laboratory however used semi-automatic software (Teclog plus own Excel sheet).

THE BUILDING

Description of the building

The building subject to the tests is located in the premises of the BBRI in Limelette, Belgium (Building X2). It is an unoccupied single family house built around 1980 and fully dedicated to research work.

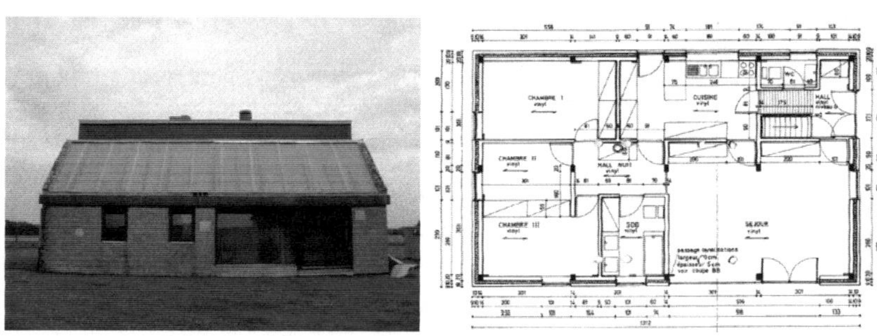

Figure 1: Southern façade and plan of the building subject to the tests

Only the ground floor was subject to the test. The basement and the attic were excluded. The internal volume of the ground floor is equal to 221 m³, while the net floor area is equal to 92 m². The building is equipped with electrical convection heaters and a fan assisted balanced ventilation system. It is also equipped with four externally mounted air transfer devices.

REPETABILITY

Results of the measurements

The 10 tests made by the BBRI under repeatability conditions resulted in air leakage rate at 50 Pa ranging from 699 to 738 m³/h for depressurisation and from 732 to 754 m³/h for pressurisation. The average value between depressurisation and pressurisation ranged from 715 to 744 m³/h. (Figure 2 - Table 1)

According to EN 13829:2001 [2], the air flow coefficient C_L and the air flow exponent n were determined using an unweighted least square regression. The air leakage rate $q_{\Delta pr}$ at a specified reference pressure difference was calculated using equation (1)

$$q_{\Delta pr} = C_L\,(\Delta p_r)^n \qquad (1)$$

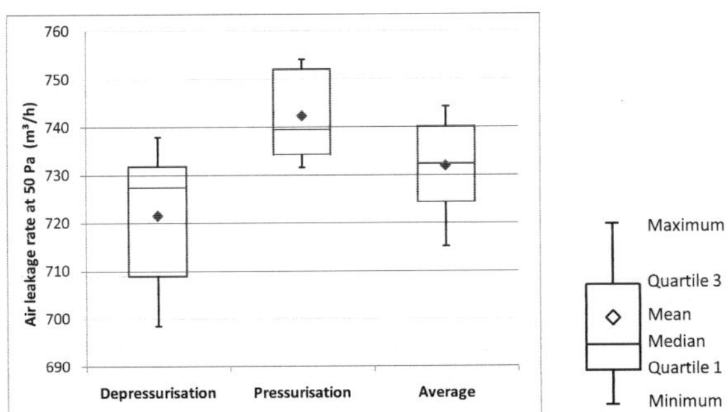

Figure 2: Variation of the air leakage rate at 50 Pa under repeatability conditions

	Depressurisation			Pressurisation			Average
	C_L	n	q_{50}	C_L	n	q_{50}	q_{50}
Replicate	m³/(h.Paⁿ)	-	m³/h	m³/(h.Paⁿ)	-	m³/h	m³/h
1	63.76	0.6155	708.5	79.43	0.5753	754.1	731.3
2	60.93	0.6277	710.2	79.31	0.5687	733.6	721.9
3	74.25	0.5865	736.4	70.58	0.6049	752.4	744.4
4	72.30	0.5918	732.1	74.87	0.5901	753.1	742.6
5	72.86	0.5890	729.8	66.80	0.6125	733.4	731.6
6	68.58	0.6028	724.9	78.42	0.5741	741.1	733.0
7	77.13	0.5749	731.0	83.97	0.5599	750.6	740.8
8	73.22	0.5906	738.0	73.54	0.5895	737.9	738.0
9	70.75	0.5880	705.9	72.00	0.5943	736.3	721.1
10	66.95	0.5995	698.6	67.29	0.6100	731.6	715.1

Table 1: Results of the 10 tests made under repeatability conditions

Variability of the results

The average air leakage rate had a repeatability standard deviation ranging from 3.5% at 4 Pa to 1.2% at 100 Pa through 1.4% at 50 Pa (Figure 3 - Table 2).

The variability of the depressurisation and pressurisation tests taken alone was slightly higher:

- From 5.2% at 4 Pa to 1.7% at 100 Pa through 2.0% at 50 Pa for depressurisation;
- From 5.1% at 4 Pa to 1.4% at 100 Pa through 1.2% at 50 Pa for pressurisation.

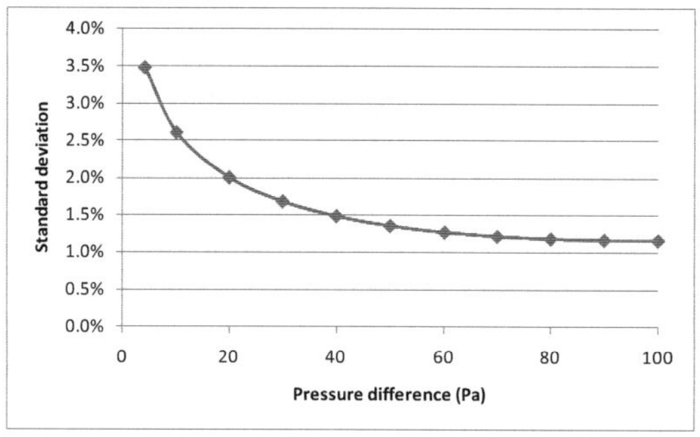

Figure 3 : Repeatability standard deviation s_r of the average air leakage rate

Δp (Pa)	4	10	20	30	40	50	60	70	80	90	100
	q_4	q_{10}	q_{20}	q_{30}	q_{40}	q_{50}	q_{60}	q_{70}	q_{80}	q_{90}	q_{100}
Replicate						**(m^3/h)**					
1	163.0	280.9	424.1	539.7	640.3	731.2	815.0	893.2	967.1	1037.3	1104.4
2	160.0	276.2	417.6	532.0	631.8	721.9	805.0	882.7	956.1	1025.9	1092.7
3	165.3	285.4	431.2	549.1	651.7	744.4	829.8	909.6	984.9	1056.5	1125.0
4	166.9	286.9	432.1	549.1	650.9	742.6	827.1	906.0	980.4	1051.0	1118.6
5	160.5	278.3	421.9	538.3	639.8	731.6	816.3	895.5	970.3	1041.5	1109.6
6	166.0	284.5	427.6	542.7	642.8	733.0	816.0	893.4	966.5	1035.8	1102.1
7	176.8	297.3	440.5	554.4	652.7	740.8	821.5	896.6	967.2	1034.0	1097.7
8	166.3	285.5	429.8	545.9	647.0	738.0	821.8	900.1	973.9	1043.9	1110.9
9	162.0	278.4	419.5	533.1	631.9	721.1	803.1	879.8	952.0	1020.7	1086.3
10	155.2	270.2	410.9	525.1	624.9	715.2	798.6	876.6	950.4	1020.5	1087.7
Mean	164.2	282.3	425.5	540.9	641.4	732.0	815.4	893.4	966.9	1036.7	1103.5
Repeatability	5.7	7.4	8.5	9.1	9.5	9.9	10.4	10.9	11.4	12.1	12.8
standard deviation s_r	3.5%	2.6%	2.0%	1.7%	1.5%	1.4%	1.3%	1.2%	1.2%	1.2%	1.2%
Repeatability	16.0	20.6	23.8	25.5	26.7	27.8	29.0	30.4	32.0	33.9	35.9
limit r[1]	9.7%	7.3%	5.6%	4.7%	4.2%	3.8%	3.6%	3.4%	3.3%	3.3%	3.3%

Table 2: Average air leakage rate for the 10 tests made under repeatability conditions

[1] Repeatability limit (r): The value less than or equal to which the absolute difference between two test results obtained under repeatability conditions may be expected to be with a probability of 95 %.
$r = 2.8 \cdot s_r$

The observed higher repeatability standard deviation s_r for low pressure differences can be explained by the combination of two phenomenons:

1. Considering all measurements, the values at low pressure difference were more variable (in relative terms) than those at high pressure difference;
2. For each regression, the standard deviation of the logarithm (ln) of the air flow rate around the regression line grows as one goes away from the mean value of the logarithm of the pressure differences achieved during the measurement (mean value was 45.3 Pa).

Note: The estimate of the standard deviation of y (y = ln(q)) around the regression line at the value x (x = ln(Δp)) is given by [2] :

$$s_y(x) = s_n \sqrt{\frac{N-1}{N} s_x^2 + (x - \bar{x})^2} \qquad (2)$$

s_n : standard deviation of air flow exponent n
N: total number of test readings
s_x : standard deviation of x
$\bar{x}$: mean value of x

For low pressure differences, the combination of higher variability with the distance from the mean pressure led to high standard deviation while for high pressure differences, the lower variability tended to counterbalance the effect of the distance from the mean pressure.

Supplementary tests

On the basis of the observations by Murphy et al. [7] stating that variability of the results correlates with the air leakage rate, 2 times 10 supplementary tests with higher and lower air leakage rate were carried out by BBRI. The same building with a slightly different preparation was used for these tests.

Note: There was a fortuitous change in the preparation of the building between the two days of test with lower leakage rate so the data were split into two series.

The average air leakage rate had a repeatability standard deviation ranging:

- From 3.0% at 4 Pa to 0.4% at 100 Pa for higher leakage (mean q_{50} = 1523 m^3/h);
- From 2.7% at 4 Pa to 0.6% at 100 Pa for lower leakage (mean q_{50} = 284 m^3/h);
- From 1.7% at 4 Pa to 0.6% at 100 Pa for lower leakage (mean q_{50} = 297 m^3/h).

These supplementary tests are probably not extensive enough to come to any conclusion about the correlation between the variability of the results and the air leakage rate. However they could not strengthen the earlier observations.

Weighted vs unweighted regression

According to EN 13829:2001 (ISO9972:1996 modified) [2], an unweighted least square regression was used for the calculation of the air flow characteristics. However experts in the field seem to be inclined to favour a weighted regression. So the air flow characteristics of the original tests were calculated again with the weighted regression of CAN CGSB-149.10-M86 [1] in order to see the difference (In this standard, ln Δp_i and ln q_i are weighted with q_i^2)

Compared to the unweighted regression (Table 2 and Figure 3), the weighted regression (Table 3 and Figure 4) showed lower standard deviations for the results at low pressure difference (4 to 30 Pa) while there was no noticeable difference at higher pressure difference

(40 to 100 Pa). As far as the mean air leakage rates are concerned, there was no noticeable difference whatever the pressure difference.

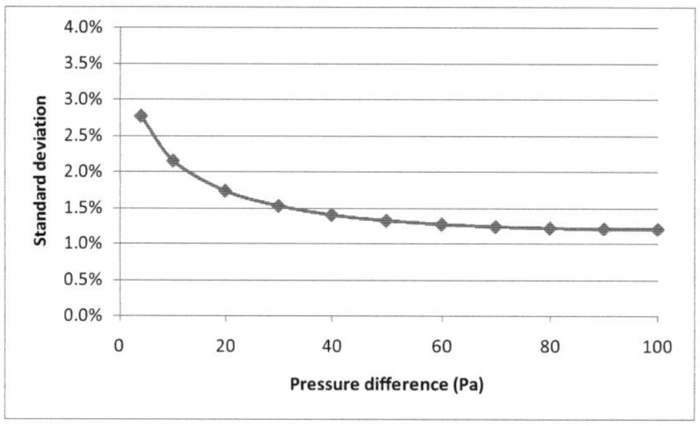

Figure 4: Repeatability standard deviation s_r of the average air leakage rate (original tests - weighted regression)

Δp (Pa)	**4**	**10**	**20**	**30**	**40**	**50**	**60**	**70**	**80**	**90**	**100**
	q_4	q_{10}	q_{20}	q_{30}	q_{40}	q_{50}	q_{60}	q_{70}	q_{80}	q_{90}	q_{100}
Replicate						**(m³/h)**					
1	165.3	283.5	426.4	541.5	641.6	731.9	815.0	892.6	965.7	1035.3	1101.7
2	162.8	279.5	420.8	534.6	633.6	722.8	804.9	881.6	954.0	1022.7	1088.3
3	164.7	284.7	430.6	548.6	651.4	744.2	829.8	909.8	985.3	1057.1	1125.7
4	168.4	288.6	433.7	550.3	651.7	743.0	827.0	905.4	979.3	1049.4	1116.5
5	161.1	279.0	422.6	538.9	640.2	731.8	816.3	895.3	969.9	1040.8	1108.6
6	164.6	283.0	426.3	541.8	642.3	732.9	816.4	894.3	967.8	1037.6	1104.3
7	173.9	294.1	437.7	552.2	651.3	740.3	821.9	897.8	969.3	1037.1	1101.7
8	167.7	287.1	431.2	547.0	647.6	738.2	821.6	899.4	972.7	1042.3	1108.8
9	162.0	278.5	419.5	533.1	632.0	721.1	803.1	879.8	952.0	1020.7	1086.3
10	157.2	272.4	412.9	526.7	625.9	715.6	798.3	875.6	948.7	1018.2	1084.6
Mean	164.8	283.0	426.2	541.5	641.8	732.2	815.4	893.2	966.5	1036.1	1102.6
Repeatability	4.6	6.1	7.4	8.3	9.0	9.7	10.4	11.1	11.8	12.6	13.3
standard deviation s_r	2.8%	2.2%	1.7%	1.5%	1.4%	1.3%	1.3%	1.2%	1.2%	1.2%	1.2%
Repeatability	12.8	17.1	20.7	23.2	25.3	27.2	29.2	31.1	33.1	35.2	37.3
limit r	7.8%	6.0%	4.9%	4.3%	3.9%	3.7%	3.6%	3.5%	3.4%	3.4%	3.4%

Table 3: Average air leakage rate for the 10 original replicates of the test (weighted regression)

REPRODUCTIBILITY

Results of the measurements

The tests made by the BBRI and the 10 external laboratories under reproducibility conditions resulted in air leakage rate at 50 Pa ranging from 699 to 772 m³/h for depressurisation and from 704 to 793 m³/h for pressurisation. The average value between depressurisation and pressurisation ranged from 713 to 772 m³/h. (Figure 5).

Detailed information about the measurements (mean values) is given in Table 4.

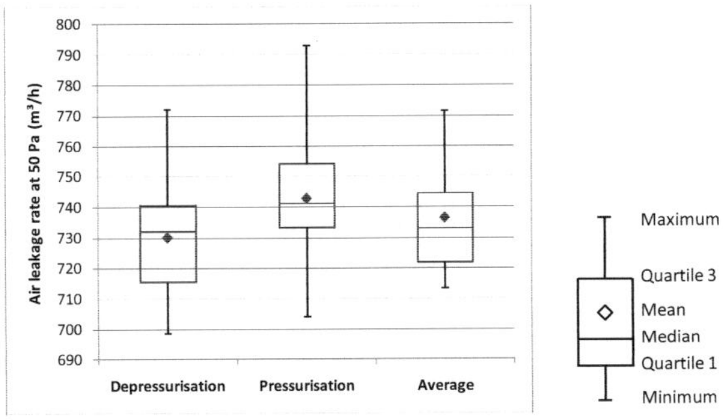

Figure 5: Variation of the air leakage rate at 50 Pa under reproducibility conditions

	Depressurisation			Pressurisation			Average
	C_L	n	q_{50}	C_L	n	q_{50}	q_{50}
Laboratory	$m^3/(h.Pa^n)$	-	m^3/h	$m^3/(h.Pa^n)$	-	m^3/h	m^3/h
1	74.94	0.5858	741.1	69.82	0.5926	709.2	725.2
2	70.21	0.6023	740.7	52.32	0.6717	724.1	732.4
3	92.37	0.5233	715.7	59.02	0.6362	711.1	713.4
4	70.07	0.5961	721.5	74.62	0.5873	742.4	732.0
5	88.70	0.5458	750.3	80.84	0.5715	756.1	753.2
6	72.97	0.5922	740.2	76.17	0.5894	764.2	752.1
7	76.86	0.5787	739.5	42.32	0.7187	704.0	721.8
8	64.60	0.6243	742.9	77.48	0.5945	792.8	767.9
9	77.40	0.5879	772.1	75.60	0.5936	771.0	771.5
10	63.46	0.6240	728.9	77.31	0.5799	747.2	738.1
11	66.73	0.6038	708.2	76.53	0.5777	733.3	720.7

Table 4: Results of the airtightness tests made by 11 different laboratories under reproducibility conditions (mean values for the 2 laboratories that made more than 1 test)

Variability of the results

The average air leakage rate had a reproducibility standard deviation ranging from 5.9% at 4 Pa to 2.6% at 100 Pa through 2.4% at 50 Pa (Figure 6 - Table 5).

The variability of the depressurisation and pressurisation tests taken alone was slightly higher at 50 and 100 Pa and noticeably higher at 4 Pa:

- From 7.9% at 4 Pa to 2.9% at 100 Pa through 2.5% at 50 Pa for depressurisation;
- From 11.1% at 4 Pa to 3.2% at 100 Pa through 2.9% at 50 Pa for pressurisation.

At 50 Pa the reproducibility limit was 6.7% which means that the absolute difference between two test results obtained under reproducibility conditions may be expected to be less or equal to 6.7% with a probability of 95 %.

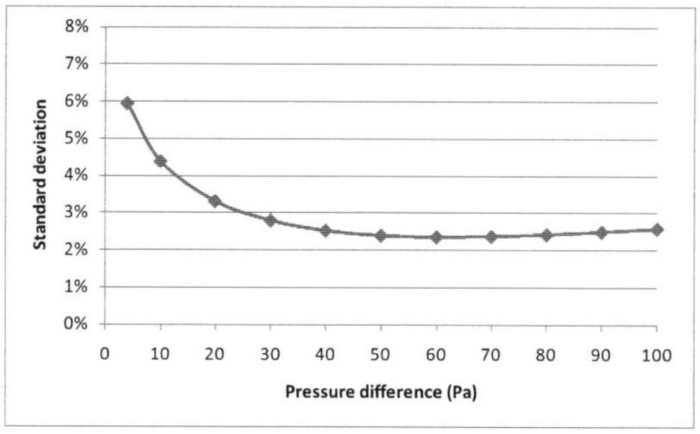

Figure 6: Reproducibility standard deviation s_R of the average air leakage rate

Δp (Pa)	**4**	**10**	**20**	**30**	**40**	**50**	**60**	**70**	**80**	**90**	**100**
	q_4	q_{10}	q_{20}	q_{30}	q_{40}	q_{50}	q_{60}	q_{70}	q_{80}	q_{90}	q_{100}
Laboratory						**(m³/h)**					
1	163.8	281.0	422.7	536.8	635.9	725.2	807.5	884.2	956.6	1025.4	1091.0
2	147.3	263.3	409.0	529.2	635.5	732.5	822.7	907.5	988.1	1065.1	1139.1
3	166.7	281.8	419.9	530.7	626.8	713.3	792.8	867.0	936.9	1003.3	1066.7
4	164.3	282.5	425.7	541.1	641.5	732.0	815.3	893.2	966.6	1036.4	1103.0
5	183.8	306.5	451.4	566.2	664.9	753.2	834.0	909.0	979.4	1046.0	1109.5
6	169.1	290.6	437.7	556.2	659.2	752.2	837.7	917.6	992.9	1064.4	1132.8
7	143.0	256.4	399.8	518.9	624.8	721.7	812.2	897.5	978.8	1056.6	1131.5
8	165.1	288.3	439.6	562.6	670.3	767.9	858.0	942.5	1022.3	1098.4	1171.2
9	173.5	298.1	449.0	570.5	676.2	771.4	859.2	941.1	1018.3	1091.7	1161.8
10	161.7	280.4	425.4	542.8	645.3	738.1	823.6	903.7	979.3	1051.3	1120.1
11	162.3	278.7	419.6	533.1	631.8	720.8	802.7	879.2	951.4	1019.9	1085.4
Mean	164.2	282.9	427.1	543.5	645.0	736.6	821.0	899.9	974.4	1045.3	1113.0
Reproducibility standard deviation s_R	9.75 5.9%	12.39 4.4%	14.12 3.3%	15.16 2.8%	16.25 2.5%	17.62 2.4%	19.31 2.4%	21.32 2.4%	23.58 2.4%	26.06 2.5%	28.70 2.6%
Reproducibility limit R[1]	27.29 16.6%	34.70 12.3%	39.53 9.3%	42.44 7.8%	45.50 7.1%	49.33 6.7%	54.08 6.6%	59.69 6.6%	66.03 6.8%	72.96 7.0%	80.36 7.2%

Table 5: Average air leakage rate for the tests made by 11 different laboratories under reproducibility conditions

DISCUSSION

The study showed that there can be a significant difference between the results in depressurisation and in pressurisation. It also showed that the uncertainty of the average value between depressurisation and pressurisation was globally lower than the uncertainty of both of them taken apart. So since natural conditions never lead to a fully pressurized or depressurized building, it seems preferable to favour tests in both modes.

Measurements taken in the range of 10 to 100 Pa showed relatively low reproducibility standard deviation (< 3%) for the results calculated at 30 to 100 Pa. The reproducibility standard deviation at 50 Pa was 2.4%. At lower pressure difference however, the standard

[1] Reproducibility limit (R): The value less than or equal to which the absolute difference between two test results obtained under reproducibility conditions may be expected to be with a probability of 95 %.
$R = 2.8 \cdot s_R$

deviation was larger. The standard test procedure from EN 13829 is therefore well suited for results calculated at 50 Pa like the well known n_{50} value but should be taken with greater care for results calculated at 4 or 10 Pa like the effective or equivalent leakage area for example.

The application of a weighted least square regression showed a possibility to reduce the standard deviation of the results calculated at low pressure difference. This option could therefore be evaluated for the revision of ISO 9972 [6].

For the tests, the mean air leakage rate at 50 Pa ranged from 284 m^3/h to 1523 m^3/h. This range matches with the airtightness of some new built single family dwellings in Belgium but other tests with higher air leakage rates should be done to cover a larger part of the building stock.

No correlation could be found between the variability of the results and the air leakage rate. The number of tests was however too low to come to a conclusion on that point. Extrapolation of the results of this study to higher air leakage rates should therefore be considered with care.

The tests of this study were made under favourable weather conditions in a low-rise building and with low temperature difference between inside and outside. It should be noticed that uncertainty of the results can increase with the wind, the height of the building and the temperature difference.

ACKNOWLEDGEMENTS
This report is based on a research project done at the Belgian Building Research Institute in the framework of the "Etanch'Air" project with the financial support of the Walloon Region.

REFERENCES

[1] CAN CGSB-149.10-M86. 1986. *Determination of the Airtightness of Building Envelopes by the Fan Depressurization Method.* Canadian General Standards Board
[2] EN 13829:2000. *Thermal performance of buildings - Determination of air permeability of buildings - Fan pressurization method (ISO 9972:1996, modified).* European Committee for Standardization
[3] ISO 5725-1:1994. *Accuracy (trueness and precision) of measurement methods and results - Part 1: General principles and definitions.* International Organization for Standardization
[4] ISO 5725-2:1994. *Accuracy (trueness and precision) of measurement methods and results - Part 2: Basic method for the determination of repeatability and reproducibility of a standard measurement method.* International Organization for Standardization
[5] ISO 5725-6:1994. *Accuracy (trueness and precision) of measurement methods and results - Part 6: Use in practice of accuracy values.* International Organization for Standardization
[6] ISO 9972:2006/Amd 1:2009. *Thermal performance of buildings -- Determination of air permeability of buildings -- Fan pressurization method.* International Organization for Standardization
[7] Murphy, W. E., Colliver, D. G., Piercy, L. R., Shipman, A. S., Penman, J., Sun, W. 1990. *A Round Robin Test of Fan Pressurization Devices.* Final report to Ashrae on research project 594-RP, University of Kentucky.
[8] Sherman, M., Palmiter, L. 1995. *Uncertainties in Fan Pressurization Measurements.* ASTM Airflow Performance Conference 10/93

PRESSURE DISTRIBUTION IN LARGE BUILDINGS DURING AIRTIGHTNESS TESTS

Stefanie Rolfsmeier, Paul Simons

BlowerDoor GmbH, Zum Energie- und Umweltzentrum 1, 31832 Springe, Germany
Tel. +49 (0) 5044 / 975-40, Fax +49 (0) 5044 / 975-44, E-Mail: info@BlowerDoor.de

ABSTRACT

When conducting airtightness tests of buildings, you must ensure that all building parts to be measured have air connection, so that the test object can be considered as one single zone. This also applies to large buildings like office buildings, schools, old people homes, indoor pools, etc. with several floors and rambling floor plans. Openings that are too small for a constant air flow from the leakages to the measuring device can prevent an even pressure distribution. According to the European measuring standard EN 13829, the pressure differentials inside the building during the airtightness test should not be less than 10% of the building pressure differential.

This article examines the different issues involved in bringing these theoretical considerations to the measuring practice. Which approaches in measuring technology can help improve knowledge of the actual pressure distribution and, consequently, the interpretation of the measuring results? What do the results look like? Which possibilities are there for conducting measurements even in critical conditions?

When conducting BlowerDoor tests in very tall buildings and/or buildings with very rambling floor plans, additional measuring points are set up in the zones considered critical. The pressure differentials between the BlowerDoor measuring device and these zones are recorded with the help of the TECLOG2 MultipleFan software. They can then be evaluated by the testing team, which can take the appropriate measures (e.g., close open windows, install an additional measuring device in the critical building part).

KEYWORDS

airtightness tests, big building, large building, high rise building; BlowerDoor, pressure distribution, pressure drop, "single-zone buildings"

INTRODUCTION

When conducting airtightness tests of large buildings, even distribution of the pressure generated by the measuring device must be ensured throughout the entire building for the duration of the test. Particularly when testing high-rise buildings and/or buildings with strongly structured floor plans, the room located the farthest from the measuring device should also be tested under the same conditions.

This article presents possibilities for ensuring an even pressure distribution. In addition, two examples – a high-rise building and a building with a rambling floor plan – demonstrate possible pressure distributions within a building.

REQUIREMENTS OF THE EUROPEAN STANDARD EN 13829 ON THE PRESSURE DISTRIBUTION IN THE BUILDING

A basic requirement for conducting airtightness tests according to the European standard EN 13289 is for the building to be considered a "single-zone building". All rooms and building parts of the test object must have air connection, so that an even pressure distribution is ensured during the test. To accomplish this, all interior doors of the test building or the tested building part are opened. Usually, the size of these openings (open interior doors) is sufficient for constant air flow.

During the measurement, the pressure differential within the building should be less than 10% of the building pressure differential between the inside and the outside.

EXAMINING PRESSURE DISTRIBUTION IN THE BUILDING

To examine the pressure distribution in the building, a building pressure differential of 50 Pascal is established with the BlowerDoor measuring device. There are only two ways to check whether the pressure distribution in the building is even:

1. Measuring the pressure differential at the building envelope in the critical zone
2. Measuring the pressure differentials within the building

Both possibilities will be presented in more detail.

Time for testing the pressure distribution

The right time for checking the pressure distribution in the building is after the test set-up and building preparation, and before leakage detection. This carries the advantage that a leakage detection test of the building envelope can be performed under the same basic conditions.

Testing the pressure differential between the interior and the exterior in "critical" zones

For the analysis, a building pressure differential of 50 Pascal is established with a BlowerDoor measuring device. In the following, the pressure differentials between the inside and the outside of the "critical" zones (e.g. most remote rooms / building parts) are checked. In order to do so, an additional portable pressure gauge is placed in the area to be tested. One connection of the measuring device ends in the zone, while the reference connection ends in the same zone outside of the building. A metal capillary tube with a tube extension can be used to transfer the outside pressure to the inside. It is possible to clamp the capillary tube in a window frame without closing its opening. The resulting pressure differentials are recorded. See Figure 1 for the measuring principle.

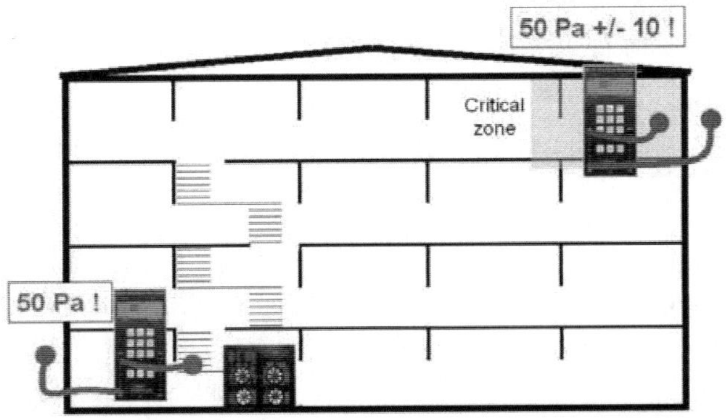

Figure 1. Measuring the pressure distribution at the building envelope

The pressure differential in the tested room must not fall below 50 Pascal minus 10%, i.e. 45 Pascal.

Testing the pressure distribution inside the building

For the analysis, a building pressure differential of 50 Pascal is established with a BlowerDoor measuring device. The pressure differentials within a building are determined by comparing the pressure near the measuring device with the pressure in "critical" zones (e.g. the most remote rooms or building parts). To do this, a long tube is led from the measuring device to the "critical" zone. See Figure 2 for the measuring principle.

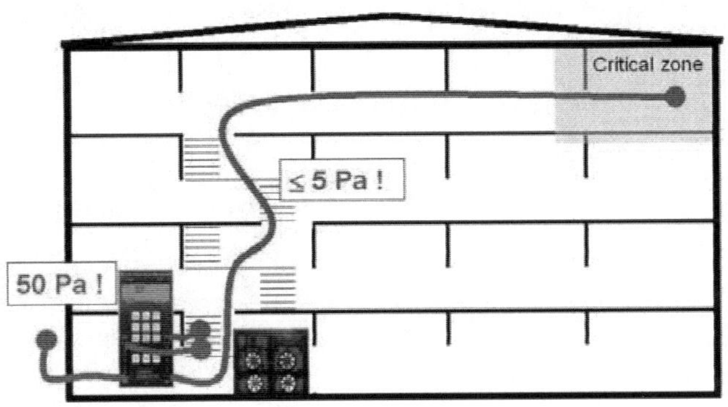

Figure 2. Measuring the pressure distribution within a building

The pressure differential within the tested room must not deviate from the building pressure differential by more than 10%, in this case 10% of 50 Pascal, i.e., a maximum of 5 Pascal.

Baseline pressure differential

Even under natural conditions, thermal forces and wind can generate a pressure differential within the building. If this is the case, the natural pressure distribution inside the building must be measured first and then taken into account when determining the pressure distribution in the building at an artificial pressure stage.

POSSIBLE SOLUTIONS FOR CASES WITH EXCESSIVELY HIGH PRESSURE DIFFERENTIALS IN THE BUILDING

If an excessively high pressure reduction is discovered in one zone when testing the pressure differentials in the building, this is caused by the following: The constant flow opening in this zone is too small or the leakages of this zone are too large for the existing constant flow opening. By setting up one or more additional BlowerDoor measuring devices in the critical zone, the same pressure level is established in all areas (Figure 3).

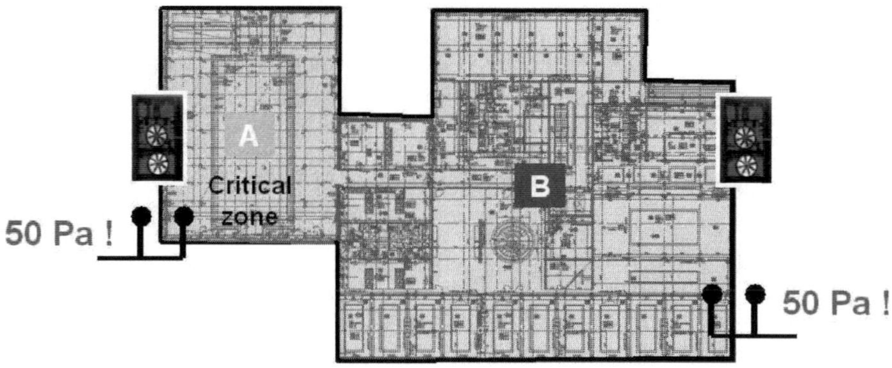

Figure 3. Distribution of measuring devices in the building in order to generate an even pressure distribution

EXAMPLE MEASUREMENTS

Demonstrations using a high-rise building and a building with a highly structured floor plan show this pressure distribution test in action. During the tests, the pressure distributions within the building were checked.

To measure the buildings, several Minneapolis BlowerDoor fans were combined. The fans were centrally controlled via the TECLOG2 MultipleFan program, which also permits the representation and recording of several building pressure differentials and internal pressure levels at once.

Example 1: Measurement of a high-rise building

Test object

In the following example, we tested a high-rise building which, once finished, will have 22 floors. A preliminary measurement of floors five to fourteen with an internal volume of 31,744 m3 was planned to determine the air change rate during construction, and to discover the existence of any severe systematic faults in the building envelope (Figure 4).

Figure 4. BlowerDoor test in the marked part of this high-rise building

Installing the measuring devices

The measuring system BlowerDoor MultipleFan with 3 fans per door frame was installed in the stairwell of the fifth floor (Figure 5). The installation opening for the measuring devices was built by the client from oriented strand boards and squared timber.

Figure 5. BlowerDoor MultipleFan (3 fans) in the stairwell

Building preparation

The tested building part was carefully separated from the non-tested areas via temporary sealing (stairwell and elevator doors were closed by oriented strand boards, end-to-end utility shafts for piping and cables were sealed, etc.).

Air connection

In order to establish a single "zone", the air connection was ensured via the continuous stairwell and the supply shafts. On each floor, the openings for the yet-to-be-installed interior doors provided the openings for constant air flow.

Basic conditions

The temperature difference between the inside and the outside was minimal. In spite of a storm forecast, the wind during the measurement remained within a tolerable range. The baseline pressure differentials were below the requirements of the European standard EN 13829.

Measuring the pressure differentials

The pressure differentials between the fifth and the fourteenth floor were measured with a highly accurate pressure gauge (DG-700). The gauge was placed near the measuring equipment on the fifth floor. One connection of the measuring channel remained free, while the other was connected to an approximately 40-meter-long tube. This was led to the fourteenth floor via the stairwell and provided the pressure from this floor (similar to Figure 2).

Measuring result

The pressure differential between the fifth and the fourteenth floor at a building pressure differential of 50 Pascal was -0.5 Pascal on average. An excerpt from the measurement program TECLOG2 MultipleFan in Figure 6 shows the result in a chart.

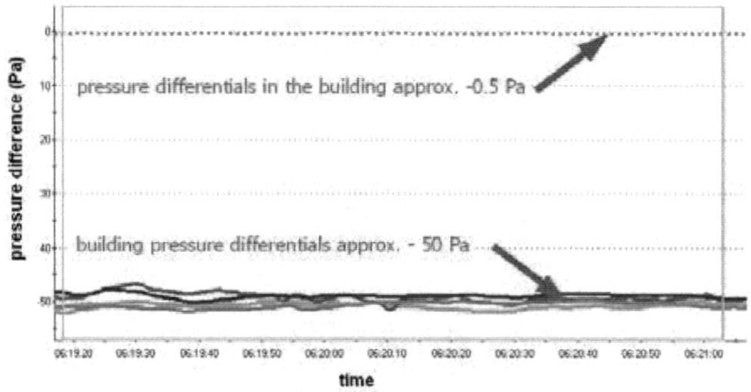

Figure 6. Excerpt from TECLOG2 MultipleFan: Building pressure differential and internal pressure distribution

The chart shows five building pressure differentials of around -50 Pascal (continuous lines), which were measured at the four different building sides and in the open stairwell. The dashed line shows the pressure distribution inside the building, approx. -0.5 Pa. It meets the requirements of the European standard EN 13289.

Example 2: Measurement of a hotel with a rambling floor plan

Test object

The next example stems from a hotel complex consisting of several building blocks of different sizes and heights. Again, only parts of the building complex (sections A, B, and C) with a volume of 23,000 m³ were measured (Figure 7).

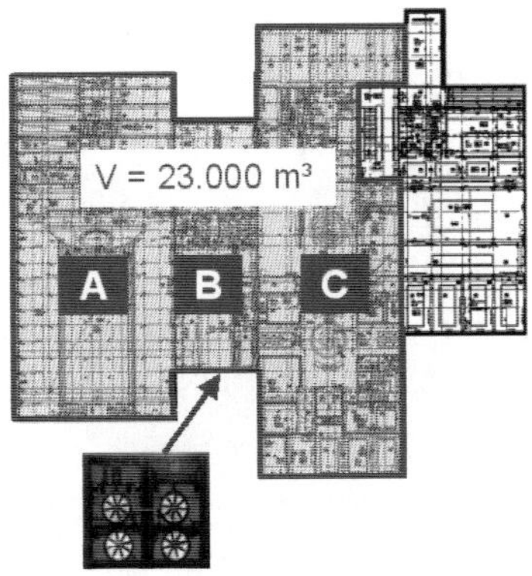

Figure 7. Partial floor plan of the hotel

Installing the measuring devices

Four BlowerDoor fans were installed in the exterior doors of Section B on the ground floor.

Building preparation

The tested zones were separated from the non-tested areas via temporary sealing (drywall in the doors of the floors and rooms, sealing of continuous shafts for piping and cables, etc.).

Air connection

Building sections A (indoor pool) and B featured sufficient air connection via diverse openings. The connection of section B to C was considered critical. There was only one air connection between these two parts of the building: a door in the stairwell on the first floor.

Basic conditions

The temperature difference between the inside and the outside was minimal, as was the wind. The baseline pressure differentials stipulated in European standard EN 13289 were not exceeded.

Measuring the pressure differentials

The pressure distribution was tested between the ground floor (section B near the measuring equipment) and the third floor (section C, most remote room). The tube length was 150 m.

Measuring result

The pressure differential between sections B and C at a building pressure differential of approx. -50 Pa was approx. 0.2 Pascal on average. An excerpt from the measuring software TECLOG2 MultipleFan in Figure 8 displays the results in a chart.

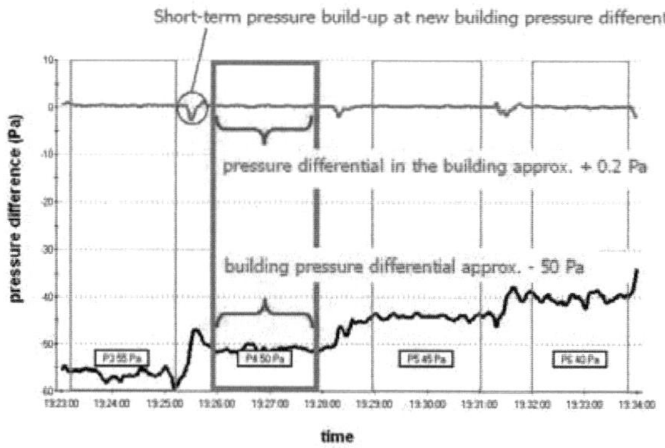

Figure 8. Excerpt from TECLOG2 MultipleFan: Building pressure differential and internal pressure distribution

The stairwell door between sections B and C, which at first had been considered too small for a constant flow opening and was thus rated critical, turned out to be sufficient. After a quick pressure build-up (see round mark) following each new setting of a building pressure differential, a constant pressure differential close to 0 Pascal was achieved. At this pressure distribution, the measurement could be conducted.

SUMMARY AND OUTLOOK

A basic requirement for air-tightness tests of large buildings is an even pressure distribution in the building. This in turn requires many openings of sufficient size for constant air flow, in order to ensure air connection in the entire tested building section ("single-zone buildings").

The pressure differentials in the building must be measured and checked, e.g., at a building pressure differential of 50 Pascal. There are two possible methods for this: either measuring the pressure differential in the "critical" zones (i.e., in the remote areas of the building) between the interior and the exterior directly at the building envelope, or by testing the pressure differential between the measuring device and the critical area within the building.

If the decline in pressure in one building section is too high, it is advisable to use one or more additional measuring devices at this location. The distribution and parallel operation of BlowerDoor fans in the building provides for an even pressure distribution.

In very tall buildings (e.g., high-rise buildings) an even temperature differential between the inside and the outside is important to ensure that the baseline pressure differential is not too high. Large buildings should also be tested at little wind or during calm weather.

If time constraints require conducting the test in spite of unfavorable weather conditions, the building pressure differential should be measured at different locations of the building and then averaged for the subsequent evaluation.

Thursday 13 October 2011

11.00 – 12.00 Parallel Session 5A with short oral presentations and posters – IAQ analysis and simulation of airflow and pollutant transport

- Performances of decentralized air handling terminals connected to building airtightness and indoor hygro-thermal climate (Gabrielle Masy, Belgium)
- Basis study about prediction of air flow environment in cross-ventilated room by neural network (Tomoyuki Endo, Japan)
- Sensitivity study for architectural design strategies of office buildings in Central Chile: effectiveness of nocturnal ventilation (Felipe Encinas Pino, Belgium)
- Nano-scale Aerosol Deposition Model for CFD in Indoor Environmental Analysis (Jun Narikawa, Japan)
- Exposure Concentration Prediction by Multi-Nesting Approach Connecting Building Space-Virtual Manikin- Nasal Airway Model (Kazuhide Ito, Japan)

11.00 – 12.00 Parallel Session 5 B - Philosophy for defining airtightness requirements

Should there be specific airtightness requirements? If so, what level is to be required? Should there be a minimum level of air leakage? The objective of this session is to review critical aspects that have to be considered to tackle such questions. It falls within the scope of a AIVC-TightVent project.

- Optimal air tightness levels of buildings (Willem de Gids, Netherlands)

12.00 – 12.15 Room Change

PERFORMANCES OF DAHT CONNECTED TO BUILDING AIRTIGHTNESS AND INDOOR HYGRO-THERMAL CLIMATE

Masy Gabrielle[1], Lebrun Jean[2], Gendebien Samuel[3], Nicolas Hansen[1], Marc Lengele[4], and Luc Prieels[5]

1 CECOTEPE
Quai Gloesener, 16
Liège, Belgium
**Corresponding author: gabrielle.masy@hepl.be*

2 JCJ Energetics
Liège, Belgium

3 University of Liège
Liège, Belgium

4 WOW Technology
Rue Pieds d'Alouette, 18
Naninne, Belgium

3 Greencom Development SCRL
Liège, Belgium

ABSTRACT

As building insulation level increases, the coupling of ventilation systems with building enveloppe airtightness becomes an important issue in order to improve buildings energy performances. A building ventilation model can be built on a set of resistances and generators in order to handle infiltration, natural ventilation as well as fan driven air flows. The model is able to assess the indoor air humidity level and the building energy balance.

Double flow ventilation can be handled through decentralized air handling terminals (DAHT), integrated in window ledges. A model of DAHT can be combined with the model of a whole building envelope, including infiltrations as well as dynamic behaviour, allowing comparisons with classical ventilation systems, such as natural or hybrid systems, or with centralized double flow systems. Results regarding energy consumptions, air humidity levels and superficial condensation risks can be analysed. Fresh air flow can be calibrated in order to meet air quality standards related to indoor humidity level and CO2 concentration.

The modelization of buiding indoor hygro-thermal climate allows a complete assessment of the seasonal heat exchanger efficiency, including heat recovery through the condensation of indoor air humidity when it flows through the exchanger.

KEYWORDS

Building ventilation, building infiltrations, heat recovery, fan models, indoor air quality.

INTRODUCTION

Double flow ventilation can be handled through Decentralized Air Handling Terminals, integrated in walls or window ledges and provided with heat recovery exchangers. Figure 1 shows a perspective and section of such a system [1].

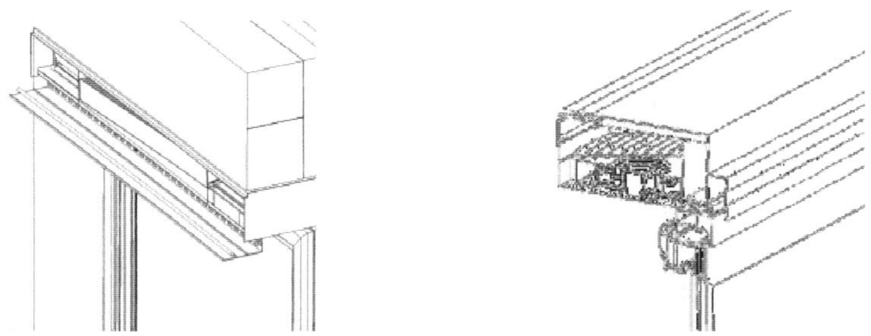

Figure 1. Decentralized Air Handling Terminals perspective and section.

Building ventilation models are interesting tools to predict the behavior of different types of ventilation systems interacting with the building envelope air tightness [2].

The aim of this paper is to define a model of DAHT including supply and exhaust fans models as well as heat exchanger model.

The efficiency of air flow control through the use of CO_2 or humidity indoor probes is assessed. The influence of the building air tightness on the efficiency of heat recovery is assessed as well.

Finally, the authority of supply and exhaust fans is checked.

DAHT MODEL DESCRIPTION

A building envelope air tightness and natural ventilation model can be built on the basis of the following generic equation regarding pressure drop through air apertures, and allowing mass air flows to be positive or negative, depending on the air flow direction [1]:

$$\Delta p = K.\left|\dot{M}\right|^{n}.\dot{M} \qquad (1)$$

Δp: Pressure drop through ventilation aperture *Pa*

$\dot{M}$: Mass air flow rate through ventilation aperture *kg/s*

n: Exponent ranging from *0* when the flow is laminar to *1* for a turbulent flow

K: Constant $Pa.(s/kg)^{1+n}$

The driving forces are due to wind pressure and to buoyancy effects:

$$\Delta p_{buo} = g.\frac{(z_2 - z_1)}{v} \qquad \Delta p_{wind} = p_c.\frac{u^2}{2.v} \qquad (2)$$

A model of a window can be built integrating the stack effet over the window height (Fig. 2).

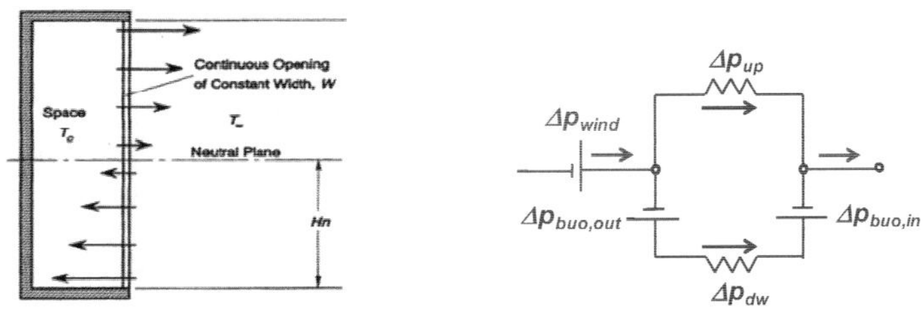

Figure 2. Window ventilation model.

DAHT supply/exhaust fans are modelled through a zero flow pressure generator followed by a pressure drop resistance. The heat exchanger adds another pressure drop resistance (fig. 3).

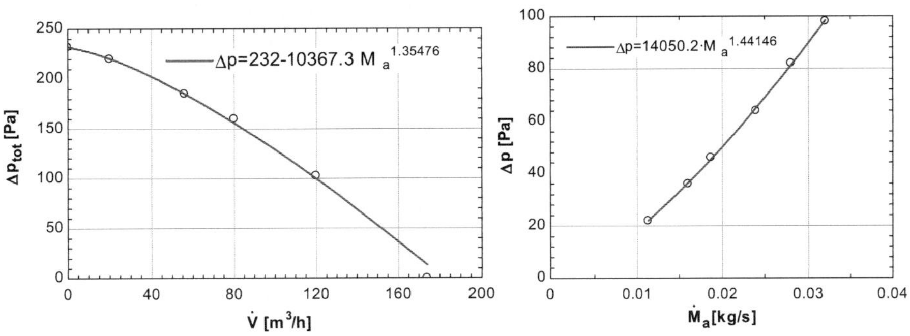

Figure 3. Regression laws: fan pressure/air flow curve (left) and heat exchanger pressure drop (right).

A model of the DAHT can be added to the window model (fig. 4).

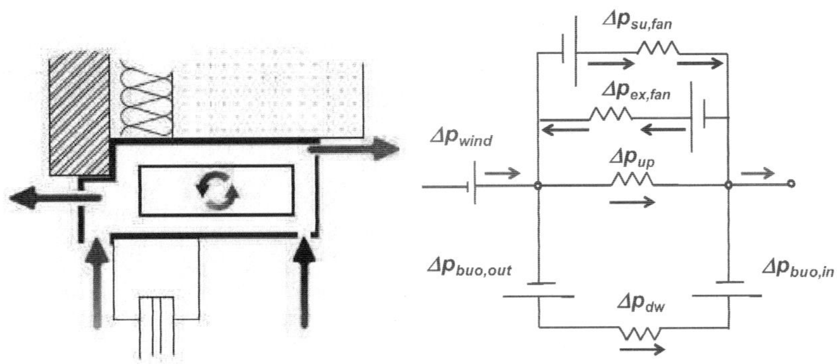

Figure 4. Model of the window provided with a DAHT.

TEST CASE STUDY

The model was used in order to assess the efficiency of decentralized air handling terminals in a semi detached house located in Belgium [1]. The house has 119 m2 floor area. It includes a living room with an open kitchen as well as a laundry at the first floor, three bedrooms and a bathroom at the second floor. It is supposed to be occupied by four inhabitants. The air tightness level is average, i.e. 12 m^3/h air flow per square meter of external wall at 50 Pa. This is equivalent to n_{50}=8.2 h^{-1}.

Zones are shared between dry zones (living room and kitchen as a whole zone on the first floor, bedrooms on the second floor) and wet zones (sanitary and laundry on the first floor, bathroom on the second floor). Humidity generation profiles are defined according to occupants presence and activities (showers, clothes drying, floor cleaning). The humidity generated by cooking is supposed to be removed through a separate kitchen exhaust fan. Annual simulations are performed with EES solver on a hourly basis with average Belgium weather data.

Simulations over a whole year with a very detailed model of the heat exchanger [3] showed that energy savings provided by latent heat recovery, due to condensation of the indoor exhaust air humidity, is only about 2 % of the whole energy saving provided by the heat recovery, so that only the sensible part of heat recovery will be considered. Measurements performed on the heat recovery exchanger showed that its efficiency is averaging 0.80 [4]. Simulations are performed with a constant 0.80 sensible heat recovery efficiency.

1. Air flow control

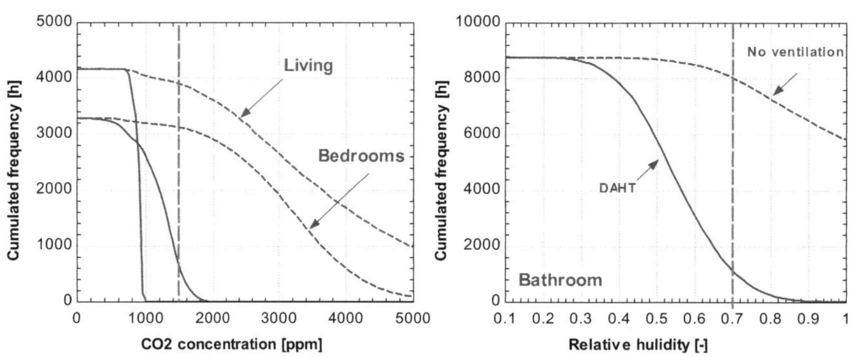

Fig. 5. Cumulated frequencies of indoor air CO2 concentration (left) and relative humidity (right) with and without ventilation through DAHT (respectively full and dotted lines).

A first simulation is performed with no ventilation system, the building air renewal beeing only provided by infiltration. Figure 5 (dotted lines) shows that CO2 concentration is exceeding 1500 ppm during most of the occupancy hours in the livingroom and in the bedrooms (the outdoor CO2 concentration is equal to 400 ppm). The relative humidity is exceeding 0.7 most of the time in the bathroom.

A second simulation is performed with DAHT for ventilation. DAHT are sized according to Belgian standards: a supply fresh air flow of 3.6 m3/h per square meter of floor area is required in dry zones, and an exhaust air flow of 3.6 m3/h per square meter of floor area is required in wet zones with a minimum of 30 m3/h per room. The living room is supposed to be occupied 14h per day during the week and 17h per day during the weekend. DAHT fans are supposed to work at 100% load during occupancy hours, and at 10% load during no occupancy hours. In bedrooms, DAHT fans work continuously at 10% load to prevent noise problems.

Figure 5 full lines show the results related to DAHT ventilation. Air quality requirements, i.e. maximum 1500 ppm CO2 concentration and 0.7 relative humidity are respected. The livingroom appears to be overventilated. Ventilation air flow control is provided in order to avoid overventilation.

Air flow control is performed through CO2 probes in the livingroom and in the bedrooms, while it is performed trough relative humidity probes in the laundry and in the bathrooms. The frequency curves of figure 5 are almost the same that those observed without control, except for the livingroom where the overventilation disappears providing energy savings as displayed on figure 6: ventilation heat losses decrease by 10 kWh per year and per square meter of floor heated area.

2. Building envelope air tightness

The influence of building airtighness level on the heat recovery energy performance is highlighted by the following simulation: the building average air tightness level i.e. 12 m3/h air flow per square meter of external wall area at 50 Pa or n_{50}=8.2 h^{-1}, is improved to reach a high air tightness level i.e. 1 m3/h.m2 at 50 Pa, or n_{50}=0.68 h^{-1}. The result is displayed on figure 6: ventilation heat losses decrease by 3.5 kwh per year and per square meter of floor heated area. The humidity level in the bathroom and the CO2 concentration in the livingroom are not affected, while the CO2 concentration in bedrooms increases beyond 1500 ppm half of the occupancy hours, suggesting an increase of the minimum DAHT fan load from 10 to 15%.

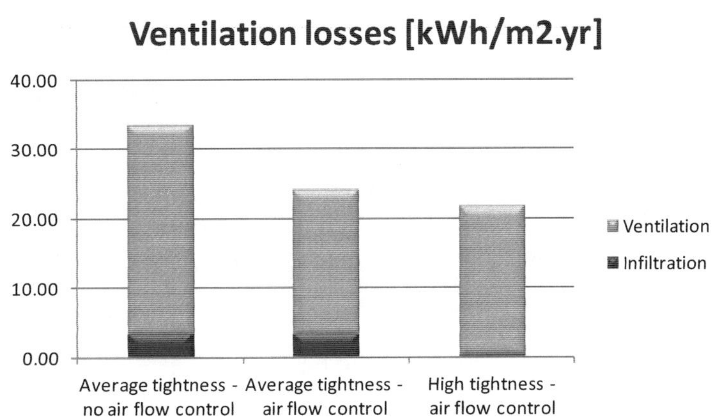

Fig. 6. Yearly specific ventilation heat losses for an average building airtightness without and with air flow control, and for a high tightness level combined with air flow control.

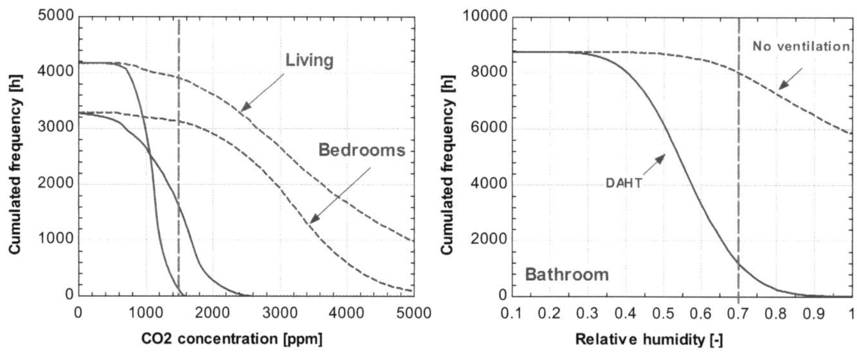

Fig. 7. Cumulated frequencies of indoor air CO2 concentration (left) and relative humidity (right) without and with controlled DAHT and high tightness level (respectively full and dotted lines).

3. Fans authority

Simulation of ventilation system interacting with the building envelope is also usefull in order to predict the level of pressure difference the DAHT fans must be able to face.

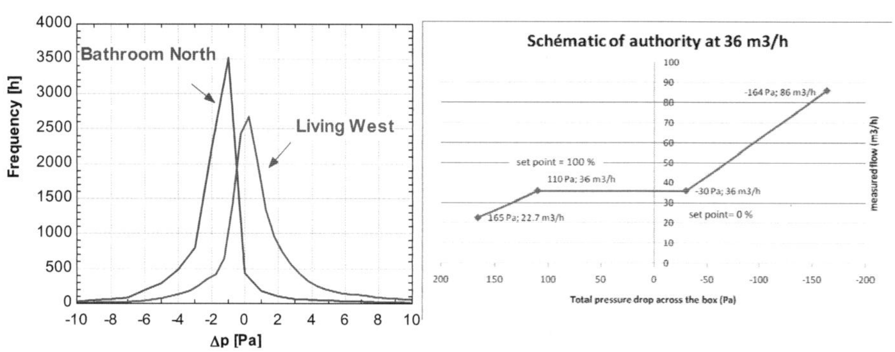

Fig. 8. Frequencies of pressure differences between DAHT fans terminals (left) and measured DAHT fan authority (right).

Results regarding a West facing DAHT in the living room and a North facing DAHT in the bathroom are presented on fig. 8, the first system being mostly pressurized, the second being mostly depressurized. The over/under pressure never exceeds + or – 20 Pa and is comprised between - and + 10 Pa most of the time suggesting that the measured DAHT fan authority is sufficient (fig. 8).

CONCLUSION

Different simulations are performed on a house case study in order to assess the energy performance of Decentralized Air Handling Terminals provided with heat recovery exchangers.

DAHT fans can be modeled through a pressure generator followed by a resistance, and coupled to a whole model of the building.

Air flow control through CO2 and relative humidity indoor probes provides a reduction of 10 kWh heat losses per year and per square meter of floor heated area. An improvement of the building air tightness level adds up a 3.5 kWh/yr.m^2 reduction.

The authority of DAHT supply and exhaust fans must be comprised between + and – 20 Pa in order to ensure that the expected fan air flow is reached.

ACKNOWLEDGEMENTS

The support of the Walloon Region for funding the Green+ project in the framework of the "Marshall Plan" to the work related in this project is gratefully acknowledged.

REFERENCES

[1] Masy, G., Lebrun, J., Prieels, L., Vincent, B. and Hansen, N. 2011. *A model of Decentralized Air Handling Terminals interacting with Building Infiltration*, Roomvent, 2011.

[2] Maria Justo Alonso, Bjarne Malvik, Hans Martin Mathisen, *Energy efficiency and indoor climate: modeling of ventilation systems using Contam W*, Roomvent, 2011.

[3] Gendebien, S., Bertagnolio, S. and Lemort, V. *Modeling and experimental validation in partially wet conditions of an air-to-air heat recovery exchanger*, BS 2011.

[4] Gendebien, S., Bertagnolio, S., Georges, B., and Lemort, V. *Investigation on an air-to-air heat recovery exchanger: modeling and experimental validation in dry conditions*, Roomvent, 2011.

BASIS STUDY ABOUT PREDICTION TO AIR FLOW ENVIRONMENT IN CROSS-VENTILATED ROOM BY NEURAL NETWORK

Tomoyuki Endo[1]

1 Kanto Gakuin University
1-50-1,Mutsuurahigashi, Kanazawa-ku,
Yokohama-city, Kanagawa,Japan
endo@kanto-gakuin.ac.jp

ABSTRACT

In many parts of Asia as typified by Japan, conditioning of the indoor thermal and air environments using natural ventilation since ancient times. When indoor thermal and air environments are predicted, the use of simulation technologies such as CFD and Heating and Ventilation Network Model has increased. Those have advantages and disadvantages. In addition, AI programs like Neural Network (NN) and Genetic Algorithm (GA) are increasingly utilized in other research areas. In architectural equipment field, there are examples of air-conditioning system models with NN. These programs are relatively easy to use, finish the calculation quickly, and even conduct assessment and prediction. However, there are few application examples of NN in simulations of thermal and air environment of cross-ventilated room. This study examines fundamental investigations in the application of NN to cross-ventilation environmental simulation. As a result, it revealed that the results of the simulations with NN tended similarly to the results of CFD under the condition that 2 openings were open. Although, by the combination of the cases with 2 openings open, the simulation of NN with 3 openings in case, which had small calculation load and high simulation accuracy, gained low accuracy and basically resulted in similar wind directions to the results of CFD.

KEYWORDS

AI, Neural Network, Cross-Ventilation, Indoor air environment, Air velocity, Vectors

INTRODUCTION

In many parts of Asia as typified by Japan, conditioning of the indoor thermal and air environments using natural ventilation in the morning and night of summer since ancient times. Although the period which conditions indoor thermal and air environments with air-conditioner continued with the spread of air conditioners, the tendency that cross-ventilation will be taken in positively is increasing again in response to gain of energy-saving momentum in recent years.

When indoor thermal and air environments are predicted, the use of simulation technologies such as CFD and Heating and Ventilation Network Model has increased. Heating and Ventilation Network Model as represented by COMIS-TRNSYS regards a space as a mass point, and that it can conduct calculation about many spaces in a short time. However, we cannot figure out the detailed indoor airflow distribution through it (Fig.1). On the other hand, CFD, which has spread rapidly because of the advance in computation technology, can show the detailed indoor airflow distribution. However, its users are required specialist knowledge and experience in grid generation and in selecting calculation algorithm, difference scheme, turbulence model and various types of boundary conditions (Fig.2). CFD also needs computation time and costs severely.

In addition, AI programs like Neural Network (NN) and Genetic Algorithm (GA) are increasingly utilized in other research areas. These programs are relatively easy to use, finish the calculation quickly, and even conduct assessment and prediction. The followings are the simple overviews about these technologies.

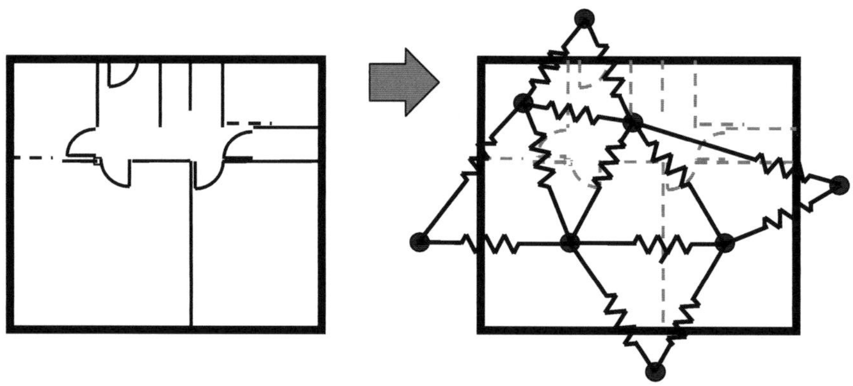

Figure 1. Network modell concept

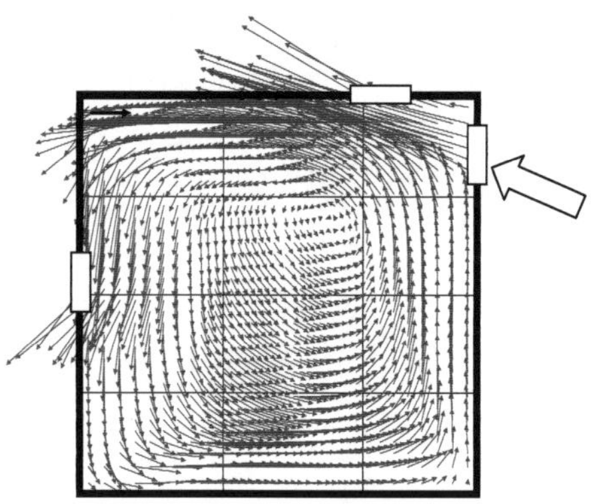

Total Cell Number, Min. y+		?
Turbulence Model		?
Algorithm		?
Discretization Scheme		?
Boundary Condition	Inlet	?
	Outlet	?
	Walls	?

Figure 2. CFD concept

Genetic Algorithm (GA)

Living things generally repeat breading and produce offspring. The offspring receive the parents' genes, so they genetically have something resemble to their parents. A mutation evolution rarely happens and a new gene which has no relation to those of the parents appears. GA includes the variable like hereditary nature, crossover, and mutation causes. Gene sequences in GA are created by only two numbers, 0 and 1. GA is an AI that explores the best appropriate gene (a combination of 0 and 1) through the repetition of mutation evaluation and crossover that are mentioned above.

Tables are numbered. The table caption is below the table.

Fuzzy Theory

People frequently use adjectives of ambiguous meaning such as *tall, hot, many, heavy*. An ambiguous (fuzzy) quantity is hard to handle as "more than" and "less than" a numerical value. We can express such quantity by using of a gently shifting scale appropriate to it in Fuzzy Theory. An air conditioner which is equipped with the "fuzzy" function appeared years ago. This function enables computers interpret those ambiguous sense like "a bit cold" or "rather cold".

Neural Network

Neural Network is AI model based on human brain function. Numerous brain cells called neuron are webbing in human brain. At each connecting part (axon), an excitement produce brain secretions and information is transmitted and learned (Fig.3). The presence or absence of information transmitting is determined by threshold level. It is the boundary of whether an excitement happens or not. In this way, AI is modelled based on human brain cells network and progresses learning by given information.

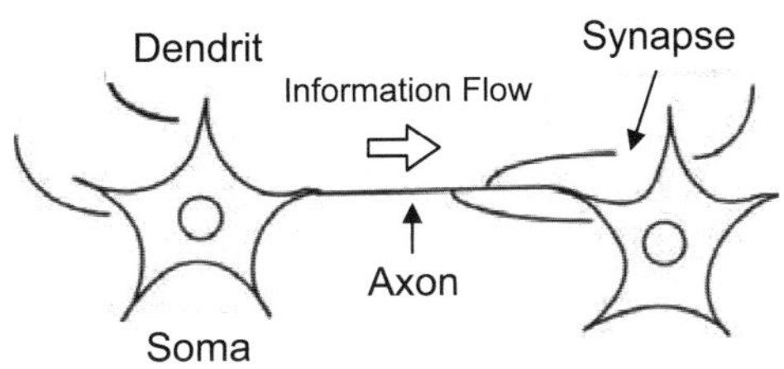

Figure 3. Neuron of Neural Network

PURPOSE OF THE STUDY

In architectural equipment field, there are examples of air-conditioning system models with NN [1]. But there are few application examples of NN in simulations of thermal and air environment of cross-ventilated room. So, this study examines fundamental investigations in the application of NN to cross-ventilation environmental simulation. Cross-ventilated indoor environment is greatly influenced by position and size of the openings, and wind direction and velocity outside of the room. Because of these variables' interaction, a high-accuracy simulation needs CFD in each individual case. But it takes large amount of time and load in calculation. In this study, the purpose is prediction of cross-ventilated indoor environment under the condition that only position of openings and wind direction are changed, but size of openings and wind velocity outside of the room are in the same condition.

OUTLINE OF THE STUDY

The examination object in this study was a room shown in Figure 4, which is modelled from second level of a house. In the calculation model, eave height (5.9 meters) and eave wind velocity were normalized as 1 respectively. This model has 14 openings in all, but this study dealt with only 3 of them (Fig.5). The indoor airflow distribution at the case when these openings are opened was simulated with NN.

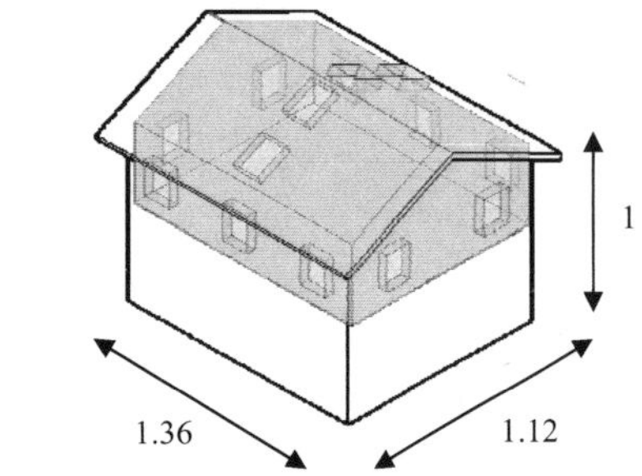

Figure 4. Object house

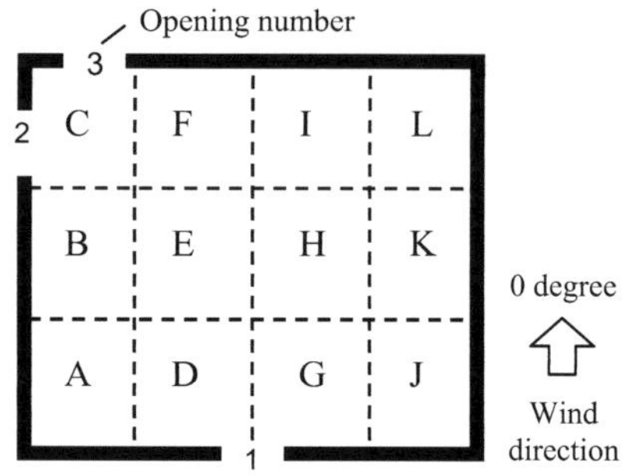

Figure 5. Object room

Some cases of wind directions were calculated in advance with CFD so that NN can learn the data. Then, the simulation of the target cases was calculated with NN. The room was divided 12 part areas as shown in Figure 5. The average value of x-way, y-way, and synthesized wind velocity (Eq.1) inside of each part were dealt with in the calculation (Fig.6).

$$V = \sqrt{V_X^{\ 2} + V_Y^{\ 2} + V_Z^{\ 2}} \tag{1}$$

V : Synthesized wind velocity, V_X : Wind velocity of X direction, V_Y : Wind velocity of Y direction, V_Z : Wind velocity of Z direction

A room which is 1 meter (normalized value: 0.17) from the wall was defined as a residential zone (Fig.7). The average synthesized wind velocity was calculated, and it was divided by wind velocity at eaves to require the velocity rate. Then, the results of CFD and NN were compared to each other.

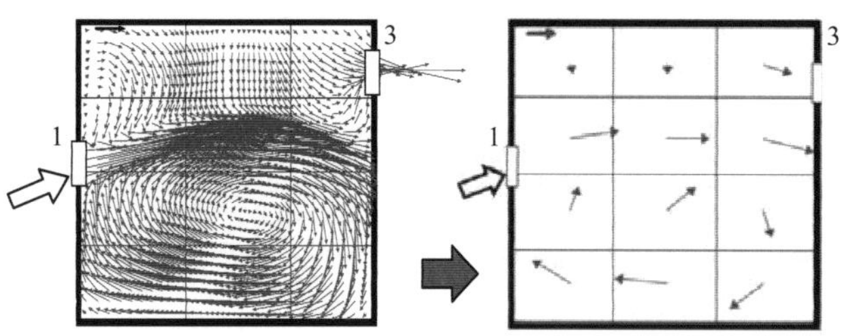

Figure 6. Wind velocity distribution in divided part areas

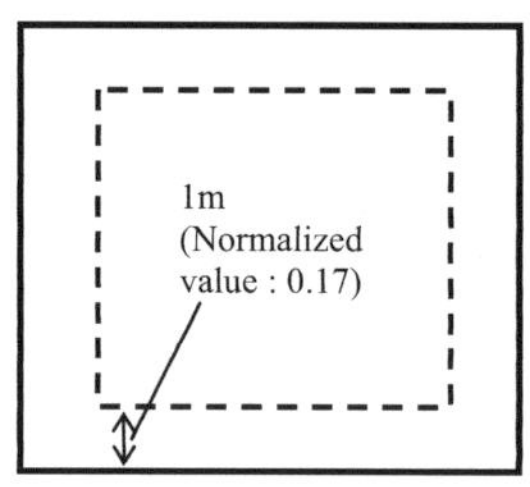

Figure 7. Residential zone

Fist, as shown in Figure 8, NN learned the indoor wind velocity distribution of 8 directions (0 to 315 degrees at intervals of 45 degrees). Then, NN predict the wind velocity distribution, wind direction, and average velocity in residential areas of another 1 direction. This case is called as Case 1.

Moreover, NN learned the results of CFD in which the opening 1 and 2, and the opening 1 and 3 were open respectively. Then, another simulation of the airflow distribution was conducted with all of the openings (1, 2, and 3) open. The learning patterns were considered with the following 3 cases (Fig.8).

Case 1: NN learned the distribution at the 8 directions, then simulate the distribution at another 1 direction.

Case 2: NN learned the distribution of 2 directions, then simulate the distribution at the in-between direction.

Case 3: NN learned the distribution of 4 directions, then simulate the distribution at the 1 direction (center of them).

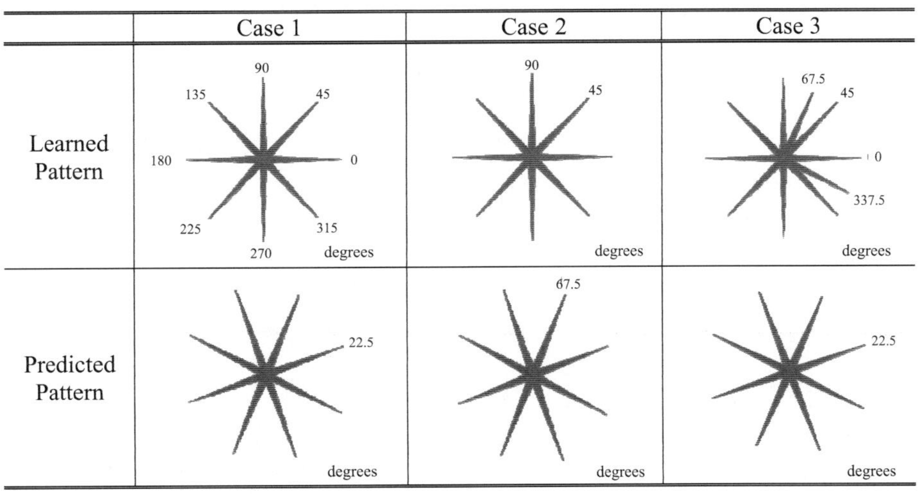

Figure 8. Learned and predicted patterns

ANALYSIS METHOD (CALCULATION CONDITION OF NN)

The connection weights between the input layer and the hidden layer were compressed as in Figure 9, and over 130,000 times repeated learning (130,000 times back propagation leaning) was conducted. In mid-course of the leaning, it was treated as convergent when the evaluative functional value was 0.001 and under. The multilayered Neural Network used in this study is applicable in many other areas, for example, a simulation of general nonlinear classification, and a tool of multiple nonlinear regression analysis. NN's leaning pace is fast enough and it is available with commonly-used personal computers.

For a highly accurate prediction, it is necessary to repeat leaning and prediction with sample answers and to reconsider the number of neurons in the hidden layer and the best value of the connection weights among the layers. After the consideration of the pretreatment with sample answers, the calculation in this study was conducted with the parameters in Table 1.

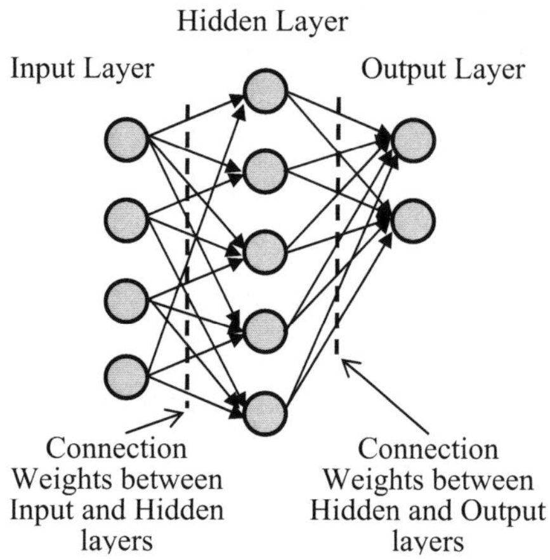

Figure 9. Layers and connection weights

The number of the hidden layers	25
The connection weights between input layer and hidden layer	0.75
The connection weights between hidden layer and output layer	0.25

Table 1. Parameteers of NN calculation

RESULTS AND DISCUSSION

Figure 10, 11, 12 show the wind velocity distributions at 22.5, 157.5, 247.5 of the wind direction degree of Case 1. The result of CFD shows slightly higher values in some parts than those of NN's simulation. However, both of them generally showed the similar areas where equal wind velocities appear, and the similar damping pattern from inlet to outlet.

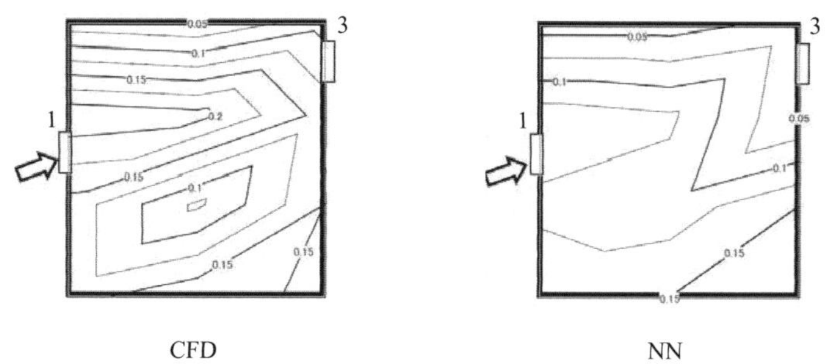

Figure 10. Wind velocity distribution at 22.5 degree of Case 1

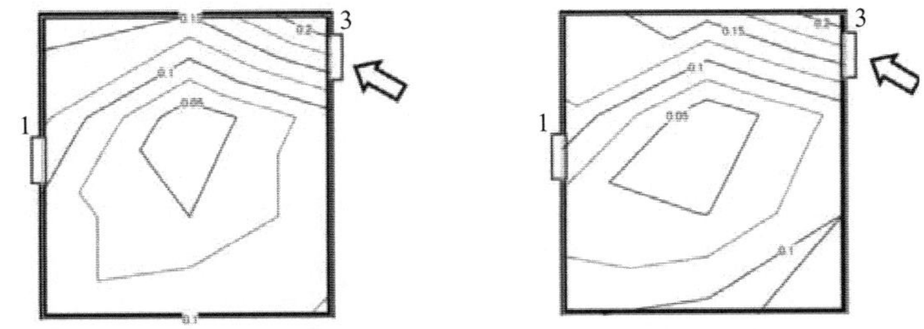

CFD NN

Figure 11. Wind velocity distribution at 157.5 degree of Case 1

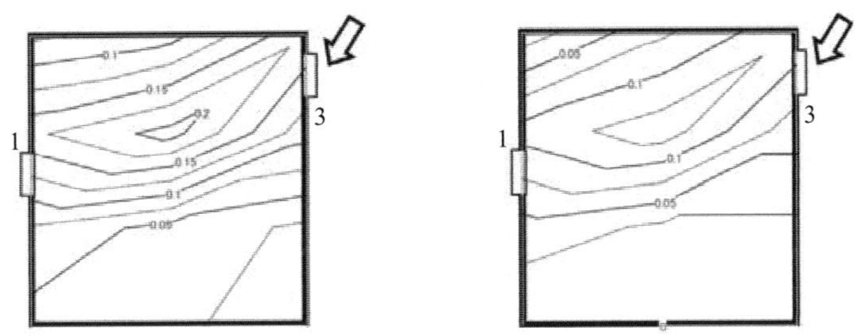

CFD NN

Figure 12. Wind velocity distribution at 247.5 degree of Case 1

Figure 13, 14, 15 show the wind velocity vectors distribution at 22.5, 157.5, 247.5 of the wind direction degree of Case 2. The wind velocity vectors at the area which seemed to be the wind pathway were simulated with very high accuracy. Besides, the condition of the indoor airflow circulation was re-created well.

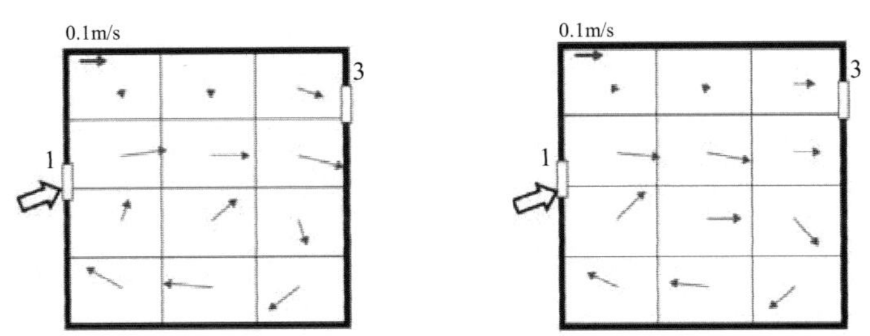

CFD NN

Figure 13. Wind velocity vectors distribution at 22.5 degree of Case 2

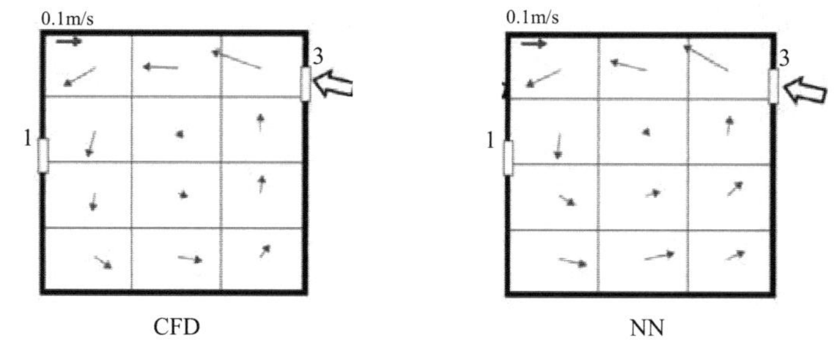

Figure 14. Wind velocity vectors distribution at 157.5 degree of Case 2

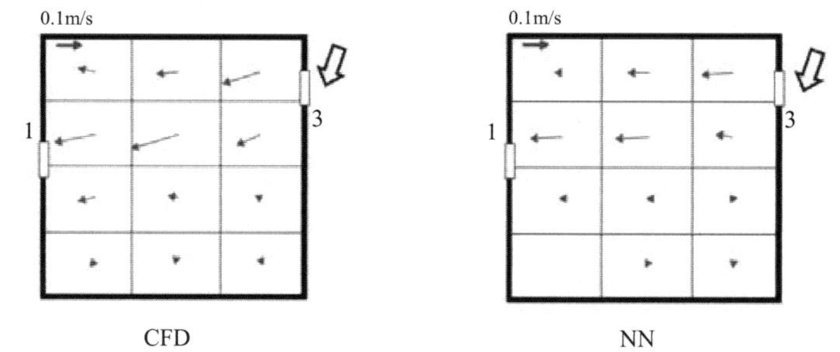

Figure 15. Wind velocity vectors distribution at 247.5 degree of Case 2

The simulation accuracy extremely lowered, when NN first learned the case with little air volume and no clear airflow pathway inside of the room such as at 90 of the wind direction degrees. It was because the value of the wind velocity was calculated exceedingly small and influenced on the simulation (Fig. 16).

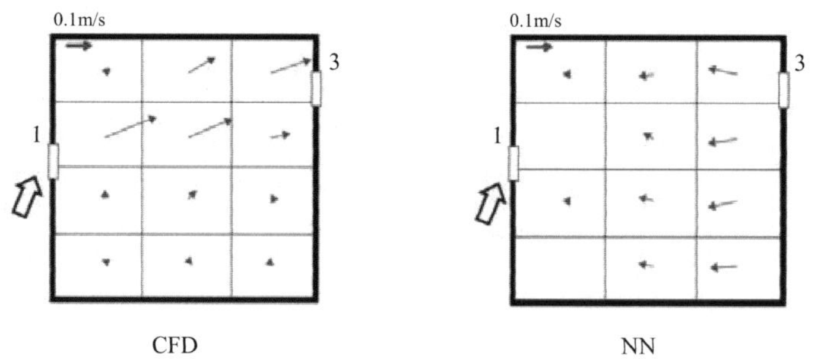

Figure 16. Wind velocity vectors distribution at 67.5 degree of Case 2

The velocity rates of NN were generally similar to those of CFD at any wind direction, and the simulation in Case 1 and 2 were highly accurate. Table 2 shows a comparison of the velocity rates in Case 2.

Degrees	22.5	67.5	112.5	157.5	202.5	247.5	292.5	337.5
CFD	0.14	0.11	0.04	0.07	0.10	0.11	0.12	0.12
NN	0.15	0.12	0.05	0.09	0.13	0.07	0.10	0.12

Table 2. Comparison of the velocity rates Case 2

So, the indoor airflow velocity distributions with all of the openings (1, 2, and 3) open were examined in Case 2, which had small calculation load and high simulation accuracy. As shown in Figure 17 and Figure 18, both of the conditions of the indoor airflow circulation were relatively similar to each other. But, there is still some problems remaining in the simulation accuracy in the small wind velocity areas.

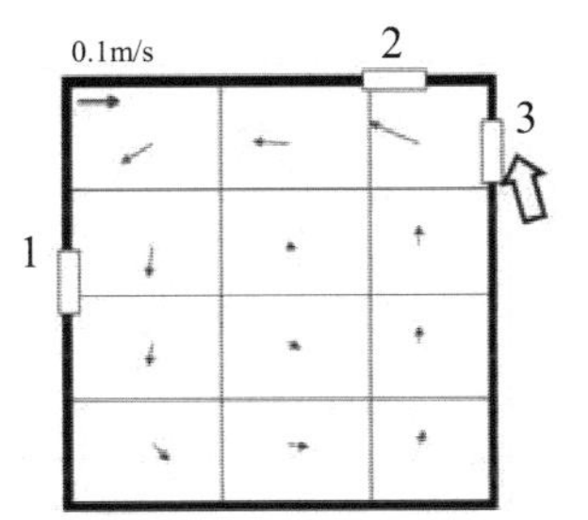

CFD

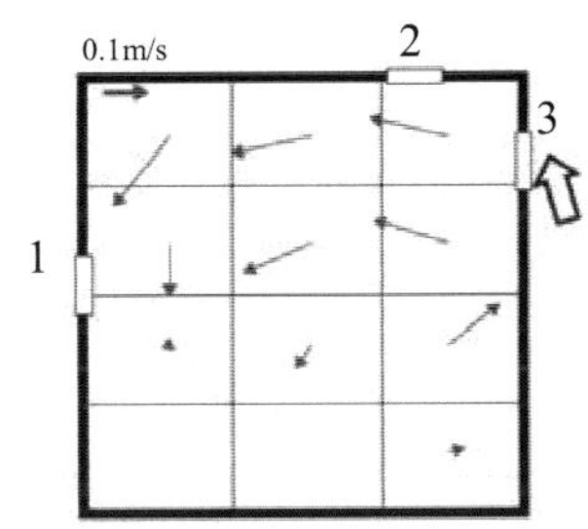

NN

Figure 17. Wind velocity vectors distribution at 112.5 degree of Case 2

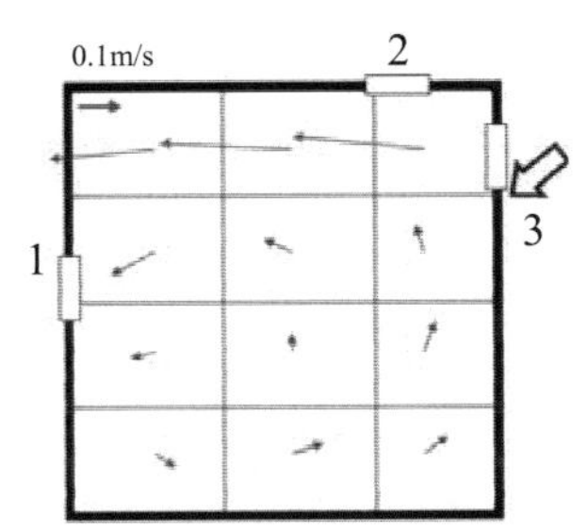

CFD

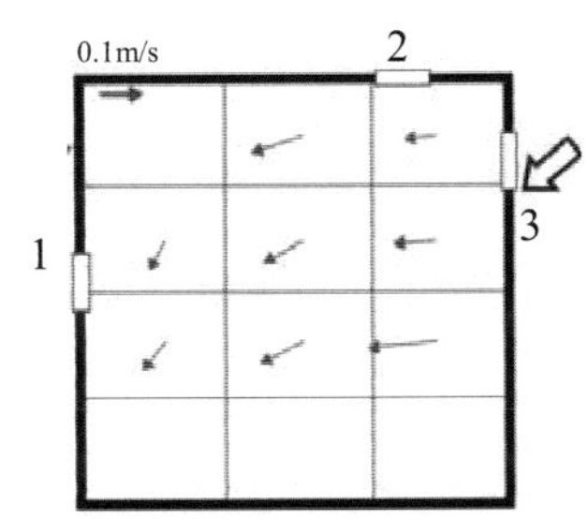

NN

Figure 18. Wind velocity vectors distribution at 202.5 degree of Case 2

CONCLUSION

This study dealt with the fundamental considerations about the cross-ventilated indoor air environmental simulations through NN. It revealed that the results of the simulations with NN tended similarly to the results of CFD under the condition that 2 openings were open in Case 1, Case 2 and Case 3.

By the combination of the cases with 2 openings open, the simulation of NN with 3 openings in Case 2, which had small calculation load and high simulation accuracy, gained low accuracy and basically resulted in similar wind directions to the results of CFD.

Further research is needed to examine other learning methods and to improve the calculation accuracy.

ACKNOWLEDGEMENTS

Kurabuchi Labolatory (Tokyo Science University) provided for CFD data which were used in this study. I appreciate Prof. Takashi Kurabuchi.

REFERENCES

[1] Yukihiro Hashimoto, Itaru Murasawa. 1998. *Study on Control of HVAC System Using Artificial Neural Networks*, The Society of Heating, Air-Conditioning Sanitary Engineers of Japan. 68, 11–19.

[2] Hiroshi Ichikawa. 1993. The Multilayered Neural Network –Application to nonlinear problem analysis-. Kyoritsu Shuppan Co., Ltd.

SENSITIVITY STUDY FOR ARCHITECTURAL DESIGN STRATEGIES OF OFFICE BUILDINGS IN CENTRAL CHILE: EFFECTIVENESS OF NOCTURNAL VENTILATION.

Waldo Bustamante [*1], Felipe Encinas [2] and Francisco Sánchez de la Flor [3]

1 School of Architecture. Catholic University of Chile
Street address
Address, Country
**Corresponding author: wbustamante@uc.cl*

2 Architecture et Climat. Université Catholique de Louvain
Place du Levant 1
B 1348 Louvain-la-Neuve, Belgium

3 Escuela Superior de Ingeniería de Cádiz
C. Chile 1
Cádiz, Spain

ABSTRACT

Office buildings in Chile show higher cooling than heating energy demand. The climate of the country show important differences between cities by the ocean and those of interior regions, located between the coastal and the Andes range. Main cities of Central Chile, where more than 40% of buildings are constructed every year are Santiago and Valparaíso, both located at around 33° S. Santiago presents a Mediterranean climate, with a high temperature oscillation between day and night during cooling period. Valparaíso, by the coast, shows lower temperature fluctuation compared with Santiago during identical period.
In order to define design strategies for energy efficiency of office buildings in mentioned cities, a sensitivity study has been made, considering variables like size of windows (window to wall ratio), type of windows (clear and selective glazing, including low e, single and double glazing), use and type of solar protection and use of nocturnal or diurnal ventilation. In opaque facades (walls and roofs), thermal insulation is considered. In case of walls, in order to increase thermal inertia external insulation is assumed.

The sensitivity analysis is developed considering a square building containing office rooms on all four orientations. This10 story building has been specially proposed and designed for this analysis. Methodology considers an evaluation of heating and cooling demand of the building in both cities. For this purpose, a simulation software under dynamic conditions has been used (TAS of Environmental Design Solutions Limited).

The lowest cooling energy demand is reached when using the lowest window to wall ratio (20%), with solar protection in east, west and north oriented glazed areas. In fact, fully glazed facades in both cities are not recommended. Nocturnal ventilation was highly effective for decreasing cooling demand in both cities. In the case of Valparaíso, due to relatively low temperature during cooling period (maximum lower than 26°C), diurnal ventilation for cooling purposes is also effective.

KEYWORDS: Office buildings, cooling demand, energy efficiency, nocturnal ventilation

INTRODUCTION

Chile shows a wide latitudinal variation (from 17°30′ S to 56°S), which generates high North-South climate variation. On the other hand, the presence of the Pacific Ocean and the Coastal and Andes mountains, generate important climate variation from East to West. Santiago (located in the foothills of the Andes) is the governmental capital and also the industrial and financial centre of the country. Valparaíso is the main port of the country, located at almost the same latitude of Santiago but by the coast.

Climate of Santiago is Mediterranean, showing high temperatures and solar radiation during spring and summer. Mean value of maximum temperature is 29,7°C and mean minimum is 13°C for the warmest month of the year (January). Mean temperature of coldest month (July) are: 3.9 °C (mean minimum) and 14.9 °C (mean maximum). A high temperature fluctuation is observed, especially in summer and intermediate seasons. Climate of Valparaiso is influenced by the Pacific Ocean, showing lower temperature oscillation than Santiago. Mean value of maximum temperature is 20.8°C and mean minimum is 13.5°C for the warmest month of the year (Jan.). For the coldest month (July), mean minimum is 9.2°C and mean maximum is 14.3°C[1].

In Chile, around 4.73 million of square meters of buildings of the Industry, Commerce and Financial Institutions sector were constructed during 2008 [2]. 53,2% was built in Santiago and 6,6% in Valparaíso. In the country there is no mandatory thermal behaviour requirements for office buildings and most of their design patterns are brought from developed countries, even if some architectural strategies, such as double skin, are not suitable –for example- in Central European countries due to the generation of overheating problems, especially when they are designed with fully glazed façades [3;4].

Normally office buildings show higher cooling than heating energy demand. Several studies in different countries have been done approaching the impact that different architectural strategies have over the energy demand [5]. A study performed in 2004 in London for an office building, showed benefits in energy use if windows size, solar protection, and internal profit, are optimized. During two representative weeks, one with hot temperate climate and the other with extreme hot climate, 23% and 40% of refrigeration energy reduction were respectively obtained, once previous modifications were applied. On the other hand, once nocturnal ventilation is applied to the optimized building, an additional reduction of 13% is reached [6]. A study performed in 1998 in Sweden (predominantly cold climate), shows how important are the selection of type of glazing and window to façade ratio for reducing cooling and heating energy demand. Glass use with low U values and solar transmittance may mitigate overheating problems but it does not solve it [7].

In order to define architectural design strategies for reaching thermal and visual comfort with energy efficiency in office buildings of the mentioned cities -with different climates- a sensibility study has been made. Variables considered are the following: size of windows (window to wall ratio), type of windows (clear and selective glazing, including low e, single and double glazing), use and type of solar protection and use of nocturnal or diurnal ventilation. In opaque facades (walls and roofs), thermal insulation is considered. In case of walls, in order to increase thermal inertia and effectiveness of nocturnal ventilation, external insulation is assumed. This paper will mainly show results of this study related to the effect of diurnal and nocturnal ventilation on cooling energy demand.

METHODOLOGY

This study aims to analyze the thermal behaviour of an office building, for different design strategies, during a whole year. For this analysis, simulations are performed with TAS (www.edsl.net), software under dynamic conditions. The sensibility study was developed with a 10 story building, specially designed for this study. Figure 1 shows the plan (16X16) building and 3D image. Each story contains 12 offices of typical dimensions (4 m x 4 m x 2.8 m height).

The building

Main specifications of the building are the following:
Walls: Reinforced concrete 150mm with external EPS 30 mm. U=1.0 W/m2°C
Roof: Reinforced concrete 150mm with EPS 60mm. U=0.59W/m^2°C (in Valparaíso) and with 80 mm of EPS in Santiago. U= 0.40 W/m2°C.
Windows: single glazing, clear. U=5.8 W/m^2°C, Lighting transmittance: 0.90 Solar transmittance: 0.87. In the case of windows, this corresponds to the initial situation. Type of glazing is changed during the sensibility process. It is necessary to mention that it is still common to find new office buildings with single glazing in the country.

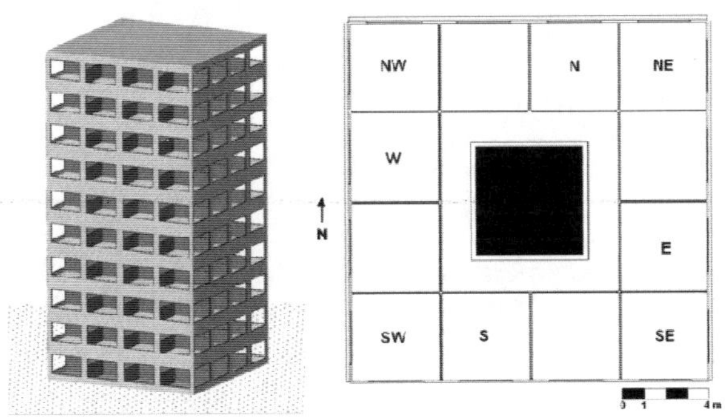

Figure 1. Plan and 3D image of the building

Internal gains and internal conditions

Internal gains of the buildings considered are the following:
People: 9.38 W/m2 (sensible) 6,88 W/m2 (latent).
Lighting: 11 W/m2.
Equipment: 11.25 W/m2.
When cooling demand was estimated, the following temperatures in the inside of each office were considered:
Week days: Maximum of 26°C from 8:00 AM till 19:00 PM.
Weekend days: No temperature restrictions.
Infiltration rate: 0.3 ach.
Ventilation rate: 1.0 ach during week days from 8:00 AM till 19:00 PM.

Sensibility analysis

For defining cases to be simulated, a factorial design was adopted. This involves a given number of samples per each input parameter and consequently running the model for all combination of samples [8]. This method is based on the sampling-based approach, where the model is repeatedly executed from the combination of input parameters sampled with some probability distribution. Since the design of this sensibility analysis consists in 4 input parameters with 3, 3, 4 and 8 parameters per each one, the total combination of samples gives a complete sample of 288 cases. Table 1 presents the different input parameters considered for this study and their associated variables.

Input parameters	Number of variables	Description of variables
Glazing ratio*	3	20% 50% 100%
Types of solar protection devices	3	Without solar protection Overhang in N orientation and blinds for E and W orientations Blinds in N, E and W orientations
Types of glazing	4	Single glazing, clear Single glazing, selective Double glazing, clear Double glazing, selective
Orientations	8	All orientations (N, NE, E, SE, S, SW, W, NW)

(*) Ratio of the glazed area with respect to the total area of

Table 1: Input parameters for sensibility analysis

The 8 orientations correspond to different office rooms showed in figure 1 (N, NE, E, SE, S, SW, W and NW). For each one of these office rooms of 6th floor, the cooling and heating demand was estimated according to variation of type of glazing, types of solar protection and glazing ratio (see Table1).

Simulations considered the following type of glazing: Clear single glazing clear (CS, 4mm) selective single glazing (SS, 6mm), clear double glazing (DGC) and selective double glazing (DGS). Properties of these types of glazing are shown in Table 2. LT: Light transmission, ST: Solar transmission

	CS	SS	DGC	DGS
LT	0,90	0,60	0,82	0,54
ST	0,82	0,50	0,68	0,41
U (W/m2°C)	5,80	5,70	2,78	2,76

Table 2: Properties of different types of glazing.

After obtaining energy demand results and selecting cases with lower heating and cooling demand, simulations considering low e glazing were made. They showed a non significant impact in lowering heating demand in both climates. As it will be observed, heating demand is significantly lower than cooling demand in office buildings in Chile.

Also, for a selective number of cases, after obtaining the 288 mentioned results, ventilation strategies (diurnal and nocturnal) were studied, in order to observe their impact on reducing cooling demand in the building.

RESULTS

First of all, we have confirmed that heating energy demand for office buildings is significantly lower than cooling demand. In the case of Valparaíso, with only diurnal ventilation for maintaining quality of air in different offices, the lower energy demand was reached with selective double glazing, a WWR of 20% and solar protection (Overhang in N orientation and blinds for E and W orientations). In this case annual cooling demand is 16.3 kWh/m2 year and heating demand reaches 3.2 kWh/m2 year. These values represent the energy demand of all 16 offices of the 6th floor of the building. In the case of Santiago, cooling demand for identical case is 31.8 kWh/m2 year and 4.5 kWh/m2 year as heating demand. When using selective and double glazing with low e, cooling demand increases in a 15% in the case of Valparaíso and around 1% in the case of Santiago (both with identical solar protection).

Sensibility analysis

Regarding the sensibility study and due to high output variability (cooling energy demand for each office) -as consequence of the input variability- the energy performance of office buildings is highly impacted by their façade glazing ratio. Differences on annual cooling demand according to window to wall ratio are significant. See figure 2 for the case of Valparaíso. Identical results were obtained for the case of Santiago. See figure 3.

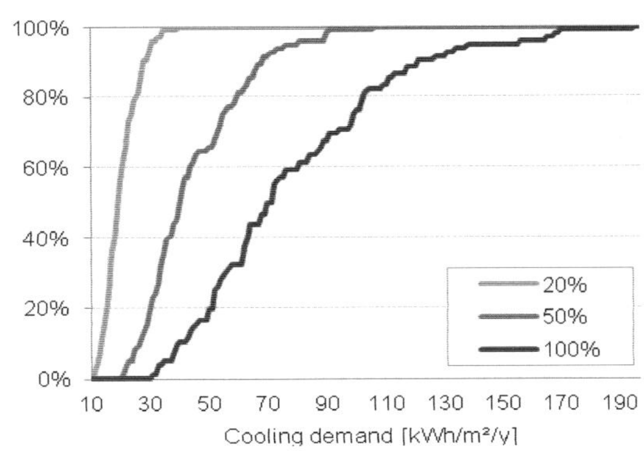

Figure 2. Cumulative frequency (%) for cooling demand with respect to glazing ratio in the case of Valparaiso.

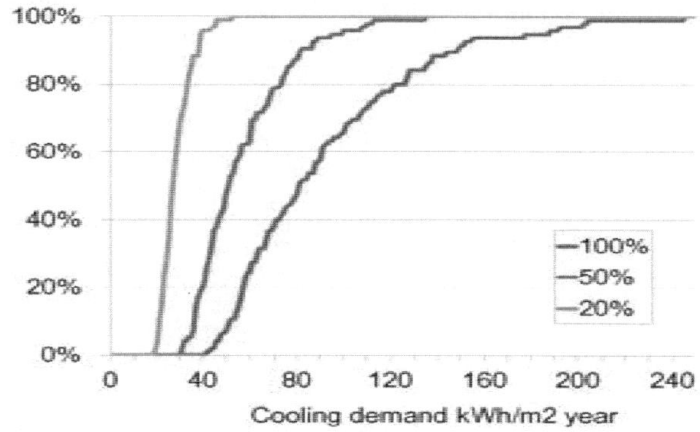

Figure 3. Cumulative frequency (%) for cooling demand with respect to glazing ratio in the case of Santiago.

On the contrary, Figure 4 shows as the variability of the output results per orientation is reduced, which is even more critical with regard to the range of low cooling demands (cases with low window to wall ratio). Figure 4 shows variability on energy demand for all office rooms of 6th floor of the building. Very similar results were observer also in the case of Santiago. According to these results, it is clear that any design strategy proposed for new office developments in Valparaiso and Santiago, the ratio of the glazed area with respect to the exposed façade should be prioritized with respect to orientation.

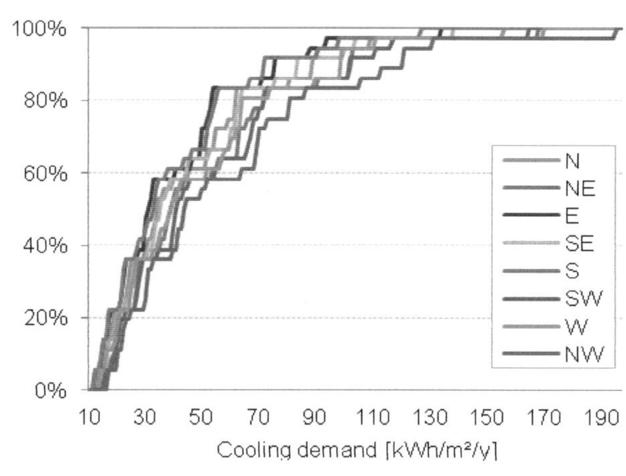

Figure 4: Cumulative frequency (%) for cooling demand with respect to orientations in the case of Valparaiso

Ventilation strategies.

As mentioned, in order to decrease cooling energy demand on office buildings in cities of Santiago and Valparaíso, strategies of ventilation were studied. In the case of Santiago, where we may observe high diurnal temperatures during spring and summer (higher than 26°C) but relatively low nocturnal temperatures (around 15°C), night ventilation was studied. In the case of Valparaíso, where higher temperatures on spring and summer are commonly lower than 26°C, diurnal ventilation for cooling was also studied.

Figure 5 shows the effect of nocturnal ventilation in cooling demand of office buildins in Santiago, when using 20% of window to wall ratio with or without solar protection (SP). Types of glazing are: Single glazing (SG), double glazing selective (DGS). It may be seen that nocturnal ventilation is very effective for getting energy efficiency in office buildings in Santiago. As we have already mentioned, envelope wall of the building is externally insulated, providing it higher thermal inertia. 8 to 10 air changes per hour for nocturnal ventilation may be recommended.

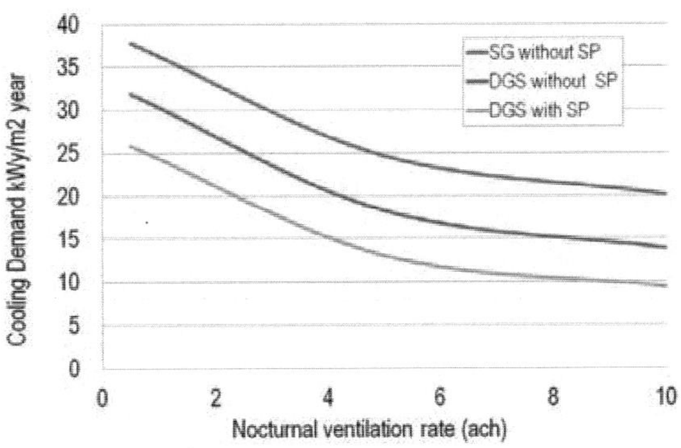

Figure 5: Cooling demand of office buildings with respect to nocturnal ventilation rate.

Figure 6 shows the case of Valparaíso, where identical cases than Santiago were studied. Nocturnal ventilation is also effective bur in this case diurnal ventilation may be recommended. Opening windows when external temperature is lower than 26°C allow reaching a cooling demand 6.9 kWh/m2 year, which decreases to 3.0 kWh/m2 year when using diurnal and nocturnal ventilation. Both cases suppose windows with double glazing selective, solar protection and envelope walls with external insulation.

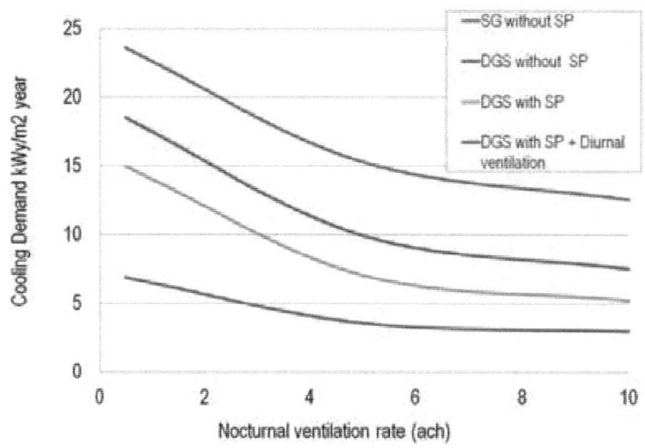

Figure 6: Cooling demand of office buildings with respect to nocturnal ventilation rate.

CONCLUSION

First of all, cooling energy demand in office buildings of Santiago and Valparaíso is significantly higher than heating energy demand. Attention on architectural design strategies for decreasing cooling demand is highly recommended. For lowering cooling demand, solar protection, size and type of windows and solar protection have been studied.

Double glazing selective may be recommended for reaching low cooling demand in both cities. Double glazing clear may also be recommended when using effective solar protection.

It has been showed that there is a high dependence between size of façade glazing area ann the cooling energy demand on office buildings in both studied cities. The lower the window to wall ratio is, a better energy performance of the building is reached. On the contrary, orientation of offices is less relevant for reaching low cooling energy demand, wich is more noticeable for lower window to wall ratio.

Finally, nocturnal ventilation is highly effective for reaching low cooling energy demand in the city of Santiago (with a Mediterranean climate). This strategy is less effective in the case of Valparaíso (with a climate influenced by the ocean). For energy efficiency in office buildings, in the case of Valparaíso, diurnal ventilation may also be applied. Nocturnal and diurnal ventilation may be combined with solar protection on windows and the lower window to floor ratio that may possible to be used.

ACKNOWLEDGEMENTS

This research has been carried out as part of the project FONDECYT N° 1090602 funded by CONICYT, Chile.

REFERENCES

[1] Instituto Nacional de Normalización. 2008. *NCh 1079 Of 2008. Arquitectura y construcción - Zonificación climático habitacional para Chile y recomendaciones para el diseño arquitectónico.* Instituto Nacional de Normalización. Santiago de Chile.

[2] Instituto Nacional de Estadíusticas. 2008. *Anuario de Edificación 2008.* Instituto Nacional de Estadísticas Santiago. Chile.

[3] Manz, H. and Th. Frank 2005. *Thermal simulation of buildings with double-skin façades.* Energy and Building 37, 1114-1121.

[4] Gratia, E. and A. De Herde 2007. *Are energy consumption decreased with the addition of a double skin?* Energy and Building 39, 605-619.

[5] Bustamante, W., Encinas F., Otárola R and Pino A. 2011. *Strategies for improving thermal performance and visual comfort in office buildings of Central Chile.* Proceedings Passive and Low Energy Architecture, PLEA 2011. Louvain la Neuve. Belgium

[6] Kolokotroni, G.I., Watkins R. 2006. *The effect of London heat island summer cooling demand and night ventilation strategies.* Solar Energy 80, 383-392.

[7] Bülow-Hube. H. 1998. *The effect of glazing type and size on annual heating and cooling demand for Swedish offices.* Proceedings of Renewable Energy Technologies in Cold Climate 98. Montreal. Canada.

[8] Hamby, DM 1994. *A review of techniques for parameter sensitivity analysis of environmental models.* Environmental Monitoring and Assessment 32, 135-154.

Nano-scale Aerosol Deposition Model for CFD in Indoor Environmental Analysis

Jun Narikawa, Cong Li, and Kazuhide Ito*

IGSES, Kyushu University
6-1 Kasuga-koen, Kasuga
Fukuoka, Japan

**Corresponding author: ito@kyudai.jp*

ABSTRACT

The overarching objective of this study was to develop a numerical model based on computational fluid dynamics to predict aerosol concentration distributions in indoor environments. Towards this end, this paper proposes a wall surface deposition model of nano-scale aerosol that can predict unsteady deposition flux of aerosol indoors; it also reports the results of sensitivity analyses for targeting a plug-flow-type chamber.

KEYWORDS

Nano-scale aerosol, Deposition model, CFD, Indoor environment

INTRODUCTION

The air quality in the overall indoor environment is called the indoor air quality (IAQ) and has attracted increasing attention with the increasing health consciousness of residents. We spend most of our time indoors and hence IAQ has a great impact on health because of the large amount of inhalation by steady breathing. In the wide spectrum of IAQ problems, this research focuses on indoor aerosol pollution issues that have a large influence on the health of indoor residents. Many studies have reported associations between aerosol in indoor environments and adverse health effects. With regard to the aerosol problems in indoor environments, a prediction method of aerosol distribution and exposure concentration level for residents is required for designs that ensure healthy indoor air quality. The aerosol concentration distribution is usually formed by various factors, for example, convection, diffusion, chemical reaction, gravitational sedimentation, thermophoresis, and electrophoresis. Concerning the particularity of indoor environments compared with outdoor environments, the ratio of wall surface area to the volume of the room or space is usually much larger. From this point of view, aerosol deposition phenomenon is one of the critical components to determine aerosol concentration level in an occupied zone or breathed air; hence, the development of an aerosol deposition model for numerical simulation is an important research issue in the field of IAQ.

The overarching objective of this study is to develop a numerical model based on computational fluid dynamics (CFD) to predict aerosol concentration distributions in indoor environments. Towards this end, this paper describes a novel wall surface deposition model of nano-scale aerosol that can predict unsteady deposition flux of aerosol indoors; it also reports the results of sensitivity analysis targeting a simple 2-dimensional rectangular duct model.

PREVIOUS STUDIES OF DEPOSITION MODELS

There have been many studies of adsorption/desorption of gaseous chemical compounds and various numerical models incorporating adsorption isotherm have been proposed. Concerning the aerosol deposition phenomenon, there have been research reports published since the

1980s [Okuyama et al., 1986, Shimada et al., 1988, Otani et al., 1989] and, in recent years, A.C.K. Lai published an elaborate and comprehensive review article that focuses on a deposition model for indoor environments. The aerosol deposition mode in an indoor environment is described by the concepts of particle loss coefficient or deposition velocity and these parameters are classified as functions of aerosol size and turbulent properties.
Deposition velocity of aerosol particles is defined by the aerosol deposition flux and reference concentration, and the wall boundary condition of aerosol deposition is described as follows:

$$J = -v_d \cdot n. \quad (1)$$

Here, J indicates deposition flux of indoor aerosol, v_d denotes deposition velocity, and n is number concentration of aerosol.
In general, the concept of deposition velocity is analogous to the mass transfer coefficient by which surface zero concentration is considered and hence incorporates not only Brownian diffusion but also the effects of turbulent diffusion and gravitational sedimentation.
Base on the previous reports, for indoor aerosol from 10 nm to 10 μm in diameter, the deposition velocity as a function of particle diameter has a concave downward profile and is minimized within the range of 0.1 to 1.0 μm. This is because the effect of gravitational sedimentation becomes predominant for larger aerosol (aerosol diameter >> 1 μm) and the effect of Brownian diffusion becomes critical for smaller aerosol (aerosol diameter << 0.1 μm).

PROPOSING A DEPOSITION MODEL THAT CONSIDERS SURFACE ASPECT

In this study, we propose a novel deposition model of indoor aerosol with a size range from gas scale to nanometer scale (< 100 nm) for wall surface boundary conditions of CFD simulation. Here we focus on Brownian diffusion and gravitational sedimentation, and other parameters, for example, thermophoresis, turbophoresis, electrophoresis, coagulation, and inertial collision, are disregarded. Although generally Brownian diffusion becomes the dominant process in deposition to a solid surface in the case of nanometer-sized aerosol, in this study, a deposition model incorporating gravitational sedimentation is developed in consideration of extendibility. For simplification, the uniform flow inside the channel cavity and isothermal conditions are assumed in the following development of a formula.

Improvement of Deposition Velocity v_d

The surface geometry of building materials must be reproduced by a numerical grid design in a range of possible resolutions, and the integration of surface roughness effect into the deposition model is needed for the microscopic phenomenon below the resolution with a numerical grid.
If the first grid point on the wall surface adopted in the indoor air flow analysis is set inside the viscous sub-layer (wall unit y+<<1), the surface geometry of the building material can be reproduced by a numerical mesh and hence the flow profile becomes linear. In this boundary condition, the flow profile at the vicinity of the wall surface can be analyzed by CFD simulation and the microscopic difference of surface characteristics of building materials must be modeled independently of flow profile information.
As shown in Figure 1, the order of deposition velocity changed in accordance with the change of building materials. Although the experimental conditions were not united for the results shown in Figure 1, there are certain amounts of rationality in separate modeling of flow information and microscopic surface characteristics of building materials when the flow field conditions are almost the same in these experiments.
In this hypothesis, deposition flux is expressed as follows:

$$J = -S'\overline{v}_d \cdot n. \quad (2)$$

Here, $\overline{v_d}$ denotes deposition velocity for a smooth surface, S' indicates the ratio of effective surface area of a rough surface to that of a smooth surface of building materials.

Novel Nano-scale Aerosol Deposition Model for CFD in Indoor Environment

In this section, constant flow along a flat plate is considered. The hypothetical deposition layer at the surface boundary between solid phase (building material) and air phase is introduced as shown in Figure 1 and it is assumed that the control volume (C.V.) of CFD consists of an imperceptible area element *dS*, thickness *dh*, and volume *dV*. The first grid point as the hypothetical deposition layer is assumed to be set inside of the viscous sub-layer (wall unit y+<<1).

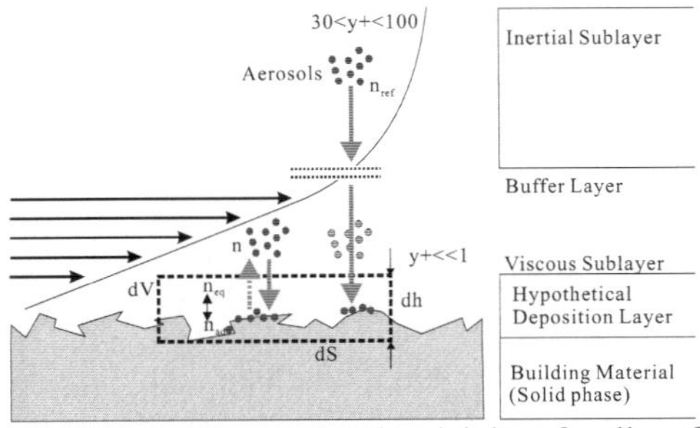

Figure 1 Transport phenomenon of aerosol at the vicinity of wall surface and hypothetical deposition layer

In this condition, the convective flow inside the C.V. is negligible and deposition phenomenon is described by Brownian diffusion. In addition, local equilibrium and one-dimensional deposition in the normal direction against the wall surface are assumed in C.V. (y+<<1 at *dh*).

The aerosol transportation in C.V. is described by the following transport equation of aerosol.

$$\frac{\partial n}{\partial t} = \frac{\partial}{\partial x}\left(D_a \frac{\partial n}{\partial x}\right) + \frac{\partial}{\partial x}\left(V_g n\right) - adv \qquad (3)$$

Here, n is the number concentration of aerosol [m^{-3}], D_a is Brownian diffusion coefficient of aerosol [m^2/s], and ρ_a indicates air density [kg/m^3]. In this formula, gravitational sedimentation (settling velocity) Vg [m/s] is also considered for enhancing to larger particles in the future. In this modeling, impacts of diffusion flux and gravitational settling flux act in parallel.

adv denotes mass transfer at the interfacial surface of air and solid surface and is expressed on the basis of the potential model approach as follows:

$$adv = -\alpha S\left(n_{eq} - n\right), \qquad (4)$$

where α is mass transfer coefficient at the interfacial surface between solid phase and air phase [m/s], S is the contact area pre-unit volume of building material [$m^2/m^3_{(CV)}$], and n_{eq} is the equilibrium concentration [m^{-3}]. The aerosol concentration in C.V. is assumed to be identical to n_{eq} by local equilibrium.

The deposition amount on the surface of a building material is expressed as follows by using deposition phase concentration n_{ad} [m^{-2}].

$$S\frac{\partial n_{ad}}{\partial t} = adv \qquad (5)$$

Here, instantaneous equilibrium in C.V. is adopted and the time change of aerosol concentration in the air phase is assumed to be negligible. When equation (3) is subjected to volume integration in C.V., equation (6) is derived and equation (7) is also derived from equation (5) in a similar way.

$$0 = D_a \frac{\partial n}{\partial x} dS + V_g n dS - adv \cdot dV \qquad (6)$$

$$S\frac{\partial n_{ad}}{\partial t} dV = adv \cdot dV \qquad (7)$$

From equations (6) and (7), the following equation is introduced.

$$-\left(Da\frac{\partial n}{\partial x}+V_g n\right)\Bigg|_B=-\left(S\frac{\partial n_{ad}}{\partial t}\right)\frac{dV}{dS}=-S'\frac{\partial n_{ad}}{\partial t} \tag{8}$$

Here, S' indicates the ratio of effective surface area of rough surface and smooth surface of building materials and the same parameter as in equation (2).

Relationship between Concentrations of Deposition Phase and Air Phase

In order to close the equations, it is necessary to introduce the relationship between deposition phase concentration n_{ad} and air phase concentration of aerosol n (= n_{eq}). In this study, a simple relationship as shown in equation (9) is introduced.

$$n_{ad}=f(n_{eq}) \tag{9}$$

The function of f in Equation (9) indicates an adsorption isotherm in the case of gas phase adsorption/desorption phenomenon.

Here, when a simple linear relationship between n_{eq} and n_{ed} is introduced, equation (10) is derived from equation (9) and deposition flux is also defined in equation (11).

$$n_{ad}=k_h\cdot n_{eq}=k_h\cdot n\big|_{B-} \tag{10}$$

$$-\left(Da\frac{\partial n}{\partial x}+V_g n\right)\Bigg|_B=-S'\frac{\partial n_{ad}}{\partial t}=-S'k_h\frac{\partial n_{eq}}{\partial t} \tag{11}$$

Here, k_h denotes a model coefficient of the linear relationship between n_{eq} and n_{ed} and corresponds to Henry's coefficient of gas phase adsorption isotherm.

Although equation (9) is a hypothetical assumption and not validated by experimental data, a higher-precision model can be developed by introducing higher-order function f in equation (9).

For example, when the sigmoid function like Langmuir-type adsorption isotherm is adopted, equation (12) is derived from equation (9) and deposition flux is also defined in equation (13).

$$n_{ad}=\frac{n_{ad0}k_l n_{eq}}{1+k_l n_{eq}} \tag{12}$$

$$-\left(Da\frac{\partial n}{\partial x}+V_g n\right)\Bigg|_B=-S'\frac{n_{ad0}k_l}{(1+k_l n_{eq})^2}\frac{\partial n_{eq}}{\partial t} \tag{13}$$

Here, k_l and n_{ad0} are the model coefficients.

In numerical analysis integrating the proposed deposition model, equation (11) or (13) is adopted as the boundary condition of a wall surface in an indoor environment.

SENSITIVITY ANALYSIS

In order to analyze the impact of model parameters of the proposed deposition model, sensitivity analysis was carried out for targeting a simple flow field.

Target Flow Field

In this numerical analysis, the flow field and aerosol concentration distribution are analyzed for a two-dimensional rectangular duct model. This target model is reproduced by the inside space of an experimental setup of a rectangular duct made of stainless steel. The outline of the experimental duct model is shown in Figure 2. This chamber has one supply inlet and one exhaust outlet. The cross section of this chamber is 0.02 m (x) × 0.1 m (y) and with a total length of 6.0 m including a running section (z=1.0 m) and a test section (z=5.0 m) with a building material.

Outline of Numerical Analysis

The modeling methodology is based on the Eulerian moment form of the dynamic equation for aerosol transport and dynamics in conjunction with solving the Reynolds averaged Navier-Stokes equations (RANS) for bulk fluid modeling. The Navier-Stokes governing equations were discretized by a finite volume method and flow fields were estimated using the low Reynolds number-type k-ε model (Abe Kondo Nagano model). The QUICK scheme was used for the convection term, and a SIMPLE algorithm was used. To analyze the flow field in the boundary layer and to enable the application of the aerosol deposition model at the wall surface in equation (11), the center of the computational cells closest to the wall surface should be at a non-dimensional distance (wall unit) of $y^{+}<1$, where $y+=u^{*}y_1/\nu$ and y_1 is the distance normal to the wall surface, ν is the kinematic viscosity, and $u^{*}=\sqrt{\tau_w/\rho}$ is the friction velocity. Here, ρ is the air density and τ_w is the wall shear stress. In general, the order of Brownian diffusion coefficient is about 10^{-5} to 10^{-6} [m^2/s] while the order of molecular diffusion coefficient of gas phase contaminant is about 10^{-8} [m^2/s]. Meeting the requirement of y+<<1 with this Brownian diffusion scale deviates greatly from the usual grid size used with CFD simulation. Hence, kinematic viscosity defined by air density and molecular viscosity was used to evaluate y+.

In order to analyze the aerosol dynamic equation numerically in conjunction with CFD based on RANS model, these equations are ensemble-averaged. For the cross-correlation function of time fluctuation of scalar concentration and wind velocity caused as a result of the ensemble-averaging operation, eddy-viscosity representation is adopted using turbulent eddy viscosity ν_t and turbulent Schmidt number σ_t. In general, a value of about 0.2-1.3 is adopted for turbulent Schmidt number and the value 0.7 or 0.9 has been used for most of the CFD studies for turbulent mass diffusion. Here, σ_t =1.0 was used. Concerning the aerosol dynamic equation, convection term, diffusion term, and surface deposition model were considered and other influential factors, for example, reaction-generation term, coagulation loss and gain term, thermophoresis, and electrophoresis, were disregarded in this study because the experimental condition of supply inlet aerosol concentration was reasonably low and nominal time constant (τ_n) was also short enough (= 5 sec).

The number of meshes was set to 52 (x) and 152 (z) and structured mesh was used for the analysis. The analysis was carried out in two dimensions. The air inlet velocity and turbulent intensity were set to U_{in}= 1.0 m/s and 10%, respectively, which are the same as those used in the experiments.

The concentration of aerosol at the inlet position was set to 1.0 [-] as the normalized concentration and kept constant.

Analytical conditions are shown in Table 1.

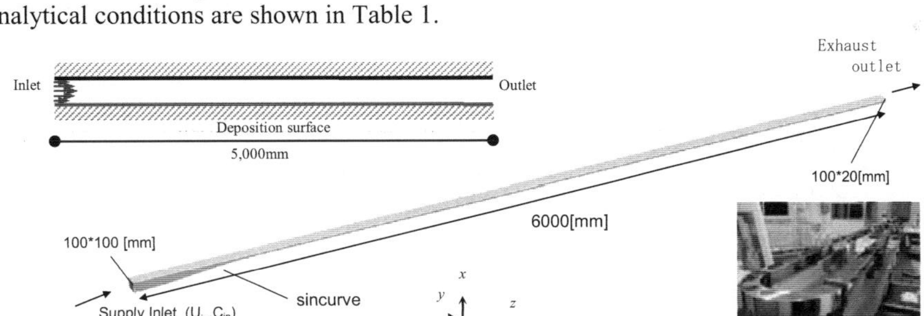

Figure 2 Model analyzed (2-dimensional flow field inside rectangular duct)

Table 1 Numerical and Boundary Conditions

Turbulence Model	Low Re k-ε model (Abe-Kondh-Nagano model, 2-dimensional Cal.)
Mesh	52(x)×152(z)
Scheme	Convection Term: QUICK
Inflow Boundary	U_{in}=1.0 [m/s], k_{in}=3/2× (U_{in}×0.1)^2,ε_{in}=$C_\mu^{3/4}$×$k_{in}^{3/2}$/l_{in}, C_μ=0.09, l_{in}=0.02[m]
Outflow Boundary	U_{out}= free slip, k_{out}= free slip, ε_{out}= free slip
Wall Treatment	Velocity, no slip; Aerosol, proposed deposition model (linear type, see eq.(11))
Aerosol	Dp=1 - 2,000 (see Table 2)

Table 2 Setting of Model Parameters and Cases Analyzed

D_p [nm]	Gas	1	2	3	7	20	70	300	2000
V_g [m/s]	0	6.4×10^{-9}	1.2×10^{-8}	1.9×10^{-8}	4.5×10^{-8}	1.3×10^{-7}	5.3×10^{-7}	4.3×10^{-6}	1.2×10^{-4}
D_a[m^2/s]	1.0×10^{-5}	5.0×10^{-6}	1.0×10^{-6}	5.0×10^{-7}	1.0×10^{-7}	1.0×10^{-8}	1.0×10^{-9}	1.0×10^{-10}	1.0×10^{-11}
$S'k_h$ [-]	1.0×10^{-5} - 1.0×10^{4}								

D_p: particle diameter, V_g: gravitational sedimentation, D_a: Brownian diffusion coefficient

Cases Analyzed

In this study, a deposition model that assumed a linear relationship between n_{ad} and n_{eq} in equation (11) was used for sensitivity analysis. In the linear model shown in equation (11), there are two model parameters: S' and k_h. $S'k_h$ was assumed to be one parameter and the sensitivity analysis was carried out. In this analysis, Brownian diffusion coefficient D_a corresponding to aerosol diameter was changed within the range from 10^{-11} to 10^{-5} [m^2/s]. The parameters of cases analyzed are shown in Table 2.

Concerning the wall surface treatment of deposition model, three cases were set: one-sided deposition (floor) of diffusion flux in rectangular model (Case 1), one-sided deposition (floor) of diffusion and gravitational settling flux in rectangular model (Case 2), and two-sided deposition (floor and ceiling) of diffusion flux and one-sided deposition (floor) of gravitational settling flux in rectangular model (Case 3).

RESULTS

The results of sensitivity analysis are shown in Figure 3. In each case, Brownian diffusion coefficient Da was changed from 10^{-11} to 10^{-5} [m^2/s] and $S'k_h$ was changed from 10^{-5} to 10^{4}. The vertical axis in Figure 4 indicates aerosol concentration decay passing through the deposition surface from the inlet to the outlet position (C_{ext}/C_{in}, C_{in} indicates supply inlet concentration, C_{ext} denotes exhaust outlet concentration of aerosol). The representative concentration distributions of aerosol in the duct model are shown in Figure 4. The aerosol concentrations in Figure 4 are normalized by supply inlet concentration.

In Case 1 for one-sided deposition of diffusion flux onto a floor, the aerosol concentration at the exhaust outlet (C_{ext}/C_{in}) decreased gradually when the aerosol diameter became small (Brownian diffusion coefficient became larger). The model parameter of linear model $S'k_h$ had high sensitivity between the range from 10^{-2} to 10^{0}, and concentration gradient on the wall surface became almost zero (adiabatic boundary condition) within the range of $S'k_h < 10^{-3}$, and wall surface concentration became almost zero (perfect sink boundary condition) at the range of $S'k_h > 10^{1}$ as shown in Figure 4 (1).

In Case 2 with one-sided deposition of diffusion and gravitational settling flux onto a floor, the change of the aerosol concentration at the exhaust outlet (C_{ext}/C_{in}) as a function of $S'k_h$ showed a similar trend to that in Case 1 for a relatively fine aerosol, that is, within the range of 10^{-8} to 10^{-5} [m^2/s] of Da, because the gravitational sedimentation was disregarded in this region. On the other hand, gravitational settling flux became larger than Brownian diffusion flux in the range of μm diameter of aerosol. In this numerical analysis, concerning the aerosol of μm size, the diffusion and gravitational sedimentation flux were assumed to act separately and in parallel and the zero gradient concentration at the wall surface was adopted for the boundary condition of gravitational sedimentation. Hence, the decrease of the aerosol concentration at the exhaust outlet (C_{ext}/C_{in}) was determined by the effect of gravitational sedimentation in the range of μm size aerosol.

In Case 3 that considered two-sided deposition of diffusion and gravitational settling flux onto a floor and one-sided deposition of diffusion onto a ceiling, the decrease of the aerosol concentration at the exhaust outlet (C_{ext}/C_{in}) became larger than that of Case 2 for the upper side deposition by Brownian diffusion.

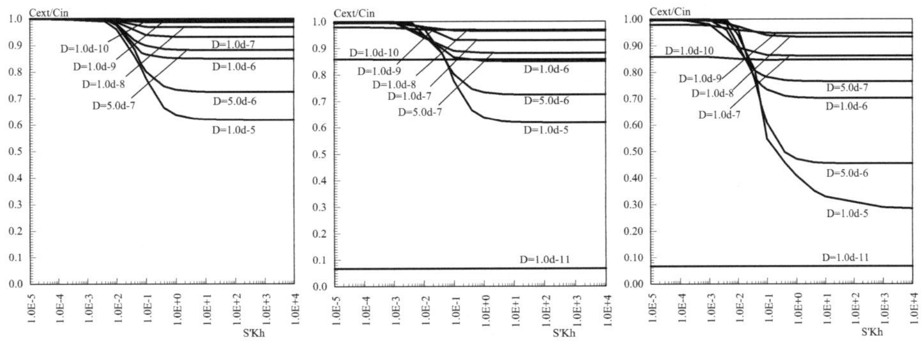

(1) Case 1 (2) Case 2 (3) Case 3

Figure 3 Results of sensitivity analysis

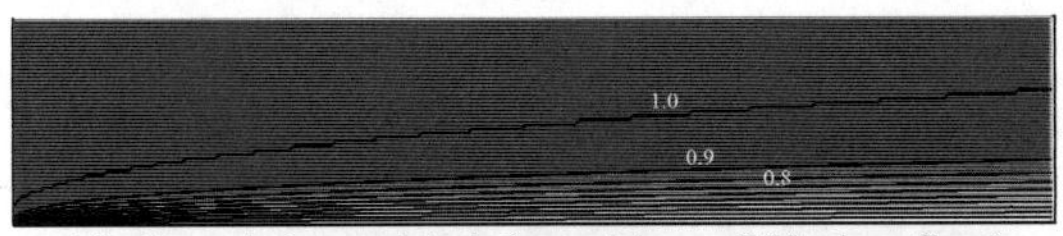

(1) Case 1 (one-sided deposition, diffusion flux)

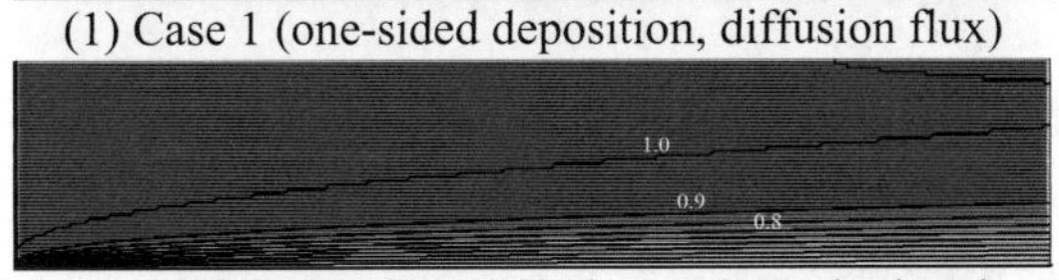

(2) Case 2 (one-sided deposition, diffusion and gravitational settling flux)

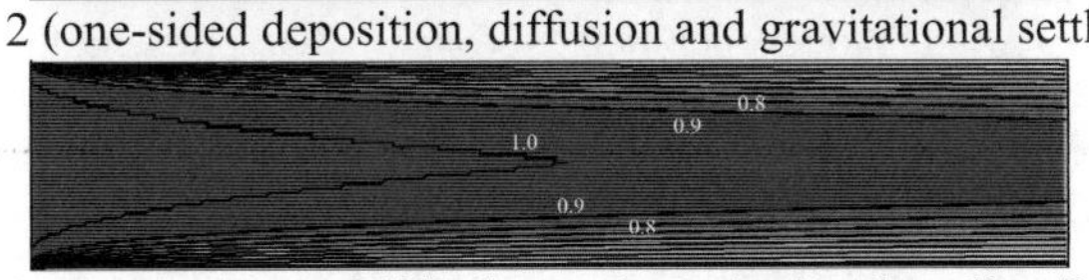

(3) Case 3 (two-sided deposition of diffusion and gravitational settling flux (on floor), and one-sided deposition of diffusion (on ceiling))

Figure 4 Results of aerosol concentration distribution in duct model
(Vertical axis in Figure 5 enlarged 50 times to facilitate visualization)

DISCUSSION

For the adsorption/desorption model for gas phase contaminant, in general, the adsorption phase concentration on a wall surface is analyzed by using the concept of adsorption isotherm. In other words, wall surface concentration of gas phase contaminant will be analyzed explicitly. On the other hand, wall surface concentration will generally be set to zero when analyzing diffusion flux. The behavior of μm size particles is known to be ruled by physical factors and hence the deposition model of previous reported study was discussed in terms of physics properties of wall surface-aerosol interaction, especially in aerosol engineering. However, the diameter of aerosol contaminants has a wide spectrum from a few nanometers to micrometers and above, and the distinction between gas phase contaminant and nano-scale aerosol will not be clear.

Although there seems to be a boundary between gas phase molecules and aerosol particles at about 2 nm for contaminants of comparatively simple structure, the macromolecules that reach a size of 100 nm and above have the characteristics of gas phase molecules. From this point of view, the development of a wall surface deposition model for numerical simulation that could apply continuously from gas phase to nano-scale aerosol (<100 nm) is necessary.

We have a research plan to report the results of two types of fundamental experiment: (i) dynamic chamber experiment with rectangular duct model, and (ii) static chamber experiment with Tedlar bag (polyvinyl fluoride).
In the Tedlar bag experiment, deposition velocity and effective surface area S' will be identified for various building materials, that is, materials of smooth surface and rough surface, by measuring time series of aerosol concentration in the bag enclosing the target size aerosol and building material.
In the rectangular duct experiment, the model parameter $S'k_h$ will be identified by measuring the concentration decrease from the supply inlet to the exhaust outlet opening and using the identification chart of $S'k_h$ and C_{ext}/C_{in} as shown in Figure 4. Then, conclusively, S' and k_h will be separately identified by the above two types of experiment.
The experimental validation of this sensitivity analysis and identification of model parameters will be reported in the future.

CONCLUSION

In this paper, a novel deposition model for nano-scale aerosol to analyze the surface concentration explicitly was proposed and the results of sensitivity analyses were also reported. The findings obtained in this work can be summarized as follows:
(1) The deposition model that incorporated a linear relationship between n_{ad} and n_{eq} was confirmed to have high sensitivity in the range from 10^{-2} to 10^{0} of the model parameter $S'k_h$.
(2) Deposition flux of relatively large aerosol (>100 nm) was determined by gravitational sedimentation and the effect of model parameter of $S'k_h$ was almost negligible for this relatively large aerosol size.
(3) In future, we will report the experimental validation of this sensitivity analysis by using two types of experiment (dynamic chamber experiment with rectangular duct model, and static chamber experiment with Tedlar bag) and identification of model parameters

ACKNOWLEDGEMENTS

This research was partly supported by a Grant-in-Aid for Scientific Research (JSPS KAKENHI for Young Scientists (S), 21676005). The authors would like to express special thanks to the funding source.

REFERENCES

[1] Murakami, S., Kato, S., Ito, K., Yamamoto, A., Kondo, Y., Fujimura, J. Distribution of Chemical Pollutants in a Room Based on CFD Simulation Coupled with Emission / Sorption Analysis, ***ASHRAE transaction*** 107, 1, AT-01-13-3, 2001

[2] Murakami, S., Kato, S., Ito, K., Zhu, Q. Modeling and CFD Prediction for Diffusion and Adsorption within Room with Various Adsorption Isotherms, ***Indoor Air***, 13, Supplement 6, pp20-27, 2003

[3] Okuyama, K., Kousaka, Y., Yamamoto, S. and Hosokawa, T., Particle loss of aerosol with particle diamteres between 6 and 200 nm in stirred tank, ***J. Colloid Interface Sci.,*** 110, pp214-223, 1986

[4] Shimada, M., Okuyama, K., Kousaka, Y., Seinfield, JH., A model calculation of particle deposition onto a rough wall by Brownian and turbulent diffusion, ***Journal of Colloid and Interface Science***, Vol. 125, No, 1, pp198-211, 1988

[5] Otani, Y., Emi, H., Kanaoka, C., Kato, K., Determination of deposition velocity onto a wafer for particles in the sice range between 0.03 and 0.8 μm, ***J. Aerosol Sci,*** Vol. 20. No. 7, pp. 787-796, 1989.

[6] Harrison, A.W., Quiescent boundary layer thickness in aerosol enclosures under convective stirring conditions, ***J. Colloid Interface Sci.,*** 69, pp563-570, 1979

[7] Lai, A.C.K., Particle Deposition indoors: a review, ***Indoor Air*** 12, pp211-214, 2002

Exposure Concentration Prediction by Multi-Nesting Approach Connecting Building Space-Virtual Manikin-Nasal Airway Model

Kazuhide Ito* and Hiroaki Asanuma

IGSES, Kyushu University
6-1 Kasuga-koen, Kasuga
Fukuoka, Japan

**Corresponding author: ito@kyudai.jp*

ABSTRACT

In this study, we developed an integrated simulation procedure for prediction of concentration of contaminant exposure using a multi-nesting method connecting building space, a Virtual Manikin, and bronchus airway in humans. On the basis of this numerical simulation, detailed information on the unsteady spacial distribution of contaminant concentration, the breathing concentration of infectious contaminant, and the non-uniform distribution of contaminant deposition in nasal airway could be provided for designers of indoor environments in the design stage and also for residents. Here, we demonstrate the application of this integrated multi-nesting approach for targeting a waiting space in a university hospital.

KEYWORDS

Airway, Virtual Manikin, CFD, Nesting, Infectious Transmission

INTRODUCTION

Indoor environmental issues, especially air-quality problems, are being addressed more and more intensively, and this trend looks set to continue. From the viewpoint of management of public health for residents or workers, the prediction and control of the concentration of contaminant in occupational exposure will be critically important for the production of healthy indoor environments. In general, concerning problems of low-concentration and long-term exposure, for example, exposure to formaldehyde and volatile organic compounds (VOCs) emitted from building materials, time- and space-averaged concentration in a whole room or occupied zone will be the controlled target of air-quality management; the necessity of evaluating and predicting time fluctuations and distributions of contaminant on a short time scale is low. On the other hand, concerning the problems of high-concentration and short-term exposure by terorrism involving nuclear, biological, and chemical weapons (NBC), gas leak accidents, and/or imperfect combustion of waste gas from burning appliances, with a local or point concentration, the breathing concentration of contaminant will present a mortal danger. More specifically, the unsteady and non-uniform concentration distribution of a target contaminant becomes a critical parameter of residential air-quality management.

In this paper, we propose a method for prediction of residential exposure concentration level from building scale to nasal airway scale inside of the human body. Here, an integrated numerical procedure of CFD technique, that is, a multi-nesting method, is applied to infectious transmission in a hospital space caused by a hypothetical acute respiratory contaminant.

OUTLINE OF TARGET SPACE AND GRID DESIGN

University Hospital (Region 1)

In this study, the waiting space of a university hospital as Region 1 is analyzed. Figure 1 shows the outline of the hospital space and the plan of the first floor, which is 42 m × 90 m and with a total floor area of 2020 m^2. This region consists of three zones: (i) reception and waiting space on the north side (3.0 m in height), (ii) hospital mall on the south side (15 m in height), and (iii) medical space on the west side; there is a heating, ventilation, and air-conditioning (HVAC) system designed and constructed in accordance with these three zones. The numerical analysis was conducted in zones (i) and (ii) because the doors of the rooms in zone (iii) are always closed and its HVAC system is also independent. Multiple fan coil units (FCU) are arranged in zones (i) and (ii) in order to control the indoor temperature. A total of 79 supply inlet openings of the air-conditioning system are installed on the ceiling and air is exhausted through lavatories in three places. The geometries of the hospital space, furniture, and the supply inlet and exhaust outlet openings were simplistically modeled in order to avoid them having predominant effects on the prediction accuracy.

This space (Region 1) was discretized by an unstructured mesh and the total number of meshes was set to approximately 600,000 for the analysis.

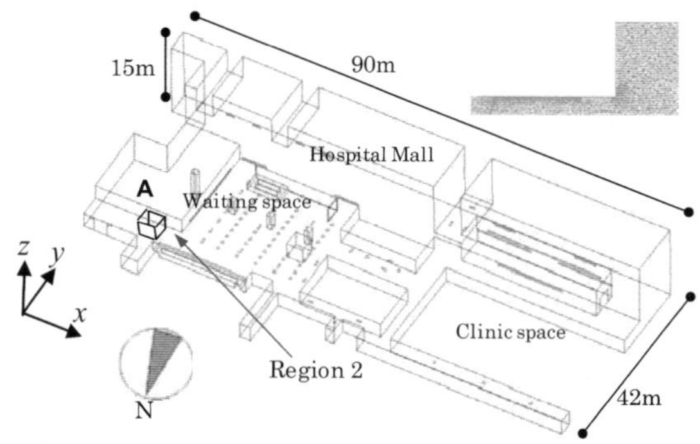

Figure 1 Perspective View of University Hospital (Region 1)

Virtual Manikin (Region 2)

Indoor environmental studies have focused on physical phenomena around the human body at the microclimate level in recent years; therefore, the need for realistic and detailed human body models, for example, Virtual Manikin or Computer Simulated Person, has been pointed out. In this study, a Virtual Manikin that reflected the average Japanese body proportions (standing adult male) was developed.

The outlines of the human body were drawn using POSER 4.0J software (Curious Labs Inc.) and the data were then read out in DXF format. The overall shape of the human body was then adjusted using three-dimensional CAD software (Vector Works and A&A Co. Ltd.). The final geometry of the Virtual Manikin and computational grids were made using GRIDGEN V15 (VINAS Co. Ltd.). The hands and feet of the Virtual Manikin were simplified in consideration of the computational load for the CFD analysis.

The Virtual Manikin was arranged at the northeast corner and in front of the lavatory in Region 1 and an analytical domain of dimensions x=3.0 m, y=3.0 m, and z=3.0 m was set by centering on a standing Virtual Manikin (see point A in Figure 1). Figure 2 denotes the outline of Region 2. The ceiling height of Region 2 corresponds to that of Region 1 and there is no supply and exhaust opening of the air-conditioning system.

The total number of computational cells in Region 2 was set to about 1.02 million. The surface mesh reproduced the complex geometry of the human body arranged with a triangular surface mesh. To resolve the boundary layer around the Virtual Manikin, four layers of prism cells were created on its surface with an equal height of 1.0 mm between layers. The tetra-meshes were then arranged from the outside of the boundary layer to the other side walls in the analytical model room. Under this numerical condition, the wall units (y^+), which express the dimensionless normal distance from the surface, met the requirement of 1.0 or less over the whole surface of the Virtual Manikin.

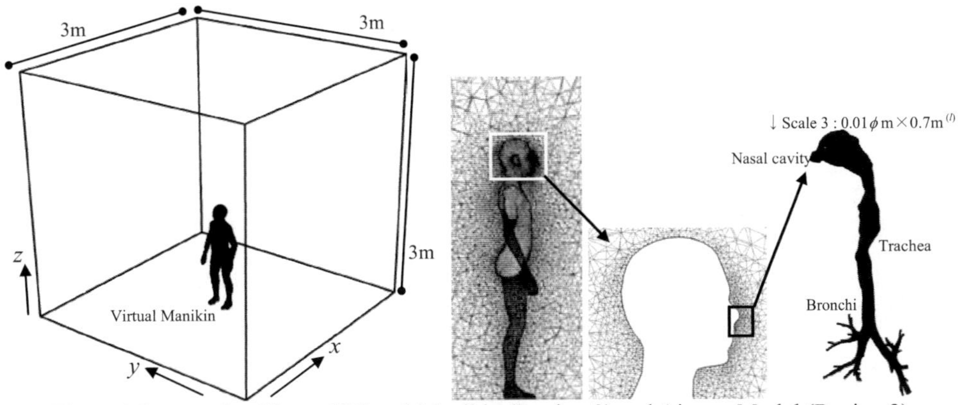

Figure 2 Perspective View of Virtual Manikin (Region 2) and Airway Model (Region 3)

Nasal Airway Inside Human Body (Region 3)

Airway model from nasal cavity to bronchial tubes (fourth bifurcation) was arranged as Region 3 and nasal cavity (nostril) was set as the boundary with Region 2. A computational airway model was created using computed tomography (CT) data (DICOM format) of a healthy adult male (average height and weight of Japanese). The smoothing of the overall shape and the creation of fluid geometries were adjusted by Mimics 4.0 (Materialise) and 3-matic (Materialise) software. The final geometry of the airway model and computational grids were made using GRIDGEN V15 (VINAS).

The total number of computational cells in Region 3 was set to about 1.0 million and the wall units (y^+) met the requirement of 1.0 or less over the whole surface of the airway model.

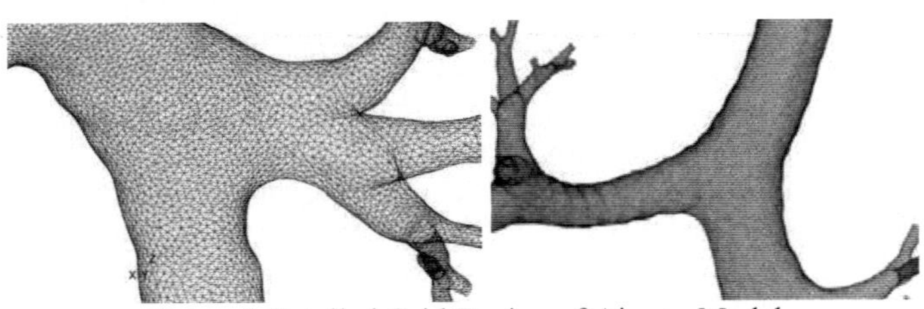

Figure 3 Detailed Grid Design of Airway Model

OUTLINE OF NUMERICAL SIMULATION

In order to connect three different analytical domains (Regions 1, 2, and 3), a one-way nesting method was applied in this analysis. From Region 1 to Region 2, distributions of average velocities (U, V, and W), turbulent properties (k and ε), and unsteady contaminant concentration distribution data were transferred. Each piece of data that passed from Region 1 to Region 2 was linearly interpolated because the size of the grid was markedly different on the boundary plane in the two regions.

From Region 2 and Region 3, the flow field in the vicinity of the nostrils is assumed to be determined almost completely dependently on the respiratory cycle of the human body; hence, the inlet velocity of the boundary plane in Region 2 and Region 3 was calculated from the steady breathing air volume and only contaminant concentration data in the breathing zone were transferred from Region 2 to Region 3.

The purpose of this simulation is to demonstrate the analytical procedure and provide an application example; then, every simulation was conducted under isothermal conditions for the reduction of computational load. The commercial CFD code ANSYS/Fluent 12 (ANSYS Co. Ltd.) was used to calculate the flow field and contaminant distributions.

Region 1

Flow fields were analyzed by the RNG k-ε model in the steady state condition [22]. The SIMPLE algorithm was used with the QUICK scheme for the convective terms, and a second-order center difference scheme was used for the others. After steady flow field analysis, unsteady aerosol contaminant concentration distributions were analyzed by solving ensemble-averaged scalar transport equation based on a Eulerian approach. Contaminant was assumed to be aerosol particles and their diameter was set to 10 μm. The contaminant was assumed to be generated through 16 points of supply inlet openings of the air-conditioning system in the zone (i) reception and waiting space constantly for two hours, and subsequently reached a concentration of zero at supply inlet positions.

Table 1 Numerical and Boundary Conditions of Region 1

Turbulence Model	RNG k-ε model (3-dimensional Cal.)
Scheme	Convection Term: QUICK
Inflow Boundary	U_{in}=0.5 [m/s], $k_{in}=3/2\times(U_{in}\times0.1)^2$, $\varepsilon_{in}=C_\mu^{3/4}\times k_{in}^{3/2}/l_{in}$, C_μ=0.09, l_{in}=0.03 [m]
Outflow Boundary	U_{out}= free slip, k_{out}= free slip, ε_{out}= free slip
Wall Treatment	Velocity: generalized log low, Contaminant: gradient zero
Contaminant	D_p=10 μm, Eulerian approach

Region 2

The flow field was analyzed three-dimensionally on the basis of the low Reynolds number k-ε model (Abe-Nagano-Kondo model) in the steady state condition. The SIMPLE algorithm was used with the QUICK scheme for the convective terms, and a second-order center difference scheme was used for the others. The no-slip condition was adopted as the wall surface boundary condition for velocity. After the steady state analysis of flow field around the Virtual Manikin, the analyses of unsteady contaminant concentration distributions were carried out under conditions that considered convection, diffusion, and gravitational sedimentation for Dp= 10-μm-size particle based on a Eulerian approach. Table 2 shows details of the numerical and boundary conditions for Region 2. In this analysis, breathing cycle (inhaled and exhaled air flow) and heat generation from the Virtual Manikin were not reproduced.

Table 2 Numerical and Boundary Conditions of Region 2

Turbulence Model	Low Re Type model (Abe-Nagano model, 3-dimensional Cal.)
Scheme	Convection Term: QUICK
Region 1 and 2 Boundary	U_{in} : Region 1 Data , k_{in}: Region 1 Data,ε_{in}: Region 1 Data
Wall Treatment	Velocity: No-slip, $k\|_{wall}$: no slip, $\varepsilon\|_{wall}=2\nu(\partial\sqrt{k}/\partial y)^2$, Contaminant: gradient zero
Contaminant	D_p=10 μm, Eulerian approach

Region 3

The flow field was also analyzed on the basis of the low Reynolds number k-ε model (Abe-Nagano-Kondo model) and no-slip condition was adopted as the wall surface boundary condition. Unsteady distribution of aerosol contaminant was analyzed using a Lagrangian model, which analyzes transient momentum equation for each particle and includes drag force and gravitational effect. Discrete random walk model was adopted to model stochastic turbulent dispersion. A total of 10,000 particles, which were assumed to have a spherical shape, were generated from the inlet plane at moment and unsteady particle trajectory was carried out under the perfect sink wall boundary condition.

Table 3 Numerical and Boundary Conditions of Region 3

Turbulence Model	Low Re Type k-ε model (Abe-Nagano model, 3-dimensional Cal.)
Scheme	Convection Term: QUICK
Inflow Boundary	U_{in}=0.92 [m/s], $k_{in}=3/2\times(U_{in}\times 0.1)^2$, $\varepsilon_{in}=C_\mu^{3/4}\times k_{in}^{3/2}/l_{in}$, C_μ=0.09, $l_{in}=10^{-3}$[m]
Outflow Boundary	U_{out}= free slip, k_{out}= free slip, ε_{out}= free slip
Wall Treatment	Velocity: No-slip, $k\|_{wall}$: no slip, $\varepsilon\|_{wall}=2\nu(\partial\sqrt{k}/\partial y)^2$, Contaminant: perfect sink
Contaminant	D_p=10 μm, Lagrangian approach

RESULTS

Region 1

The results of flow field and contaminant concentration distribution in Region 1 are shown in Figure 4. Stagnant flow field was formed except for in the vicinity of the supply inlet opening and lavatory space, which was designed as an exhaust outlet. As for the contaminant distribution, aerosol contaminant was distributed within the zone (i) waiting space of the hospital in accordance with the zoning of the air-conditioning system and contaminant generation points.

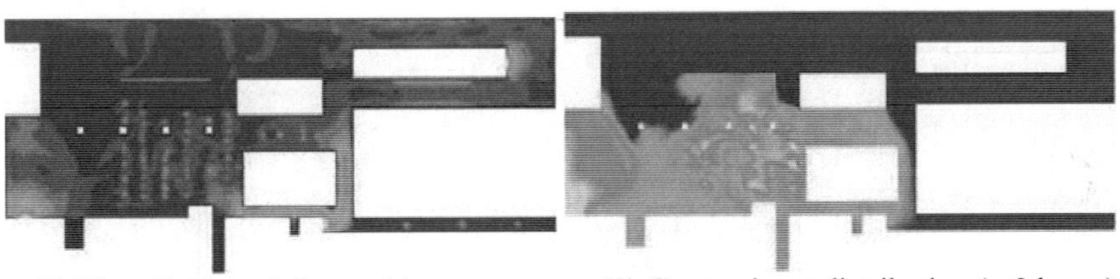

(1) Flow fluid (y=1.6 m, m/s) (2) Contaminant distribution (t=2 hours)

Figure 4 Results of Flow Field and Contaminant Distribution in Hospital (Region 1)

The flow field in the nesting region from Region 1 to Region 2 (see point **A** in Figure 1) is shown in Figure 5(1); a stagnant flow field with air velocity below 0.15 m/s and flow from the center of the waiting space to the lavatory space was formed. Figure 5(2) denotes the concentration distribution of D_p= 10-μm-size aerosol contaminant in the same nesting region.

Region 2

The Virtual Manikin faced the east side of the hospital building.

Figure 5(3) indicates the concentration distribution of the aerosol contaminant in Region 2 with the Virtual Manikin. This is the result from two hours after the start of analysis, and the concentration in Figure 5(3) was normalized by the supply inlet concentration of the air-conditioning system on the ceiling. The existence of the Virtual Manikin influenced the flow field and contaminant distribution in Region 2.

In Figure 6, the estimation results of time histories of breathing concentration and total inhalation dose of the Virtual Manikin are shown. Here, the pulmonary ventilation rate of the Virtual Manikin was assumed to be 1.67×10^{-4} [m^3/s] for estimation of total inhalation dose.

Air change rate of Region 1 was 4.36 [h^{-1}] and hence contaminant concentration in the breathing zone reached a steady state from the beginning of contaminant generation within tens of minutes as shown in Figure 6(1). The breathing concentration immediately decreased after the cessation of contaminant generation (t=2 hours).

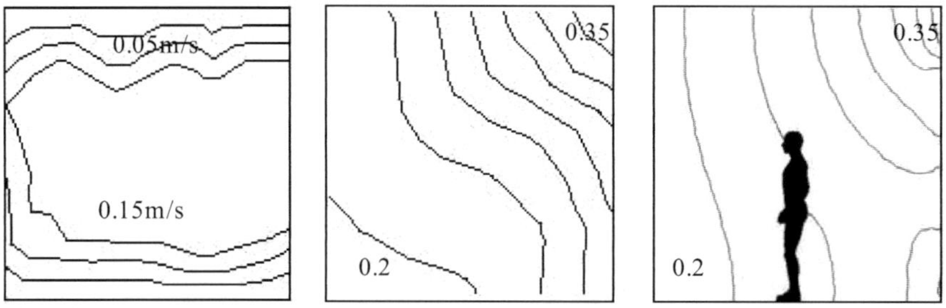

(1) Flow field (without Virtual Manikin) (2) Concentration distribution (without Virtual Manikin) (3) Concentration distribution (with Virtual Manikin)

Figure 5 Flow Field and Contaminant Distribution in Region 2

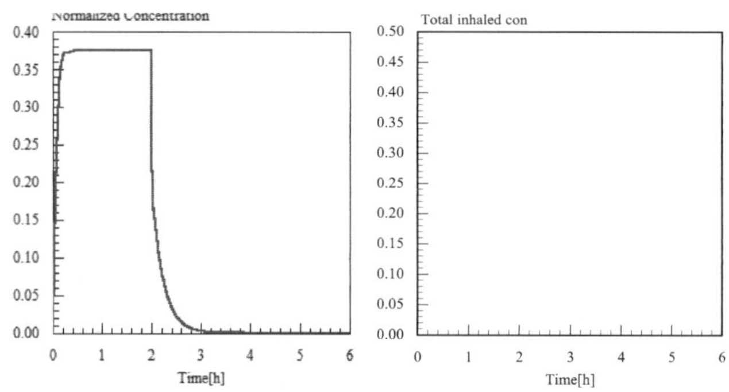

(1) Breathing concentration (2) Total inhalation dose

Figure 6 Results of Time Series of Exposure Concentration (Region 2)

Region 3

The results of flow field and path line of aerosol particles are shown in Figure 7. The results of relatively large particles (Dp= 50 μm) as well as fine particles (Dp= 10 μm) are introduced for reference.

The flow inside the airway model was analyzed under the hypothesis of constant inhalation and steady state. A complicated flow field was formed in the airway model because of the complex geometry. Concerning the 10-μm-size particles, 66% of total generated particles at the boundary of the nares were deposited on the surface of the nasal cavity and the rest were transported until the third bifurcation of bronchial tubes through the pharyngis. On the other hand, 100% of the 50-μm-size particles were deposited at the inner wall of the nasal cavity and were not transported as far as the bronchus.

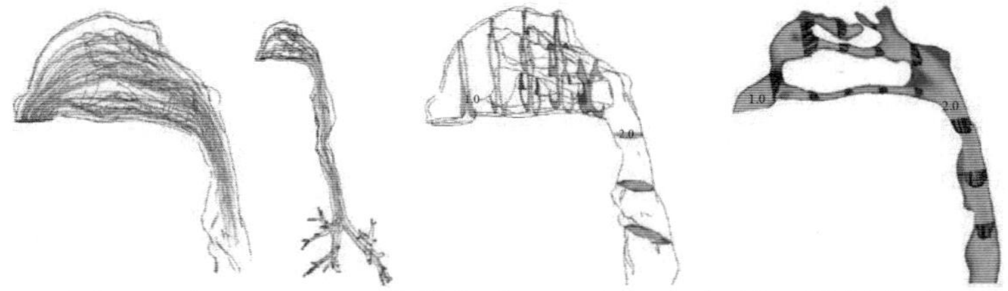

(1) Stream line (2) Velocity maginitude (m/s) (3) Velocity vector (m/s)

(4) Path line of Dp= 10 μm particles (5) Path line of Dp= 50 μm particles

Figure 7 Results of Stream Line and Path Line of Contaminant (Region 3)

DISCUSSION

A previous research paper reported that particles of micrometer order or above were deposited and captured in the upper airway region and particles of the order of 0.1 μm or below reached deep into the lungs. Our simulation results show the possibility of 10-μm-size particles passing through the nasal airway. However, the reproducibility of flow and contaminant distribution in the airway model will be strongly dependent on the degree of roughness of the surface geometry. Fluid resistance of nasal hair and interaction between particles in air phase and the surface of the airway model, and also the respiration cycle, will also be important factors to improve prediction accuracy.

Here, we proposed an integrated simulation procedure for prediction of concentration of contaminant exposure using a multi-nesting method connecting building space, a Virtual Manikin, and bronchus airway in humans and showed an application example of this procedure. The nesting from Region 2 to Region 3 was insufficient in this study and the development of a seamless nesting method to connect breathing information of the Virtual Manikin and the boundary of nasal airway is an important research issue for the future.

CONCLUSION

A comprehensive numerical prediction based on one-way nesting method that directly connects the boudary conditions from hospital building scale to nasal airway inside a human by the intermediate of a Virtual Manikin was proposed in this study.

The findings obtained in this work can be summarized as follows:

(1) One-way downscaling analysis of residential exposure concentration for a large-scale hospital waiting space was demonstrated. The possibility of an engineering approach was represented for the estimation of breathing contaminant concentration and non-uniform distribution of aerosol deposition inside nasal airway.

(2) In the future, the experimental validation and improvement of prediction accuracy of this proposed numerical procedure will be needed.

ACKNOWLEDGEMENTS

This research was partly supported by a Grant-in-Aid for Scientific Research (JSPS KAKENHI for Young Scientists (S), 21676005). The authors would like to express special thanks to the funding source.

REFERENCES

[1] Ma, B. and Lutchen, K.R., CFD simulation of aerosol in an anatomically based human large-medium airway model, Annals of Biomedical Engineering, Vol.37, No.2, 2009, pp.271-285

[2] Longest, P.W. and Vinchurkar, S., Validating CFD predictions of respiratory aerosol deposition: Effects of upstream transition and turbulence, Journal of Biomechanics, 40, 2007, pp305-316

[3] Deschamps, T., Schwartz, P., Trebotich, D., Air-flow simulation in realistic models of the trachea, Proceedings of the 26th Annual International Conference of the IEEE EMBS, 2004, pp3933-3936

[4] Stapleton, K.W., Guentsch, E., Hoskinson, M.K., Finlay, W.H., On the suitability of k-ε turbulence modeling for aerosol deposition in the mouth and throat: A comparison with experiment, Journal of Aerosol Science 31 (6), 2000, pp739-749

[5] .Schroeter, J.D., Garcia, G.J.M., Kimbell, J.S., 2011. Effects of surface smoothness on inertial particle deposition in human nasal models, Journal Aerosol Science 42, 2011, pp52-63

[6] Asanuma, A., and Ito, K., Integrated Approach of CFD and SIR Epidemiological Model for Infectious Transmission Analysis in Hospital, AIVC 2011 conference proceedings (submitted)

OPTIMAL AIR TIGHTNESS LEVELS OF BUILDINGS

Willem de Gids
VentGuide
Netherlands

1. Introduction

The air tightness of building has been a serious problem over the last 30 years. In 1979 the international Air Infiltration Centre (AIC) was erected within the International Energy Agency (IEA) platform. Infiltration of cold air into buildings needs to be heated to reach to a comfortable indoor climate. But the energy penalty due to that should be minimized. The AIC (later AIVC) had as one of their tasks to find solutions for good air tight buildings and to promote the knowledge about building construction to reach acceptable level of air tightness of buildings. Many publications were produced. The slogan was and is "Built tight and Ventilate right" The problem of air tightness is still important in the building practice of these days. Last year in 2010 TightVent Europe was formed to cope with the problem of air infiltration due to building leakages in low energy buildings. The big question is are buildings still too leak ? To find an answer on the above question the reasons for airtight buildings should be considered.

2. Why air tightness of buildings

The air tightness of buildings is necessary because several aspects are influenced by uncontrolled leakage of air through buildings. The most important aspects are:

- Health
- Building damages
- Comfort
- Disturbance of ventilation
- Energy

2.1 Health

From the health point of view a building must be heated at least at a level that the heating system in the building may reach temperature by which people can perform normal tasks. This is partly a design problem for the heating system but to high local leakages may cause problems of to cold areas in buildings. In the developed countries of the world nowadays the level of air tightness of buildings may not any longer cause problems of to cold areas in buildings. But in very cold climate eastern countries this effect is still eminent.
Uncontrolled infiltration of outside air has as unavoidable effect also exfiltration of warm some times humid air through the building fabric. Microbiological growth of all kind of species can take place in the building construction itself. Because the wind direction on

buildings is not constant and may vary after some time the spread of the microbiological species into a building may happen due to a period of infiltration of air. See figure 1.

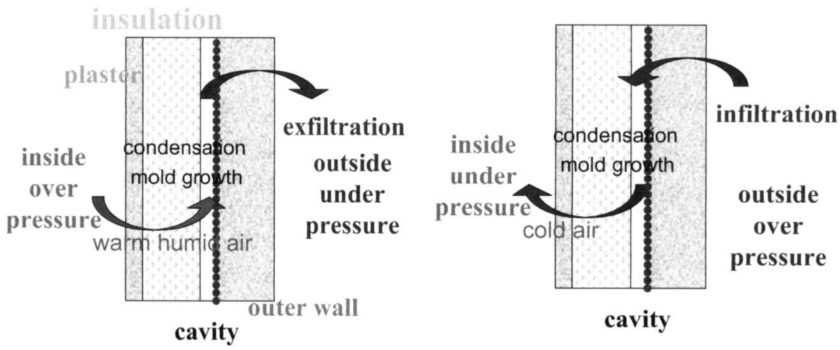

Figure 1 Infiltration and exfiltration related to mold growth

Some people have the hypothesis that this effect is one of the reasons for lung related health problems by sensitive persons such as young children.

2.2 Building damages

The same phenomenon as mentioned above namely condensation in the building construction may cause detrimental effects on the building structure itself. The construction may loose its function and may in extreme cases collapse. Especially in roof constructions in very cold climates this may happen and is a very serious reason to protect the construction for condensation in voids spaces such as cavities.

2.3 Comfort

The distribution of the air leakage over the building envelope play an important role related to comfort. Although the whole building may have achieved a reasonable air tightness level during its construction, local leakages in such cases may cause comfort problems. A cold air stream is coming through seems or joints in the construction a persons in the room are complaining about draught problems. Some times this phenomenon appears even through electrical sockets. A hole of about 4 mm and a jet of relatively cold air is blowing along the neck of a sitting person.
As described in paragraph 2.1 under health infiltration of could air may also case in less extreme cases comfort problems because the distribution of heat is disturbed.

An completely other aspect of comfort the hindrance due to too high under pressures in case the exhaust system is working but the designed supplies are closed. A inhabitant may experience this in two hindrances:

1. noise problems such as whistling and fluttering sound through the building envelope
2. to high under pressures in the building and as a cause of this, slamming doors.

2.4 Disturbance of ventilation

In case the infiltration is so high through one façade that the incoming flow rate is higher than the exhaust fan capacity exfiltration may occur. Exfiltration always means an energy penalty. Some rooms in the building may have due to this phenomenon during a lot of hours bad indoor air quality conditions. Polluted air from windward situated rooms is transported to leeward side rooms. The supply of outside air which was intended with the designed ventilation system is completely disturbed.

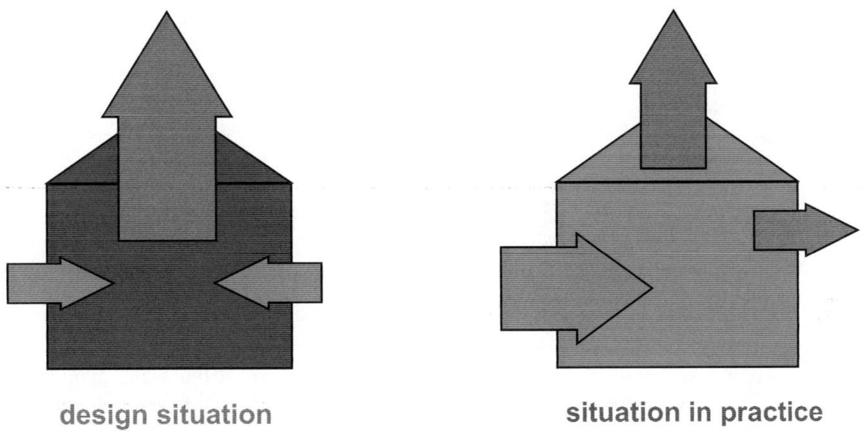

Figure 2 Disturbance of infiltration on the designed ventilation

2.5 Energy

It is quite clear that the lower the air tightness level of a building the higher the energy use for heating. (see figure 3) Because the cold incoming air will be heated to a temperature at which the occupants experience thermal comfort.

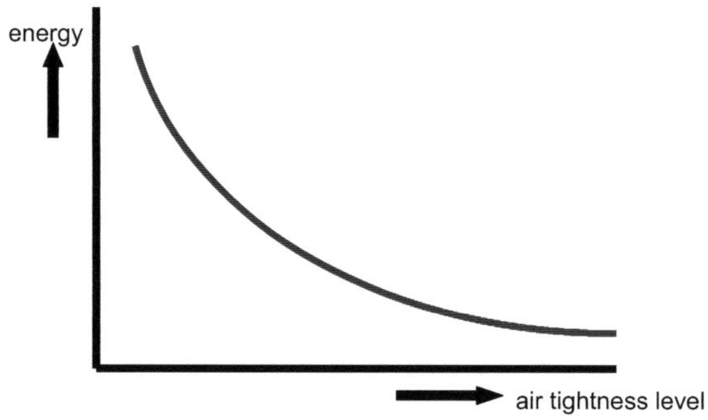

Figure 3 Relation energy use for heating and air tightness level of buildings

Nevertheless not all infiltration may be seen as a loss. In some demand controlled ventilation systems with for instance CO_2 control part of the infiltration may be used to dilute human effluents for which CO_2 is used as a marker. The CO_2 sensor may not see the difference of purposed provided air through the ventilation system and the infiltrated air through leakages in the building. So the question here is how much of the required air should come from the ventilation system and how much is allowed to come through leakages.
So there must be an air tightness level above which it is not any longer very efficient to make a building airtight.
All electrical and electronic appliances in homes have a lot of led signals which are burning and cannot be switched off. So they are using electrical energy 24 hours a day and 365 days a year. In homes the number of this signal leds of electronic equipment may be about 30 to 40 using in some cases together more than 7 to 10 W. Why should you take a lot of effort to increase the air tightness level of your building while all electronic equipment is using is using even more energy?

3. Is there an optimum ?

Considering the cost for energy and the cost to make building better air tight, there should be an optimum. Indeed normally people don't know about exact data for the relation given in figure 3 nor about the relation between the costs to realize the air tightness level of a building.

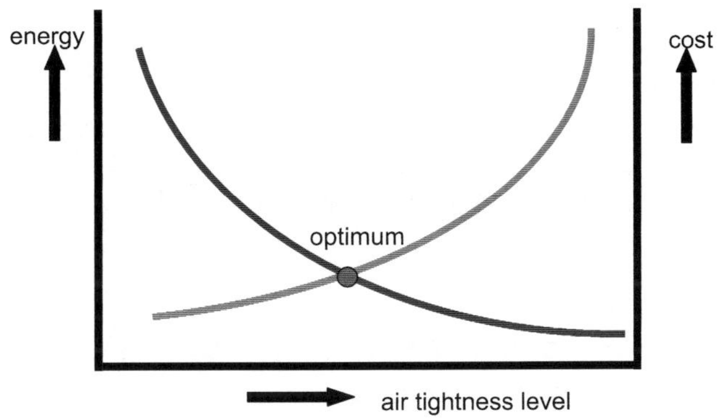

Figure 4 Possible optimum for air tightness of buildings

4. Discussion

- To find an optimum for air tightness of buildings you need good and precise data for the relation with energy and for the relation with cost to realize the air tightness. This data is generally lacking. More information is needed. Who is starting gathering this information? There is a need to come to defendable levels of air tightness

- For passive housing a N_{50} value of 0.6 ACH is often used or even required. Without the knowledge of the relations given in figure 4 is looks strange to put forward such high levels of air tightness. What can be realized during building construction practice may not automatically be the target figure for all low energy buildings.

- Who is able to justify these high levels of air tightness? The building and energy sector are waiting for a real rational and a good motivation.

Thursday 13 October 2011

12.15 – 13.15 Parallel Session 6A - DCV and sensor technology - CLEAR UP project

Demand Controlled Ventilation (DCV) is usually seen as an effective way for reducing the energy consumption. The aim of this workshop is to present new sensors for DCV developed in the framework of the EU project Clear-up and to discuss with ventilation experts their possible applications in buildings.

- Introduction to DCV - Willem de Gids, TNO, the Netherlands
    - What is DCV and why its application
    - The need for demand control in dwellings
    - Indoor air pollutants in general and the most dominant ones for dwellings
    - A strategy for controlling most domestic pollutants
- Introduction to DCV – Anne-Marie Bernard, Allie Air, France
    - Potential of energy saving
    - Different types of sensors

- Success of hygro regulated DCV in France
- Impact of cost reduction

- Energy-efficient Demand-Controlled Ventilation using Micromachined Metal Oxide Semiconductor Gas Sensor Technology - Simone Herberger, Applied Sensor, Germany

- Panel discussion on the potential applications of these new sensors

12.15 – 13.15 Parallel Session 6B – Uncertainties in airtightness measurement - field data

- Modernizing ISO, EN and ASTM air leakage standards ... more accuracy in less time (Colin Genge, Canada)
- Evaluation of selection criteria of an air tightness measurement method for multi-family buildings (Bassam Moujalled, France)
- Improvement of air tightness of communities (Markku Hienonen, Finland)

13.15 – 14.00 Lunch break

MODERNIZING ISO, EN AND ASTM AIR LEAKAGE STANDARDS

... more accuracy in less time

Colin Genge

Retrotec Inc.
1060 East Pole Rd.
Everson, WA 98247 USA

ABSTRACT

A building was tested the equivalent of over 1,000,000 times under windy conditions where each test satisfied the conditions of ASTM, CGSB, ISO, EN, ATTMA and USACE testing standards in every respect. The air flow measurements made at lower reference pressures, such as 4 and 10 Pa, varied over a wide range of +87% to -45% from the average, while the results at 50 Pa varied 15% from the average. The basis for a mathematical model was created to evaluate the error potential of existing standards and to advance new test procedures that would achieve the desired level of accuracy for results referenced to any pressures from 4 to 75 Pa, in the least amount of testing time over the widest possible range of test conditions.

Variations in flow measurements at 4 Pa were reduced to 7%, while at 75 Pa they were reduced to 5.7%. The 95% confidence interval calculation was checked against several batches of tests but the actual confidence interval was no better than a 60%. The new mathematical model could be used to accurately predict the potential variation for any result.

KEYWORDS

Air leakage measurement, baseline, USACE, Induced Variation, baseline variation, confidence interval

SUMMARY

The data in Table 1 shows the variation in results that occurred when repeated tests were done on the same building: according to existing standards, and using the test procedure proposed in this paper.

Test in strong wind	Existing standards				Proposed 1 direction		Proposed 2 directions	
	ASTM	CGSB	ISO, EN & ATTMA	USACE	1-1d	2-1d	1-2d	2-2d
Total test-fan time, seconds	60	90	100	320	300	300	240	240
Results at 4 Pa	+78% -42%		+87% -45%		+15- 13%		±7%	
Results at 10 Pa		43%			+12- 11%		±6%	
Results at 50 Pa	14%	15%	15%		6.5%	6.5%	±6%	±6%
Results at 75Pa				9.3%		6.1%		±5.7%

Table 1. Tests were performed in compliance with the standards shown.

INTRODUCTION

The existing paradigm for air leakage measurement in buildings has remained relatively unchanged for 25 years but during that time, modern electronics has allowed the operator to effortlessly collect thousands of readings in an unbiased fashion. Since 1985, gauge accuracy has improved at least 8 times and test fan accuracy has increased threefold but existing standards take no consideration of that improvement. Standards designed to test residential houses are now being ineffectively applied to large buildings, where stack and wind pressures can be 5 to 10 times greater than testing houses, making those standards obsolete in some geographic areas (over 75% of the time). Clearly, large building specific testing standards must be developed to be robust in most weather conditions. House testing standards will benefit from a fresh look at their procedures as well.

All of today's air leakage testing standards evolved from the CGSB and ASTM standards dating back to the early 1980s. Many standards still require gauge accuracies to be within 2 Pa, however wind fluctuations could add another 2 Pa of uncertainty. With a desire to get results at low reference pressures, such as 10 and 4 Pa, and it becomes clear that large uncertainties can exist. Currently committees are revising ISO and ASTM standards in efforts to standardize not only the test methods but also the test results themselves.

Air leakage test results are typically referenced to pressures such as 75, 50, 10 and even 4 Pa. Data is taken over a range of test-fan induced pressures, and a curve fit provides information on the building's air leakage at any specific reference pressure. Since baseline pressures of 1 Pa or more are virtually always present, results that are referenced to low pressures such as 4 Pa are affected to a relatively large degree. When baseline pressures increase to 3, 5 and even 10 Pa, obtaining stable results at 4 Pa becomes impossible. This effect has never been quantified, either theoretically or experimentally for results at low reference pressures. Essentially, a strategy is required to determine a stable result at 4 Pa where the baseline pressure represents a significant percentage of the reference pressure.

METHODOLOGY

The same building was tested the equivalent of over 1,000,000 times under windy conditions where each test satisfied the conditions of ASTM, CGSB, ISO, EN, ATTMA and USACE testing standards in every respect. A calibrated three phase 2 HP test fan, driven by a variable frequency drive for stability, was used for this project. The pressure pickups were set up in two different locations: a dual pickup with one open tube (with a rain cover) on the same balcony as the fan (west side), and a second open tube (with cover) in the top end of the south east door; a single pickup at the bottom end of the east sliding door (we expected this pickup to display approximately -1 Pa of stack pressure due to the temperature difference between inside and outside of ~8C).

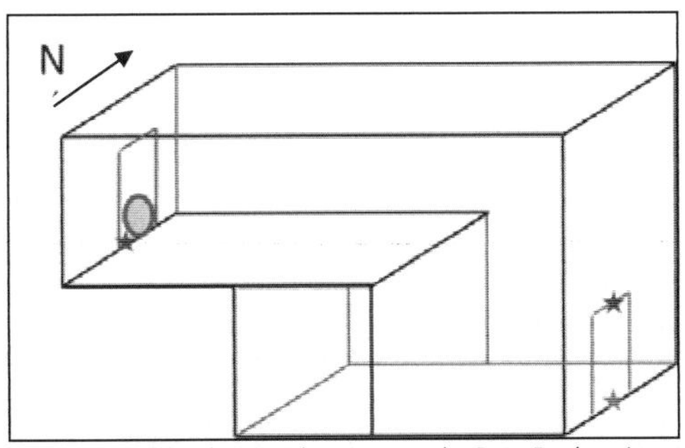

Figure 1. Test schematic. The yellow circle and red rectangle represent the Door Fan location; green stars represent the dual pressure pickup locations, and the orange star represents the single pressure pickup location.

Figure 2. Test Building: the fan setup is visible in the balcony door of unit 611/610 (above). The test fan was always on the Leeward side of the building.

Tests were completed using Retrotec's FanTestic software, allowing for a variety of testing configurations. Options to change the testing configuration included: number of baseline pressure points, number of test fan induced pressure points, range of test pressures, and number of readings within each baseline/induced test pressure point (based on time for collection). Several combinations of these configurations were used. In the final round of testing, data was collected continuously for 600 seconds at each point and broken up into a series of individual tests for analysis later.

Data was analyzed in Excel; of specific importance were ranges of maximum and minimum baseline pressure values, over several averaged "bins". A collection or "bin" of data readings is called a "point". Assuming that FanTestic was set to deliver one reading per second, the bins were averaged over 5, 10, 15, 20, 30, 40, 60, 120 and

240 seconds for both the induced building test pressures and the baseline pressures. Each raw induced building test pressure reading was adjusted to account for the overall average baseline pressure taken during the test. The maximum pressure variation was then determined from the overall average baseline for each binned dataset; this was used to create the four points of a trapezoid, in which all data should fall. This model is discussed further.

Understanding the effect of wind

It became clear after analyzing all the above variables that the mechanics of the equipment could be easily quantified and understood, but that wind was the true chaos generator and its effect had to be well documented before progress could be made. For example, many of the standards require a baseline pressure to be taken, usually for about 10 seconds. Figure 3 shows a continuous stream of baseline pressures that were taken for 10 seconds each. Depending on when the pressure was observed, the value read could have been somewhere between -0.3 to -5.5 Pa. Each one of these values would have appeared legitimate to the operator but would have meant very different corrections applied to the induced building pressures. For example, if the operator unwittingly measured the baseline as being -1 Pa, it is extremely likely that by the time the test fan was turned on to create an induced envelope pressure, this baseline could easily have shifted to -5 Pa. This would mean that the operator was actually testing at 6 Pa when the operator mistakenly thought that the real corrected test pressure was 10 Pa.

Data set 918, of baseline averages shown on the next few figures, were done under especially windy conditions and without the benefit of shielding and wind averaging, in order to amplify the effects. A robust methodology should be developed to handle these highly fluctuating baseline pressures; In addition, the existing standards only refer to average baseline pressures and say nothing about fluctuations. They require the magnitude of the average to fall within certain limits but we discovered in the course of the study that the magnitude of the baseline has had little to do with the variation of results from test to test. The truly important factor is, "what is the range of fluctuation that could occur when induced envelope pressures are being measured?" It is possible to have a baseline average of zero but the fluctuations could have swung 10 Pa in both directions, making induced pressure readings dependent upon which way the baseline pressure had swung at a particular instance in time. In the past, it has always been assumed that taking more test points than the minimum 5 or 7 would somehow handle these wind fluctuations. What we have experienced over the past 30 years in analyzing this type of data is that only one of the test points, taken during the fluctuations, can wildly affect the results. Often it was blamed on the lowest test point but in fact none of the test points were safe from the effect of these fluctuations since there was no telling when they would occur. It was clear that intermediate points in the typical curve fit only acted as pivot points and did not constrain the data from the effects of fluctuations at the lowest test points. What did have an effect, was taking sufficient readings at the lowest test point so the effect of the fluctuations was reduced to insignificance. In order to understand how long this lower test point must be taken, a clear picture must be obtained by carefully examining the fluctuations that were occurring over the baseline pressure measurement period.

It turned out that these fluctuations were the dominant effect that dictated how accurate and/or repeatable a particular test was. Fluctuations in baseline pressure test points were used to predict the variation that was likely to be occurring while test fan readings were taken.

Test ID #918 baseline pressures

The following graphs show a rolling average over 10s and 240s. The red marker indicates the overall baseline average for all readings taken over 600 seconds.

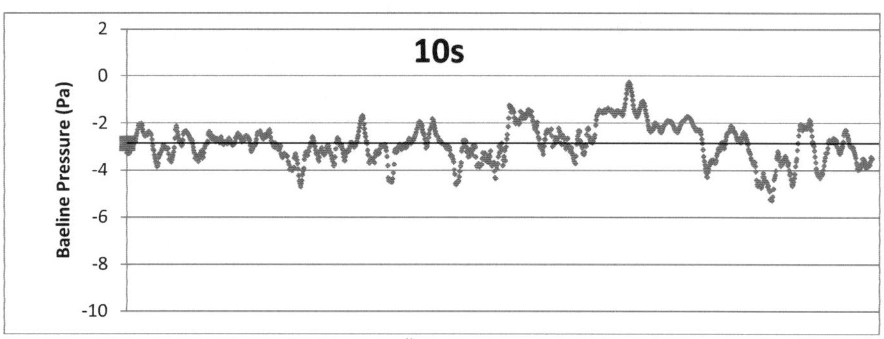

Figure 3. Rolling average over 10s.

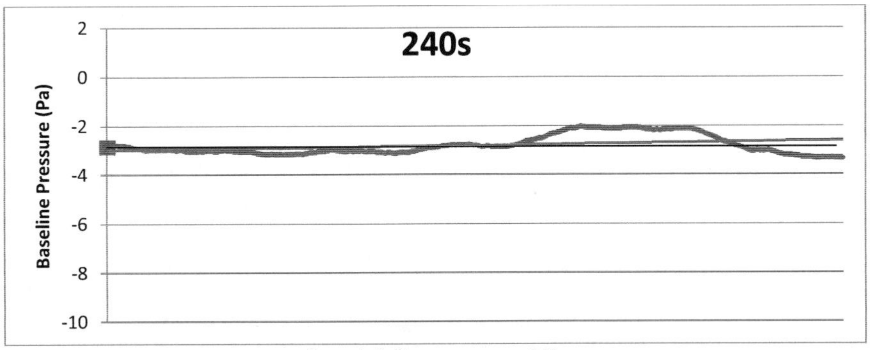

Figure 4. Rolling average over 240s.

THE MODEL

How the model works

A mathematical model was developed in an attempt to predict the effect of wind, how to eliminate it and how other factors influence accuracy and repeatability. The model's real value comes from having a tool that will look at a wide range of data collection possibilities in order to determine the most effective model. First, the model was checked against real data and then by making some realistic assumptions, an evaluation of existing standardized test methods was performed. This would be the mathematical equivalent of performing millions of tests on the same building under different test conditions. Next, all the Factors affecting accuracy were adjusted to optimize the accuracy of air leakage tests at different reference pressures.

Figure 5 illustrates the mathematical model, based on a trapezoidal shape. The lowest test point is used to create the base of a trapezoid which is centered on a control value, considered to be the true result. The width of the trapezoid's base is defined on one side by the sum of the squares of the baseline variation plus the gauge error around zero, plus the sum of the squares of the induced variation plus the gauge error in the range of induced pressure. This results in a pressure that is added on both sides of the centered control value. Perpendicular wings are added to the space in the direction of the y-axis and are defined by the test fan flow error at that test pressure. This process is repeated for the top of the trapezoid.

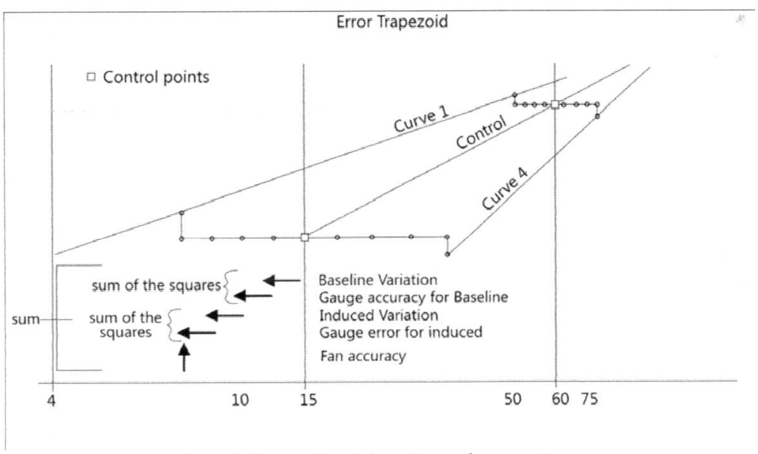

Figure 5. Trapezoid model used to verify assumptions.

The model uses the variables listed below to calculate the maximum range of errors for a variety of reference pressures. The model is validated by plotting actual test data against the model's predicted space in which test data should fall (within the trapezoid). In the Figure 6, the rolling average of the data was created based on 600 seconds of streaming data that would emulate a series of approximately 600 test points. In other words, for 5 second averaging, data was converted into a series of test points that contained all the data taken during a 5 second interval. Since a rolling average was used, a continuous stream of 5 second test points, updated every second, was created that would have a similar effect to having taken the point 595 times.

Baseline Variation	This is the maximum baseline variation that could occur between any complete 'sets' of baseline readings. (For example, if 12 baseline points had to be taken for 10 seconds each for a total of 120 seconds, this 120 second increment is one set of baseline readings.) This set is compared to every other set to determine the maximum variation between any two sets. It is conceivable that any one of these sets could have been measured as the true baseline. This variation is a measure of how wrong that reading could possibly have been. The true average was considered to be the average of all 600 readings for that particular test. The variations for each set were measured from this average.
Gauge accuracy for baseline	Gauge specification that will influence the accuracy of the baseline measurement. It is added to the above variation using sum of the squares.
Induced Variation	There will be a variation in the baseline pressure that is occurring during test fan operation that is creating the induced pressure on the envelope. This fluctuation will be invisible to the operator since all they will see is the magnitude of the induced pressure created by the test; but it is reasonable to assume that the range of variations that occurred during test fan operation are similar to the range of variations that occur during the baseline measurement. By looking at the time taken to take induced pressure test points, the range of potential variations that could be occurring during the measurement of the induced pressure can be established. This value cannot be measured but must be predicted from baseline points.
Gauge error for induced	Gauge error specification that will influence the magnitude of the induced pressure measurement. Gauge error at a much higher induced test pressure is usually different, and may even be in the opposite direction, of the gauge errors for baseline.
The previous 4 Factors can all be additive. For this analysis the first two were combined as a sum of the squares, the second two were combined separately as a sum of squares, and then these two sums of squares were added to establish the maximum movement in the positive and negative direction along the X-axis from the control point.	
Test fan flow accuracy	This variable is added to the previous 4 Factors in such a way as to create the widest possible trapezoidal envelope.

Table 2. Factors that affect accuracy and repeatability used in a mathematical model.

Minimum number of Induced Points	This value is tracked (not used in calculations). Experimentation has shown that the difference between the lowest and highest induced pressure points has no impact on the results as long as sufficient data is collected at the lowest and highest test points.
Total time to get Induced Pressure	This is shown for information only (not used in calculations).
Reference pressure for results	The model produces a maximum potential error based on the reference pressure. Typically the lower the reference pressure, the greater the error associated with that value.

Table 3. Other Factors recorded in a mathematical model.

Checking the model against real data

This process requires thinking clearly about how data can be used to verify the model. While all the Factors listed above will affect results in the field, because a wide range of equipment can be used, data from our study is not affected in the same way because the same equipment was semi-permanently installed and used in that way for every test. Each Factor is discussed, as reference to it influences the spread of data points we get at each induced pressure. Our data set was manipulated in order to produce hundreds of induced pressure test points at each nominal test pressure. This would be the equivalent of performing hundreds of simultaneous air leakage tests and should clearly show the amount of variation caused by the Factors below.

Baseline Variation	Used to determine the range of induced variation, but not part of error trapezoid model calculation used for comparing data.
Gauge accuracy for baseline	This is irrelevant in checking the model because we do not know what the correct value for the baseline is.
Induced Variation	The degree of variation was the prime Factor that displaced the induced test pressures to the left and right. In each case, the amount of variation in the baseline was used to estimate the amount of variation that was, in all likelihood, occurring over the same time average interval for the induced test pressure.
Gauge error for induced	Gauge errors will not be apparent since the true value is unknown but errors in repeatability and zero drift will be causing some movement in the test data. A value of 0.2 Pa was chosen.
Test fan flow accuracy	Overall test fan flow accuracy will not be apparent either since the true value is unknown but again, errors in repeatability appear to be in the range of 2% which would be a combination of gauge repeatability, test fan flow signal noise and wind on the test fan itself.

Table 4. Factors that affect accuracy and repeatability used in a mathematical model.

The following graphs in Figure 6 to 8 show the true or control value using a dashed line. The sides of the trapezoid are defined by straight lines which are affected by the factors described above, and determine the width of the trapezoid's base and top. As the amount averaging period for baseline pressure readings increases, the variation in the dataset decreases. Various standards require different minimum time intervals and will produce similarly different results if the test is repeated hundreds of times. While these graphs do a good job of showing how well the model performs, they do not represent the full extent of results variation that would be expected when the effect of the major Factors are added on to the variations displayed.

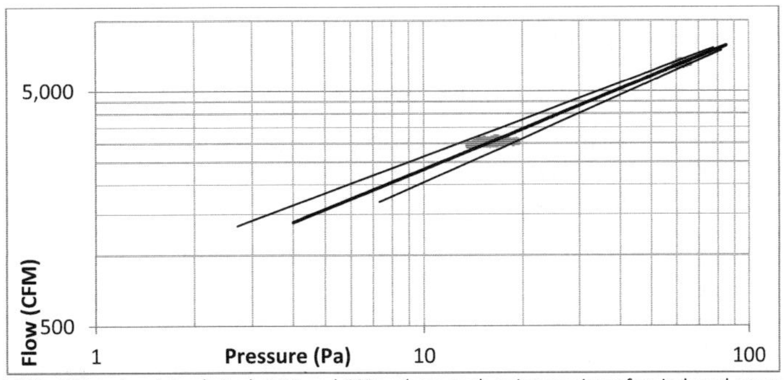

Figure 6. **Test ID 918**. 595 test points plotted at 15 and 60Pa where each point consists of an induced pressure and the flow required to induce that pressure, averaged over 5s. Note how the trapezoid contains this data fairly well.

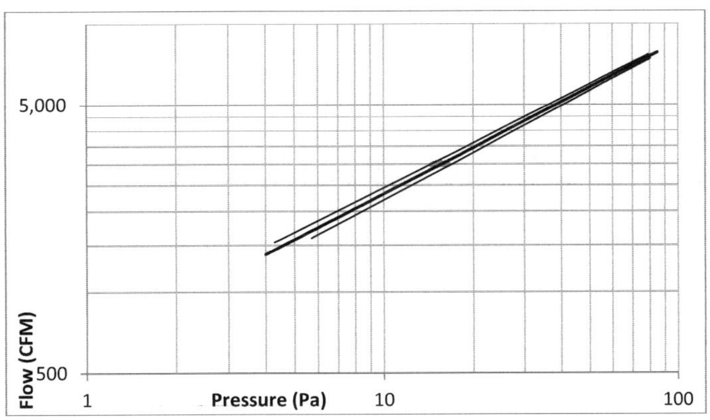

Figure 7. **Test ID 918**. 480 test points plotted similarly but averaged over 120s. Note how the trapezoid contracts with the data because the variation in baseline is reduced because of the longer time interval.

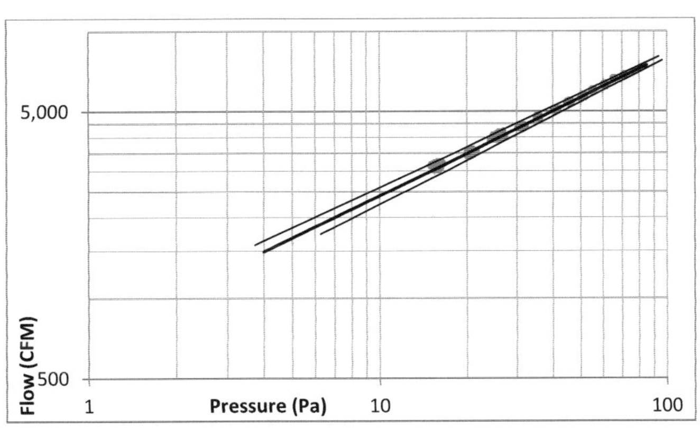

Figure 8. **Test ID 801**. 110 test points for a third data set at each of 12 different induced pressures, from 15 to 70Pa over 10s for each point. Data for each induced pressure was taken for 120 seconds but plotted as 110 ten second chunks.

APPLYING THE MODEL TO DIFFERENT STANDARDS

In this section, we are using the Model to show how much variation would have occurred using the Factors from different standards under the conditions that data set 918 was collected under. Had 500 or so tests been conducted over 40 minutes on the same building, much variation would occur in the results if all Factors were set to the widest ranges of values allowed by those standards. Each Factor was looked at individually and added cumulatively from one table to the next, in order to understand their respective effects. Factors are highlighted in green when they are first added onto the table.

Baseline variations cause errors

Data for Test ID 918 consisted of a continual sampling of baseline pressures for 600 seconds. A series of rolling averages were created for 10, 30, 60, 120 and 240 seconds to quantify the variation within each group against the 600 second average. Examining Figure 6, which is a running 10s average of baseline data for 600 seconds, we notice that we could have recorded a baseline average between zero and -5 Pa, and this variability is recorded in Table 5 under NEBB and ASTM. The variability in the baseline for ISO, EN and ATTMA is 3 Pa (from -4 to -1) based on 30 s averages and USACE is 2 Pa based on 120 s averages. The average over 600 seconds is subtracted from the highest and lowest baseline in order to find the maximum Baseline Variation from the average.

Test ID #918	Test Standard				
	NEBB	ASTM	CGSB	ISO & EN & ATTMA	USACE
Number of Baseline test points	1	1	1	30	12
Time to take each point	10	10	10	1	10
Total Time to get Baseline	10	10	10	30	120
Highest Baseline based on Total Time	-5	-5	-4.5	-4	-3.5
Lowest Baseline based on Total Time	0	0	-0.5	-1	-1.5
Average over 600 seconds	-3	-3	-3	-3	-3
Baseline Variation from the Average	2.6	2.6	2.5	2.0	1.5

Table 5. Effect of all the above Factors on Test ID 918.

Testing in both directions

Testing in both directions allows accuracy requirements for baseline measurements to be relaxed because errors in determining the correct value largely cancel out when the same baseline value is added to a test in one direction and then subtracted from the test in the other direction. For example, if the baseline was measured incorrectly as being +5 Pa when it should have been zero, then the result would be that an induced test pressure of +25 Pa as read on the gauge would be improperly corrected to +20 Pa. During the test in the opposite direction, an induced test pressure of -25 Pa, as read on the gauge, would be corrected to -30 Pa. Both resulting curve fits would be wrong because of the baseline error, but when the curves are combined, that error would effectively disappear.

Since both baseline variation and gauge error essentially cancel out when testing in both directions, the potential error is dramatically reduced. It is also entirely probable that there would be some reduction in the induced variation since it is unlikely that the same magnitude of variation would occur in the same but opposite directions when testing both ways.

These results also indicate that the common practice of taking the lowest test point at a higher pressure, to reduce the effects of wind, actually has the opposite effect. Errors are drastically increased by increasing the lowest test point. It appears as if the best scenario may be testing from 10 to 60 Pa when results at 4 Pa are required.

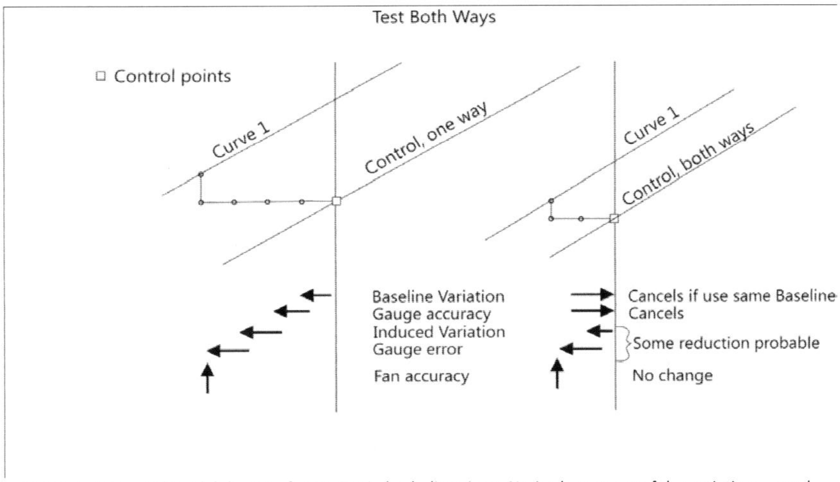

Figure 9. Trapezoid model diagram for testing in both directions. Notice how many of the variations cancel.

Errors using existing standards for test ID 1540 were reduced to half from when tested in both directions.

Test ID #1540 - test in both directions	Existing standards					New Factors		
	NEBB	ASTM	CGSB	ISO & EN ATTMA	USACE	1	2	3
Number of Baseline test points	1	1	1	30	12	30	30	12
Time to take each point	10	10	10	1	10	90	30	30
Total time to get Baseline	10	10	10	30	120	2700	900	360
Allowable variation, 1 Baseline pt						1	2	2
Max variation in any 1 Baseline pt						1	1.89	1.89
Baseline Variation over total time	3.03	3.03	2.5	1.89	3.03	0.4	0.5	0.50
Gauge accuracy near zero	2.4	0.25	2	2	0.25	0.25	0.25	0.25
Induced Pressure Variation	3.03	3.03	2.5	3.03	2.45	0.6	0.85	0.85
Time to take 1 Induced Point	10	10	10	10	20	360	120	120
Gauge error for induced	2.4Pa	5%	2Pa	2Pa	1%	1%	1%	1%
Test fan flow accuracy	4%	5%	7%	7%	5%	3%	5%	5%
Lowest test point, at least	10	10	15	10	25	15	15	25
Highest test point, at least	75	50	50	50	75	60	60	75
Minimum # of Induced Points	5	5	8	5	10	2	2	2
Results at 4 Pa		+35% -25%		+45% -31%		7% -7%		
Results at 10 Pa			24%				9%	
Results at 50 Pa		10%	11%	11%			6%	
Results at 75Pa	+6%				7%			5.9%

Table 6. Effects of all variables when testing in both directions, including new Factors.

New confidence interval calculation

A series of air leakage tests were used to see if the confidence interval generated by the ASTM standard was wide enough to include successive tests that were done under exactly the same conditions. Tests were grouped in batches that had exactly the same set of conditions. It doesn't appear as if this method of statistical analysis can accurately predict the type of variation that was witnessed due to variations in baseline pressure and their effect on the induced pressure readings.

The uncertainties elsewhere in this paper were calculated using the Model in Excel that could be converted into an equation for use in standards. If the procedure is conducted as described using 30 baseline points for at least 10 seconds and the induced pressures points are taken for 120 seconds then the maximum uncertainty can be calculated using: the maximum variation between baseline points, the averaged baseline, number of test directions, gauge repeatability around zero, gauge error for induced pressure points, test fan accuracy at the tested flow, the lowest induced point and the highest induced pressure point.

Test number	Set	Version	Wind	Pr/Dp	Pts	Span	Time	Q4	Lower	Upper	Setup						
													1445	1380	1380	1385	1430
	1	FT	Calm	Pr	4	10 to 50		1445	1304	1604			0	0	0	0	0
2010-12-14-2316	2	FT	Calm	Pr	4	10 to 50		1380	1251	1542			0	0	0	0	0
	1	FT	10mph	Pr	12	15 to 70	10/12s	1380	1295	1473			0	0	0	0	0
	2	FT	10mph	Pr	12	15 to 70	12s	1385	1344	1430			1	0	0	0	0
2010-12-18-2046	3	FT	10mph	Pr	12	15 to 70	12s	1430	1402	1458			0	1	1	1	0
												Avg 4					
												20%	25%	25%	25%	25%	0%

Test number	Set	Version	Wind	Pr/Dp	Pts	Span	Time	Q4	Lower	Upper	Setup						
													1600	1850	1640	1780	1523
2010-12-18-1009		DEV	10mph	Dp	12	10 to 70	30s	1600	1438	1776			0	1	0	1	0
2010-12-18-1349		FT	10mph	Dp	12	15 to 70	10/12s	1850	1691	2046			1	0	1	0	1
	1	FT	10mph	Dp	12	15 to 70	10/12s	1640	1508	1789			0	1	0	0	0
2010-12-18-1659	2	FT	10mph	Dp	12	15 to 70	12s	1780	1688	1881	B-1		1	0	1	0	1
2010-12-20-1001		FT	5mph	Dp	12	15 to 70	30s ??	1523	1471	1574			1	1	1	1	0
												Avg 4					
												37%	43%	43%	43%	29%	29%

CONCLUSIONS

1. The majority of batches of tests fell outside the confidence interval more than 30% of the time even though no variation would be expected from one test to another, as exactly the same gauges and test fans are used each time. Results will fall outside the confidence interval more when equipment inaccuracies are taken into account.
2. The variation in baseline pressure test points was a useful indicator in determining how much variation there would be in induced pressure readings; 10 baseline points did not appear to be sufficient while 30 baseline points appeared to be sufficient to determine the variation in induced pressure readings.
3. Each baseline test point must be averaged over at least 10 seconds.
4. A maximum allowable baseline variation of 0.5 Pa appeared to be optimum when testing in one direction. (Gauge accuracy around zero was important)
5. A maximum allowable baseline variation of 1.0 Pa appeared optimum when testing in both directions. (Gauge accuracy around zero was less important)
6. Baseline variation can be reduced by 75% or more by locating pressure pickups away from windward locations.
7. Taking one 300s baseline was optimum in the windiest conditions and yielded less variation than taking two 150s baselines.
8. No reduction in uncertainty was achieved by taking more than the lowest and highest induced pressure test point. Points in between did not yield better results.
9. For testing in one direction, induced pressure test points taken for 300 seconds produced the least variation in results and 120 seconds was sufficient when testing in both directions.
10. The ratio between the lowest induced pressure test point and the largest baseline point must be 3.3 or greater.
11. As long as the magnitude of the largest baseline test point does not violate the previous condition, the variation between the maximum and minimum baseline points is the best indicator of the variation that will occur in the leakage measurements made at the induced test pressures.
12. Results referenced to 50 and 75 Pa had virtually the same leakage measurement variation.
13. Baseline variations caused the greatest degree of uncertainty with the existing ASTM ISO and EN standards, very little with the USACE procedure and slightly less with the proposed methodology.
14. Using the proposed procedure the uncertainty created by Baseline variations was reduced to only 7 or 8% when testing in one direction and reduced to only about 1% when testing in both directions.
15. Uncertainty in the proposed method is dependent primarily on gauge accuracy, test fan accuracy, and the number of test directions.
16. The proposed model could be utilized to establish a meaningful confidence interval.

PROPOSAL TO IMPROVE ACCURACY AND REPEATABILITY

Some of the foregoing proposal is based on conclusions that are well documented by this paper while other conclusions can easily be induced by our operation of the model or by further manipulations of the data taken during the study. No statistical gurus have so far contributed to the study and I'm sure that additional contributions can be made to polish this way of analyzing enclosure air leakage testing data. We welcome their contributions.

Number of baseline test points

The more test points that are used, the more likely it is that the maximum variation will be uncovered. Twelve is probably the minimum number and there may not be any value in going beyond 30. Either way, evaluating the difference between these points is an essential way of determining the amount of variation that will probably be occurring later during induced envelope pressure test points.

It is proposed that when testing in both directions, an initial baseline pressure be taken and then a final baseline pressure taken only after the 2nd direction is completed, assuming the change in test directions occurs before any environmental conditions change too drastically. The reason for this proposal is that if intermediate baseline pressures are taken after the first test and before the 2nd test, a different baseline value may be applied to the first test when compared to the 2nd. Since baseline pressure measurements tend to go through a lot of variation not much additional accuracy is gained by taking a total of 4 sets baseline pressure test points versus only 2. What is more likely is that accuracy will be lost because an incorrect baseline will be subtracted from a test in one direction versus the other. As long as the baseline pressure does not change from one test direction to the other, it is more accurate to apply the baseline correction to both test directions equally. In fact, when testing in both directions a good case can be made for not correcting the data at all, however, the important function of using baseline variations to determine induced pressure variation is lost as is the ability to evaluate the accuracy of the test using this method.

Baseline pressure readings should be measured at the lowest elevation of the building which is generally away from the maximum impact of the wind. Determining pressure differentials at the top of the building can best be performed from inside the building rather than exposing pressure probes to the high velocity winds at the top. Averaging pressures on all 4 sides of the building will reduce the impact of wind and may be done electronically or pneumatically.

Minimum time required to take each baseline point

Each baseline point must be averaged over a minimum time period in order that the total time to acquire all points is long enough to detect any wind gusts that may be occurring. Time required for testing in one direction is at least double the amount of time required for testing in both directions because in the latter case, variations almost completely cancel out.

Allowable variation, one baseline point

This is the maximum allowable variation between any one baseline point and the average of all points and is that at 1 Pascal for testing in one direction except for results at 4 Pa where it should be a maximum variation of 0.5 Pa. For testing in both directions these values are doubled because this variation has very little impact on the results. As the total time used to acquire a baseline point increases this variation tends to decrease.

Gauge accuracy near zero

The magnitude of baseline pressures do not present nearly as much of a problem as the variation in the baseline over time. When successive baselines are taken under mildly windy conditions over a period of 10 seconds for example the resultant test points vary substantially. By taking 30 baseline test points for 30 seconds each, an accurate picture of the variability can be obtained. As the time taken to measure each baseline point is increased, the variation from one baseline point to another is decreased. Even under windy conditions taking 30 baseline test points for 30 seconds each, where the maximum variation between baseline pressure test points and the overall average 1.0 Pa or less will yield a baseline average that is within 0.25 Pa of the true value.

Minimum time to take one induced pressure test point

Since the total number of induced pressure test points has been reduced down to only 2, it makes sense to take more time on those 2 points. The overall testing time can be reduced and accuracy will increase. For testing in one direction, take each induced pressure test point for one half the time required to acquire all baseline pressure test points.

Gauge error for induced pressure test points

The baseline variation can be used as a guide to how long the induced pressure test point should be taken. It is only necessary to take test points at the lowest and highest test pressures, since those in the middle add little to the accuracy of the overall result. It is better to spend more time taking one accurate point then to waste time taking test points at mid-test pressures which have little impact on the overall result. It is reasonable that if the variation in the baseline pressure test points is only 0.5 Pa, than the time taken for induced pressure test points can be cut in half to 180 seconds each with no loss of accuracy. Similarly, if the variation is only 0.25 Pa, then the induced pressure test point can be cut down to 90 seconds, again with no loss of accuracy.

Modern digital gauges will have accuracies of +/ -1% or 0.25 Pa, whichever is greater. Many air leakage standards were written in the 1980s, when accuracies of +/-2 Pa were all that could be expected.

Test fan accuracy

Calibrated test fan accuracies are commonly 5%, which would be an appropriate requirement for most testing. Where test results are required at a reference pressure of 4 Pa, accuracy requirements of 3% is recommended. The model can be used to test different scenario accuracies and evaluate the impact on the desired accuracy of the end result.

Lowest test point, at least

The lowest induced pressure test point must be at least 15 Pa for results that are referenced at pressures up to 50 Pa. For reference pressure is above 50 Pa there is no need for this test point to go below 25 Pa.

Highest test point, at least

The highest test point must be at least as high as the reference pressure. There must be a ratio between the lowest and highest induced pressure must be greater than the ratio between the lowest test point and the reference pressure. If the test is conducted between 15 and 60 Pa which is a ratio of 4 to 1 then the lowest reference pressure could be one quarter of 15 Pa which 3.75 which would accommodate a reference to 4 Pa.

Minimum number of induced points

Taking only 2 induced pressure test points is proposed because it appears to be considerably more accurate to take more readings over longer period of time to firmly establish to test points at the lowest and highest is pressures that it does to take numerous test points between these 2 extremities. By reducing the number of test points but increasing the amount of data taken for each point, the overall accuracy of the test is improved while the time to complete the test may be decreased.

Confidence interval calculation

The majority of batches of tests fell outside the confidence interval more than 30% of the time even though no variation would be expected from one test to another, as exactly the same gauges and test fans are used each time. Results will fall outside the confidence interval more when equipment inaccuracies are taken into account.

A new method of calculating uncertainty based on the instability of the baseline pressure readings and equipment accuracy shows a lot of promise in improving the quality of his calculation. Once variation was reduced by taking one longer Baseline reading and two induced pressure points the primary factors in determining confidence of results came from gauge and test fan accuracy. Uncertainty due to Baseline variation was reduced from 7% or 8% to below 1% when testing in one and two directions respectively. Baseline variations caused the greatest degree of uncertainty with the existing ASTM ISO and EN standards, very little with the USACE procedure and slightly less with the proposed methodology.

Proposal for testing in one, and testing in both directions

The major advantage of testing in both directions is that 2 of the major sources of error virtually cancel out and others are reduced. It was showed that testing times can be vastly reduced, and accuracy maintained, by simply testing in both directions. If induced pressure tests occur under the same conditions, it is better to perform only one initial set of baseline pressure test points and then only one set of final baseline pressure test points after the 2nd induced pressure test is completed.

Factors	Proposal 1 direction		Proposal 2 directions	
	1-1d	2-1d	1-2d	2-2d
Number of Baseline test points	30	30	30	30
Minimum time to acquire each baseline point	10	10	10	10
Total time to acquire all baseline points (s) ≥	300	300	300	300
Allowable maximum variation between Baseline pts (Pa)	0.5 Pa	0.5 Pa	1 Pa	1 Pa
Gauge accuracy near zero cancels for both ways	0.25	0.25		
Gauge repeatability and drift (Pa)			0.1	0.25
Time to take 1 Induced Point (s) ≥	300	300	120	120
Gauge error for induced Points ≤	1%	1%	1%	1%
Test fan flow accuracy ≤	5%	5%	5%	5%
Lowest Induced Point must be (Pa) ≥	10	25	10	25
Lowest Induced Point must be (Pa) ≤	15	50	15	50
Lowest Induced Point ≥3.3 times the largest baseline point (in magnitude)	3.3 X	3.3 X	3.3 X	3.3 X
Highest Induced Point (Pa) ≥	50	75	50	75
Number of Induced Points =	2	2	2	2
Reference pressure for results (Pa)	4 to 50	50 to 75	4 to 50	50 to 75
Minimum total time to complete test	900	900	1080	1080
Maximum uncertainty at 4 Pa	+15-13%		7%	
Maximum uncertainty at 10 Pa	+12-11%		6%	
Maximum uncertainty at 50 Pa	6.5%	6.5%	6%	6%
Maximum uncertainty at 75 Pa		6.1%		5.7%

Table 7. Proposals 1 and 2 for testing in single and both directions.

Alternatively, the testing requirements could state the degree of confidence required and the model could be used to determine the uncertainty.

Factors that must become constants:

- 10 x 30s baseline points (it would be possible to reduce the amount of time required to take each point under stable conditions but such a relationship has not yet been determined and it may be far simpler to require the 300 s baseline to be taken in every case which would be useful to determine stability).
- 2 -300s induced pressure points for testing in one direction or two 120s induced pressure points for testing in two directions
- Lowest and highest Induced Point values prescribed.

The remaining Factors would be used to calculate the uncertainty to be expected:

- Maximum Variation in Baseline Points
- Gauge and test fan accuracy that would have to be verified for the range of pressures and flows measured.
- Lowest and highest induced pressures
- Reference pressure

Largest baseline point (in magnitude)- possibly. We are unsure of the relationship between the largest baseline point and the uncertainty. As long as it's within the prescribed limits it would be small for testing in two directions but potentially more error for testing in one direction.

REFERENCES

Sherman and Palmiter. 1995. *Uncertainties in Fan Pressurization Measurements.* ASTM Airflow Performance Conference, April 13, 1995.

Murphy, Colliver and Percy. 1991. *Repeatability and Reproducibility of Fan Pressurization Devices in Measuring Building Air Leakage.* AHSRAE Transactions, Volume 97(II).

EVALUATION OF SELECTION CRITERIA OF AN AIR TIGHTNESS MEASUREMENT METHOD FOR MULTI-FAMILY BUILDINGS

Bassam Moujalled[*1], Fabrice Richieri[1], Rémi Carrié[2], and Andrés Litvak[3]

1 CETE du Sud-Ouest
Rue Pierre Ramond, CS 60013
33166 Saint-Médard-en-Jalles, France
**Corresponding author:*
bassam.moujalled@developpement-durable.gouv.fr

2 INIVE
Lozenberg 7
BE-1932 Sint-Stevens-Woluwe, Belgium

3 CDPEA
58 rue Jean Duvert, Ecoparc, Le Fiducia
33295 Blanquefort Cedex, France
andres.litvak@cdpea.fr

ABSTRACT

There are often practical limitations to measure the airtightness of a multifamily building as a whole as described in EN 13829. The building may be too large; the floors may not be connected with an internal airflow path; or there may be large leaks in the stairway. In such cases, the measurement is performed on a sample of apartments for compliance check purposes, which raises a number of questions especially as legal disputes may arise.
Therefore, our objective was to evaluate the limitations of several sampling methods and suggest improvements based on a field data from ten new multi-family buildings, representing 208 units. In each building, we measured a) the air tightness on the whole building, b) in each apartment, and c) in the common areas of the building. The envelope area was found to be the most reliable parameter as a selection criterion for the sampling method, i.e., it is the best parameter we found to correlate airtightness with. Our analysis also confirms that the leakage in the common areas can have a significant impact on the air permeability of the whole building, especially in the presence of with lift shaft and/or basement parking.

KEYWORDS

Airtightness measurement, multi-family buildings, sampling method.

INTRODUCTION

European standard EN 13829 [1] describes the measurement method of air permeability of buildings. This measurement is meant to be performed on the whole building. In the case of multi-family buildings, there are often practical limitations to measure the air permeability of the whole building. The main reasons we found are: the building is too large; the floors are not connected with an internal airflow path; or the stairway is very leaky, e.g. due to a lift shaft or a fire access door. For these buildings, it is common to measure the airtightness of individual apartments separately. Note that although there exist protocols to better evaluate the leakage to the outside (i.e., avoiding double-counts of leakage to interior spaces), they are seldom used in practice.
Because these measurements (if they are performed) are usually necessary for compliance checks, some organizations or regulations propose specific rules a) to choose the units that must be tested; and b) to extract the criteria that will be used. Walther and Rosenthal [2] give an overview of different sampling methods in use in Europe. In Germany, at least 20% of the

total number of apartments should be tested, with at least one tested apartment at the top floor, one at an in-between floor and one at the ground floor. In UK, zone testing should cover at least 20% of the building's envelope area. In France, 3 apartments have to be measured if the building has 30 units or less, and 6 apartments otherwise. The apartments must have the largest ratio of floors and windows length per floor area and must be located at the top, intermediate and ground floors. The French method has been included in the French application guide GA P 50-784 of EN ISO 13829 [3]. However, to our knowledge, there has not been any careful evaluation of the relevance of these rules.
The research project MININFIL has been conducted since 2008 with the support of ADEME and the French ministry of ecology in order to enhance the knowledge of professionals on the air tightness and its impact on the energy performance in buildings. Under the task 3 of the project, an extensive campaign of airtightness field measurements has been carried out in ten new multi-family buildings. In each building, the air tightness was measured for all the apartments and for the whole building. This paper presents the approach used and analyses performed to compare several airtightness assessment methods in muti-family buildings based on our field data.

METHOD

The air permeability measurements have been performed with the fan pressurization method according to the standard EN ISO 13829 [1]. The aim of the measurements was to identify separately the air permeability of each apartment, the common areas, and the whole building. Therefore, three types of air permeability measurements have been carried out in each building:

1. Individual measurements of the air leakage rate at 4 Pa "Q_{4Pa_apart}" (m^3/h) of each apartment of the building with a blowerdoor positioned on the entrance door of the apartment. The entrance door of the building is fully opened.
2. A measurement of the air leakage rate at 4 Pa "$Q_{4Pa\_whole\ building}$" of the whole building, including leakages in apartments and common areas. This measurement is realized with a blower door using a single or double fan. The blower door is located at the entrance door of the building. The doors between apartments and common areas are fully opened.
3. A measurement of the air leakage rate at 4 Pa "$Q_{4Pa\_common\ areas}$" of the common areas. This measurement is similar to the previous, but this time the doors between apartments and common areas are closed, in order to eliminate the leakage in apartments from the measurement. This requires that the doors are air-tight. Additional tightening of the doors was done if necessary.

Table 1 presents the multi-family buildings characteristics. The number of apartments per building varies between 12 and 38, and the number of the levels between 2 and 7. The volume of buildings varies between 2365 m^3 and 5704 m^3, with 2 buildings larger than 4000 m^3.

Building code	B01	B02	B03*	B04	B05*	B06	B07	B08	B09	B10
# of levels	4	3	2	4	3	5	5	4	7	5
# of flats	17	12	17	20	16	17	17	16	38	36
Area (m^2)	1325	956	1266	1486	1150	1455	1248	1375	2256	2246
Volume (m^3)	3280	2365	3150	3700	2893	3544	3446	3031	5704	5175

Table 1: The description of the assessed buildings.

The measurements have been done at the end of the building construction. All of the apartments were unoccupied, in order to facilitate the access to all the parts of the building. However, the global measurements of buildings B03* and B05* have been disturbed by the presence of workmen during the tests. Therefore, the global measurements for these building will not be considered in the analysis.

RESULTS

In France, the air permeability "$Q_{4Pa\text{-}Surf}$" ($m^3/h/m^2$) is calculated as the ratio between the air leakage rate at 4 Pa and the envelope area of the building except the floors area "A_{TBAT}". The new thermal regulation sets the limit value required for air permeability to 1.0 $m^3/h/m^2$ for multi-family buildings. This value is based on the French regulatory low-energy building standard (BBC-Effinergie). We present the results here with the French air permeability indicator.

The ten buildings represent a total of 208 apartments. For the individual measurements, more than half of the apartments (52%) show lower results than the limit value of 1.0 $m^3/h/m^2$. For the measurement of the whole buildings, only three buildings (over eight) are lower than the limit value. The major part of the leakage in the apartments (40%) has occurred across the fenestration (joints at window sash, window sill, and shutter box), while 30% of the leakage occurs at the joints of hatch and ducts, and 25% across the electricity plugging. The leakage across the joints between walls and slabs are negligible.

Figure 1 presents the results of the individual and whole measurements for each building. Figure 1 shows that that the individual measurements of air permeability are very heterogeneous between buildings, and between apartments in the same building in some cases. Based on our observations, the buildings can be classified into two major categories:

- Buildings B05*, B06, B07 and B08 having the whole measurement and the individual measurements globally below the required limit value (1.0 $m^3/h/m^2$). For these buildings, the individual measurements are uniform and vary in a narrow range.
- For the other buildings, both the whole measurement and the median of the individual measurements are greater than the limit value. The individual measurements in each building are very heterogeneous and vary in a wide range. In B09, the upper value of the individual measurements is almost ten times greater than the lower value.

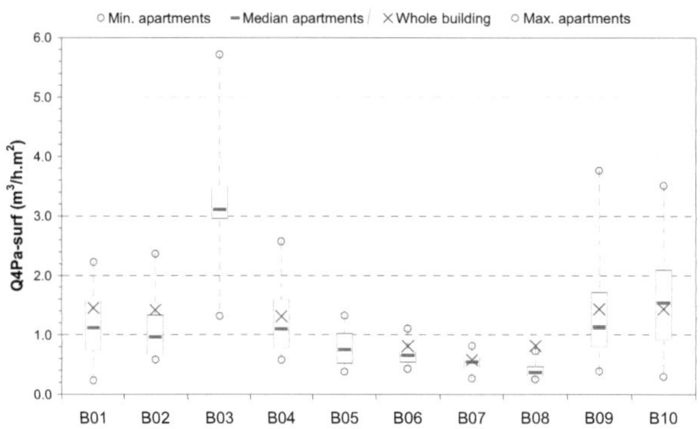

Figure 1: Box plot of the measured air permeability values in each building: the box lines indicate the statistic results of individual measurements and the red marks indicate the measurement of the whole building in each case. The whole measurements of B03* and B05*have been excluded.

Analysis of the selection criteria

GA P50-784 evaluates the air permeability of the whole building through the weighted average of the sample of apartments. Besides it does not impose any requirement on the individual measurements. However, it requires compliance of the sample to a selection criterion meant to avoid samples heavily biased towards favourable units. This criterion is based on the ratio of the length of floor and windows per unit of floor area "(PVl+Pl)/Shl". The method requires the selection of apartments with the largest value of this ratio, as they are considered to be potentially the leakiest apartments.

Figure 2 shows on the left hand-side the variation of the air leakage rates at 4 Pa versus this ratio. It shows no significant correlation between these two parameters (r^2=0.02); in fact, if anything, the air leakage rate seems to decrease with this ratio. Consequently, this criterion "(PVl+Pl)/Shl" seems inappropriate to select the leakiest apartments. We have analysed the correlation with a number of parameters with the help of "principal components analysis". In turn, we found that the correlation was more significant (r^2=0.20) although it remained weak with the envelope area "A_{TBAT}" (see right hand-side of Figure 2). This suggests that although not ideal, the envelope area is more relevant as a selection criterion than the ratio of the length of floor and windows per unit of floor area.

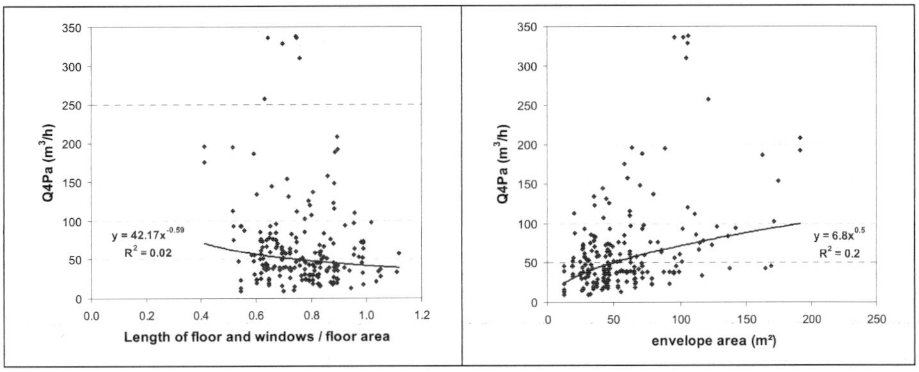

Figure 2: The variation of the measured air leakage rates at 4 Pa as a function of the sampling criteria (the GA P50-784 sampling criteria on the left panel and the envelope area excluding floor on the right panel).

Comparison of the GA P50-784 sampling method against the measurements

GA P50-784 method based on a sample of units has been compared to the results of the measurements on all units. The left panel of Figure 3 presents a comparison between the weighted average air permeability of the sample of apartments and the weighted average of all the apartments for each building. For the buildings with uniform individual measurements lower than the limit value (B05*, B06, B07, and B08), the results of the samples are very close to those obtained with all the apartments. For the other buildings, the difference is more significant.

The right panel of Figure 3 presents the comparison of the weighted average air permeability of the samples of apartments (both the sample of GA P50-784 method, and the sample of all the apartments) against the measurement of the whole building. For both samples, the weighted average air permeability of apartments is always lower than the air permeability of the whole building as it doesn't account for the leakage in the common areas caused by the lift shaft, the parking basement and other shafts and hatches. The greatest difference was found in the case of buildings B08, B09 and B10 with lift shaft and basement parking in the common areas.

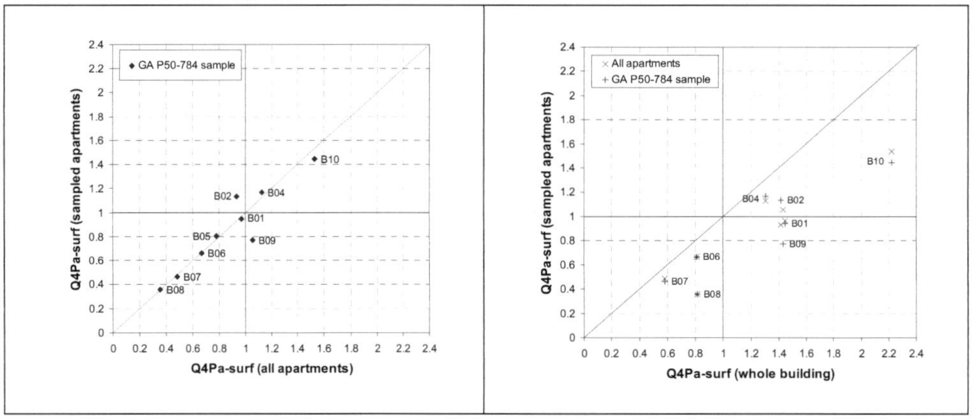

Figure 3: Comparison of the weighted average air permeability of the samples against the weighted average of all the apartments on the left panel, and against the whole building measurement on the right panel.

Analysis of the air leakage in the common areas

In order to evaluate the airtightness of the common areas, we have realised a whole building measurement with the Blower Door positioned on the entrance door of the building. The doors between the apartments and the common areas were all closed and airtight, except for the building B09 where the doors were very leaky at the moment of measurement (the results of this building will be excluded from the analysis).

The measured air flow rate accounts for all the leakages in the common areas. It may include parasitical flow through leaks to adjacent apartments of the common areas. The part of the parasitical flow depends on the airtightness of the walls separating the apartments and the common areas.

The left panel of the figure 4 presents for each building a comparison between the measured air leakage rate of the whole building (including the apartments and the common areas) against the sum of the measured air leakage rate for each apartment and the measured air leakage rate of the common areas.

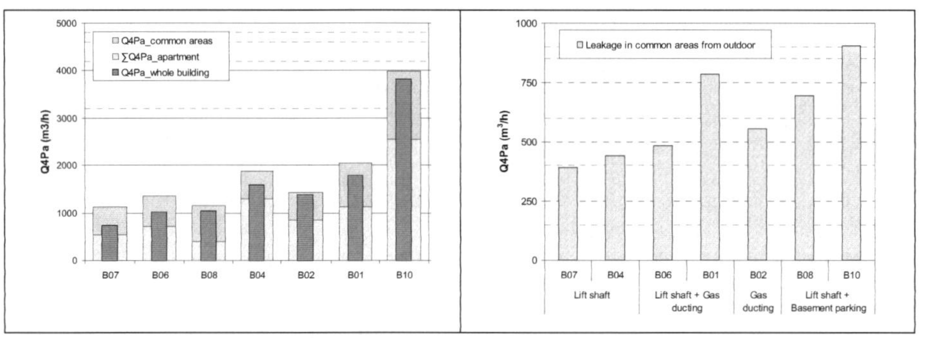

Figure 4: Comparison between the measured air leakage rate of the whole building and the sum of the measured air leakage rate of the apartments and the common areas (left panel). Air leakage in the common areas from the outdoor (right panel).

Ideally if there were no leaks between the apartments and the common areas, the air leakage of the whole buildings should be equal to the sum of the air leakage of the apartments and the common areas. As we can see on the figure, it is always lower than the sum of the apartments and the common areas air leakages. This is due to the air flow through leaks between the apartments and the common areas, which is accounted twice in the apartments and the common areas measurements. Hence, the air leakage of the whole building can be written as in equation 1, where "$Q_{4Pa\_common/apart}$" is the air flow through leaks between the apartments and the common areas. The air leakage in the common areas can also be written as the sum of two parts : the air flow from the apartments "$Q_{4Pa\_common/apart}$" and the airflow from the outdoor "$Q_{4Pa\_common/outdoor}$" (equation 2). From these two equations, we can calculate the air leakage in the common areas from the outdoor where "$Q_{4Pa\_whole\ building}$", "Q_{4Pa_apart}", and "$Q_{4Pa\_common\ area}$" are measured values.

$$Q_{4Pa\_whole\ building} = \Sigma Q_{4Pa\_apart} + Q_{4Pa\_common\ areas} - 2*Q_{4Pa\_common/apart} \quad (1)$$

$$Q_{4Pa\_common\ area} = Q_{4Pa\_common/apart} + Q_{4Pa\_common/outdoor} \quad (2)$$

The results of the calculation are given on the right panel of figure 4. The amount of the air leakage at 4 Pa through lift shaft or gas ducting in the common areas is about 500 m^3/h. In the case of lift shaft with basement parking, it becomes more important (between 700 and 900 m^3/h). The air leakage in the common areas represents 24% up to 67% of the air leakage of the whole building.

CONCLUSION

A detailed campaign of air permeability measurements has been carried out in ten multi-family buildings in France. For each building, the air permeability of individual apartments and the whole building have been measured. This represents a total of 208 units on which sampling method of the implementation guide GA P50-784 has been evaluated. The results show that the sampling method gives good results only in the case of buildings with uniform individual measurements. Moreover, the selection criterion of the GA P50-784 sampling method does not identify the apartments with greatest risk of leakage. The use of another criterion based on the envelope area appears more relevant. The results have shown that the leakage in the common areas are significant and can have an important impact on the air permeability of the whole building in the case of common areas with lift shaft and basement parking. These leakages should be considered in the measurement method if it extrapolates the individual measurements to the whole buildings.

ACKNOWLEDGEMENTS

This work has been conducted under the task 3 of the research project MININFIL that is financed by the French Mininstry of Housing and the French Agency for the Energy Efficiency and the Environment ADEME.

REFERENCES

[1] NF EN 13829 : February 2001 - *Thermal performance of buildings - Determination of air permeability of buildings - Fan pressurization method.*

[2] Walther, W, Rosenthal, B. 2009. *Airtightness testing of large multi-family buildings in an energy performance regulation context,* ASIEPI information paper#165.

[3] GA P50-784 : February 2010 – *Thermal performance of Buildings – Implementation guide for NF EN 13829 : 2001.*

IMPROVEMENT OF AIR TIGHTNESS OF COMMUNITIES

Markku Hienonen*[1], Timo Kauppinen[2], Erkki Vähäsöyrinki[3]

1 Building Supervision Office of Oulu
Solistinkatu 2
90100 Oulu, Finland
markku.hienonen@ouka.fi

2 VTT Technologies and Services for Buildings
PO Box 1100
Oulu, Finland
timo.kauppinen@vtt.fi

3 VTT Expert Services
PO Box 1100
Oulu, Finland

ABSTRACT

From the beginning of year 2007 the buildings in Finland must have energy efficiency calculations, which requirements are now part of Building Codes, based on European Performance of Buildings Directive. According the renewed code, being into the force from July 2012, air tightness number q_{50} cannot be more than 4 $m^3/(h*m^2)$. Better air tightness can be shown by measurements. The air infiltration must be calculated in compensation calculations based on air tightness number 2.0 $m^3/(h*m^2)$. The energy efficiency requirements caused an immediate response in the building sector. Air tightness's have been measured since 70`s but new requirements launched a real boom. For example, in the city of Oulu building supervision authorities connected air tightness in their quality control programs of buildings - before that the authorities demanded the moisture management and control plan for multi-story houses and commercial both public buildings.

More than 30 years ago, the typical air leakage number n_{50} in one-family houses varied in the level of 6 – 7 1/h (changes/hour). In the turn of millennium the level was 2-3 1/h, but for instance in the city of Oulu the air tightness of new one-family houses has elicited to improve to the level 1 1/h or even below that. The best result since now is 0, 1 1/h, in the target where special attention has paid in air tightness, measured by three measurers and four different Blower Doors.

In this presentation the progress of air tightness especially in one-family buildings has been considered, and how a city can effect on the quality of new buildings. The best case of 0, 1 1/h is introduced, also the structural details. Air tightness is just one part of energy efficiency control – when air leak number will be in the level of 0, 5 – 0, 6 1/h – in the level of passive house – the effect of air tightness is not so crucial dealing with energy consumption and energy savings. Also calculations of the influence of air tightness for energy consumption in various cases will be introduced.

KEYWORDS

Air Tightness, Blower Door, Energy Efficiency of Buildings, Building Performance

1. INTRODUCTION

The function of the building envelope and building services is to maintain the goals of indoor environment and performance set for the building. Air tightness of buildings means how air-tight the exterior walls of the building are – how the incontrollable air flows through the structures has been prevented. Air-tightness of the structures of exterior walls interacts significantly with thermal comfort, and, if the building is intight enough, also with energy consumption.

Uncontrolled air leaks through building envelope can cause also health hazards and structural damages. This air leak flow comes about pressure differences caused by wind and temperature differences and also caused by insufficient function of ventilation system. The location of the building, the height of the building and the condition of building envelope has an effect on air leak flow.

Air tightness of a building is measured by specific device or set of devices; also the own ventilation system of building can be utilized. Tracer gas-method can also be used, especially when the controlled rate of ventilation must be separated from air leaks (concentration decay method). Air tightness measurements based on external fans demand measurable pressure difference across building envelope, and the air flow of fan will be determined. The air leak number is presented based on building air volume (n_{50}) or/and on area of building envelope (q_{50}) mainly at 50 Pa pressure difference.

2. AIR TIGHTNESS AND ORDERS OF AUTHORITIES

2.1 Air tightness and building codes

There has not been numerical scale-based requirements dealing with air tightness in Finland before 2008; the only indirect mention was in ventilation codes, in which the recommendable air leak number n_{50} was 1,0 changes/hour mostly intended to ensure the proper function of mechanical ventilation system. Also the air tightness of ventilation ducts was determined and classified. When requirements of energy performance calculation came into the building codes in 2008 (first version, the recent version from 2010), the interest of different parties of building trade grew rapidly in air tightness measurements. According the new code, being into the force from July 2012 [1], air tightness number q50 cannot be more than 4 m3/ (h*m2). Better air tightness can be shown by measurements. The air infiltration must be calculated in compensation calculations based on air tightness number 2.0 m3/ (h*m2). The valid standard SFS-EN 13829 is presented in the building code [2].

In the finnish building code [3] dealing with the evaluation of energy consumption and heating need typical air leak numbers of building enevlope and ratings has been presented:

Typical values (n_{50}) of air tightness			
Class	**Details**	**Building type**	**Typical values, n_{50} [1/h]**
Good		One-family houses	1....3
		Apartrments, offices	0,5....1,5
Average		One-family houses	3...5
		Apartrments, offices	1,5...3
Poor		One-family houses	5....10
		Apartrments, offices	3...7

Table 1. Air leak numbers by building type

In the new indoor quality classification [4] recommendation for the maximum value of one family houses $n_{50} < 1 - 2$ 1/h and for other buildings $n_{50} > 0,5 - 0,7$ 1/h. Air tightness for apartments is recommended $n_{50} < 0,5 - 0,7$ 1/h (including external and both internal leaks through exterior walls, floors and intermediate walls).

2.2 Air tightness of one-family houses – previous results

VTT started air tightness measurements 1980. Air tightness of buildings has improved during 30 years – one can evaluate that the air tightness of one-family houses is more than halved. Nowadays the best houses will go under the limit of passive houses (n_{50} = 0, 6 1/h). Typical values of new houses are between 1-2 1/h. There are still problems.

VTT studied in 1981 air tightness level of some buildings [5]. The data included 42 one-family houses with various materials. Insulation material was mineral wool (32) and sawdust (12). Table1. Shows the results. The sample was relatively low and the results are therefore only suggestive. The air leak numbers concentrated between 7-9 1/h.

Type of building	Targets	n_{50} (mean value)
One-family houses, built before 1973	7	7,9
One-family houses, built after 1973	9	6,6
Row houses , before 1973	1	9,3
Row houses, after 1973	15	9,8
Log houses	10	9,7

Table 2. Air leak numbers according to building type (1981)

Type of building	Targets	n_{50} (mean value)	Lowest	Highest
One-family houses	56	5,3	1,6	18
Row houses	102	5,6	1,7	14,9
Log houses	13	10,7	5,3	14

Table 3. Air tightness of one-family houses 1981-1988

Table 2 shows the results of 171 one-family houses during 1981-1998 [6]. Main part of the houses has been reclamation cases. This means that the results can be worse than average or distribution of leak points has concentrated in that way, that it caused draft problems. Age of the buildings varies. Biggest group of one-family houses was between 3 - 4 1/h and of row houses between 4 – 5 1/h. There are also data available for Finnish housing fair which shows [7] that air-tightness has been improved and more attention has been paid to that topic.

Air tightness is totally depending on the solutions made during the construction and installation stage – to improve air tightness afterwards is extremely difficult. According to some single tests, air leak number will increase in the long run approx. 0, 5 1/h – 1, 0 1/h. By tightening the air leak number can improve 0,5 – 1,0 1/h, equal to the decrease of air tightness in course of time, depending on how the repair works has carried out and how wide the repairs have been.

3. MEASUREMENT OF AIR TIGHTNESS AND CERTIFICATION PROCEDURE

Air-tightness measurements of buildings in Finland have been done from early 70's and more systematically from 1980 on. When building codes have been renewed, lot of serviced providers sprang up and also the measurements were certified – air-tightness measurer certification procedure and courses started 2008. Now it is also part of building thermographer-certification procedure. Building thermographer-certification started 2003. Many manufacturers and contractors have included air tightness measurements in their quality control procedures [8].

4. THE ROLE OF BUILDING SUPERVISION OFFICE (BSO) IN FINLAND.

4.1 Background

Every new building in Finland needs building permit and building supervision office to give that permit. In other words building supervision has contact to every builder when they start to design their project and before they can start to build. BSO has excellent possibility to give that kind of information they conceive important. We can fairly speak about some kind of momentum.

Almost in every municipality (ca. 330) has own building supervision office. Responsibility of the BSOs in Finland is to control, that houses to be built will be carried out according to law, rules and city plan, also taking care of the environment.

Recommendable task is also, that building supervision give advices and take care how builders can get better energy efficiency, better sustainability and longer life cycle. Normally building supervision offices in Finland do the minimum or only a bit more.

Building permit in Finland includes calculated energy certification, including among others goal of airtightness. Energy certification must be dated before dwellers will move in the house.

4.2 The role of building supervision office in Oulu, "BSO – Oulu - model"

BSO-Oulu tries to use this momentum effectively. The goal is that BSO-Oulu should do:

- Produce measureable added value to our customers and to City of Oulu,
- help customers,
- to have the courage give advices,
- to have willingness to be co-operative,
- to use public media,
- do development work together with designers and builders,
- To create network with local and national actors.

BSO-Oulu has done all those things during last ten years and the organization has employed a quality manager too for these tasks.

The main network of organizations consists of Oulu University of Applied Sciences (OUAS), VTT`s (Technical Research Centre of Finland) research team in Oulu, four biggest companies producing single-family houses in Finland and the Ministry of Environment. With BSO of City of Umeå (Sweden) BSO-Oulu had employee's exchanges in the year 2007 and 2008.

Last few years main focus has been to improve new buildings more energy-efficient and sustainable. In the near future BSO-Oulu will work with existing buildings to achieve the same. In the year 2010 started one project concerning existing buildings in co-operation of Ministry of Environment. During last ten years BSO-Oulu has arranged together with network of companies many seminars for professionals, designers, public authorities, responsible managers, foremen, builders and families together ca. 10 000 personal education days. For every builder of single family house in Oulu BSO arranges two or three times per year a course that includes minimum of 12 hours info sessions during six evenings. That happens after they have got the building site and before they start to design their project.

4.3 Examples of tools BSO-Oulu has produced and usde in everyday work

- Interactive application in Internet, www.pientalonlaatu.fi [9] (first price 2006-RIL-Finnish Association of Civil Engineers),
- Quality cards - http://www.ouka.fi/rakennusvalvonta/oppaat/laatukortit.htm [10] (figure 1, figure 2).

4.4. International activities and projects

BSO-Oulu is a partner in "Increasing Energy Efficiency in Buildings"-project (IEEB). The consortium of this EU-project constitutes OUAS, Luleå University of Technology (LTU,

Sweden), Umeå University (UU, Sweden) and NORUT Narvik (Norway). Total value of IEEB-project is 1, 7 million €, see http://www.interregnord.com/fi.aspx [11].

In the near future the challenges will be big - passive houses and 0-energy houses. We all need new studied and documented information that will be suitable in northern Finland, Sweden and Norway.

In IEEB-project air-tightness is in very important role. BSO-Oulu's task is to help exporting the results of this project go into practice. When choosing the results going into practice one shall always be critical. Only those things/methods who are important, influential (energy-efficient, costs of life-cycle) and riskless (for example moisture risk) are acceptable for common use to all builders. Only in few "test-buildings" is acceptable to prove something unsure.

4.5. Examples of the results in Oulu –economical evaluation

In the year 2008 single-family houses in Oulu were more than 30% more energy-efficient in total average value compared with general regulations in Finland. In bigger dwelling-houses comparable value was over 25% and in other buildings approx. 20%. We have calculated that if we put money ca. 0, 1 M€, our customers will get back 20 M€ during next 50 year life cycle. Income value/investment is ca. 200. In the next few years this kind results are very difficult (maybe impossible) to achieve, because regulations will tighten/go forward very fast in Finland too.

5. RESULTS OF AIRTIGHTNESS IN OULU REGION.

In Oulu region air-tightness is now ca. four times better than seven years ago (4…6 --> 0.4...1, 9 1/h) and clearly better than average value in Finland. This variation of airtightness means ca. 20…25 % lower energy consumption. In Finland we calculate, that a change of 1 1/h in air leak number value means about 7 % in energy consumption (heated rooms in house). Cost efficiency is very good, in most cases the cheapest way to get houses more energy efficient.

Good airtightness is one of the most important factors, when constructing very energy efficient or passive houses. Poor air-tightness will cause moisture risks, increase energy consumption and indoor climate is uncomfortable. In city of Oulu, if air tightness in calculated energy certificate is 4 1/h at 50 Pa pressure difference inside-outside, measurement or other clarification are not needed. If goal of airtightness in energy certification is 2…4 1/h clarification according airtightness card or measurement is needed. If goal of airtightness is < 2 1/h, measurement is required.

Figure 1 Air-tightness in Oulu Region

Figure 2 Air-tightness card /BSO-Oulu

6. CASE STUDY HOUSE LINNAKANGAS IN KEMPELE ECO-QUARTER, KEMPELE FINLAND

6.1 Demonstration building

This house is situated in north Finland ca. 15 km distance from city of Oulu in so called Kempele Eco-quarter (kempele is a neighboring community south of Oulu). The whole Eco-quarter is not connected in normal electrical network like almost all finnish single-family houses do. Eco-quarter consists ten single-family houses and "Powerhouse", where both electric energy and heat energy is produced for the whole quarter by CHP and a bit by windmill. Fuel is mainly wood chips and partly biofuel. Eco-quarter has won first prize in the year 2010 given by RIL - Finnish Association of Civil Engineers, see e.g.

http://www.ril.fi/en/main-page.html [12] ,
http://www.ril.fi/fi/etusivu/ajankohtaista-2/ril-palkinto-2010-kempeleen-ekokylalle.html [13]

Figure 3. Energy quarter

Figure 4. CHP "Powerhouse"

Figure 5. House Linnakangas in Kempele Eco-quarter

6.2 Structural details of House Linnakangas

Floor, ground based:

- coating (tile)
- reinforced concrete slab 100 mm
- polyurethane 160 mm, alongside ext. wall 190 mm
- washed coarse gravel with radon ventilation pipes
- U-value 0.103…0.107 W/m2K

External wall:

- stainless reinforced white concrete 80 mm
- polyurethane 170 mm with ventilation grooves
- reinforced concrete 100…120 mm
- U-value 0.15 W/m2K
- calculated life cycle 100 year

Roof:

- bitumen
- plywood board 15 mm
- wooden frame and ventilated space
- blowed heat insulation 600 mm
- bitumen coating
- reinforced concrete slab 200 mm
- U-value 0.07 W/m2K

Windows:

- partly fixed in ext. wall, U-value 1.0 W/m2K
- partly opening, U-value 0.85…1.0 W/m2K

Doors:

- partly sliding doors with window, U-value 1.0 W/m2K
- partly normal doors with window, U-value 0.8 W/m2K

Effiency of ventilation recovery (year): 73 %
Airtightness: 0.08 1/h. (0.07…0.08 1/h) -ca. 7x better than passive house limit 0.6 1/h

Heat transmission inside the house: waterpipes in floor/concrete slab

House automation:

- Ouman Oy, Kempele, www.ouman.fi [14]
- heating-, ventilalation-, moisture damage-, fire carbon monoxcide-warning/control.

Ventilation equipment:

- Enervent Pandion 2 pieces, SFP-value 1,6
- preheating/cooling with pipes 400 m in the ground
- all ventilations pipes inside envelope

Other equipments:

- hood with direct outflow (using only when preparing food)
- sauna stove, "fuel" /propellant piece of wood

Wideness, area and volume:

- 347 brm^2 (gross area)
- heated air volume 897 m^3

Calculated energy efficiency: ET-value 115 kWh/brm^2/year, Energy class A

Figure 6 Sliding window in living room

Figure 7 Fixed window in living room

6.3 How good airtightness has been achieved

Builder/owner has good experience in desingning constructions over 25 years. Based this experience the owner had designed and done himself all the most important details conserning air-tightness. Some examples of details:

- joint between floor and external wall tightened with elastic mass.
- almost all pipes inside envelope
- all perforations(some pipes and electrical cables) of envelope tightened with elastic mass.
- all joints between window/door frame and ext. wall tightened with polyurethan and elastic mass.

- all opening windows adjusted in building site after installation (sealings were not airtight)
- premeasuring of airtightness and repairing the critical points, result of premeasuring was 0,1 1/h.

6.4 Measuring of airtightness in house Linnakangas

Because the premeasured result was so good, it was important to check the final result not only by one measurer and one Blower Door. The final measurement decided to carry out with four different calibrated Blower Door and three different measurer. The result in all four measurements was the same 0, 08 1/h. (0, 07...0, 08 1/h). In spite of that the house is quite big (897 m^3), the air leakage was so low, that all the adjustments in Blower Door must be done manually. In the end of the tests was found out, that there was a small air leakage in Blower Door, see Fig. 12. In normal house it does not matter, but in so tight house the result should be better than 0, 07 1/h, if that leakage has been found before final measuring.

Figure 8. Company , 8.6.2010

Figure 9. Installation

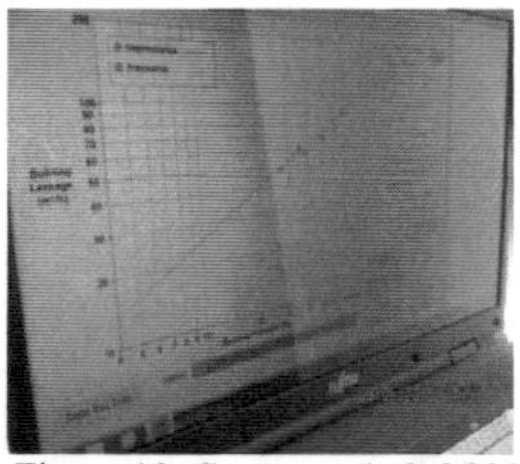
Figure 10. Company 1, 8.6.2010

Figure 11. VTT 14.6.2010

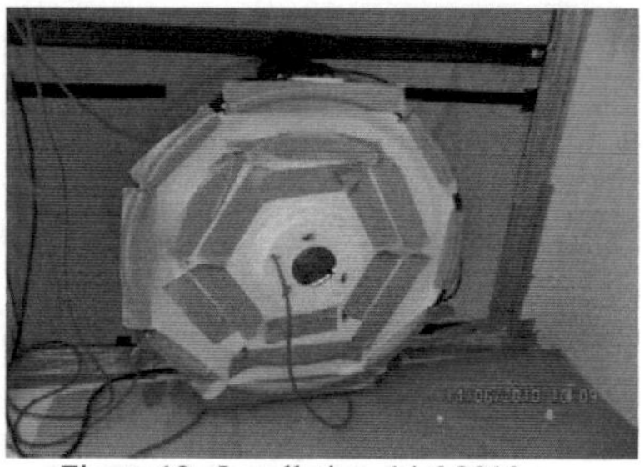
Figure 12. Installation 14.6.2010

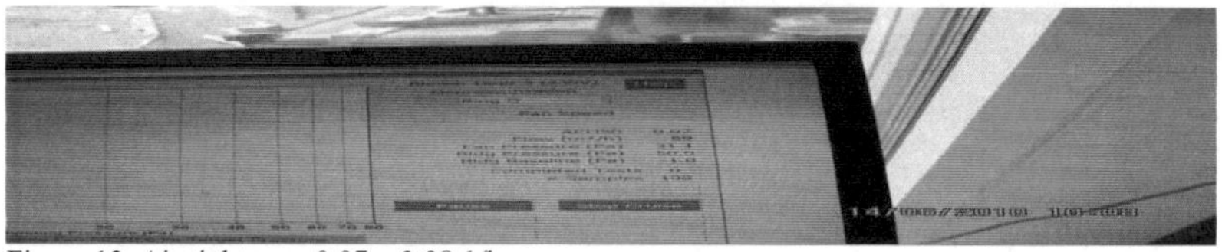
Figure 13. Air-tightness 0,07...0,08 1/h

7. CONCLUSIONS

It is possible to reach better airtightness by systematic work not only in pilot houses, but in every new house. To design details in advance is necessary, it is too late to design all the details in building site. Educating and training designers, responsible foremans, carpenters and others workers in building site is needed. Details shall not be too complicated and many different details in the same house shall be avoided. Measuring both air-tightness and thermographic survey is needed. Without the location of possible leak points it is impossible to make any repairs. The fact, that the house will be measured can influence a lot, even so that air-tightness sharpen from 4 --> 2 1/h. That kind of experience was in Oulu year 2005. When building passive houses, it is necessary to know in advance, that good airtightness (<0.6 1/h) is possible to reach. Many contractors have already reached the level 1…1, 5 1/h in Oulu region. Now the next step is <0, 6 1/h as wide as possible in normal or low energy house. Anyway, some new challenges will exist:

- house automation should be designed/developed so that using fireplace and hood is possible.
- in some parts of Finland there are no professionally skilled measurers nearby
- if outgoing air is 100, how much should incoming air be in very airtight house ?
- when house is very airtight and ventilation system break in some reason at night, do we need carbondioxide alarm ?

ACKNOWLEDGEMENTS

Mr. Risto Linnakangas (the owner and designer of the pilot house)

REFERENCES

[1] D5. Suomen rakentamismääräyskokoelma. Ympäristöministeriö. Rakennuksen energiankulutuksen ja lämmitystehontarpeen laskenta . Ohjeet 2007 (D 5, Finnish building codes. Ministry of Environment. Calculation of energy consumption and heating capacity of buildings. Instructions 2007. Ministry of Environment 2007)

[2] SFS-EN 13829

[3] D3. Suomen rakentamismääräyskokoelma. Ympäristöministeriö. Rakennusten energiatehokkuus. Määräykset ja ohjeet 2012. *(D 3, Finnish building codes. Ministry of Environment. Energy efficiency of buildings. Requirements and instructions 2012. Ministry of Environment 2011)*

[4] Sisäilmastoluokitus 2008 *(Classification of Indoor Environment 2008). www.sisailmatieto.fi*

[5] Rantamäki, Jouko, Kauppinen, Timo. Suomalaisten rakennusten ilmanpitävyys mittausten perusteella. SIY Raportti 13. Sisäilmastoseminaari 1999. Säteri, Jorma & Hahkala, Harri (toim/ed.). Sisäilmayhdistys SIY. Helsinki 1999 *(Air tightness of finnish buildings on the grounds of measurements). Indoor Air Association, report 13, pp.329-336. Sisäilmatieto Oy 1999.*

[6] Hekkanen, Martti, Kauppinen, Timo. Rakennusfysiikka 2007. Uusimmat tutkimustulokset ja hyvät käytännön ratkaisut 18-19.10.2007 Tampere. Toim (ed.). Juha Vinha & Minna Korpi (2007), pp. 277 - 286 *(Thermography and air tightness measurements in evaluation of building performance – air leak numbers of one-family houses in Oulu Housing Fair), Seminar of Building Physics, Tampere 2007. Tampere University of Technology*

[7] http://www.asuntomessut.fi/en/english-home

[8] Kauppinen, T., Ojanen, T., Kovanen, K., Laamanen, J. and Vähäsöyrinki, E. Rakennusten ilmanpitävyys (Air tightness of buildings) in Indoor air seminar, Helsinki. Säteri, J, Backman, H (ed). *Indoor Air Association, report 27, ISBN 978-952-5236-35-8, Loimaa, Finland 2009.*

[9] www.pientalonlaatu.fi

[10] http://www.ouka.fi/rakennusvalvonta/oppaat/laatukortit.htm

[11] http://www.interregnord.com/fi.aspx

[12] http://www.ril.fi/en/main-page.html

[13] http://www.ril.fi/fi/etusivu/ajankohtaista-2/ril-palkinto-2010-kempeleen-ekokylalle.html

[14] www.ouman.fi

Thursday 13 October 2011

14.00 – 15.30 Parallel Session 7A - Evaluation of ventilation strategies

- Numerical validation for natural ventilation design (François Demouge, France)
- Performance of low pressure mechanical ventilation concept with diffuse ceiling inlet for renovation of school classrooms (Søren Terkildsen, Denmark)
- Definition of occupant behaviour patterns with respect to ventilation by means of multivariate statistical techniques (Felipe Encinas Pino, Belgium)
- Control and performance of innovative Ventilation systems in Low Energy Buildings: a case study (Juslin Koffi, France)
- Shelter-in-place strategy: CONFINE, an airtightness level calculation tool to protect people against accidental toxic releases (Gaëlle Guyot, France)
- Liabilities of Vented Crawl Spaces And Their Impacts on Indoor Air Quality in Southeastern U.S. Homes (Jonathan Coulter, USA)

14.00 – 15.30 Parallel Session 7B - IAQ and energy impacts of envelope leakage

- Laboratory investigation of timber frame walls with an exterior air barrier in a temperate climate (Jelle Langmans, Belgium)
- State of Art of Non-Residential Buildings Airtightness and Impact on the Energy Consumption (Valérie Leprince, France)
- Investigations on the effects on airtight performance improvement and energy consumption of insulation retrofit in detached houses (Hiroshi Yoshino, Japan)
- The influence of air permeability and type of underlay on the hygrothermal performance of an inclined roof (Paul Steskens, Belgium)
- Impacts of Airtightening Retrofits on Ventilation and Energy in a Manufactured Home (Andrew Persily, USA)

15.30 – 15.50 Room Change

NUMERICAL VALIDATION FOR NATURAL VENTILATION DESIGN

François Demouge[1,*], Nicolas Le Roux[1,2], Xavier Faure[1]

1: Université Paris-Est, CSTB
11 rue Henry Picherit
44300 Nantes, France
**Corresponding author: francois.demouge@cstb.fr*

2: Université de La Rochelle, LEPTIAB
Avenue Michel Crépeau
17042 La Rochelle, France

ABSTRACT

The aim of this study is to check the accuracy of a nodal model to predict correctly the flow fields involved inside a building by wind-induced pressure. The model is confronted to experimental tests involving a one-storey dwelling of 84 m² at a reduced scale of 1/10 placed in a wind tunnel facility. Different configurations are tested considering openings of different sizes for outside openings as well as for internal doors. For each configuration, various wind incidences are studied. The confrontation shows that the model is able to reproduce the experiment with a relative error of ±20% for situations involving velocity ratios of outside openings greater than 0.2. For lower velocity ratios, higher discrepancy is observed. From an engineering point of view, the results of the model are found acceptable for the evaluation of the potential of wind-driven ventilation of a given building.

KEYWORDS

Wind-driven ventilation, model scale experiment, nodal modelling

INTRODUCTION

Ventilation is an important source of energy consumption since fresh air comes generally from the outside environment. The air mass flow needs to be adjusted so that thermal comfort in summer and the air quality in winter are ensured. These two objectives are to be held while reducing the global energy consumption of the building [1].

The use of heat ventilation and air conditioning systems (HVAC) has been widely used until now. It offers the possibility to reach the two first objectives mentioned but it can increase dramatically the energy needs. Concerning dwellings, HVAC coupled with heat recovery system can reduce the energy need in an important manner. Energy efficient building labels as Minergie [2] are good examples of what can be done using HVAC systems. In temperate climates, one can obtain high reduction of energy consumption by considering the potential of natural ventilation for designing hybrid ventilation systems [3]. This potential depends greatly on the site wind velocity and wind direction, but also on the internal architecture and on the areas of internal and outside openings [4].

In the design process of ventilation systems, nodal models can give detailed information on the flow distribution and the air quality inside the building. The simplicity of using nodal models and the low resources required allow one to test easily a large number of configurations (internal distribution, opening sizes, mechanical flow rates, internal temperatures, usage scenarios, etc.) [5] [6]. However, there is a need to validate the ability of nodal models to reproduce the various physical phenomena involved.

The present paper constitutes a first step towards the validation of the nodal model MATHIS, developed at CSTB, for the design of hybrid ventilation systems and focuses on the ability of the tool in reproducing the interaction between wind-induced pressure and internal architecture of buildings. In a first step, the main assumptions of the nodal model used are briefly presented. The model allows to predict the internal pressure, temperature and species fields and openings flow rates depending on internal heat sources, outside wind and temperature conditions. In a second step, the test apparatus used for confrontation is presented. The model represents a one-storey dwelling of 84 m² composed of 8 internal volumes at a reduced length scale of 1/10 placed in a wind tunnel facility. Different configurations were tested considering openings of different sizes for outside openings as well as for internal doors. For each configuration, various wind incidences were studied. In a third step, numerical and experimental results are compared. This comparison is made for the various configurations tested in terms of pressure field inside the dwelling, openings flow rates and wind driven ventilation potential for a given geographic site.

NUMERICAL MODELLING

The numerical tool MATHIS (***M**odélisation de l'**A**éraulique, de la **Th**ermique et de l'**H**ygrométrie **InS**tationnaire d'un bâtiment*) is currently under development at CSTB. The aim of the tool is to unify the different nodal models developed along the past years in the different departments of CSTB and dedicated each one to specific studies (mechanical ventilation design, air quality assesment, evaluation of the interaction between heating systems and ventilation systems, fire safety engineering, etc.).

MATHIS is a nodal model similar to other tools as COMIS or CONTAM. It is based on the breackdown of the configuration studied into nodes and branches. Each node represents a room or a portion of the ventilation network of a building. Each branch represents the aeraulic components (openings, ducts,...) between those volumes. Mass and energy conservations applied in nodes allow one to access to the pressure, temperature and species mass fractions. For the ith node, they can be expressed by the following ordinary differential equations :

$$\frac{dP^i}{dt} = \frac{Cs_P^i/Cs_V^i - 1}{V^i}\left(\dot{E}^i\right) \quad (1)$$

$$\frac{d\rho^i}{dt} = \frac{1}{V^i}\left(\dot{m}_{in}^i - \dot{m}_{out}^i\right) \quad (2)$$

$$\frac{dY_k^i}{dt} = \frac{1}{\rho^i V^i}\left(\dot{m}_{k_{in}}^i - \dot{m}_{k_{out}}^i\right) \quad (3)$$

Temperature is deduced using the perfect gaz law :

$$T^i = \frac{P^i}{\rho^i\left(Cs_P^i - Cs_V^i\right)} \quad (4)$$

Mechanical energy conservation applied in branches gives the flowrates for each aeraulic component. It takes the form of the steadystate generalized bernoulli equation (which means neglecting the branch inertia). For a branch connecting the node i at height z^i to node j at height z^j, it is expressed as follows:

$$P^i - \rho^i g z^i + \rho^* g\,(Z_a^i + z^i) = P^j - \rho^j g z^j + \rho^* g\,(Z_a^j + z^j) + f(\rho^*, \dot{m}) \quad (5)$$

with $f(\rho^*, \dot{m})$ the pressure loss associated to the branch. The mass flux through the branch is deduced from (5) as:

$$\dot{m} = sgn(\Delta P).\rho^*.K.\left(\frac{|\Delta P|}{\rho^*}\right)^n \quad (6)$$

with: $$\Delta P = (P^i - \rho^i g z^i) - (P^j - \rho^j g z^j) + \rho^* g[(Z_a^i + z^i) - (Z_a^j + z^j)] \quad (7)$$

$$\rho^* = \begin{cases} \rho^i \; if \; \dot{m} > 0 \\ \rho^j \; if \; \dot{m} < 0 \end{cases} \quad (8)$$

Given the aim of the modelling, different optional models can be linked to this aeraulic model (heat diffusion through walls, thermal radiation, combustion, sources and sinks of heat and species, etc.).

TEST APPARATUS

The experiment used for the confrontation of the numerical tool as been done in the framework of studying the potential of wind-driven ventilation for dwellings. We give herein a short description of the apparatus. One will find in ref. [7] full details of the experiment. The apparatus represents a model of a classical one-storey dwelling. Full scale dimensions of the dwelling are 8.5 m large, 12 m long and 3 m high. Figure 1 presents a view of the dwelling in the wind tunnel and a schematic representation of the internal architecture with the functionality of each room. Each room communicates with the corridor through internal doors and with the outside environment through windows (outside openings). 10 different configurations are studied. They are classified as "case n°i - xx%" where i corresponds to different sizes of internal doors and xx to the percentage of the maximum sections of outside opening (see ref. [7] for details of the configurations).

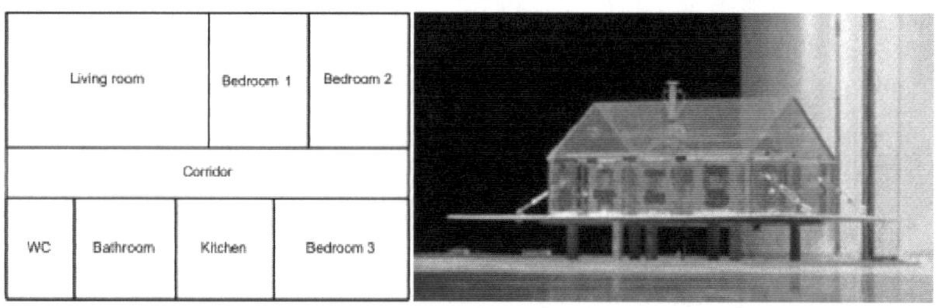

Figure 1: Internal architecture (left) and view of the dwelling in the wind tunnel (right)

COMPARISON BETWEEN EXPERIMENT AND PREDICTION

The model tests are simulated with MATHIS at full scale. The openings are modelled through the orifice equation, based on the Bernoulli's assumption of steady incompressible flow [8]. Thus, the caracteristics of the orifice might be included in equation (6) as :

$$\begin{cases} n = 0{,}5 \\ K = C_z.A.2^n \end{cases} \quad (9)$$

For sharp-edge orifices, high aspect ratios and normal wind incidence, the discharge coefficient C_z is generally taken to 0.6 [8]. However, as presented in several studies, this parameter can take different values depending mainly on wind incidence and aspect ratios [9][10][11]. The C_z variations of the openings of the scale model have been calibrated with and without wind on a specific bench [7] and are implemented in the numerical model. External pressure coefficients measured in the wind tunnel are applied on the outside openings of each room. The internal pressure coefficients predicted inside each room are then compared to the measured ones. Figure 2 presents an example of results obtained for one of the configuration tested. Figure 3 presents the correlation obtained between predicted and measured internal pressure coefficient for all the cases, rooms and wind incidence. The coefficient of determination r^2, defined as the square of the linear correlation coefficient between prediction and experiment, is close to 0.99. However, it is obvious that the general trend of the numerical model is to give higher C_P than the experiment. Moreover, discrepancy is higher for small values of C_P.

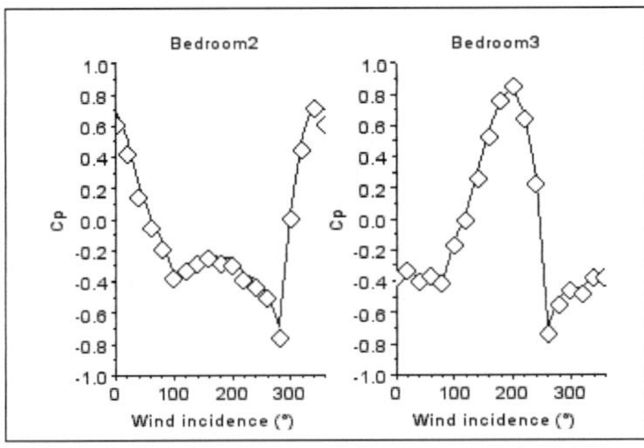

Figure 2: internal pressure coefficients in bedrooms 2 and 3 as function of wind incidence - case n°3 - 50% (◇: experiment; —:prediction)

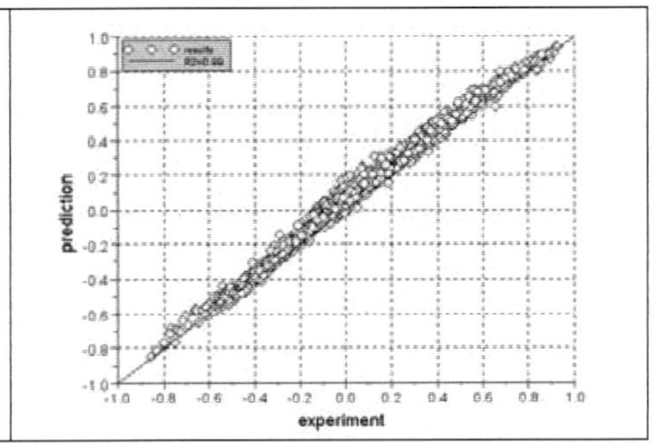

Figure 3: correlation between predicted and measured internal pressure coefficient

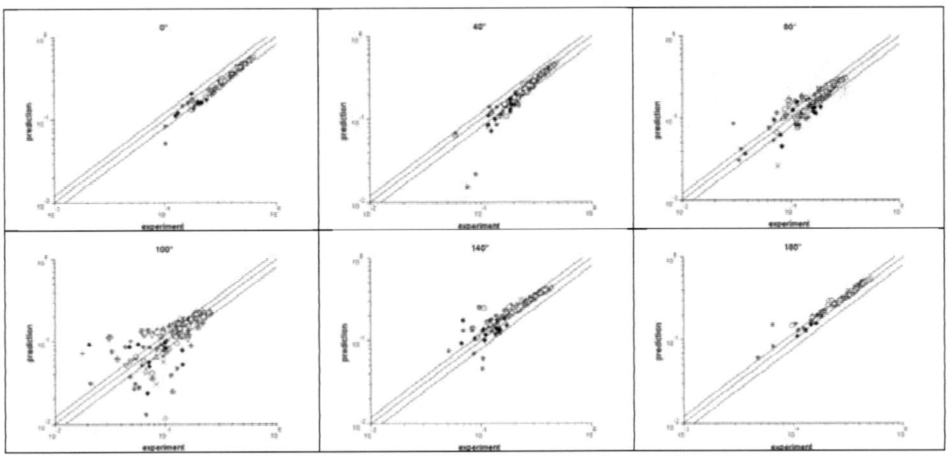

Figure 4: comparison of measured and predicted velocity ratio $R = W/U$ for 6 different relative incidence
○: case n°1 - 10% ; ✳: case n°1 - 25 % ;
+: case n°3 - 10% ; ×: case n°3 - 25 % ; ⊕: case n°3 - 50% ; ◆: case n°3 - 75% ;
Δ : case n°5 - 10% ; ◇: case n°5 - 25% ; ∇: case n°5 - 50% ; ⊕: case n°5 -75%

Figure 4 presents the correlation between predicted and measured velocity ratio R at outside openings defined as $R = W/U$. Results are presented for 6 different relative wind incidences (wind incidence relative to opening orientation). Each set of dots represents a configuration.

The lines represent an error of ±20% relative to the experiment. Those results shows the same general trend for the different relative wind incidences: for R greater than 0.2 the numerical results are within 20% from the experiment. For R lower than 0.2, high discrepancy is observed between the numerical and the experimental results.
Situations with low velocity ratio R are mainly obtained for relative wind incidence near 90°. They are also obtained for configurations with large outside openings and comparatively small internal doors such as for exemple case n°1 - 25% or case n°3 - 75%. On the one hand, this high discrepancy observed for low velocity ratios might be explained partly by measurement uncertainities on mean external pressure coefficient C_P, particularly for wind incidence near 90°. On the other hand, the effect of the differences of geometry between the dwelling and the bench used for openings calibration might also be important on the C_Z variation law for incidences higher than 90°. On this last point, research efforts are currently being done at CSTB with the aim to better understand the effects of wind incidence, Reynolds number orifice, aspect ratio and position of openings on C_Z variations.

Test Case for a real site consideration
The potential of wind driven natural ventilation has then been evaluated for a specific site and orientation using both the experimental results and MATHIS software. The evaluation consists in calculating a satisfaction rate. It is defined as the percentage of time for which the air change rate satisfy the french standard recommendation of $135m^3/h$ within the kitchen for dwellings of more than 5 rooms. Figure 5 and 6 present the wind rose of the considered geographic site (Le Bourget, FRANCE) and the results of the potential of wind driven ventilation for a specific orientation (with the living room facing north) respectively. The residus in Figure 6 are defined as the difference between both approachs on the satisfaction rate.

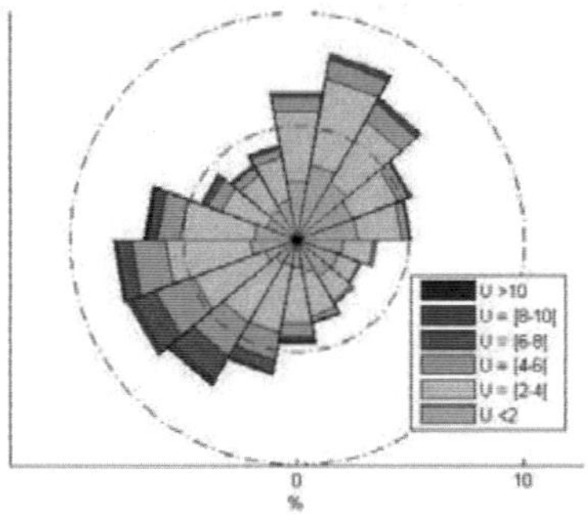

Figure 5: Wind rose of Le Bourget (FRANCE)

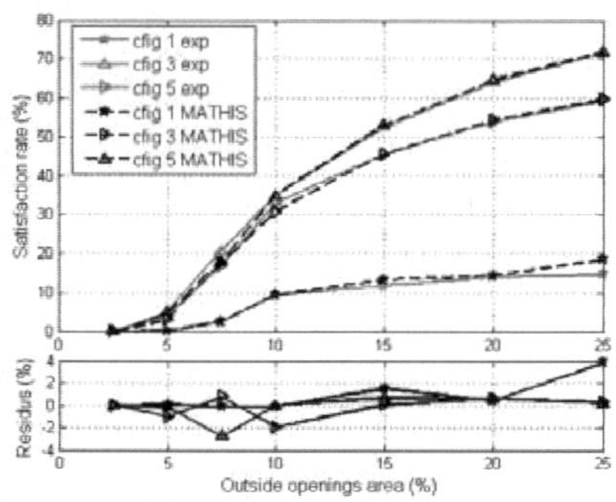

Figure 6: Satisfaction rate for the air change rate in the kitchen depending on the outside openings area

Even if some discrepancies have been previously seen, the global results, considering an engineering point of view, are very good. A maximum of 4% of error is identified on the satisfaction rate between the experiment and MATHIS results.

CONCLUSION

The present study allows to validate the use of a nodal model like MATHIS for the design of wind-driven ventilation systems. In the design process, while wind tunnel tests or thoroughly validated CFD modellings are still needed to obtain the external pressure coefficients to apply on the different façades of a given building, a nodal model can be used to test the effect of various configurations of internal architecture and openings sizes. Here, the tests involved an isothermal one-storey building only. Others configurations involving several storey and stack effect should be tested in order to complete the validation.

Nomenclature

A	orifice area (m^2)	V	volume (m^3)
C_P	pressure coefficient	W	opening velocity ($m.s^{-1}$)
Cs_P	specific heat at constant pressure ($J.kg^{-1}.K^{-1}$)	Y	mass fraction (kg/kg)
Cs_V	specific heat at constant volume ($J.kg^{-1}.K^{-1}$)	z	heigth from the ground (m)
C_z	discharge coefficient	Z_a	height from reference altitude (m)
$\dot{E}$	net heat flux inside node (W)	ρ	density ($kg.m^{-3}$)
K	aeraulic resistance ($kg^n.m^{-3(n-1)}.Pa^{-n}$)	ρ^*	density of the gas through a branch ($kg.m^{-3}$)
$\dot{m}$	mass flux ($kg.s^{-1}$)		
n	flow equation exponent	*subscripts*	
P	total pressure at the ground of a node (Pa)	*in*	inward
R	velocity ratio	*out*	outward
T	temperature (K)	k	k^{th} specie
U	wind velocity ($m.s^{-1}$)		

REFERENCES

[1] French Ministry of Building, Transport and Environment. (2010, July) Reglementation Thermique 2012. [Online]. http://www.rt-batiment.fr/batiments-neufs/reglementation-thermique-2012/presentation.html

[2] F. Beyeler, N. Beglinger, and U. Roder, "Minergie: The Swiss Sustainable building Standard," *Innovations*, vol. 4, no. 4, pp. 241-244, 2009.

[3] H. Chang, S. Kato, and T. Chikamoto, "Effects of outdoor air conditions on hybrid air conditioning based on task/ambient strategy with natural and mechanical ventilation in office buildings," *Building and Environment*, vol. 39, pp. 153-164, 2004.

[4] F. Allard, *Natural Ventilation in Buildings. A Design Handbook.*: James & James, 1998.

[5] J. Laverge, N. Van Den Bossche, N. Heijmans, and A. Janssens, "Energy saving potential and repercussions on indoor air quality of demand controlled residential ventilation strategies," *Building and Environment*, vol. 46, pp. 1497-1503, 2011.

[6] J. Koffi, F. Allard, and J.-J. Akoua, "Numerical Assessment of the Performance of Ventilation Strategies in a Single-family Building," *International journal of Ventilation*, vol. 9, no. 4, pp. 337-349, March 2011.

[7] X. Faure and N. Le Roux, "On the potential of wind-driven ventilation for dwellings and low rise buildings," *Roomvent 2011 - Trondheim, Norway*, 2011.

[8] D. Etheridge and M. Sandberg, *Building Ventilation. Theory and Measurement.*: John Wiley & Sons Inc, 1996.

[9] P. Karava, T. Stathopoulos, and A.K. Athienetis, "Wind driven flow through openings – A review of discharge coefficients," *International Journal of Ventilation*, vol. 3, no. 3, 2004.

[10] Y-H. Chiu and D. W. Etheridge, "External flow effects on the discharge coefficients of two types of ventilation opening," *Journal of Wind Engineering and Industrial Aerodynamics*, vol. 95, pp. 225-252, 2007.

[11] Chia R. Chu, Y-H. Chiu, Yan-Jhih Chen, Yu-Wen Wang, and C-P. Chou, "Turbulence effects on the discharge coefficient and mean flow rate of wind-driven cross-ventilation," *Building and enivronment*, vol. 44, pp. 2064-2072, 2009.

PERFORMANCE OF LOW PRESSURE MECHANICAL VENTILATION CONCEPT WITH DIFFUSE CEILING INLET FOR RENOVATION OF SCHOOL CLASSROOMS

Søren Terkildsen[*1], Svend Svendsen[1],

1 Technical University of Denmark
Department of Civil Engineering, Section of Building
Physics and Services
Brovej, Building 118
2800 Kgs. Lyngby, Denmark
**Corresponding author: sterk@byg.dtu.dk*

ABSTRACT

In a great portion of Danish primary schools the mechanical ventilation systems is outdated or simply rely on opening of windows to ventilate the classrooms. This leads to high energy consumption for fans and/or ventilation heat losses and poor indoor environment, as the ventilation systems cannot provide a sufficient ventilation rate. A recent study with 750 Danish classrooms show that 56 % had CO_2-concentrations over a 1000 ppm, which is the recommended limit by the Danish working environment authority and this adversely affects the performance and well being of the pupils. This paper describes a mechanical ventilation concept to lower energy consumption and improve the indoor environment, developed for refurbishment of school classrooms. The performance of the concept is investigated through computer simulations and measurements of energy consumption and indoor environment. The measurements are made at a pilot project at Vallensbæk primary school, where a system is installed in two classrooms used by the 6th grade. The system is designed with an oversized air handling unit and duct system to reduce the pressure loss and thus fan power required to operate the system. The supply air to the rooms is distributed through diffuse ceilings inlets, where the air is supplied over a large part of the ceiling area through small perforations. The flow rate to the rooms is determined by motion and CO_2 sensors and controlled by a new type of flow control dampers increasing the flow rate according to the demand. The measurements and simulation results show that the indoor environment fulfils indoor environment category II in EN 15251 and the CO_2-concentration is kept below 800 ppm. The heat recovery efficiency is 85 % and the SFP-value of the system is measured to 495 J/m^3 at the design flow rate, lower than 1/4 of the maximum requirement in the Danish building code. This result in a yearly fan power consumption of 4.0 kWh/m^2, included a primary energy factor of 2.5. The results show that the indoor environment is improved to an acceptable level with little energy use, making the concept applicable for implementation in future energy renovation project of school classrooms.

KEYWORDS

Low pressure, diffuse ceiling, mechanical ventilation, energy renovation, performance, school children.

INTRODUCTION

Large parts of suburban Copenhagen was developed and build through the 1960-70. The schools build in those years rely on opening of windows or mechanical ventilation systems that are now outdated to provide fresh air. The recommended maximum CO_2-concentration by the Danish working environment authority is 1000 ppm [1], but a recent study with 750 Danish classrooms showed that 56 % had CO_2-concentrations above that [2]. This shows poor indoor air quality in the classrooms which is due to an insufficient ventilation rate and this adversely affects the school children learning ability and health. Wargocki *et al* (2007) [3] has shown that school children performance can be increased up to 15 % by improving the indoor air quality. Solutions on how to renovate and improve the ventilation systems in existing schools and thereby improve the indoor environment is therefore needed. However, this must be done without excessive energy use, due to the increasing climate concerns. Traditional ventilation system can have specific fan power (SFP) values of 5.5-13 kJ/m^3 while new systems have SFP-values of 2-2.5 kJ/m^3 [4]. The heat recovery in traditional ventilation systems is typically cross flow heat exchangers with efficiencies around 50 % if any heat recovery at all, while new systems have efficiencies of 80-85 %. The energy requirements on buildings in Denmark will be lowered by 75 % in the following years up to 2020 [5]. Current ventilation systems cannot comply with these requirements and new solutions that have minimal fan power consumption and heat loss are therefore needed, although without compromising the indoor environment. This paper describes a pilot mechanical ventilation system that is installed in two classrooms at Vallensbæk primary school, as part of an overall renovation of the ventilation system at the school. The system design is based on a low pressure concept and the objective was to determine the performance of different design solutions and components, and investigate how a conventional oversized air handling unit operates at low pressure. The system has high efficient heat recovery, diffuse ceiling inlet and demand controlled flow rate controlled by a new type of flow control damper. The goal is to keep the indoor CO_2-concentration below the recommended 1000 ppm by the Danish Working Environment Authority and reach a SFP-value of 0.65 kJ/m^3, equalling 1/3 of the current requirement of 2.1 kJ/m^3 in the Danish building code.

CASE

General description of pilot system

The pilot system is installed in class room 42 and 44 in a wing of Vallensbæk primary school, see Figure 1 left. The rooms are presently occupied by the 6th grade (age 11-12) and there are 24 pupils in each class. Room 42 is 80 m^2 and room 44 is 60 m^2, with a floor height of 2.5 m in the main part of the rooms.

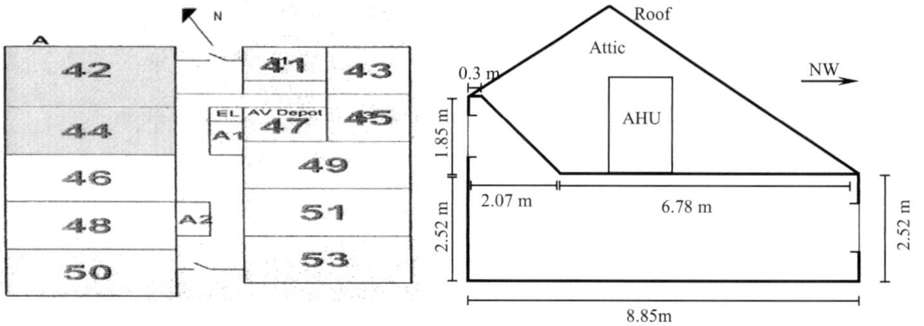

Figure 1. Plan of rooms used for pilot project (left) and section of the rooms (right)

The main facade with windows is oriented North-West. Therefore is the solar gain during the occupied hours limited and only fabric curtains are installed as shading device. The air handling unit (AHU) is placed in the attic above room 42 to reduce duct lengths, see Figure 1 (right). The fresh air to the rooms is supplied through a diffuse ceiling in the horizontal part of the ceiling and the extract is placed in the sloped part of the ceiling.

Indoor environment requirements

The main purpose of the new ventilation system was to improve the indoor air quality, and maintain the CO_2-concentration below 1000 ppm, which is the recommended limit by the Danish working environment authority. The system was only designed to provide fresh air and not cool the air, as the need for cooling is expected to be limited. This is due to the orientation of the windows, limited internal heat loads and varying occupancy during the day. The rooms are used for maximum 1.5 hours at the time before a break of at least 20 minutes, where most of the pupils leave the room, which is providing a break to recondition the room. The room temperature should however comply with the recommended temperature range in EN 15251 category II of 20-24 °C in the winter and 23-26 °C in the summer [6]. The heating demand is to be covered by the heating system, but the supply air is preheated in the heat recovery to decrease ventilation heat loss and avoid draught problems. The design flow rate for the system is set to 5 l/s per person resulting in a flow rate of 450 m^3/h for one room with 24 pupils and one teacher. This corresponds to an air change rate of 2.0 h^{-1} and 2.6 h^{-1} for room 42 and 44, respectively.

Air handling unit

The AHU is a conventional unit with a heat exchanger radial fan and filters. The unit is oversized compared to standard dimensioning to reduce pressure losses in filters and heat exchanger. The unit has a minimum flow of 450 m^3/h in order to function properly according to the producer, this correspond to the design flow rate for one room and will therefore be tested if only one room is occupied. The maximum flow rate is 2500 m^3/h which is well above the expected demand of 900 m^3/h for the two rooms. The fan will therefore operate at low speeds the entire time, minimizing the fan power consumption. The heat recovery is a counter flow exchanger and has a design efficiency of 85 %. Higher efficiency is not expected to provide further energy savings as the ventilation heat loss will be covered by internal gains from occupants, lighting system and equipment [7]. A heating coil has been omitted as studies of diffuse ceiling inlets show that it is possible to supply cold air ($\Delta T>8$ °C) without causing draught problems [8]. The heat recovery will therefore be sufficient to preheat the air making a heating coil dispensable and an easy way to reduce cost. The unit has compact filters on both the supply- and extract side with efficiencies EU7 and EU5, respectively.

Diffuse ceiling inlet

The supply air is blown into the room via a suspended ceiling with porous plates. The ceiling consists of passive and active plates, where only the active plates are permeable. Above the plates is a 15 cm cavity that functions as a pressure chamber making the supply air distribute evenly through the active plates, see Figure 2.

Figure 2. Picture of room 42 with principel airflow pattern of the diffuse ceiling inlet indicated with arrows.

There are eight active plates of 0.6x2.4 m giving a total inlet area of 11.5 m^2 in each room. This large inlet area reduces the inlet air velocity to a minimum, thereby preventing draught in the occupant zone. The low air velocities also allow for lower inlet temperatures, in [8] there was found no lower temperature limit that caused draught. Another advantage of diffuse ceiling inlet is the low pressure loss induced. In Hviid et al (2010) [9] pressure losses of 1.5-2.5 Pa was recorded and for the product used the producer states 2 Pa. The extract diffusers are the existing from the old ventilation system, which are placed on the sloped part of the ceiling. This was done to lower cost as the original diffusers had reasonable size and pressure loss. The placement of the extract diffusers high in the room is also favourable to remove warm and polluted air.

Duct system

Reducing the pressure loss in the duct system is an essential and relatively easy part in reducing the total pressure loss and thereby fan power needed. By increasing the dimension by one size, the pressure loss in straight duct can be reduced by 35 % and larger reductions can be achieved on single losses like bends, enlargements etc. In the design of the duct system a pressure gradient of 0.1 Pa/m was aimed for, resulting in air velocities of 1-2 m/s, at the design flow rate of 450 m^3/h to each room.

Control system

The air flow to the rooms is controlled by a new type of dampers (LeanVent®) see Figure 3, the damper has an aerodynamic design that minimize turbulence generation and thereby pressure and it can operate at low air velocities.

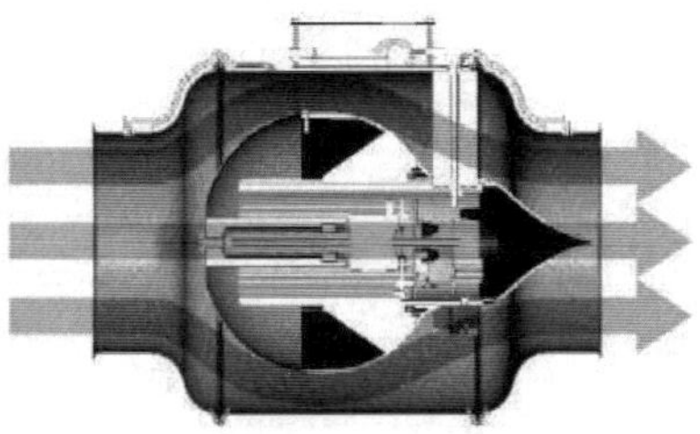

Figure 3. Principle sketch of airflow pattern in LeanVent® dampers (left) and picture a damper installed at Vallensbæk school (right).

The dampers calculate the flow rate by measuring the pressure across the damper and regulate it by adjusting the position of droplet shaped head in the flow direction. The ventilation demand is determined by a CO_2-sensor in each room. The sensor transmits a 0-10 V signal to the inlet dampers corresponding to CO_2-concentrations of 0-2000 ppm. The damper then regulate the flow rate according to the diagram shown in Figure 4.

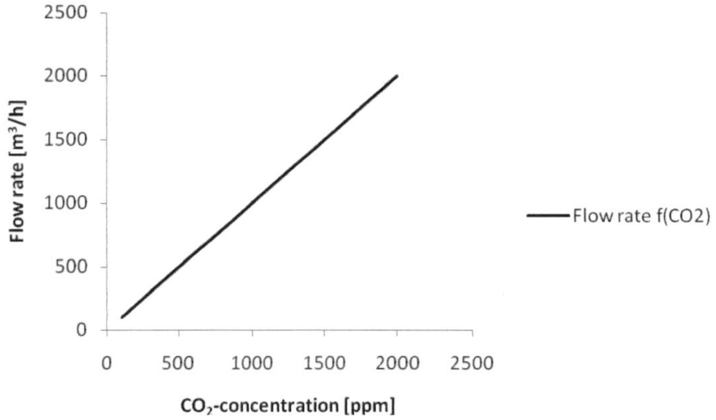

Figure 4. Theoretical ventilation flow rate to one room as a function of the CO_2-concentration.

The CO_2-concentration in the rooms in the morning and after breaks is expected to be close to the outside concentration of 360-400 ppm, and the flow rate will therefore be lower than the design value. However as the CO_2-concentration build up the flow will increase and maintain the concentration below 1000 ppm. The AHU has a time control, and motion sensors are installed in each room. The time control is set to start the system at 7.30 and shut down at 15.30, to avoid unnecessary ventilation outside occupied hours. The use of the classrooms is highly irregular as the pupils have classes at other locations, therefore motion sensors was installed to shut down the flow rate to the room if it was not occupied. This does however not function correctly yet, as motion in one room will start the ventilation in both rooms. The fan is set to maintain a constant static pressure in the duct system that is controlled by a static pressure sensor placed in midway in the supply duct. The optimal set point for static pressure was determined through the setup process of the system and the aim was to lower it as much as possible to reduce fan power.

Measurements and calculations

A series of measurements and calculations was carried out to determine the performance of the new ventilation system and individual components or design solutions. The performance is evaluated based on the indoor environment, energy consumption, pressure loss and general applicability. The following measurements and calculations were carried out.

- Measurements and theoretical calculation of the pressure loss in the entire system and individual components, to determine the pressure loss characteristic of the system.
- Measurements and theoretical calculation of energy consumption to determine SFP-value of the system.
- Measurements of CO_2-concentration, temperature and relative humidity in the classrooms before and after installation.
- Calculations of indoor environment and annual fan power consumption in the simulation tool BSim [10].

RESULTS AND DISCUSSION

Pressure loss characteristics and fan power consumption

In Table 1 is listed the design- and measured pressure loss for the different components in the ventilation system at a flow rate of 900 m^3/h (corresponding to the design flow rate of the two rooms).

Component	Design pressure loss – supply system	Design pressure loss – exhaust system	Measured pressure loss – supply system	Measured pressure loss – exhaust system
Intake/exhaust	5	5	-	-
Heat exchanger	55	55	32	32
Filter	14	3	0	1
Duct system	32	39	12	21
Leanvent damper	20	20	19	23
Diffuse ceiling/ extract diffuser	2	10	1	8
Sum	128	132	64	85

Table 1. Design and measured pressure losses for components in pilot ventilation system at a flow rate of 900 m^3/h.

The total measured pressure loss was lower than the design values, more than 100 Pa for supply and exhaust combined. The difference is mainly due to lower pressure loss in the heat exchanger and duct system. The lower pressure losses in the duct system are due to lower pressure losses in both the fire dampers and silencers than calculated. The Leanvent® dampers are the key to this low operating static pressure as they can regulate the air flow by inducing pressure losses of only 14-30 Pa depending on the flow rate. The static pressure setpoint to control the fan speed was set to 25 Pa, well below the standard values of 150-200 Pa recommended by the AHU manufacturer. In Figure 5 is shown the pressure loss characteristic for system along with the SFP-value characteristic depending on the flow rate.

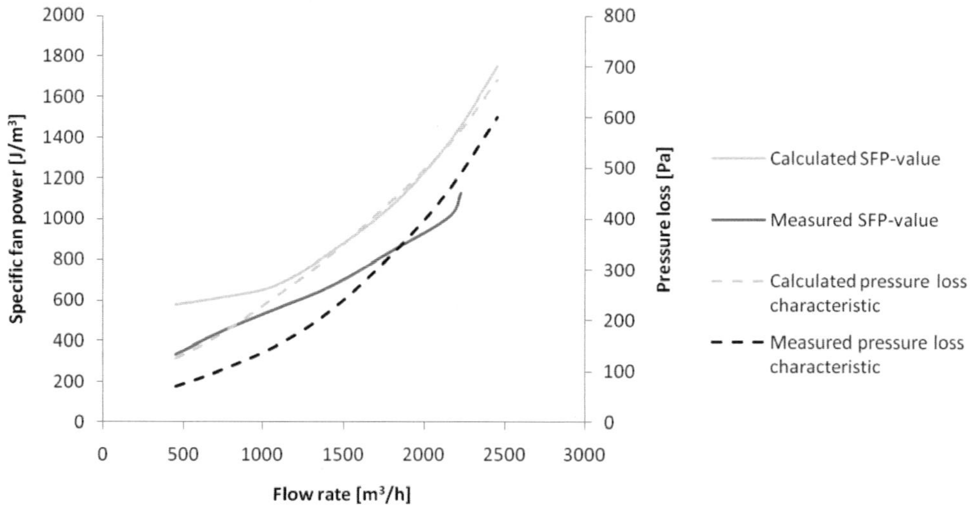

Figure 5. Calculated and measured SFP-values for the new ventilation system along the measured and calculated pressure loss Characteristics of the system.

The measured pressure loss characteristic is 100-150 Pa lower than the calculated. This is due to higher calculated pressure losses in the producers' software tool, compared to the measured values. The difference in pressure loss affects the SFP-values, where the measured value is lower than the calculated value. For the design flow rate of 900 m^3/h, the measured SFP value is 495 J/m^3. The maximum flow rate provided by the AHU during the measurements was 2230 m^3/h, which is 200 m^3/h lower than the producer specification. The yearly fan power consumption was calculated to 219 kWh in the simulation tool, equalling 4.0 kWh/m^2 including a primary energy factor of 2.5. In the simulation model the system was operating eight hours per day, five days a week and the system was turned off during school holidays.

Indoor environment

Figure 6 show the temperature and CO_2-concentration readings for week 6 in january and february before the new ventilation system was installed.

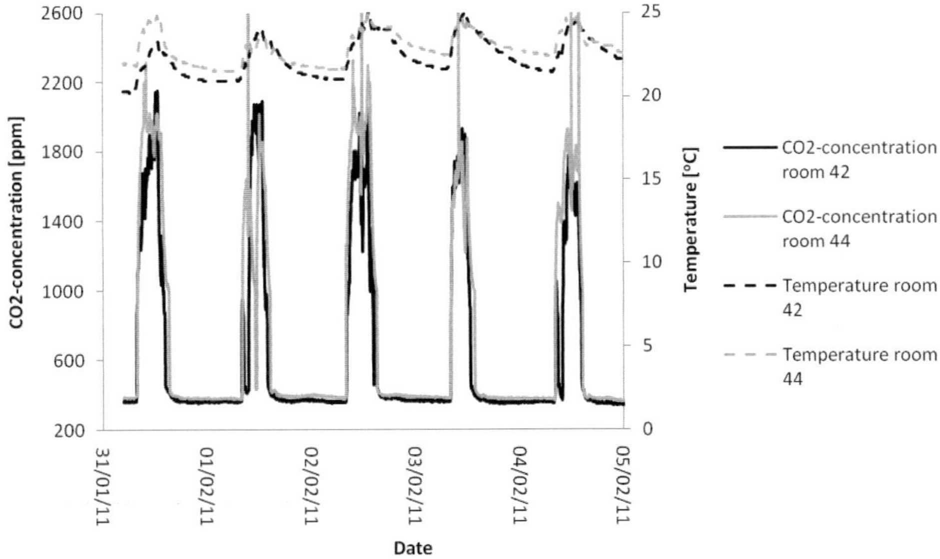

Figure 6. Indoor environment measurements in the two classrooms used in the study, before installation of the new ventilation system.

The figure show that the temperature varied between 21-25 °C in the two rooms, which is 1 °C higher than the design temperature range for the winter period. The maximum temperature will probably be decreased when the ventilation rate is increased with the new ventilation system. Otherwise it should be fixed by optimizing the heating system in the classrooms, to improve comfort and lower energy use. Figure 6 also shows that the CO_2-concentration increased from below 400 ppm at night to around 1800 ppm during occupied hours every day with maksimum values of 2600 ppm in room 44, values well above the recommended value by the Danish working environment authority of 1000 ppm. In Figure 7 is shown the temperature and CO2-concentration readings after installation.

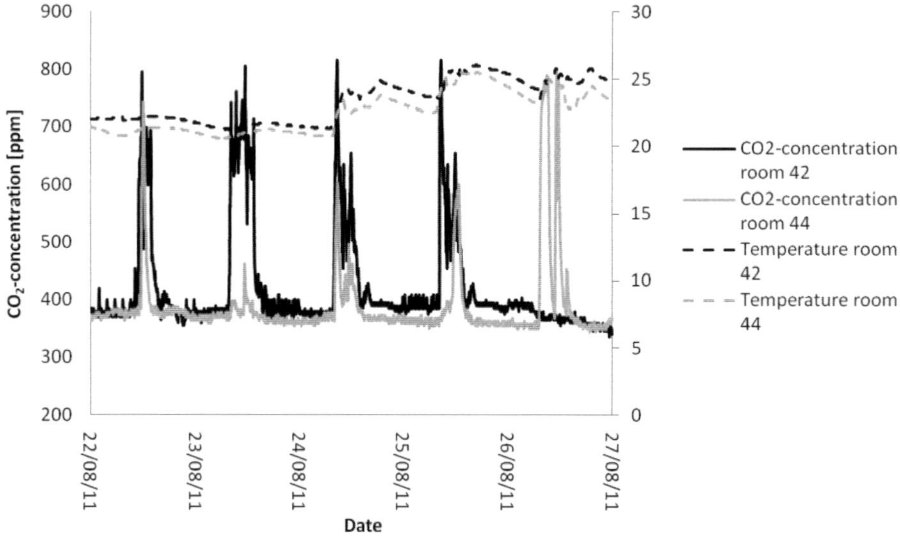

Figure 7. Indoor environment measurements in the two classrooms after installation of the new ventilation system.

The measurements after installation show that the temperature readings varied between 21-26 °C, this is 2 °C lower than the minimum design temperature. This is however assumed to be acceptable as the pupils should be dressed for colder (>20 °C) temperatures in the morning. The temperature increase during the day is maximum 3 °C in the two room, but if needed the pupils can adjust their clothing level during the day. The highest measured CO_2-concentration was around 1000 ppm lower after installation of the new system with peaks around 800 ppm. This is well below the requirement recommended value from the Danish working environment authority. The measured CO_2-concentrations had peaks higher than the results from the simulation model, which has a maximum concentration of 750 ppm, see Figure 8.

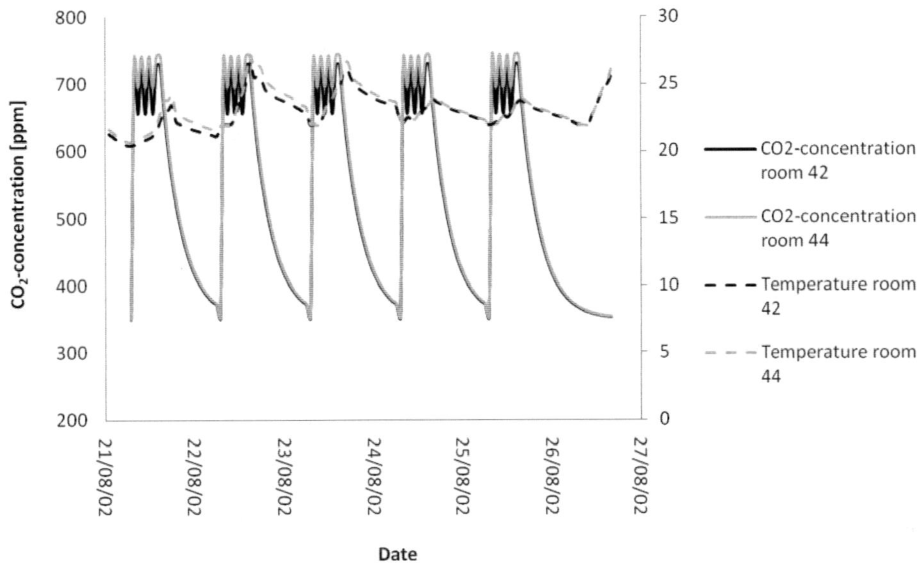

Figure 8. Simulated CO_2-concentrations and air temperatures in the two classrooms

The simulated CO_2-concentrations give the same uniform pattern for each day, that does not correlate with measured results. This is due to the irregular use of the school classrooms that is diffucult to model in a simulation tools occupancy schedule, but overall the results show that the CO_2-concentration fulfills the set requirements.

CONCLUSION

The measurements on the pilot ventilation system showed that it is possible to make school renovation that improve the indoor air quality while lowering the energy consumption for ventilation. The average CO_2-concentration was lowered from around 1800 ppm to 800 ppm, which means that the indoor environment now meet the Danish working environment authority requirements. After the new ventilation system was installed the indoor air temperature has only been measured in a part of the summer period that was somewhat mild. In this period the temperature range fulfils EN 15251 category II, but further measurement are necessary to determine thermal indoor environment under warmer conditions and in the winter as well. If further measurement show a need for cooling, the ventilation rate can be increased during occupied hours and/or cool the rooms by ventilating at night. The pressure losses throughout the system was reduced by increasing the size of the AHU and duct components, which was easy on paper, but did make it more complicated to fit the system into the attic. How the installation and manual work can be carried out should therefore always be concidered. The diffuse ceiling inlet has a pressure loss of around 2 Pa. This is a huge reduction compared to standard ceiling diffusers that induce pressure losses of 30 Pa or more, and help to lower the fan power needed to operate the system. Furthermore has Nielsen *et al* (2009) [8] showed that it is an efficient method to supply the fresh air without causing draught. The new type of damper was able to control the air accurately at air velocities of only 1-2 m/s. Therefore was the pressure loss across the damper only 19-23 Pa at the design flow rate and even lower at smaller flow rates. The SFP-value of the system was measured to 0.5 J/m^3 at the design flow rate, lower than the calculated from the producers' software. The reason is difference in pressure loss and shows the importance of reducing pressure losses to improve energy efficiency. The energy consumption has only been measured for two months but the simulation showed a yearly energy consumption of 4 kWh/m^2 for the system. This makes the concept suitable for use in energy renovations to fulfil the energy requirements for 2020.

Further work

Further measurements, experiments and optimizations on the system are planned to investigate the performance of the components more in depth.

- Trazor gas measurements to determine the ventilation efficiency and mean age of air of the diffuse ceiling inlet.
- Air velocity and temperature measurements in a room to investigate local discomfort by calculating the draught rate.
- Performance test on the pupils under different ventilation flow rates to investigate how improved air quality affect their performance and well being.
- Measurements have only been carried out in the summer, but it is necessary to investigate how the system operates without a heating coil – does the heatexchanger freeze, is the inlet temperature to low and cause discomfort due to low temperature or draught.
- Implementation of a static pressure setpoint reset (SPSR) control agorithm, that ensures one damper is fully open and thereby reduce fan power needed. The energy savings will in this case not be great as the fixed static pressure already is below 25 Pa, it will however be interesting to investigate whether a SPSR control algorithm can operate at such low pressure.

ACKNOWLEDGEMENTS

Thanks to the partners involved in the project; Vallensbæk municipality and Vallenbæk school for making it possible to realize this project, Exhausto A/S, LeanVent A/S and LL ventilation A/S for good co-operation in the design and installation process and lastly Plan C for facilitating the project. Plan C is a project under Gate 21, which is funded by European regional development fund.

REFERENCES

[1] Danish Working Environment Authority. 2011.

[2] Menå, H.R., and Larsen, E. 2010. *Målinger peger på dårligt indeklima I danske skoler,* HVAC magasinet, 10, 78.

[3] Wargocki, P., and Wyon, D.P. 2007. *The effects of outdoor air supply rate and supply air filter condition in classrooms on the performance of schoolwork by children*, HVAC&R Research. 13(2), 165-191.

[4] Blomsterberg, Å., Delmotte, C.H., Barles, P., Pennycock, K., de Gidds, W., Hardegger, P., and Maldonado, E. 2000. *Design methods for ventilation systems in residential and commercial buildings*, EC JOULE-TIP-Vent project.

[5] Danish Enterprise and Construction Authority. *Danish building code*. 2010.

[6] EN 15251. 2007. *Indoor environmental input parameters for design and assessment of energy performance of buildings addressingindoor air quality, thermal environment, lighting and acoustics, European Standard*, European Committee for Standardization.

[7] Jacobs, P., and Knoll, B. 2009. *Diffuse ceiling ventilation for fresh classrooms*, Proceedings of the 30th AIVC Conference.
[8] Nielsen, P.V., and Jakubowska, E. 2009. *The performance of diffuse ceiling inlet and other room air distribution systems,* Proceedings of cold climate HVAC, Sisimiut.
[9] Hviid, C.A., Svendsen, S., 2010. *Experimental and numerical analysis of perforated suspended ceilings as diffuse ventilation air inlets*, Ph.D. theses, Depertament of civil engineering, Technical University of Denmark.
[10] Aggerholm, S., and Grau, K. 2005. *Buildings energy demand – PC-program and calculation guideline*. Danish Building Research institute. SBI-guideline 213.

DEFINITION OF OCCUPANT BEHAVIOUR PATTERNS WITH RESPECT TO VENTILATION BY MEANS OF MULTIVARIATE STATISTICAL TECHNIQUES

Felipe Encinas Pino*[1], Francisco José Sanchez de la Flor[2] and André De Herde[1]

1 Architecture et Climat
Université catholique de Louvain
Place de Levant, 1
1348 Louvain-la-Neuve, Belgium
**Corresponding author: f.encinaspino@uclouvain.be*

2 Escuela Superior de Ingenieros
Universidad de Cádiz
C/ Chile, 1
11002 Cádiz, Spain

ABSTRACT

It has been demonstrated that there is a strong relationship between occupant behaviour and the thermal performance of dwellings. At the same time, some aspects of this behaviour, especially with respect to natural ventilation, constitute some of the most important sources of uncertainty in the field of building energy simulation. A survey about perception of thermal comfort and occupant behaviour was carried out in Santiago de Chile during December 2009 and January 2010 in a pilot case study corresponding to an apartment building. This paper proposes a methodology based on the systematic application of multivariate statistical techniques which were applied to the collected data of the survey. The results of the analyses show that daytime ventilation is not strongly correlated to the perception of thermal comfort, probably because it is mainly oriented to hygienic purposes. On the contrary, nigh ventilation appears as a very significant predictor for the same dependent variable. The final objective of these models corresponds to the definition of occupant behaviour profiles which can be used as hard data to make calculations of thermal behaviour of dwellings more accurate and reliable. At the light of the results of the energy building simulations, night ventilation presents a high potential as passive cooling technique, considering also the climatic conditions of Santiago de Chile.

KEYWORDS

Summer thermal comfort, occupant behaviour patterns, energy building simulation

1. INTRODUCTION

It has been demonstrated that there is a strong relationship between occupant behaviour and thermal performance of dwellings. Indeed, according to Macdonald et al. (1999), some variables related to occupant behaviour constitute some of the main sources of uncertainty in the field of energy building simulation [1]. In that sense, depending on the variability of aspects such as scheduled internal gains or natural ventilation (by means of manually operable windows), a wide range of variation in the energy consumption of dwellings may be expected. This also can be observed through various studies, where large differences in the thermal behaviour of similar buildings have been observed, which suggests that occupant behaviour exerts a strong influence [2].

Uncertainty and sensibility analyses frequently deal with this situation, since they can generate a great range of forecast values based on the distribution of the input variables. For example, in the case of the physical properties of building materials, this variability has been studied and may be obtained from references as Clarke et al. (1999) or Lomas & Bowman (1987) [3] [4]. However, Hyun et al. (2008) explain that the widely varying occupant influences - especially related to operable windows - have not been directly measured or investigated [5].

At the same time, most of the building energy simulation programs are deterministic, rather than probabilistic and consequently their results frequently are not expressed in probabilistic terms. Additionally, a considerable difference between the standard values of ventilation used for simulations (based on average occupant behaviour) and the ventilation patterns in real occupied dwellings may be expected. Therefore, if the aim is to represent a wide range of cases based on a more real approach (instead of a singular and/or standard case study), it is necessary to characterize the occupant behaviour in terms of patterns to be used as input data in energy building simulations.

Due to the link between occupant behaviour and energy consumption, it is important to define it from the interaction with the control mechanisms of windows during both day and night, and also establishing the reasons for that specific behaviour, as is recommended by IEA (1988) [6]. Andersen et al. (2009) indicate that most of the energy building simulation programs provide possibilities of regulation of control systems (such as opening/closing windows), but there are no guidelines for how the simulated environment should be managed by the software [7]. Similarly, Yu et al. (2011) explain that it is difficult to completely identify the influences of occupant behaviour through simulation due to users' behaviour diversity and complexity; current simulation tools can only imitate patterns in a rigid way [8]. Accordingly, the definition of a set of behaviour patterns – based on the quantification of real inhabitants' behaviour – would significantly improve the validity of the outcomes of the simulations.

Consequently, this study tries to establish occupant behaviour patterns for apartments from the real estate market of Santiago de Chile based on field data and also determining their impact on the summer thermal comfort. According to this, four main research questions were formulated:

(1) Do patterns of occupant behaviour exist with respect to perception of thermal comfort and ventilation (by means of windows opening)?
(2) Do different occupant behaviour patterns lead to significant differences in the thermal comfort of the apartments in summer?
(3) Which are the most important physical variables (e.g. wind conditions, difference between indoor and outdoor temperatures) in terms of their influence with respect to the ventilation rate in each one of the occupant behaviour patterns?

2. METHODOLOGY

2.1 Research data

Due to the importance of the occupant behaviour and ventilation on the thermal behaviour of apartments it is necessary to collect data about these aspects based on real sources. Nonetheless, due to the lack of references in the national state of art of Chile, a survey to obtain this information is required.

The pilot case study corresponds to the *Edificio Don José*, located in the Santiago borough, city of Santiago. This is an apartment building, constructed in 1993-1994, with 22 floors and 8 apartments per floor. The building is situated in an urban environment, near to the city centre. The survey was applied to 91 randomly selected apartments in two summer months (December 2009 and January 2010). The sample size corresponds, consequently, to 91 cases over a population of 166 apartments. The margin of error and the confidence level are 6% and 90%, respectively. It is important to remember that the scope of the survey is related to the indoor environment and occupant behaviour in apartments of Santiago de Chile based on a pilot case study. Due to this, the survey frame was considered as appropriate. A most ambitious experience may be proposed as further research. In that case, the population of the survey can be extended to various apartment buildings in Santiago de Chile.

2.2 Approach

The seminal articles of van Raaij & Verhallen – published in the 1980s – distinguish two different occupant behaviours as determinant of energy house in the home. According to them, purchase and maintenance–related and usage–related energy behaviour can be identified [9].

In this study, occupant behaviour was defined in the sense of the usage–related behaviour involved the day–to–day energy–conscious behaviour of use of ventilation (by means of windows opening) and passive/active strategies for thermal conditioning (such as external solar protection to avoid overheating or the use of heating systems in winter). It is important to notice that in most households, energy behaviour generally does not constitute a separate type of behaviour but is contingent on other behaviour associated with, for example, household work, childcare, in–home entertainment and sleeping [9]. This situation is highly consistent with the findings of this study, especially with respect to daytime ventilation, as it will be explained latterly.

Fig. 1 presents the methodology of this study, which is based on the systematic application of multivariate statistical techniques to the collected data of the survey. As a first step, the definition of behaviour variables was carried out (Fig. 1a) with the aim of obtaining factors of behaviour by means of a factor analysis (Fig. 1b). Principal component analysis – a kind of factor analysis according to the procedure to extract factors – can be used for exploratory purposes, in order to know the relations structure on a specific set of variables.

Afterwards, a discrete choice model was implemented to determine the level of incidence of the factors of behaviour on the overall predicted thermal comfort (Fig. 1c). If these independent variables have enough explanatory capacity, they will be able to predict the behaviour of the dependent variable (perception of thermal comfort) in the analyzed apartments.

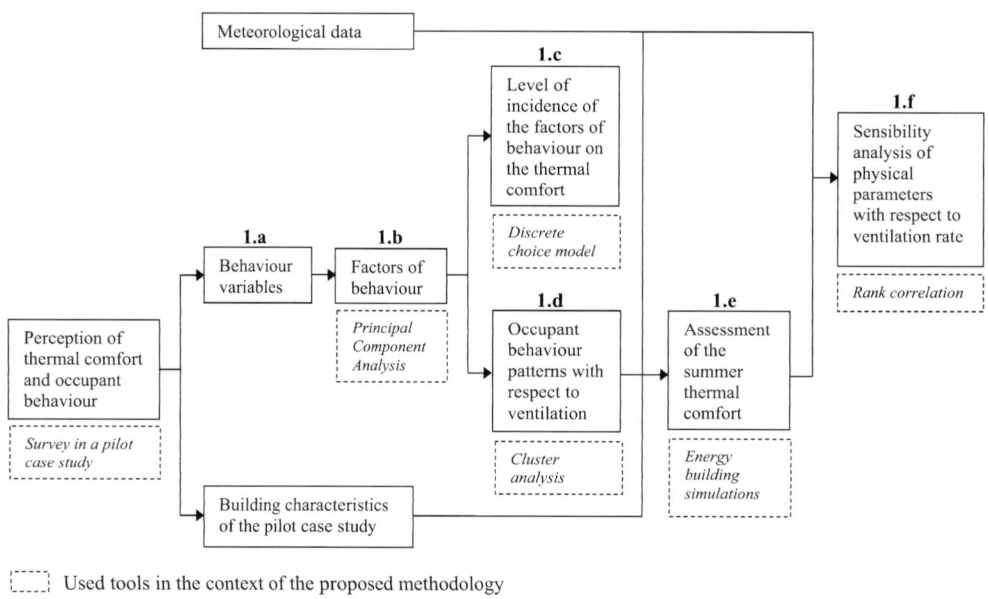

Figure 1. Methodology

In the fourth step the occupant behaviour patterns with respect to natural ventilation were established by means of a cluster analysis (Fig. 1d). Indeed, four different profiles were identified (P1, P2, P3 and P4) which were defined as *cool*, *conscious–moderate*, *comfort* and *conscious–warm* respectively. With this information, energy building simulations were performed using the TAS software [10], with the aim of assessing the summer thermal comfort in the studied cases (Fig. 1e).

Finally, by means of a sensibility analysis, the influence of physical variables (e.g. wind conditions, difference between indoor and outdoor temperatures) with regard to the obtained ventilation rate in each one of the occupant behaviour patterns was established (Fig. 1f).

3. RESULTS

3.1 Definition of factors of behaviour by means of a Principal Component Analysis

Principal Component Analysis was used to identify factor underlying occupant behaviour, since this technique allows the detection of subjacent dimensions that, in this context, can be understood as drivers of behaviour.

Table 1 presents the seven variables that were considered to carry out this analysis. These questions were selected in order to represent the different aspects related to the perception of thermal comfort, natural ventilation and strategies and systems that affect the thermal behaviour of apartments. Additionally, the obtained ratio of 12.5 between the number of observations (N=88) and the number of variables can be considered as appropriate [11].

Table 2 shows the rotated component loadings, which give information about the strength of the relationships between the variables and the obtained components. These loadings are expressed in terms of correlation coefficients (with values between 0 and 1). According to the Kaiser's criterion (Eigen values >1), four components were extracted, which account for the 73.2% of variance.

Subject	**Survey questions**	**Behaviour variables**
Perception of thermal comfort	Q6	Thermal sensation in winter*
	Q10	Thermal sensation in summer*
Ventilation	Q16	Daytime ventilation in winter
	Q20	Daytime ventilation in summer
	Q24	Use of night ventilation in summer
Strategies and systems	Q13	Presence of external solar protection
	Q27	Use of heating systems in winter

* To define the categories of these questions, the 7 point subjective scale of EN15251 [12] was considered

Table 1. Definition of behaviour variables for the Principal Component Analysis

Behaviour variables	**Components**				**Communalities**
	C1	C2	C3	C4	
Q6		0.85			0.76
Q10			0.65		0.52
Q13				0.90	0.82
Q16	0.76				0.75
Q20	0.86				0.81
Q24			0.84		0.74
Q27		-0.71			0.72

Rotation method: Varimax with Kaiser normalization

Note: component loadings <0.6 are suppressed according to the sample criterion established by Hair et al. [11]

Table 2. Component loadings and communalities based on the Principal Component Analysis (N=88)

In consequence, the four defined components of the rotated matrix with their proportions of explained variance are:

C1: *Daytime ventilation*, both in winter and summer [21.3%]

C2: *Perception of favourable thermal comfort in winter and avoiding the use of heating appliances*. Both situations can be related to a good thermal behaviour since occupants declare they generally do not feel cold during winter and at the same time they minimize the use of heating [18.7%]

C3: *Perception of unfavourable thermal comfort in summer and use of night ventilation*. This situation can be represented for occupants that describe their apartments in summer as "warm" or "hot" and consequently, open windows during night time [17.4%]

C4: *Presence of solar protection* [15.8%]

Fig. 2 presents perceptual maps per orientation, based on the factor scores obtained by means of the Principal Component Analysis. In this case, perceptual maps are the graphical expression of the associations between two components that compose the solution and where their observations are clustered by a specific criterion.

Fig. 2 (left) shows the perceptual map of C3 vs. C4 (horizontal and vertical axis, respectively). According to this, both north and south orientations were characterised as comfortable in summer and they do not present solar protection, which can be considered as expected. Indeed, in Santiago de Chile (33°28'S; 70°78'W), direct solar radiation can be easily controlled in the north orientation by means of an overhang (which can be provided, for example, by an upper balcony) and there is no need of any kind of solar protection for southern facades. On the contrary, western orientations (W, NW and SW) may present and unfavourable thermal behaviour in summer, since they were characterized as "not comfortable". At the same time – and probably due to this – these orientations concentrate the presence of solar protection, which can also be considered as expected.

Fig. 2 (right) presents the perceptual maps of C3 vs. C1, which means that the perception of summer thermal comfort is compared to daytime ventilation. However, it is not possible to identify a correlation pattern between them. For example, there is not a direct relationship between orientations that declare comfortable indoor conditions in summer and a specific behaviour with respect to ventilation (ventilate or not ventilate). This situation can be understood from the idea that daytime ventilation of C1 may be mainly oriented to a hygienic purpose (instead of cooling). However, to test this hypothesis is necessary other kind of multivariate analysis technique which be able to predict the perception of thermal comfort.

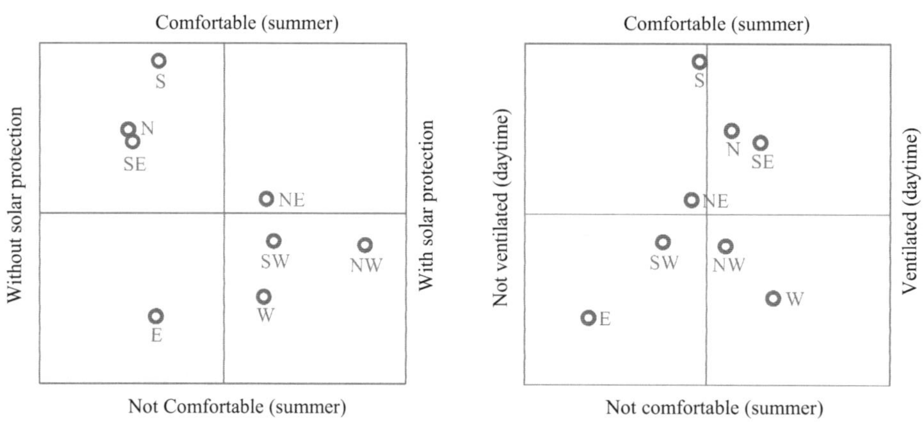

Figure 2. Perceptual maps of C3 vs. C4 (left) and C3 vs. C1 (right) per orientation

3.2 Incidence of the factors of behaviour on the overall thermal comfort by means of a discrete choice model

The logistic regression analysis is a mathematical model with the aim of predicting the behaviour of a dependent variable as function of one or more independent variables. The objective of this model is to predict the probability of occurrence of an event with a dependent variable that assumes the value of 1 when the event occurs and zero in the absence of the event. The prediction is made from a group of independent variables with explanatory capability with respect to the dependent variable

With the aim of predicting the level of incidence of the variables that determines the perception of thermal comfort, one question of the survey (Q41) was expressed as: "Do you feel comfortable in your apartment in terms of thermal comfort?" The two possible answers were "yes" or "no", which convert this variable in dichotomical. Therefore, the probability of occurrence of the answer "yes" ($Y=1$) in the question 41, can be expressed through the following logistic regression:

$$\boldsymbol{P}_{(Q41=\frac{1}{C_{i,i=1...4}})} = \frac{1}{1+e^{-(a+b_1C_1+b_2C_2+b_3C_3+b_4C_4)}} \quad (1)$$

where C_1, C_2, C_3 and C_4 are the factor scores that were obtained through the Principal Component Analysis, b_1, b_2, b_3 and b_4 are the coefficients for these variables, a is a coefficient of the model and e is the base of natural logarithms.

Table 3 presents the obtained coefficients for the logistic regression model proposed for the Q41 of the survey. According to the obtained solution, b_1 is a not significant coefficient. This means that component C_1 is not significant to predict the probability of occurrence of Q41. This situation can be understood from the previously mentioned hypothesis that daytime ventilation is mainly oriented to a hygienic purpose, instead of cooling.

Another important aspect regarding the obtained coefficients of Table 3 is the sign of b_3. As can be observed, this is negative, which means that while the value of C_3 is higher, the probability that Q41 can be answered as "yes" is lower. In other words, if the thermal sensation of the occupants during summer is hot, there are more possibilities that the people feel uncomfortable in their apartments from a wider sense.

Considering the obtained coefficients, the logistic regression applied can be expressed as:

$$\boldsymbol{P}_{(Q41)} = \frac{1}{1+e^{-(1.97+0.30\,C_1+0.62\,C_2-0.90\,C_3+0.97\,C_4)}} \quad (2)$$

In the equation (2), it can be observed that the most important aspects related to the overall perception of thermal comfort in the apartment (Q41) are solar protection (C_4) and perception of summer thermal comfort and night ventilation (C_3). These weightings can be considered as expected since the survey was taken during summer. Indeed, it is not surprising that with the presence of solar protection, a not excessive indoor temperature and night ventilation most of the people considered their apartment as comfortable.

	Coefficients	Standard error	Z value	p-value
a	1.97	0.48	4.09	4.38E-07***
b_1	0.30	0.30	1.01	0.313
b_2	0.62	0.35	1.77	0.076*
b_3	-0.90	0.45	-2.00	0.046**
b_4	0.97	0.49	1.99	0.046**

Significant variables for: *** $p<0.01$; ** $p<0.05$; * $p<0.1$

Goodness-of-fit: The number of cases correctly predicted was 76 (86.4%). Also, the McFadden's R^2 was 0.293 above the minimum value of 0.28 (for a proportion of 80/20 in the answers of the dependent variable) [13]

Table 3. Obtained coefficients from the multivariate logistic regression

3.3 Definition of occupant behaviour patterns by means of a cluster analysis

The seminal study carried out by Punj & Stewart (1983) established that cluster analysis is a statistical method for classification. According tho them, the essence of classification is that certain things of as related in a certain way. Indeed, the objective of this can be defined as the identification of a group of entities that share certain common characteristics [14]. According to Vivanco (1999), the use par excellence of cluster analysis is the generation of typologies. A typology is a group of cases that present a strong similarity [15]. In this context, cluster analysis appears as an appropriate tool to identify occupant behaviour patterns from the collected data of the survey. Van Raaij & Verhallen (1983) attempt to apply clustering procedures to both original variables and factor scores obtained by means of a Principal Component Analysis [9]. However, natural groupings were not found probably due to the inclusion of numerous types of occupant behaviour, where energy behaviour represents in practice just one aspect of them.

For the purposes of this research, the hierarchical technique was chosen as clustering method. This classifies by stages through a process that follows the structure of a tree and where each stage of the process generates a new branch. In this context, the selection of the factor scores of the Principal Component Analysis as variables for the procedure is justified since it allows correcting the interdependencies. Also, the non-equivalence of metrics between the original variables suggests the use of this procedure.

In the range of solutions proposed for the model, the alternative of 4 clusters was selected as the most representative, since their groups are consistent and well defined. Therefore, four profiles (P1, P2, P3 and P4) were characterized as function of all the originally considered behaviour variables. Fig. 3 presents the quantitative characterization of these behaviour profiles in terms of the number of hours of ventilation per day (excluding the period between 24:00 and 08:00) in summer. As can be observed, P2, P3 and P4 present a similar behaviour (in terms of the shape of their curves) with respect to the different times of the day, excepting for the period between 21:00-24:00. On the contrary, P1 presents a very particular behaviour, since the highest level of ventilation occurs at noon (the hottest period time of a summer day in Santiago). This situation is consistent with the qualitative characterization of the behaviour profiles in terms of thermal sensation in summer vs. summer daytime and night ventilation that is presented in the Fig. 4. According to this, the four profiles were described as *warm* (P1), *conscious-moderate* (P2), *comfort* (P3) and *conscious-cool* (P4).

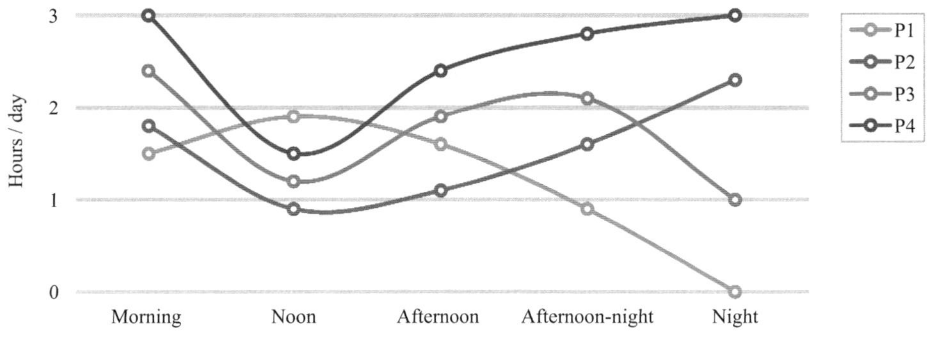

Note: 3-hours periods are defined as morning (9:00-12:00), noon (12:00-15:00), afternoon (15:00-18:00), afternoon-night (18:00-21:00) and night (21:00-24:00)

Figure 3. Quantitative characterization of the behaviour profiles in terms of number of ventilation hours (by means of manually operable windows) per day in summer (mean values)

P1 can be identified as *warm* due to the thermal sensation in summer is perceived as "neutral" and also they do not use night ventilation (probably they think that do not need to ventilate during nigh-time). This is also consistent with the reasons for opening windows in summer that they declare since a high percentage of cases are associated with hygienic purposes (and not for cooling) (Fig. 5). On the contrary, P4 is related to the concept of *conscious-cool* because its category for thermal sensation is "hot" and both ventilation regimes (daytime and nigh-time) are identified as high. This situation suggests that night ventilation may be motivated for cooling reasons, which can also be observed from the reasons for ventilating of Fig. 5. Similarly, P2 was identified as *conscious-moderate* because the thermal sensation in summer is perceived also as "hot", but the level of night ventilation is lower than in the case of *conscious-cool* (P4). However, its level of summer daytime ventilation is low, which represents a key aspect concerning the summer thermal comfort that this behaviour profile can latterly obtain. Finally, in an intermediate level, P3 was defined as *comfort* since to the thermal sensation in summer is perceived as "warm" (not too hot) and probably due to this, its level of night ventilation is medium.

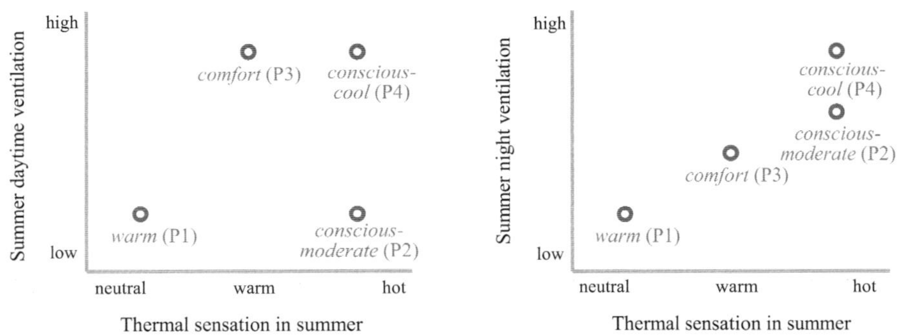

Figure 4. Qualitative characterization of the behaviour profiles in terms of thermal sensation in summer vs. summer daytime ventilation (left) and summer night ventilation (right) (median values)

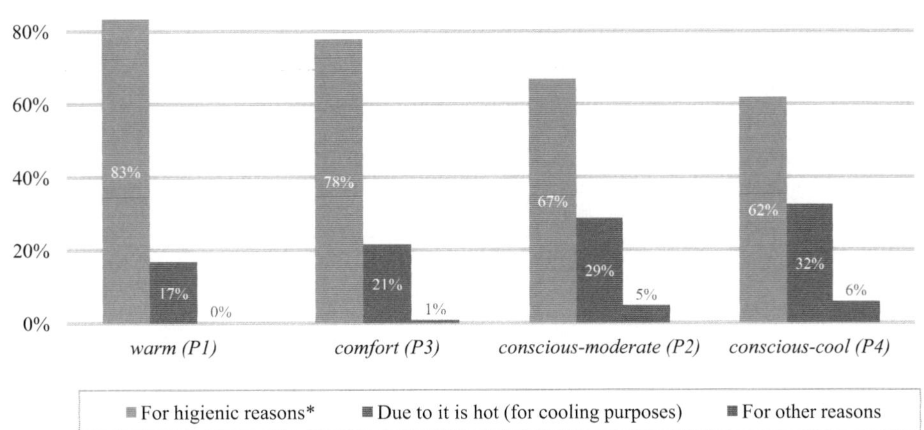

Note (*): Corresponding to the answers of the survey: "for entering fresh air" and "while the rooms are cleaned"

Figure 5. Column chart of results for the question Q22 (percent with regard to the total number of mentions): "Which are the reasons for opening windows in your apartment in summer?" by behaviour profiles

3.4 Assessment of the summer thermal comfort applying the behaviour profiles

After the definition of occupant behaviour patterns in terms of profiles, energy building simulations were carried out. These numeric simulations were performed using TAS software [10], applying the obtained behaviour profiles as input data (in terms of window opening regimes). The four profiles were applied to a floor layout of the pilot case study, using also the collected data of the survey to define internal gains of the different apartments (between 105 and 115 Wh/m²/day, including occupation, lighting and equipments). Hourly meteorological data for the year 1989 in Santiago de Chile were taken from ASHRAE (2001) [16], which were also compared and validated with respect to the monthly values of the NCh 1079 national standard (based on a period of 30 years of meteorological observations) [17] .

The aim of these numeric simulations is to find a relationship between occupant behaviour patterns, ventilation rates and thermal behaviour. The proposed profiles, at the moment, just represent an intention of ventilation (since they are defined as function of windows opening), but they need to be characterized in terms of their impact on the thermal comfort of the apartments. Fig. 6 presents the number of summer overheating degree hours according to the adaptive model of EN 15251 [12] for *warm* (P1) and *conscious-cool* (P4) profiles per orientation. These results are highly consistent with regard to the perception of summer thermal comfort per orientation in the perceptual maps of the Fig. 2.

According to this graph, it is possible to observe the favourable impact of the night ventilation strategy for reducing overheating. However, this thermal behaviour (when night cooling is applied) probably is also influenced by windows operation during daytime, as can be observed by the simulation outputs (in terms of overheating degree hours) of the different occupant behaviour patterns. Anyway, these results can also be explained considering the climatic characteristics of Santiago (because of the difference between outdoor and indoor temperatures that it is possible to reach during night-time). It is important to notice that the recommended climate conditions for the application of night cooling according to the literature (e.g. a range between 30-36°C as maximum allowable daytime temperatures and 8°C as minimum diurnal temperature swings [18] [19]) can be perfectly applied to the Mediterranean climate of Santiago de Chile as it has been observed in previous studies.

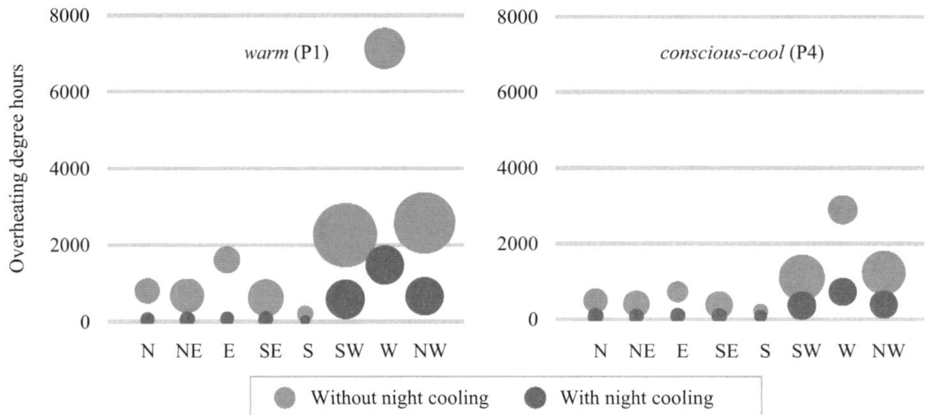

Note: The centre point of each bubble is the extent of overeating measured in degree hours (mean value for the different spaces). The area of the bubble represents the standard deviation for the distribution of values including the same rooms.

Figure 6. Bubble plots for overheating degree hours in summer according to the adaptive comfort model of the EN15251 [12] for *warm* (left) and *conscious-cool* (right) profiles per orientation

4. DISCUSSION

4.1 Sensibility analysis of physical parameters with respect to ventilation rate

The previously observed behaviour may respond that the idea that different ventilation regimes (in terms of windows opening) generate different air change rates, which also may be influenced by external variables such as wind conditions. This situation suggests that windows operation may be correlated with ventilation rates, which could be determined by means of a sensibility analysis. The sensibility analysis assesses the contribution of the inputs parameters (in this case corresponding to physical variables) to the total uncertainty in analysis outcomes (here defined as the ventilation rate).

Fig. 7 shows the rank correlation coefficients of different physical parameters with respect to the air change rate (in terms of h^{-1}) for the different occupant behaviour profiles in the main bedroom per orientation. All the parameters (windows operation, difference between indoor and outdoor temperatures, wind direction and wind speed) appear as significant variables with respect to the ventilation rate ($p<0.05$), which modify their relative level of importance as function of the higher exposition to opened windows, as can be observed through the comparison between the different profiles. If we consider also the notorious difference between the overheating degree hours in summer by profiles (as it was presented in Fig. 6), the time at when windows are opened/closed represents a key aspect with respect to both air change rate and summer thermal comfort. Additionally, wind direction and wind speeds show important differences in each profile with regard to orientation, presenting considerable higher correlation coefficients in the wind–oriented facades (S, SW, W and NW).

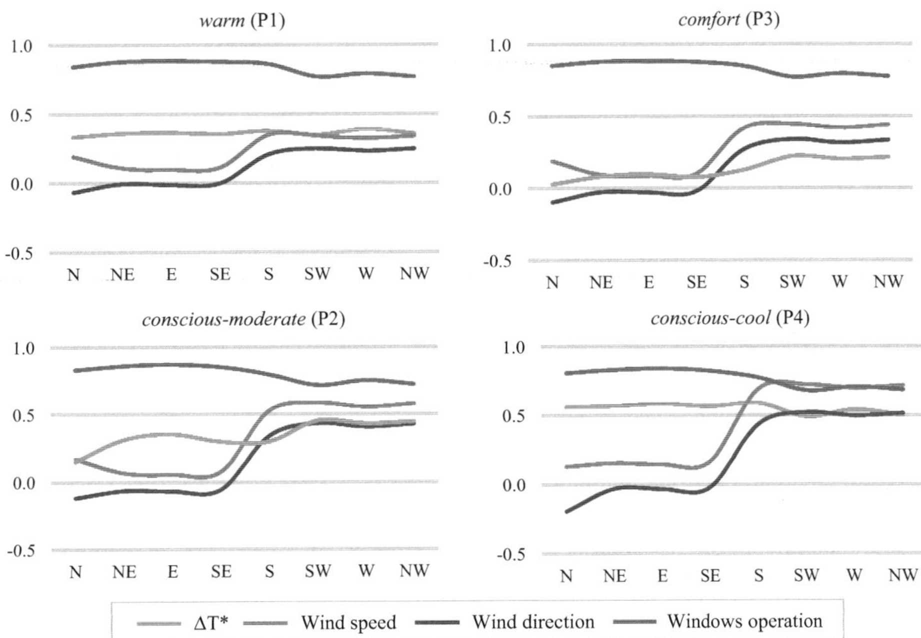

Note (*): ΔT = Difference between indoor and outdoor temperatures [°C]

Figure 7. Rank correlation coefficients for different physical parameters for *warm* (left) and *conscious-cool* (right) profiles with respect to air change rate [h^{-1}] in the main bedroom per orientation

5. CONCLUSIONS

The explanatory analysis carried out through the Principal Component Analysis and the multivariate logistic regression established the relative importance of the different variables that determine the perception of thermal comfort of an apartment in Santiago de Chile. Through these multivariate statistical techniques, the role of ventilation in the thermal sensation of the occupants was identified, associating daytime ventilation with hygienic purposes, while night ventilation appeared directly related to passive cooling.

These observations are useful to understand the perception of occupants about the different aspects related to the thermal comfort of apartments in Santiago de Chile, especially during summer. However, if one of the declared objectives of the survey is to provide information for energy building simulations based on a more real approach, it is absolutely necessary to generate hard data from the collected information. As it was explained, the definition of ventilation regimes is one of the main sources of uncertainly in an energy building simulation, mainly due to its dependence on the inhabitant's behaviour (by means of windows opening). Nonetheless, if these ventilation regimes are defined based on hard data collected directly from the inhabitants, the results of the simulation are more representative and reliable and in consequence, uncertainty decreases. This process was carried out by means of a cluster analysis, obtaining four different behaviour profiles (P1, P2, P3 and P4). These occupant behaviour patterns were characterized as function of all the originally considered variables. According to this, the four profiles were identified as *warm*, *conscious-moderate*, *comfort* and *conscious-cool*, respectively. At the same time, it was observed as these different occupant behaviour patterns lead to significant differences with respect to the summer overheating degree hours, obtained by means of energy building simulations.

The sensibility analysis carried out with the obtained results of these numeric simulations indicate that the difference between outdoor and indoor temperatures, wind direction and wind speed appear as sensitive physical variables with respect to the ventilation rate. Also, these variables increase their relative level of importance regarding ventilation profiles as function of the higher exposition to opened windows. In that sense, the time at when windows are opened / closed represents a key aspect with respect to both air change rate and summer thermal comfort.

At the light of these results, night ventilation presents a high potential as passive cooling technique, considering also the climatic conditions of Santiago (because of the difference between outdoor and indoor temperatures that it is possible to reach during night-time). However, this thermal behaviour (when night cooling is applied) probably is also influenced by windows operation during daytime, as can be observed by the simulation outputs (in terms of overheating degree hours) of the different occupant behaviour patterns.

6. ACKNOWLEDGEMENTS

This study was carried out as a part of a PHD Thesis at the *Architecture et Climat* research centre from the *Université catholique de Louvain* in Belgium. The study is funded by the *Bourse de coopération au développement* from the same university.

The author would like to thank to *Pilar San Martin Vila* and *Carlos Aguirre Núñez* for their support for the development of the questionnaire survey and the statistical models, respectively.

7. REFERENCES

[1] Macdonald, IA, Clarke, JA & Strachan, PA 1999, 'Assessing uncertainties in building simulation', Proceedings of Building Simulation 1999, Paper B-21, Kyoto, Japan

[2] Guerra, O 2011, 'Behavioural Patterns and User Profiles related to energy consumption for heating', Energy and Buildings, vol. 43, pp.2662-2672

[3] Clarke, J, Yaneske, P & Pinney, A 1999, 'The harmonisation of thermal properties of building materials', Report CR59/90 of the Building Research Establishment, Watford, UK

[4] Lomas, K, & Bowman, N 1987, 'Developing and testing tools for empirical validation', Chapter 14, Vol. IV of SERC/BRE final report, An investigation in analytical and empirical validation techniques for dynamic thermal models of buildings, Building Research Establishment, Watford, UK

[5] Hyun, S, Park, C & Augenbroe, G 2008, 'Analysis of uncertainty in natural ventilation predictions of high-rise apartment buildings', Building Services Engineering Research and Technology, vol. 29, no. 4, pp. 311-326

[6] IEA 1988, Inhabitant Behaviour with Respect to Ventilation - a Summary Report of IEA Annex VIII, International Energy Agency, Energy Conservation in Buildings and Community Systems Programme, AIVC - Air Infiltration and Ventilation Centre

[7] Andersen, RV, Toftum, J, Andersen, KK, & Olesen, BW 2009, 'Survey of occupant behaviour and control of indoor environment in Danish dwellings', Energy and Buildings, vol. 41, pp.11-16.

[8] Yu, Z, Fung, B, Haghighat, F, Yoshino, H, Morofsky, E 2011, 'A systematic procedure to study the influence of occupant behaviour on building energy consumption', Energy and Buildings, vol. 43, pp.1409-1417

[9] van Raaij, WF, Verhallen, TMM 1983, 'Patterns of Residential Energy Behaviour', Journal of Economic Psychology, vol. 4, pp.85-106

[10] EDSL 2010, EDSL TAS, http://www.edsl.net

[11] Hair, J, Anderson, R, Tatham, R, & Black, W 2005, Análisis Multivariante (Quinta edición), Pearson Educación, S.A., Madrid

[12] CEN 2007, EN15251:2007 Indoor environmental input parameters for design and assessment of energy performance of buildings addressing indoor air quality, thermal environment, lighting and acoustics, European Committee for Standardization, Brussels

[13] Ortúzar, J 2000, Modelos econométricos de elección discreta, Ediciones Universidad Católica de Chile, Santiago

[14] Punj, G & Stewart, DW 1983, 'Cluster Analysis in marketing research: review and suggestions for application', Journal of Marketing Research, vol. 20, no.2, pp.134-138

[15] Vivanco, M 1999, Análisis Estadístico Multivariante. Teoría y práctica, Editorial Universitaria S.A, Santiago

[16] ASHRAE 2001, IWEC - International Weather for Energy Calculations, American Society of Heating, Refrigerating and Air-Conditioning Engineers, Atlanta

[17] INN 2008, NCh1079–2008: Arquitectura y construcción – Zonificación climática habitacional para Chile y recomendaciones para el diseño arquitectónico, Instituto Nacional de Normalización, Santiago

[18] Givoni, B 1994, Passive and low energy cooling of buildings, John Wiley & Sons, Inc. Canada

[19] Givoni, B 2009, 'Indoor temperature reduction by passive cooling systems', Solar Energy, vol. 85, pp.1692-1726

CONTROL AND PERFORMANCE OF INNOVATIVE VENTILATION SYSTEMS IN LOW ENERGY BUILDINGS: A STUDIED CASE

Juslin Koffi *, Francis Allard

LEPTIAB, University of La Rochelle
Avenue Michel Crépeau, 17042 La Rochelle Cedex 1, France
**Corresponding author: kkoffi01@univ-lr.fr*

ABSTRACT

As part of a project aiming at assessing ventilation in low energy buildings, this study analyses the performance of innovative ventilation systems used in a single-family building. Five ventilation systems are investigated by simulation using SIMBAD Building and HVAC Toolbox. The results then show better performance in terms of energy demand and indoor air quality (IAQ) for balanced ventilation systems, either permanent or intermittent management. The investigated demanded-controlled ventilation (DCV), which strategies based on CO_2 level and on combined CO_2 and humidity, levels lead to less good but acceptable IAQ. They also allow important reduction of energy demand compared to exhaust-only ventilation system.

KEYWORDS
Innovative ventilation, low-energy building, indoor air quality.

INTRODUCTION

Low-energy buildings are built with well-insulated envelope in order to reduce the energy demand. In France, the conventional primary energy consumption should be inferior to 50 kWh/m^2/year for the residential low-energy buildings. This consumption takes into account the energy demand for heating, which includes the energy demand of ventilation and infiltration, space lighting, air-conditioning, ventilation auxiliaries and hot water production. The up-coming French thermal regulation RT2012 will apply this specification for the new built buildings. Nevertheless, the regulation on ventilation does not specially deal with low-energy buildings. One then can wonder about the energy impact of ventilation in such buildings. This concept of buildings brings out additional questioning on the link between innovative ventilation systems and indoor air quality (IAQ). The main concern is: which ventilation systems are suitable for low-energy buildings? The adequate ventilation system should meet the energy requirements while providing acceptable indoor air quality.

Researchers are trying to elaborate proper answers to that question. Huynh [1] recently analyzed this dilemma and Maier et al. [2] showed the energy and IAQ benefits of mechanical ventilation in low-energy buildings. Besides, Karlsson et al. [3], investigating residential buildings, noted that the set-point temperature, the building orientation and U-values as well as the local climate are factors that can influence the energy demand. Mahdavi et al. [4] compared passive and low-energy residential buildings: according to the authors, both type of buildings can achieve good indoor CO_2 level and meet the energy requirements. Moreover, Koffi et al. numerically assessed the IAQ and energy performances of different ventilation systems in a single-family building [5] and a low-ernergy apartment [6].

In the same framework, the present study aims to assess the performance of ventilation systems supposed to bring accurate responses to these questions. Final propositions of the project would give indications about the suitable way of the management of ventilation in low-energy buildings according to the use of the building. For this purpose, existing mechanical exhaust-only and balanced ventilation systems are designed in a single-family building. In addition, three demanded-controlled ventilation (DCV) systems are studied. They are respectively based on: 1) CO_2 concentration and humidity level, 2) CO_2 concentration and presence detection, and 3) day and night airflow rates management. The simulations are carried out in a French local climate. This paper analyzes the impact of these ventilation strategies on the indoor air quality and the energy demand.

METHODS

The simulated building

The study is about a single-family low-energy house presented by Figure 1; we used the weather data of Trappes, a French city near Paris. SIMBAD Building and HVAC Toolbox [7] is used for the simulations. This tool implements multizone and nodal building models in MATLAB/Simulink environment and combines heat and mass transfer equations; the airflow model is described by Koffi et al. [5]. Table 1 presents the thermal properties of the building walls which are designed to achieve the energy requirements of low-energy buildings. Besides, the envelope air leakage is set to 1.70 ach at 50 Pa pressure difference (i.e. 0.6 m^3/h per square meter of envelope under 4 Pa) for limiting air infiltration.

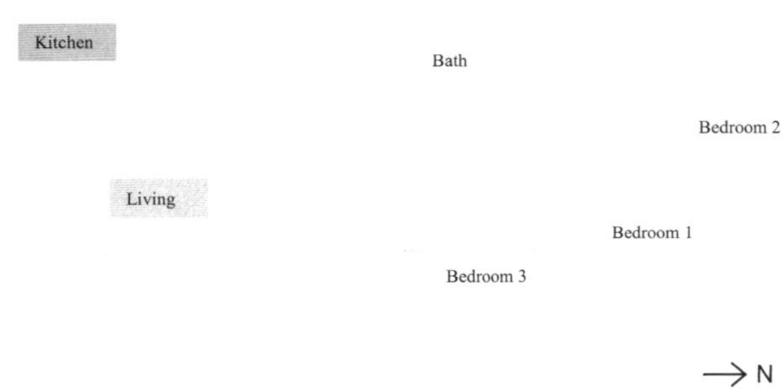

Figure 1. The studied building.

Layer	U (W/m^2.K)	R (m^2.K/W)
External walls	0.16	6.1
Ceiling	0.10	10.1
Floor (over a crawl-space)	0.17	5.6
Floor (on ground)	0.23	4.3
Inter-storey floor	0.36	2.5
Internal walls	0.20	4.8
Windows	1.2	-

Table 1. Thermal properties of the wall layers.

The house is occupied by four persons according to occupancy schedules similar to those described by Mansson et al [8]: two adults sleep in bedroom 1 and two children in bedrooms 2 and 3. Each occupant releases in the occupied space some amounts of water vapour, carbon dioxide and sensible heat depending on its age and metabolism. In addition, water vapor is generated during cooking breakfast (50 g/pers), lunch (150 g/pers) and dinner (300 g/pers), as well as shower (300 g/pers for 20 minutes), clothes washing (200 g for 2 hours) and drying (1000 g for 20 hours). Besides, some emission models of volatile organic compounds (VOCs) due to building materials and to activities like cooking, cleaning, smoking and incense burning are simulated using extisting database PANDORE [9] and IA-QUEST [10].

The studied ventilation systems

Five existing or newly designed ventilation systems are studied in the defined building in order to achieve both energy requirements and acceptable indoor air quality in low-energy building:

1) System MI-0 is the permanent exhaust ventilation commonly is in French homes. It deals with mechanical air exhaust from the kitchen (45 m^3/h), the bathroom (30 m^3/h) and the toilets (15 m^3/h). The fresh air enters the living-room and the bedrooms through self-regulated air-inlets. During cooking, the airflow rate is raised to 120 m^3/h in kitchen for half an hour; then, the total exhaust flow rate is raised from 105 m^3/h to 180 m^3/h.

2) MI-1 is a balanced ventilation system with both mechnical supply and exhaust ducts. The exhaust flow rates are the same as for strategy MI-0. The input airflow rate is 20 m^3/h in each bedroom and 45 m^3/h in the living-room. During cooking, the airflow rate is set to 120 m^3/h in the kitchen and the living-room. Furthermore, in order to reduce the energy demand, strategy MI-1 deals with 0.85 heat recovery efficiency on the exhaust air for pre-heating the supply air.

3) MI-2: this exhaust system deals with CO_2-sensors in the main rooms and combined CO_2 and humidty-dependant exhaust devices in the service rooms.

4) MI-3: we simulate this exhaust system using again CO_2-dependant air-inlets in the living-room and the bedrooms as well as strategy MI-2. However, MI-3 is distinguished from MI-2 by the use of presence detection sensors in the exhaust rooms. In fact, the airflow rate is set to a minimum value in each exhaust room until it is occupied; then, the system extracts the maximum airflow rate for half an hour once the occupant has lived the room.

5) MI-4 is a balanced demanded controlled ventilation strategy based on occupant presence in the whole building. During daytime, the maximum airflow rate, 70 m^3/h, is supplied in the living-room while only 8 m^3/h is input in each bedroom. At night, the system supplies 30 m^3/h per bedroom and 10 m^3/h in the living-room. If there is nobody in the house, the global input and output airflow rates are reduced to 25 m^3/h. As well as strategy MI-0, this system deals with heat recovery which efficiency is set to 0.85.

During lunch and dinner, the exhaust airflow rate in the kitchen is boosted to 120 m^3/h for 30 minutes. Systems MI-0 and MI-2 distribut this additional airflow along with the pressure differences on the façades. Moreover, strategies MI-3 and MI-4 use a balanced hood in the kitchen during cooking, supplying and exhausting 105 m^3/h airflow rate for 30 minutes. In order to prevent the dispersion of the released pollutants, the removal efficiency of the hood is set to 2 and the heat exchanger efficiency is 0.35.

Table 2 summarizes the control factors of the studied ventilation systems.

Syst.	Type	Control factor	Cooking airflow rate (m^3/h)
MI-0	Exhaust-only	Permanent	120
MI-1	Balanced	Permanent	120
MI-2	Exhaust-only	CO_2 + humidity	120
MI-3	Exhaust-only	CO_2 + presence detection	45+105 (balanced hood)
MI-4	Balanced	Time clock + presence	45+105 (balanced hood)

Table 2. Control factors of the studied ventilation systems.

ENERGY CONSUMPTION

Table 3 and Figure 2 present the final energy consumptions for the studied ventilation systems. This energy demand includes the heating energy in which contains the energy loss due to ventilation and infiltration. The fan energy, which is a funtion of the total airflow rates, is the electrical consumption of fans and eventually hoods: then, it has been multiplied by 2.58 when calculating the final energy demand of fans. Thus, balanced ventilation systems MI-1 and MI-4 engender the highest demand. Besides, the consumption of strategy MI-3, using a hood, is silghtly superior to that of systems MI-0 and MI-2.

The maximum energy demand is obtained with exhaust-only ventilation, strategy MI-0, which is considered as the reference system in this study. The heating energy demand for the latter system is 2733 kWh, i.e. 20.7 kWh/year per square meter of surface (in final energy). In addition, without any airflow reduction or heat recovery, the energy demand due to ventilation is 1980 kWh (i.e. 15 kWh/m^2): then, ventilation accounts 72% for of the energy demand. This part is very high as far as the building is well-insulated and emphasizes the necessity of using heat recovery or reducing as well as possible the airflow rates by means of the control of ventilation.

	Final energy (kWh)		
Syst.	Heating	Ventilation	Fan
MI-0	2733	1980	115
MI-1	1263	390	313
MI-2	2061	1211	93
MI-3	1688	865	118
MI-4	1228	350	309

Table 3. Final energy demand (kWh).

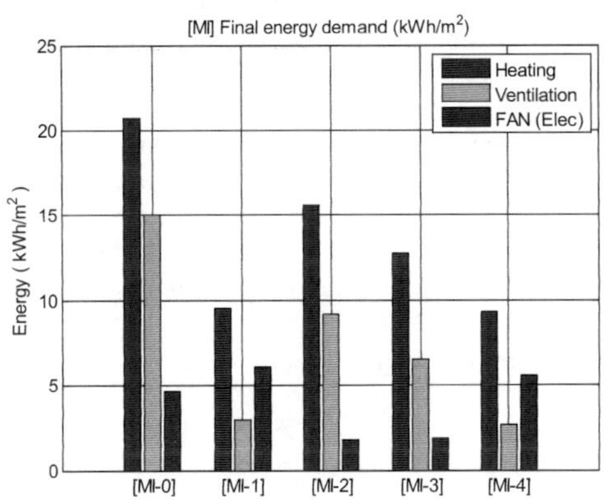

Figure 2. Comparison of final energy demand (kWh/m^2).

Benefits of ventilation heat recovery

Balanced ventilation strategies MI-1 and MI-4 are used for assessing the benefits of heat recovery. The results show a better performance since the ventilation energy is limited to 390 kWh with strategy MI-1 and 350 kWh for MI-4, representing more than 80% savings compared to the ventilation energy of system MI-0. As consequence, the heating energy demand is highly reduced to 1263 kWh with MI-1 and 1228 kWh for MI-4; the savings due to heat recovery represent more than 53% of heating energy demand of the reference system. But in fact, both systems lead to quit similar savings but one can note a slight advantage to use strategy MI-4, due to the airflow control.

Impact of ventilation control

When using demanded-controlled ventilation strategies based on CO_2 and presence (MI-3), and on both CO_2 and humidity (MI-2), one can also expect to decrease greatly the total energy demand through the control of ventilation airflow rates. MI-2 generates 2061 kWh energy demand against 1688 kWh for MI-3: the savings due to these control strategies are respectively 24.5% and 38.2%. Besides, the energy part of ventilation will represent from 51% (MI-3) to 59% (MI-2) of the corresponding heating energy demand, that less than ventilation impact of MI-0 but more than that of balanced strategies.

In the present study, CO_2-DCV MI-3 results in a better energy performance than the combined CO_2 and humidity strategy MI-2. The main reason of this difference is a difference in the design of these strategies. In fact, system MI-2 uses, in the exhaust rooms, the maximum airflow rate between that generated by CO_2 level and that due to humidity ratio. Not at all, strategy MI-3 deals with presence-dependant airflow control in these rooms. This situation results in higher energy expense by MI-2.

Finally, according to the requirements, the studied ventilation systems seem to be suitable for use in such a low-energy building for the considered local climate. Balanced ventilation strategies MI-1 and MI-4 are the strategies offering the best performance in terms of energy demand. Even strategy MI-0 leads also to acceptable results. Nevertheless, we can note high losses through ventilation and infiltration which is, most of the time, difficult to control. One should then read this conclusion carefully. In fact, the energy analysis does not consider the use of electricity. Therefore, the conclusion is not valid if electricity is used as energy for heating the space. The energy demand could have also been influenced by the set up temperature: we used 20°C, without any schedule, instead of 19°C as usually recommended. This change was done in order to get maximum levels of the energy demand.

ANALYSIS OF INDOOR AIR QUALITY

CO2 level and exposure to VOCs

Figure 3 and Figure 5 present the daily CO_2 concentrations in bedroom 1 and the living-room. The first room is generally the most polluted one: in fact, two adults occupy it during the night. In addition, Figure 4 and Figure 6 present the cumulative occurrence of CO_2 in these rooms, by calculating the ppm-hours. This index represents the product of the CO_2 concentrations, respectively superior to guideline values of 1000, 1500 and 2000 ppm, and the total duration over the heating period. This index is calculated only when the rooms are occupied in order to deal with exposure.

Using the reference system MI-0, the concentration in the bedroom 1 can exceed 1500 ppm and 1200 ppm in the living-room during the night. As shown by the CO_2 index, the concentration are lower than 2000 ppm. The second system, MI-1, brings out better air quality compared to MI-0 (and also CO_2 DCV and CO_2 and humidity-controlled ventilation). With this strategy, one can expect to ensure CO2 concentrations lower than 1500 ppm. The main benefits with system MI-1 reside in adequate and permanent air input in the bedrooms and the living-room. In addition, due to air exfiltration occurring from these rooms, a noticeable part of the emitted pollutant is exhausted outdoors without crossing the building, so that the pollutant levels are somewhat lower than with MI-0.

Strategy MI-4 is designed to always bring the fresh air where necessary according to a night/day schedule. This concept brings the best indoor air quality during the occupancy periods as shown by the CO_2 indexes. Globally, the carbon dioxide level in the bedrooms rarely reaches 1200 ppm with this system. The concentration difference with the other systems can reach 300 to about 1300 ppm. There is a real benefit to increase the airflow rate in the occupied rooms.

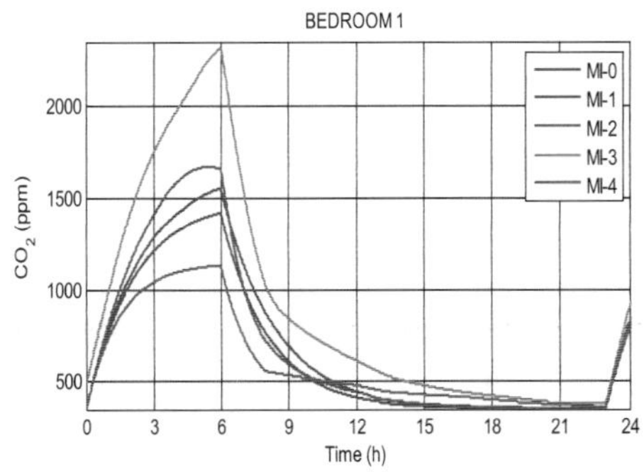

Figure 3. CO_2 concentrations in bedroom 1.

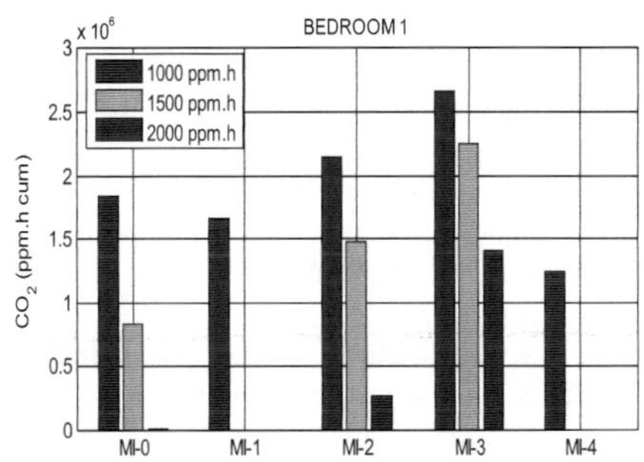

Figure 4. CO_2 index in bedroom 1.

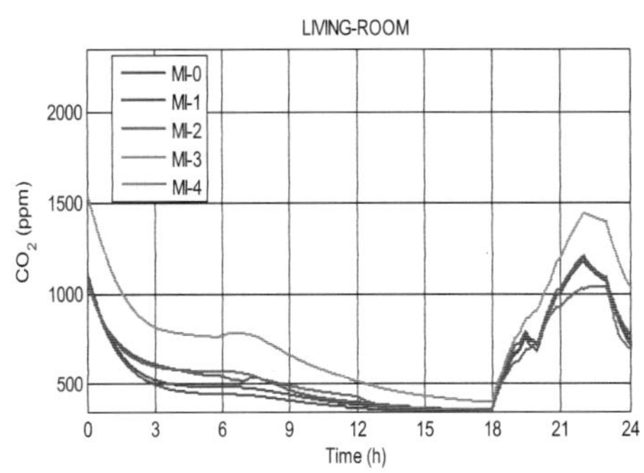

Figure 5. CO_2 concentrations in the living-room.

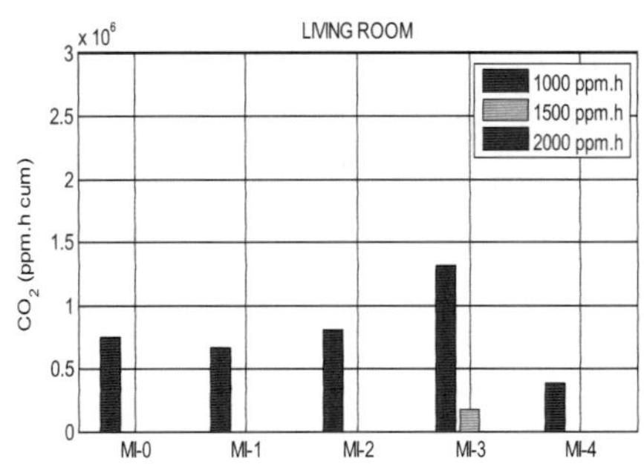

Figure 6. CO_2 index in the living-room.

On the contrary, the CO_2 and presence DCV MI-3 leads to the highest pollution in this study; the concentrations frequently exceed 1500 ppm in the living-room during the week-ends and 2000 ppm in bedrooms. This result seems to be inconsistent with the objective of that system. In fact, during the night, strategy MI-3 only depends on CO_2 concentrations in the bedrooms; the exhaust airflow rates are set to their minimum values as the corresponding rooms are not occupied. Then, the increase of CO_2 concentrations has a very reduced effect on the renewal airflow rate of the building. What happen are an enhancement of the flow rates through the air inlets and a diminution of the infiltrations in the bedrooms. During the day, the airflow rates are increased only when one or more exhaust rooms are occupied. This strategy should have integrated CO2-dependence in the exhaust rooms.

The combination of humidity control and CO_2 control in the exhaust rooms helps to improve the air quality. Then, with strategy MI-2, one can note a real decrease of the CO_2 concentration and index in the rooms. The occurrence of concentrations higher than 2000 ppm is thus divided by a factor greater than five compared to MI-3; this solution seems to be more adapted. Globally for the studied systems, the 1000 ppm guideline seems impossible to avoid. One solution can be the increase of the demanded airflow rate, for instance using system MI-4; however, this way of doing is likely to penalize the energy savings.

A parallel can be made with VOCs through the analysis of formaldehyde concentration (Figure 7 and Figure 8). The level of this contaminant is kept almost constant in the bedrooms when using strategy MI-1 as the airflow rates are constant. Some slight variations are visible with MI-0 during cooking periods.

As well are for CO_2, strategy MI-4 leads to the lowest pollution level in the bedrooms during. The concentrations decrease to a minimum value when the occupants are in these rooms. When they are of out the building during the day, the VOCs level increases until reaching the maximum level among the studied systems: however, this does not matter as far as the main concern of IAQ should be the occupants' exposure. Finally, we can note that strategy MI-3 fails to provide air quality as good as the other systems according formaldehyde concentrations; the level of this pollutant increases during the night and reaches levels more than three times compared MI-0.

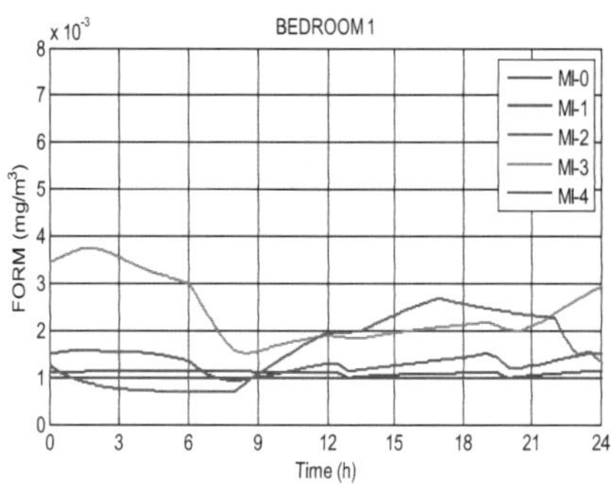

Figure 7. Formaldehyde concentrations in bedroom 1.

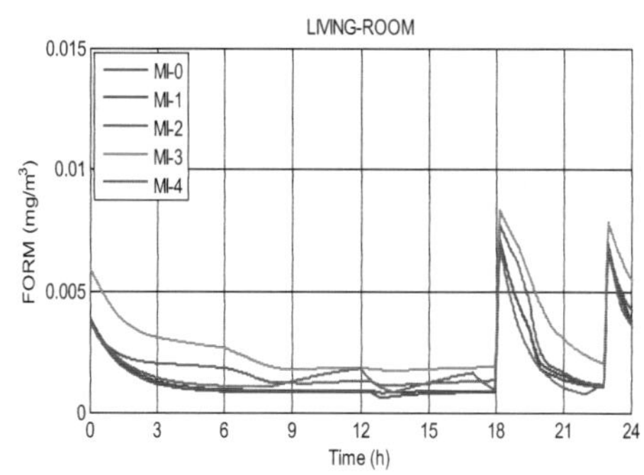

Figure 8. Formaldehyde concentrations in the *living-room*.

Humidity level

The daily evolution of relative humidity in the kitchen is illustrated by Figure 9. Figure 10 presents the cumulated number of hours over the heating period for which the relative humidity is higher than 75%. Only the bathroom and the kitchen are mainly concerned by this index; the maximum RH can reach 100% especially during cooking and shower. In the bedrooms, the humidity is most of the time lower than 60%, but can sometimes exceed this value especially with strategy MI-3; in winter, the lower values can touch 20% mainly due to outside low absolute humidity.

In the kitchen, strategies MI-0, MI-1 and MI-2 lead to comparable RH index values, representing about one hour per day. The observed differences may result from difference of internal air transfer and mainly from the humidity ratio in the living-room which is less important with MI-1. On the contrary, MI-3 and MI-4, using higher pollutant removal efficiency, allow a noticeable decrease the humidity level too.

In the bathroom, the humidity level remains higher with strategies MI-3 and MI-4 than with the other systems. This is mainly due to reduced airflow rates when the occupants are out of the building while water vapour is still released from clothes washing and drying. For MI-3, this happens as soon as nobody is in the concerned exhaust rooms. The airflow rate reduction lasts for 23 hours per day with MI-3 against 7 hours with MI-4; that difference in the basic factor of the RH index difference, about 270 hours, between these systems. Finally, the humidity ratio is kept at minimum level with strategy MI-2: the RH remains superior to 75% for about 630 hours against more than 950 with hours with MI-0 and more than 1600 hours for MI-3 and MI-4. Due to control by humidity and CO_2 concentration, MI-2 can in fact bring airflow rate up to 45 m^3/h in the bathroom while the other strategies are limited to 30 m^3/h.

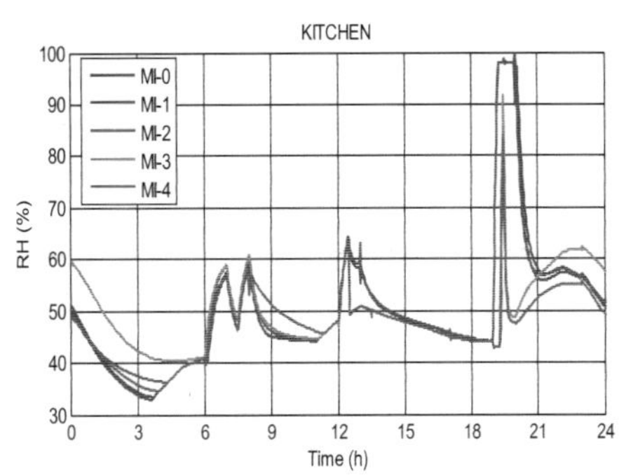

Figure 9. Relative humidity in the kitchen.

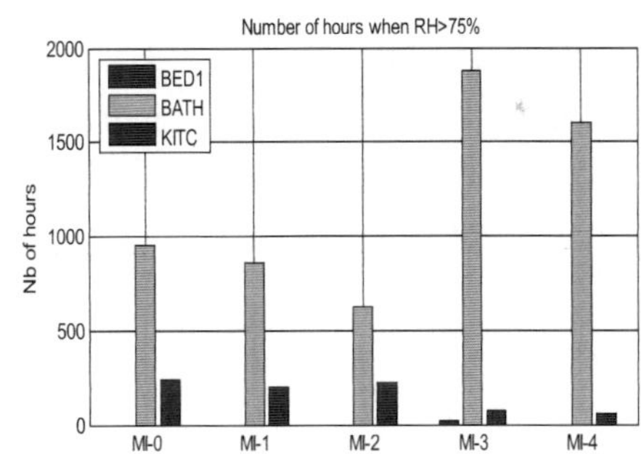

Figure 10. Relative humidity index (RH>75%) in the exhaust rooms.

CONCLUSION

This study deals with the assessment of innovative ventilation systems in a low-energy house. The analysis of the results brings out many differences among the studied strategies. The energy consumption, which is the first criterion for evaluating low-energy buildings, seems to be acceptable in the present study. In almost all the studied cases, the energy allowed for heating is kept in the range of the requirements as far as no electricity is used for this purpose.

Another observation is that, for the studied cases, the impact of ventilation can represent a considerable part of the total energy demand, about 30% for balanced ventilation systems, and from 50 up to 72% with exhaust-only strategies. This is mainly due to the combined effect of a well-insulated envelope and uncontrollable infiltration in spite of a supposed airtight envelope; it may be very interesting to investigate the energy consequence of the envelope air leakage. Besides, the results showed that 0.85 heat recovery efficiency can lead to more than 80% savings in energy demand with balanced ventilation strategies. In the same way, using demanded-controlled ventilation systems based on CO_2 and presence detection (MI-3), and on both CO_2 and humidity (MI-2), can help make savings from 24 to 38% on the energy demand.

The analysis of the indoor air quality clearly brings out differences in the performances of the studied systems according to the control factors. Exhaust-only DCV strategies (MI-2, MI-3) fail at times to bring the expected air quality. The difficulty with in their operating mechanism is that the input airflow rates depend on parameters, which values do not, most of the time, match with those of the exhaust rooms. Then, when the exhaust airflow rate is maximum, and if the CO_2 concentrations in the bedrooms or the living-room are not high enough to provide the corresponding airflow rates, these strategies can promote a lot air infiltration.

For the analyzed parameters (CO_2, water vapor and VOC), balanced ventilation systems MI-1 and DCV based on presence MI-4 bring the better performances during the occupancy periods compared to the exhaust systems. The ccontrol of ventilation through presence carried out in strategy MI-4 shows that it is possible to reduce the energy demand while performing good indoor air quality. However, the performance of this strategy would depend on the influence of the occupants who should be aware of the operating of the ventilation system.

The results thus outline the importance of ventilation control: if adequately performed, ventilation control appears as a good way of providing good air quality in order to prevent damages on the occupants' health. The most important way of controlling air quality seems the adjustment of ventilation to the demand. Nevertheless, this analysis does not consider the impact of all the simulated pollutants. Further studies would bring more detail about this through the use a built of indoor quality criteria.

ACKNOWLEDGEMENTS

This study is carried out with the financial support of ANR, the French National Research Agency, and the assistance of AirH, CSTB, CETIAT, INERIS and Allie'Air.

REFERENCES

[1] Huynh, C. K. (2010), *Building energy saving techniques and indoor air quality – a dilemma*, The International Journal of Ventilation, ISSN 1473-3315 Volume 9 No 1, June 2010.

[2] Maier, T., Krzaczek, M., Tejchman, J. (2009), Comparison of physical performances of the ventilation systems in low-energy residential houses, Energy and buildings 41 (2009) 337-353.

[3] Karlsson, J. F., Moshfegh, B., (2006), Energy demand and indoor climate in a low energy building – changed control strategies and boundary conditions, Energy and Buildings 38 (2006) 315-326.

[4] Mahdavi,A. and Doppelbauer, E.-M. (2010), *A performance of passive and low-energy buildings*, Energy and Buildings 42 (2010) 1314-1319.

[5] Koffi, J., Allard, F., Akoua, J.-J., (2011), *Numerical Assessment of the Performance of Ventilation Strategies in a Single-Family Building*, The International Journal of Ventilation ISSN 1473-3315 Volume 9 No 4 March 2011, pp 337-350.

[6] Koffi, J., Allard, F., (2011), *Methodology for assessing Ventilation Systems in Low Energy Buildings*. Proceedings of Roomvent conference 2011, 19-22 June 2011, Trondheim, Norway.

[7] Husaunndee, A., Lahrech, R., Vaezi-Nejad H., Visier J.-C., (1997), *SIMBAD a simulation toolbox for the design and test of HVAC control systems*, Proceedings of the 5th IBPSA Conference, Prague, Czech Republic, Vol. 2, pp269–276.

[8] Mansson, L.-G., (2001). Evaluation and demonstration of domestic ventilation systems, IEA-ECBCS Annex 27 Final report.

[9] PANDORE, (2010), *An Indoor pollutant sources database.* LEPTIAB, University of La Rochelle, France. http://leptiab.univ-larochelle.fr/Presentation-PANDORE.html.

[10] Sander, D., Won, D. and Magee, R.J., (2006), *IA-QUEST, Indoor Air QUality Emission Simulation Tool, Version 1.1*, Users' Guide National Research Council Canada, April 2006.

SHELTER-IN-PLACE STRATEGY: *CONFINE*, AN AIRTIGHTNESS LEVEL CALCULATION TOOL TO PROTECT PEOPLE AGAINST ACCIDENTAL TOXIC RELEASES

Gaëlle Guyot[1*], Olivier Gentilhomme[2], Rémi Carrié[3]

1 CETE de Lyon
46 rue Saint Théobald. BP128
38081 L'Isle d'Abeau Cedex, France
**Corresponding author:*
Gaelle.Guyot@developpement-durable.gouv.fr

2 INERIS
Parc Technologique Alata, BP2,
60550 Verneuil-en-Halatte, France
3 ICEE
93 rue Molière, 69003 Lyon, France

ABSTRACT

Accidental releases occurring in industrial platforms or during transportation of hazardous materials can entail the dispersion of toxic gas clouds. In case of such an event, the best protection strategy for people is to identify a shelter in a nearby building and stay in this room until the toxic cloud has finally been swept off. In addition to seeking refuge in an airtight room, this strategy called "passive shelter-in-place" also includes closing all external openings and turning off all mechanical ventilation systems and openings.

Following the AZF chemical accident (Toulouse, 2001, 30 deaths), a French law was adopted in 2003 that can compel public and private building owners to adopt such a shelter-in-place strategy. To prove that the shelter airtightness is sufficient and that the occupants will not be exposed to irreversible effects, simulations are required using for instance the modeling tool CONFINE. Originally developed by CETE de Lyon, this software is a pressure code able to model the infiltration of a pollutant inside a 3 zone - building (shelter, attic and rest of building).

This paper aims at giving an overview of CONFINE (governing equations, modeling hypotheses...) and will illustrate its application on one example of shelter-in-place strategy for a public building.

This paper will also present some unexpected results about the impact of wind velocity on shelter-in-place effectiveness. If a higher wind velocity results in a better dilution of the toxic gas outdoor, this situation does not necessarily lead to a lower concentration inside the room, and can conduce to more severe shelter airtightness.

KEYWORDS

Air infiltration, envelope, leakage, shelter-in-place, ventilation, airflow calculation, vulnerability, toxic risk, land-use

INTRODUCTION

Two strategies can be implemented to protect people against toxic risk: shelter-in-place vs. evacuation[7]. In France, shelter-in-place is the sole protective measure recommended, even close to industrial platforms. Following the AZF chemical accident (Toulouse, 2001, 30 deaths), a French law was adopted in 2003. It establishes a new tool around all SEVESO II (high level) classified establishments [9]: the technological risk prevention plan (PPRT) [1]. This local land-use tool specifies in particular protective construction works for new and existing buildings, including implementation of a shelter-in-place system against toxic risk. Such a system includes: 1- general constraints for the whole building design (e.g. system to quickly stop all voluntary airflows) and for a room used as a shelter (minimum size per occupant, presence of sanitary) ; 2- airtightness requirement for this room, with the objective to protect people during 2 hours against irreversible effects[2]. Since 2005, we have been developing the CONFINE software to calculate the minimum airtightness level required for a shelter in order to maintain the internal concentration under a given limit. CONFINE has been designed as a practical tool for operational studies on exposed buildings. It is also used as a research and development tool, to work out regulations and help in decision-makings.

CONFINE'S OVERVIEW: THEORETICAL BASIS OF AN ORIGINAL APPROACH

CONFINE is a pressure code, which considers that each building can be simplified into 3 aeraulic zones (shelter, attic space and rest of the building) delimited by 10 different types of surfaces (Table 1). Each zone is considered having the following homogeneous characteristics: temperature, reference relative pressure and concentration.

Shelter surfaces	Other surfaces
Surface A: outdoor, upwind	Surface F: Attic/outdoor
Surface B: outdoor, at roof	Surface G: Attic/ Building
Surface C: outdoor, downwind	Surface H: Building /outdoor, upwind
Surface D: attic	Surface I: Building /outdoor, at roof
Surface E: building	Surface J: Building /outdoor, downwind

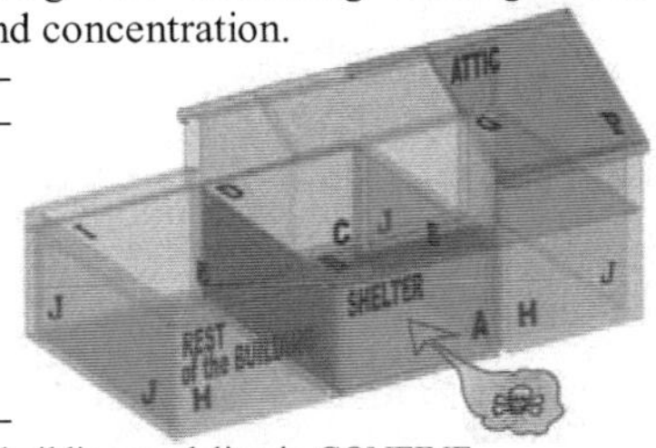

Table 1 and Figure 1 : 10 Types of surfaces used for the building modeling in CONFINE

CONFINE supposes also that all voluntary airflows are stopped and that the initial interior concentration is null. The calculation takes into account climate data, as well as aeraulic and geometric characteristics of the walls. Under these conditions, infiltration airflows are only due to wind pressure and stack effects, according to equations (1) and (2).

$$p_{w,surf,i} = C_{p,i} 0.5 \ \rho_o v_{build}{}^2 \tag{1}$$

$$P(h) = P_0(h_{ref}) + p(h) = P_0(0) - \rho_0 \ g \ h_{ref} + p(h_{ref}) - \rho g \ (h - h_{ref}) \tag{2}$$

$$\rho \approx \frac{P_0}{RT} \tag{3}$$

With:

$p_{w,surf,i}$: wind pressure on surface i (Pa)
$C_{p,i}$: wind pressure coefficient of surface i (Pa). Source: EN 15242 [4]
ρ_o: outdoor air density (kg/m^3)
v_{build}: wind velocity on building (m/s).
P: absolute pressure (Pa)
p: relative pressure (Pa)
g: acceleration of gravity (= 9.81 m/s²)
h: height of a leakage default compared with ground (m)
h_{ref}: reference height of the zone (m),
$_0$: dash referring to atmospheric characteristics
ρ: air density (kg/m^3)
T: temperature (K), T_{indoor} = 293.15 K
P_0: atmospheric pressure in normal conditions
R: universal gas constant (287.055 J kg^{-1} K^{-1}).

Wind velocity impacting the building is based on the meteorological wind velocity (usually measured at 10 m) corrected according to the logarithmical Businger relation [5] with a Monin-Obukov length [6]. This relation takes into account building height, roughness length (relief) and atmospheric stability. This same relation is used by SEVESO industrials for their own previous atmospheric dispersion calculations. Common weather conditions are D5 and F3. The first letter corresponds to the atmospheric stability based on the Pasquill scale (from A: very unstable to F: very stable) while the second figure is the meteorological wind velocity.

Airflows through each surface are calculated using the power law equation (4) and by solving the system (5) of mass balance equations for each of the 3 zones. Results are the 3 reference pressure $p(h_{réf})_i$, from which airflows can be calculated.

$$q_{vl,\Delta Pi,j} = C_{i,j} \ \Delta P_{i,j}{}^n = C_{i,j} < P_{t,j} - P_{t,i} >^n \tag{4}$$

$$\text{For k =1,2,3:} \qquad \sum_{i,j} q_{m,i,j,k} = 0 \tag{5}$$

With:
$q_{vl,\Delta P}$: volumic airflow through an opening with a pressure difference ΔP across it ($m^3.s^{-1}$)
C: flow coefficient of the opening (airtightness defect) ($m^3.s^{-1}.Pa^{-n}$)
P_t: total pressure at both sides from the opening, including wind and stack effect (Pa)
n: pressure exponent. Fixed to 2/3 in CONFINE (-)
$_{i,j}$: subscripts referring to zones at both sides of the opening
$\langle a\rangle^n$ = sign(a).$|a|^n$ by convention, depending on the direction of the flow
q_m: mass airflow through the opening ($kg.s^{-1}$)

Airtightness of each zone is modeled as a single central path located in the center of each surface listed in Table 1. The flow coefficient of this path C_{ij} is calculated with equation (6), distributing leakage index $Q_{4Pa\_Surf}^{i}$ of the zone or of the adjacent zone proportionately to the area $S_{i,j}$ of this surface. Leakage index of zones "attic" and "rest of the building" are inputs of the CONFINE model: their values are given in tables and quite conservative.

$$C_{i,j} = S_{i,j} * min(Q_{4Pa\_surf,i}, Q_{4Pa\_surf,j}) * (1/4)^{2/3} \quad (8)$$

$$Q_{4Pa\_surf,i} = q_{vl,i,\Delta P=4Pa} / \sum_j S_{i,j} \quad (9)$$

Once all airflows have been calculated, CONFINE calculates indoor concentration in each zone with the equation (8).

$$\text{For } i=1,2,3 \quad V_i \frac{dC_i}{dt} = \sum_j (q_{v,j->i} C_j) - \sum_j q_{v,i->j} * C_i \quad (10)$$

With:
C_i, C_j: concentrations in zones i et j (kg/m^3)
$q_{v,j->i}$: volumic airflow from zone j to zone i (m^3/s)
$q_{v,i->j}$: volumic airflow from zone i to zone j (m^3/s)
V_i: volume of the zone i (m^3)

The limit indoor concentration in shelter, usually the French threshold of irreversible effects, allows to calculate the minimum airtightness level required for the shelter, which is expressed as the air exchange rate at 50 Pa (9).

$$n_{50,i} = \frac{\sum_j S_{i,j}}{V_i} \left(\frac{50}{4}\right)^{2/3} * Q_{4Pa\_surf,i} \quad (11)$$

The tool CONFINE was validated through test cases with CONTAM 2.4b [3].

AN OPERATIONAL TOOL : CASE STUDY WITH A SCHOOL

The school "Pasteur" is located about 1 kilometer away from an SEVESO classified establishment AS. Since 2009, a PPRT constrains such a building to set a shelter-in-place system, in order to protect occupants from a toxic chlorine cloud (Table 2).

Duration (min)	Concentration (ppm)	Wind velocity (m/s)	Atmospheric stability	Outdoor temperature (°C)	Roughness length (m)
60	110	5	D	20°C	0.95

Table 2. Characteristics of the chlorine toxic cloud

A vulnerability diagnostic of the building led to identify a shelter composed of 3 classrooms and a part of a central corridor. It can accommodate all 164 children and adults of the school with all needed characteristics: a floor area of 248 m^2, more than the recommended 1.5 m^2 per head; a volume of 960 m^3, more than the recommended 3.6m^3 per head; no external surface directly exposed to the industrial site; only one sanitary and 2 doors should be installed. In the classrooms, closing windows can stop ventilation. Whereas, since the ventilation of sanitary is ensured by a mechanical system, this room requires the installation of additional elements: an emergency circuit breaker and devices to close rapidly the air inlets. Table 3 lists all input data finally used in the CONFINE model.

Surface A,B,D (m^2)	0	Surface H (m^2)	1359	$V_{rest\ of\ the\ building}$ (m^3)	11821
Surface C (m^2)	175	Surface I (m^2)	838	$H_{building}$ (m)	15
Surface E (m^2)	411	Surface J (m^2)	1144	Slope of the roof (°)	0
Surface F (m^2)	1122	$V_{shelter}$ (m^3)	960	$Q_{4Pa\_Surf,attic}$ ($m^3/h/m^2$)	30
Surface G (m^2)	869	V_{attic} (m^3)	1469	$Q_{4Pa\_Surf,building}$ ($m^3/h/m^2$)	10

Table 3. Input data

To protect people from irreversible effects of chlorine during the 2 hours, indoor concentration in the shelter should stay below 14 ppm. For this shelter in this school, CONFINE calculated that the airtightness level should be lower than $\mathbf{n_{50}=2.3\ h^{-1}}$.

A DEVELOPMENT TOOL: STUDY OF WIND VELOCITY IMPACT

To assess the climate influence on the minimum airtightness required for a shelter, two accidental scenarii are simulated on a building located only 200 m far away the loss of containment. First, the toxic gas cloud atmospheric dispersion is studied using a integral-type model named PHAST, and then the infiltration of the cloud within the building is modeled using CONFINE.

1. Description of the investigated scenarii

The two investigated scenarii are: a pipe failure connected to an ammonia vessel - the pipe diameter is 2" and the vessel is storing 40 ton of ammonia - and a catastrophic rupture of a wagon filled with 60 ton of chlorine. Both substances are pressurized liquefied gases initially stored at ambient temperature (20°C). All input data are listed in Table 4.

Scenario		Scenario a: Failure of the pipe connection on a ammonia storage	Scenario b: Catastrophic rupture of a chlorine vessel
Released product	[-]	Ammonia (NH_3)	Chlorine (Cl_2)
Maximum quantity likely to be released	[ton]	40	60
Product phase	[-]	Liquid	Liquid
Stored temperature	[°C]	20	20
Stored pressure	[bar abs.]	8,5	6,8
Type of release	[-]	Pipe connection failure	Vessel rupture
Orifice diameter	[mm]	50	-
Release height	[m]	1	1
Release direction	[-]	Horizontal	-

Table 4. Input data for the investigated scenarios

2. Investigation on toxic cloud dispersion

Both scenarii result in a two-phase outflow with the emission or "source term" characteristics detailed in .

Scenario		Scenario a (NH_3)	Scenario b (Cl_2)
Released mass flow rate	[kg/s]	11,1	-
Final velocity of the release	[m/s]	221	102
Final temperature of the release	[°C]	-33,4	-34
Released duration	[s]	3600	Instantaneous
Liquid fraction	[-]	0,85	0,84
Mean diameter of the droplets	[mm]	$7,3\times10^{-3}$	$2,0\times10^{-1}$

Table 5 : Source term" characteristics

The atmospheric dispersion of this "source term" was investigated using the integral-type model PHAST v6.4 supplied by DNV. The integral-type model is based on solving the governing fluid equations on a parametric way. It can handle the atmospheric dispersion of lighter-than-air products, heavier-than-air products or passive products. However, there are several drawbacks in the integral-type model. It assumes that the ground, over which the cloud is dispersing, is perfectly flat and presents a uniform roughness. In addition, the weather conditions are considered invariant during the whole release (in magnitude and in direction).
The toxic cloud profiles concentration that will penetrate the building where is located the shelter were obtained for 3 different weather conditions: D5, D10 and F3. The external temperature is 20°C. The roughness is equal to about 1 m. The results are obtained at 1 m above the ground. All results are summarized in Table 7.

Weather condition	**Scenario a (NH_3)**		**Scenario b (Cl_2)**	
	Maximum concentration (ppm)	Profile duration (s)	Maximum concentration (ppm)	Profile duration (s)
F3	6300	3600	21000	440
D5	4700	3600	24000	180
D10	3580	3600	34000	52

Table 7. Main concentration profiles characteristics 200m far from the accident

Although one should not oversee the possible influence of atmospheric stability, it can be seen that a higher wind velocity tends to increase gas dilution. Yet, it does not necessarily mean that the toxic effects observed on a person located just outside the building will be less severe. This is the case for scenario (a) since the maximum concentration is reduced while the exposure duration remains constant (= 3600 s). But, in scenario (b), since the exposure duration increases while the maximum concentration reduces, due to this longest duration, a person may be more sensitive to lower gas concentrations.

3. Calculation of the n50 of the shelter

The building considered is an individual house with following characteristics (Table 8). The tool CONFINE was used to calculate the maximum n50 required for the shelter. Results are given in Table 9.

Surface A,B,I (m^2)	0	Surface G (m^2)	100.2	$V_{rest\ of\ the\ building}$ (m^3)	250.5
Surface C (m^2)	7.5	Surface H (m^2)	37.8	$H_{building}$ (m)	4.18
Surface D (m^2)	10.8	Surface J (m^2)	69.25	Slope of the roof (°)	10-30
Surface E (m^2)	26.3	$V_{shelter}$ (m^3)	27	$Q_{4Pa\_Surf,attic}$ ($m^3/h/m^2$)	30
Surface F (m^2)	163.8	V_{attic} (m^3)	97.5	$Q_{4Pa\_Surf,building}$ ($m^3/h/m^2$)	2

Table 8. Input datas

Weather condition	V_{build} **(m/s)**	$P_{wind,surfaceH}$ **(Pa)**	n_{50} **(vol/h) - Scenario a**	n_{50} **(vol/h) - Scenario b**
F3	1.87	1.06	**23**	**11.3**
D5	3.45	3.58	**6.5**	**5.8**
D10	6.90	14.32	**2.0**	**4.6**

Table 9. Maximum n50 of the shelter (in vol/h at 50 Pa)

For this case study, a shelter-in-place system with an "easy to obtain" airtightness requirement for a room, will be efficient to protect people from irreversible effects caused by the dispersion of both scenarii.
In addition, it is important to note that if a higher wind velocity may increase gas dilution in the atmosphere, it can also entail a higher indoor gas concentration, and so, requests a more severe airtightness level for the shelter. In fact, a higher velocity increases the wall pressure on the building, which increases infiltrations (Table 9). Given this finding, and if relevant, it may be worth calculating the airtightness requirement even for the highest wind velocity.

CONCLUSION

In France, the basic strategy of a prevention program to efficiently protect people from accidental toxic clouds is based on sheltering-in-place. In the vicinity of dangerous industrial sites, buildings owners also have to adapt their building with shelter-in-place systems, including an airtight room.

We have developed CONFINE to evaluate the needed airtightness level to maintain in a shelter room toxic concentration under a given limit, usually lower than driving to irreversible effects. This tool can be used as a research and development tool, for guiding regulations and decision-making. For instance, it was used to demonstrate that even if higher wind velocities lead to lower outdoor concentration profiles, they can also increase indoor concentrations. This tool has also been developed as a practical tool for operational studies on exposed buildings. In 2010, the French Ministry for Ecology funded INERIS and CETE de Lyon for developing CONFINE as a free web application and for training private research consultancies, in order to stimulate a market transformation in this field.

Note however that a shelter-in-place system will be efficient only if people know how to use it. Therefore, such schemes should be accompanied by training and communication schemes to raise awareness among the potential end-users.

ACKNOWLEDGEMENTS

The authors are very grateful to the French ministry for ecology, sustainable development, transport and housing (MEDDTL), and in particular to the DGPR department, for its support throughout this work. The sole responsibility for the content of this publication lies with the authors. It does not necessarily reflect the opinion of the Ministry.

REFERENCES

[1] French ministry of Ecology and Sustainable Development, Technological risk prevention plan (PPRT)-Acting together to control risks, 3p., 2007

[2] CETE de Lyon, CERTU, INERIS, Guide PPRT, Complément technique relatif à l'effet toxique, 108 p., 2008

[3] Guyot, G., Carrié, F-R., CETE de Lyon, Modélisation des transferts aérauliques en situation de confinement. Bases théoriques et éléments de validation, 38 p., 2009

[4] EN 15242, Ventilation for buildings. Calculation methods for the determination of air flow rates in buildings including infiltration, 2007

[5] Businger, J. A., Wyngaard, J. C., Izumi, Y., and Bradley, E. F., Flux-Profile Relationships in the Atmospheric Surface Layer, J. Atmos. Sci. 28, 181–189,1971

[6] Havens, J., and Spicer, T.O., Development of an atmospheric dispersion model for heavier-than-air mixtures (DEGADIS model), Vols. 1- 3, University of Arkansas, 1985

[7] Montoya, M.I, Guyot, G., Planas, E., French and Catalan approaches to assess shelter in place effectiveness in the event of a toxic gas release, 32nd AIVC Conference, 1st Tightevent, 6p., 2011

[8] Penelon, T. et Antoine, F., INERIS, Simulation de la dispersion toxique d'ammoniac et de phosgène selon plusieurs scénarios – Courbes de concentration en fonction du temps à différentes distances de la source, rapport interne de modélisation n°55774, 2006.

[9] European Commission : Environment. Chemical Accidents (Seveso II) - Prevention, Preparedness and Response. Available at: http://ec.europa.eu/environment/seveso/index.htm

---

[i] Q_{4Pa_Surf} is the airtightness indicator in French Thermal regulation. See Eq. (9).

LIABILITIES OF VENTED CRAWL SPACES AND THEIR IMPACTS ON INDOOR AIR QUALITY IN SOUTHEASTERN U.S. HOMES

Jonathan Coulter[1]*, Bruce Davis[1], Cyrus Dastur[1], Melissa Malkin-Weber[1] and Tracy Dixon[1]

[1]*Advanced Energy*
909 Capability Drive, Suite 2100
Raleigh, North Carolina 27523, USA
**Email: jcoulter@advancedenergy.org*

ABSTRACT

This study documented that houses in the southeastern United States built on typical wall-vented crawl spaces possess the following characteristics: 1) bulk water, water vapor and associated moisture issues, 2) mold spores, 3) measured holes between the crawl space and living space and 4) measured transmission of mold spores from the crawl space to the living space. When these characteristics exist together, the study indicates that contaminants (mold spores and moisture vapor) present in the crawl space are being transmitted through holes in the house floor and heating, ventilation and air conditioning (HVAC) system into the livable parts of the home, thereby exposing occupants to potentially harmful crawl space contaminants. These results confirm vented crawl spaces as important sources of mold species in the home environment. In order to reduce this exposure, closed crawl spaces in combination with thorough house and duct air sealing are recommended. Such a system was found to be a robust intervention that reduced the moisture and indoor air quality problems associated with typical wall-vented crawl spaces.

INTRODUCTION

Approximately 20% of new homes in the United States (200,000 per year) are built on vented crawl space foundations according to the U.S. National Association of Home Builders (NAHB). An estimated 26 million existing homes have vented crawl space foundations. Because adults and children today spend increasing amounts of time indoors, home environmental health is important for people's well being. Mounting evidence suggests that exposure to mold in damp buildings is an important risk factor for childhood asthma [1]. The strongest identifiable risk factor for the development of asthma appears to be exposure to environmental allergens, including indoor and outdoor pollutants [2].

Because vented crawl spaces in the mixed-humid climate of the southeast experience periodic high levels of moisture, they are very likely building areas where mold can be found. One remedy, in the form of properly closed crawl space standards, demonstrated houses will be notably drier, more energy efficient and support less mold growth compared to houses built over vented crawl spaces [3]. However, it was previously unclear whether the presence of crawl space mold species resulted in exposure to occupants in houses. The overall purpose of this study was to evaluate the importance of typical wall-vented crawl spaces as sources of mold species in the livable parts of the home environment.

Forty-five homes in North Carolina were selected for mold species sampling and a building science evaluation to characterize the conditions of typical wall-vented crawl spaces. This report will explore the influences of this foundation construction technique on mold species growth, indoor air quality and house durability in houses located in the southeastern United States. From the results of these studies, we created a subsequent study of 36 new houses to validate the improved energy and moisture performance of the closed crawl space protocol established above compared to traditionally vented crawl spaces. Results from this study became available in June of 2007.

Methods

Mold species sampling

Prior to mold species sampling, the HVAC system was kept off for four hours. In each home, a minimum of two sets of samples were taken during the test using a Wilcoxon matched-pairs signed rank test. First, before the HVAC system fan was turned on, three samples were taken. One was taken near the return grill for the HVAC system, one in the crawl space and one outside the house. Then, the system fan was turned on and allowed to run for at least five minutes before two additional samples were taken, one near the return grill and one at the closest supply air diffuser (or register) to the system fan. The supply diffuser sample air was isolated from the potential contaminant sources within the house, thus allowing characterization of the relative contribution of the HVAC system to the total bioburden within the house. The sampling was conducted by two trained indoor air quality technicians using Andersen two-stage cascade impactors, which collect and separate both non-respirable and respirable sized particles. The sampler was connected to a vacuum pump calibrated to collect air samples at the rate of 0.5 liters per second. Equipment calibration was conducted at the beginning of sampling, at mid-day and at the end of the day. A sampling period of 3.5 minutes was used for the outdoor air sample and all samples collected within the houses. The sampling period for the crawl space samples was one minute. The collection medium used for impaction of mold spores was Malt Extract Agar, an aciduric mycologic medium designed for the collection of environmental fungi. After sampling, the culture plates were incubated at ambient temperature for 96 hours prior to enumeration and identification. Mold identification was accomplished by macroscopic examination of colony morphology and microscopic examination of fungal elements.

Characterization protocol

Data was collected to better understand the thermal, moisture and air leakage data associated with each ventilated crawl space:

- Homeowners interviews about how they operate in house and crawl spaces to determine any potential indoor air quality related health issues
- Air leakage and zone pressure testing was performed to quantify the "holes" between the house and outside, the crawl space and house and the HVAC system and crawl space

- House characteristics such as house measurements, topography, HVAC and other equipment and existing moisture control strategies
- Crawl space characteristics such as evidence of past moisture problems (wood rot, condensation, mold growth, puddles on vapor barrier, etc.), wood moisture content to evaluate the potential of wood for supporting current mold growth, temperature measurements of the ground, water pipes, ductwork, air handling cabinet and floor framing to assess surface condensation potential
- Long-term temperature and relative humidity data from the central area of the crawl space compared to outside between July 2004 and August 2005

Building pressure diagnostics

Detailed air leakage was measured using a multi-pressure testing system connected to three different systems – house envelope leakage measuring system, crawl space to house leakage measuring system and HVAC duct leakage measuring system. The testing order was as follows:

1) Baseline-HVAC system off, all windows and doors closed

2) House leakage test only

3) House, crawl space and duct leakage measuring systems run together

4) House and duct leakage measuring systems run together

5) Baseline-HVAC system on

House description and moisture history

In order to document past or current moisture problems, we conducted a 100-point inspection which included house air temperature and relative humidity, crawl space air temperature and relative humidity, outside air temperature and relative humidity, crawl space surface temperatures, house framing wood moisture content, crawl space construction details, crawl space and exterior grading conditions and drainage systems. Data was collected using commercially available and calibrated wood moisture meters, spot radiometers, digital thermometers and relative humidity meters.

RESULTS

The average number of vents per house was 13, with the maximum being 22 and the minimum being four vents. Sixty-seven percent of all vents were found open, 26% were partially open and 7% were closed at the time of the data collection.

Moisture

In 33% of the homes, moisture was present on the ground vapor retarder, duct and plumbing systems located in the crawl spaces. Seven percent of the homes had a leaking condensate drain for the HVAC system. Active plumbing leaks were found in 31% of the houses. Water was found inside 15% of the duct systems.

Although moisture was not always visible at the time of the testing, the presence of recent moisture accumulation in the crawl space was visible by the following means in Table 1.

Drip line visible on ground	22%
Absence of ground vapor retarder	27%
Absence of full coverage of ground vapor retarder	100%
Discoloration on walls	49%
Termite tunnels	4%
Animals and insects	36%
Dryer exhaust terminating in crawl space	16%
Visible mold growth	62%
Wood moisture readings at mold supporting levels (≥19 %)	67%
Wood moisture meter readings at wood rot supporting levels (≥ 25 %)	36%

Table 1. Moisture indications and percent of frequency found inside crawl spaces.

Mold species

Mold species sampling provided an evaluation of the total number of breathable mold spores, reported in colony forming units per cubic meter of air and the most common species of mold growth found. Table 2 shows the summary of bioaerosol results (in colony forming units per cubic meter) by the possibility of transmission and sample location. Figure 1 displays the bioaerosol levels for houses with the possibility of transmission. Figure 2 illustrates bioaerosol levels for all houses by location.

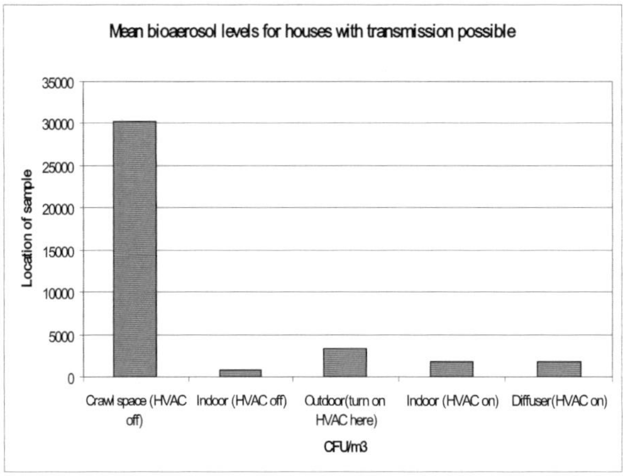

Figure 1. Mean bioaerosol levels for houses with transmission possible.

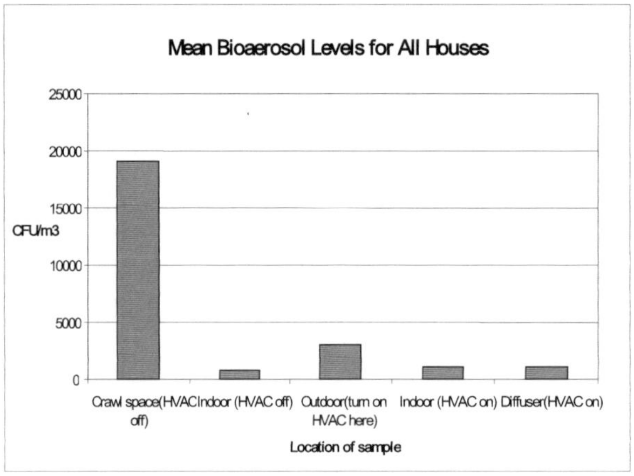

Figure 2. Mean bioaerosol levels for all houses.

Sample	# Houses	Mean	Std. Dev.	Max	Min
Transmission possible	In CFU/m3				
Crawl space (HVAC off)	21	30,163	16,230	41,146	1,348
Indoor (HVAC off)	21	861	1,233	5,802	146
Outdoor(turn on HVAC here)	21	3,235	3,862	11,756	349
Indoor (HVAC on)	21	1,761	2,425	11,756	373
Diffuser(HVAC on)	21	1,822	2,607	11,756	166
Transmission not detectable	In CFU/m3				
Crawl space(HVAC off)	10	161	508	1,607	0
Indoor (HVAC off)	10	55	173	548	0
Outdoor(turn on HVAC here)	10	2,033	2,524	8,418	40
Indoor (HVAC on)	10	176	556	1,759	0
Diffuser(HVAC on)	10	1,415	2,282	11,756	0
No transmission	In CFU/m3				
Crawl space(HVAC off)	14	16,041	15,144	41,146	105
Indoor (HVAC off)	14	1,323	3,045	11,756	71
Outdoor(turn on HVAC here)	14	3,427	4,630	11,756	146
Indoor (HVAC on)	14	645	765	3,219	124
Diffuser(HVAC on)	14	556	1,101	4,326	71
All homes	In CFU/m3				
Crawl space(HVAC off)	45	19,102	18,179	41,146	0
Indoor (HVAC off)	45	825	1,911	11,756	0
Outdoor(turn on HVAC here)	45	3,027	3,836	11,756	40
Indoor (HVAC on)	45	1,061	1,837	11,756	0
Diffuser(HVAC on)	45	1,062	2,011	11,756	0

Table 2. Summary bioaerosol results.

Measured transmission

Initial assessment of transmission of crawl space air and its contaminants, including mold spores and moisture vapor, into the living space was determined to be present if two conditions held true. First, the concentration of the mold samples had to be higher in the living space once the HVAC system was turned on compared to the level of spores with the HVAC system off. Second, the mix and rank order of the indoor samples with the HVAC system running shifted to reflect the dominant mold species present in the crawl space sample and the rank order of species was different from the outdoor sample. If only one condition held, the house was classified as "transmission not detectable", and if neither condition held true, the house had no transmission.

Transmission of air and its contaminants was possible in 21 (47%) of the houses characterized. In ten (22 %) houses, transmission was not detectable, or rather, only one of the two conditions held true. No transmission was found in 14 (31%) of the houses.

Measured holes between the crawl space and living space

Three leakage paths were measured: total house air leakage, air leakage between the living space and the crawl space and air leakage between the HVAC duct system and the crawl space. See Table 3 for total house leakage testing documented.

CFM 50 per ft² of surface area *	M³/h/m² at 50 Pascals	Classification	Percent of houses tested
<0.25	<4.6	Minimal	0
0.26-0.45	4.7-8.2	Limited	24
0.46-0.60	8.3-10.9	Moderate	42
0.61-0.75	11-13.7	Excessive	20
>0.76	>13.8	Major	13
* Cubic Feet per minute at 50 Pascals			

Table 3. Measured house leakage.

The majority of the homes (69%) had 11% and 30% of the total house air leakage coming from the crawl space. The measured leakage between the HVAC duct system and the crawl space are shown in Table 4. Five homes were not classified for this test because they were unable to reach their target pressure.

CFM 25 per ft² of conditioned floor area as a percentage	M³/h/m² at 25 Pascals	Classification	Percent of houses tested
< 3%	<0.55	minimal	0
3.1-5%	0.56-0.91	limited	4
5.1-8%	0.92-1.46	moderate	9
8.1-12%	1.47-2.19	excessive	18
> 12%	>2.20	major	65

Table 4. Classification of duct leakage.

See Table 5 for the mean equivalent hole size for air leakage across the floor between the house and crawl space and house duct system and crawl space.

Equivalent hole size in ft² (m²)	Mean	High	Low	NA*
House to crawl space	0.5 (0.046)	2 (0.19)	0.0	
Crawl space ducts	0.4 (0.04)	1.5 (0.14)	0.1 (0.01)	2
*NA indicates numerical data could not be calculated due to difficulty in reaching target pressure.				

Table 5. Equivalent hole size by location.

CONCLUSION

This study, conducted during typical 12-month conditions, documents moisture characteristics of typical wall-vented crawl spaces in the southeastern United States and measures the impact on living space mold sources. In some situations, the use of foundation vents to dry a crawl space may cause additional moisture. Of the houses in this study:

- 49% had moisture-induced wall discoloration
- 62% had visible mold growth
- 67% had wood moisture meter readings at mold-supporting levels
- 36% had wood moisture meter readings at wood rot-supporting levels

Indoor air quality is compromised when moisture conditions exist in combination with air leakage between the house and crawl space and between the HVAC duct system and crawl space, as mold species can be delivered into the house through the air leaks. Therefore, both a moisture management strategy for the crawl space and an air sealing plan to reduce house and duct leakage should be incorporated into new and existing homes.

To demonstrate a protocol that will eliminate crawl space moisture problems and stop the total air leakage between the house and the crawl space, Advanced Energy also tested an intervention protocol on 12 similar-sized homes in southeastern United States. This intervention study compared the standard vented crawl space design with a closed crawl space design. The closed crawl space design included a sealed ground vapor retarder that extended up the perimeter walls of the crawl space, air-sealed the perimeter wall between the crawl space and outside, air-sealed penetrations between the house and the crawl space, provided a source of conditioned air to the crawl space and monitored the results [3]. The data from this study demonstrated that this closed crawl space protocol is a robust measure producing substantially drier crawl spaces (reducing conditions for mold, wood decay and insects). The data also demonstrated reduced house space conditioning energy use by 15% to 18% annually as compared to the standard vented crawl space houses. Utilizing the results of these studies, we then created a study with 36 new houses to validate the improved energy, moisture and indoor air quality performance of the closed crawl space protocol established above compared to traditionally vented crawl spaces. See conference paper entitled FORMALDEHYDE AND RELATIVE HUMIDITY IN HIGH-PERFORMANCE HOMES WITH OUTDOOR AIR INTAKES AND EXHAUST VENTILATION.

ACKNOWLEDGEMENT

This investigation is co-funded by Advanced Energy, Duke University, and the U.S. Department of Housing and Urban Development. HUD Grant No. NCLHH0096-01.

REFERENCES

[1] Etzel, Ruth and Rylander, Ragnar. 1999. Indoor Mold and Children's Health. *Environmental Health Perspectives*. Vol. 107 (Supplement 3): 463-468.

[2] National Institutes of Health. 1995. Global Initiatives for Asthma. National Heart and Blood Institute Publication Number 95-3659.

[3] Davis, Bruce and Dastur, Cyrus. 2004. Moisture performance of closed crawl spaces and their impact on home cooling and heating energy in the southeastern United States. U.S. D.O.E./ Oak Ridge National Laboratory and ASHRAE Performance of the Exterior Envelopes of Whole Buildings IX International Conference.

LABORATORY INVESTIGATION OF TIMBER FRAME WALLS WITH AN EXTERIOR AIR BARRIER IN A TEMPERATE CLIMATE

Jelle Langmans[*1], Ralf Klein[2], Staf Roels[1]

1 Catholic University of Leuven, Building Physics Section
Kasteelpark Arenberg 40 - bus 02447
BE-3001 Heverlee, Belgium
** jelle.langmans@bwk.kuleuven.be*

2 Catholic University College Sint-Lieven,
Sustainable Building Research Group,
BE-9000 Ghent, Belgium

ABSTRACT

Recently, the requirements regarding global building airtightness to reduce the exfiltration losses became more severe as result of the trend towards very low energy buildings and Passive Houses. These very strict requirements regarding airtightness are currently achieved with an interior air barrier, which is labour intensive and consequently expensive. At the same time it is observed that new wind barrier solutions - to reduce windwashing of the insulation - can have a major contribution to the global airtightness of timber frame constructions. Consequently, it is questioned whether the labour intensive interior air barrier will still be necessary in practice when the global building airtightness can be guaranteed by an improved wind barrier only.
However, moving the air barrier from the interior to the exterior of the building envelope can imply an increased moisture load, and thus, higher risks for interstitial condensation against the exterior sheathing in cold and moderate climates.
The current paper presents the results of a laboratory experiment to study the hygrothermal behaviour of light weight timber walls with an exterior air barrier only. Four independent test walls (2.3m by 0.5m) are placed between a newly developed hot and cold box, operating at controlled temperatures, humidities and air pressures. All four walls are insulated with 30 cm of standard mineral wool to which OSB is applied as interior sheathing. The test walls differ from each other by the physical properties of applied exterior air barrier; airtightness, moisture buffer capacity, vapour permeability and thermal resistance.

KEYWORDS

Hygrothermal analysis, hotbox-coldbox, exterior air barrier, natural convection, laboratory

INTRODUCTION

The importance of preventing air leakages through building enclosures is well known and documented in literature. Air convection through building envelopes can result in unwanted effects such as draught, interstitial condensation and excessive heat losses.
For timber frame constructions an airtight building envelope is commonly realised by an interior air barrier system. The term ‘air barrier’ refers to the material layer which prevents air leakage between inside and outside through the building envelope. Consequently, the most important property of this layer is the overall continuity, which leads to the requirement of sealing all the joints and intersections in this layer. In cold and moderate climates, such as North-West European areas, the air barrier function is often combined with that of the vapour retarder. Realising a good airtightness with an interior barrier however, is very labour-intensive due to many internal joints, intersections and perforations [2-3].

On the other hand, to protect the insulation layer from unwanted infiltration of outside cold air (so called windwashing), a 'wind barrier' is provided at the outside of the insulation. In addition, this exterior layer also serves as drainage plane to prevent water infiltration into the structure. The performance criteria for wind barrier systems regarding air permeance are less severe than for air barriers, and thus, are the joints in the wind barrier usually left unsealed. However, results of field tests [6,8] and numerical investigations [7] emphasise the importance of improving the continuity of the wind barrier layer to reduce heat losses. As a result of these studies, today more and more building companies start to improve the airtightness of the wind barriers by sealing the joints.

In situ measurements [1] show how the air permeance of the wind barriers can be significantly improved with minor modifications. The case study discussed demonstrated that with good workmanship and appropriate materials, an airtightness level lower than 1 ACH at 50 Pa can be reached with the wind barrier only.

Moreover, it is noticed that in Norway the wind barrier evolves more and more towards a secondary air barrier [4,9]. In the Nordic countries it is becoming common practise to measure the global buiding airtightness twice; during the windtight stage and after the building is finished. However, when improving the wind barrier to such levels it becomes impossible to control the continuity of the interior air barrier with pressurisation test, because only the air resistance of the global building envelope is measured. This means that situations can occur where the exterior wind barrier is more airtight than the inner air barrier. In this case the exterior sheathing acts as a wind and air barrier and the inner sheathing only acts as vapour barrier/retarder. Given such a non-continuous interior vapour barrier, concentrated vapour diffusion and moist air by natural convection may enter the building envelope through the gaps in the vapour barrier.

Consequently the question rises to which level this additional moisture load has an impact on the risk for interstitial condensation against the exterior sheathing.

The current paper presents the results of a comprehensive laboratory investigation in which the hygrothermal response of light weight walls with an exterior air barrier is studied.

Four highly insulated test walls enclosed between two climate chambers to simulate in and outdoor winter conditions in a temperate climate have been analysed. The test walls are based on the configuration currently used in Belgian timber framed Passive houses using Oriented Strand Board (OSB) as interior sheathing and insulated with 30 cm of insulation. The walls differ from each other by the physical properties of the applied exterior air barrier; airtightness, moisture buffer capacity, vapour permeability and thermal resistance. The investigation is performed in five consecutive stages with increasing importance of air transport in the test walls. A detailed description of the test setup and preliminary results, mainly focussing on the thermal behaviour during the first measuring step, were already presented in [10].

EXPERIMENTAL METHOD

Hot box/ cold box equipment

For the current study a new vertical calibrated hot box/ cold box was constructed. Here only the main features are discussed. A more detailed discription can be found in [10].

The test setup consists of three major parts; a test frame to install the studied building component enclosed between two climate chambers to simulate in and outdoor conditions. The warm climate chamber has a cubic inner volume with sides of 2.4 m and is completely insulated with 60cm of PUR insulation panels. The test frame, which was constructed in the same way, has a measuring area of 2.4 m by 2.4 m and a depth of 0.6 m. The cold box on the other hand is only insulated with 0.1 m polyurethane boards.

A controlled IR-bulb in the middle of the warm chamber creates the desired temperature conditions. The cold chamber on the other hand is provided with a convector accompanied with a fan system to control and distribute the temperature. As a consequence of the fan system, a small under pressure in the cold box is unavoidable. The humidity in both the warm and cold chamber is conditioned with free evaporation of salt solutions. To create a total air pressure difference across the test specimen a small ventilator is installed at the back wall of the warm chamber.

Wall configurations and sensor positioning

To investigate the hygrothermal consequences of exterior air barrier systems in timber frame construction, four highly insulated test walls (test area: 2.3m by 0.5m) were tested. All four test walls are insulated with 30 cm of standard mineral wool to which OSB is applied as interior sheathing. The test walls differ from each other by the physical properties of applied exterior air barrier; airtightness, moisture buffer capacity, vapour permeability and thermal resistance. Both the exterior sheathing of the first test wall (further referred to as REFERENCE) and the second test wall (referred to as FIBREBOARD 1) consists of bituminous impregnated soft fibre board with an exterior top layer which increases its airtightness. For the third test wall (FIBREBOARD 2) a similar bituminous impregnated soft fibre board but without top layer is applied. This means that the first three test walls are provided with a hygroscopic and capillary exterior sheathing. Contrary, the fourth wall (FOIL) is executed with a spunbunded foil at the outside. The applied foil is extremely airtight but has no water buffer capacity. The configuration of the test walls studied is shown in Figure 1 and the most important material properties are summarised in the following section.

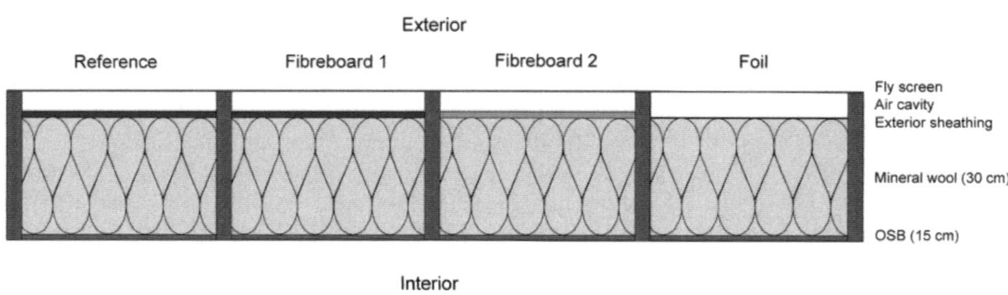

Figure 1. Wall configurations studied.

Each test wall is provided with 18 thermocouples placed at three heights: (1) 20 cm from the bottom, (2) middle height and (3) 20 cm from the top, at every material interface, in the ventilation cavity and the middle of the insulation layer. Also 15 relative humidity sensors were installed at the same positions except from the interface OSB-mineral wool (warm side). Additionally, the heat fluxes were measured at this interface at the three heights. For each wall also a pressure gauge was installed at the middle height.

Besides this continuous logging system, the wind barrier layer was constructed in such a way that each part of the wall contained three removable specimens (12cm by 12cm). The specimens are used to quantify the moisture evolution of the fibreboard on a two-weekly basis. Special care was given to the airtightness of the perimeter of each specimen. After every measurment the joinst were sealed with airtight tape.

Boundary conditions and test sequence

The experiment, which lasted for about four months, was subdivided in five main consecutive measuring steps. In the first four steps the test walls were exposed to typical winter conditions with increasing importance of natural and forced air convection.

During the first step both the interior and exterior sheathing is airtight. In the second step, gaps are introduced in the interior barrier of all walls except for the REFERENCE section. The gaps correspond with slits of 1 cm at 20 cm from the top and bottom of the OSB and cover the full width of each test wall to maintain the two dimensional situation. In step 3 and 4 an increasing overpressure was created in the hotbox.

Finally, in the last step the conditions in the cold box were adapted to create drying conditions inside the walls. Table 1 summarises the boundary conditions in both the warm and cold chamber.

Steps	Days	T_{HB} (°C)	$P_{v,HB}$ (Pa)	T_{CB} (°C)	$P_{v,CB}$ (Pa)	P_a (Pa)
1	35	20.1	1180	3.0	652	1.4 - 2.7
2	28	20.1	1185	3.4	684	0.6 - 1.6
3	24	20.1	1228	3.2	687	5.4 - 6.6
4	11	20.2	1292	3.3	698	10.1 - 12.5
5	32	24.2	1618	22.5	2007	1.6 - 1.8

[1] Overpressure depends on the position of the test wall an varies between these values

Table 1. Boundary condition in hotbox (HB) and coldbox (CB) during the consecutive measuring steps

It should be noted that the current climate conditions represent an averaged typical winter month in a temperate climate, such as Belgium.

Material properties

Most important material properties are summarized in Table 1 and Figure 2. Apart from the heat capacity all material properties were measured at the laboratory. Special care was given to the air permeability and in particular to the mineral wool. For the permeability perpendicular to the fibres ($K^{\perp}$) and parallel to the fibres ($K^{\parallel}$) seven specimen were tested. The results found are in agreement with Økland (1998) who lists an overview of measured air permeability of mineral wool from literature. However, it was observed that $K^{\parallel}$ was very sensitive to the installation of the specimen in the test setup.

Material	d (mm)	ρ (kg/m³)	c_p (J/(kgK))	λ^1 (W/m/K)	$K^{\perp}$ (m²)	$K^{\parallel}$(m²)
Fibreboard 1	18	285	2068	0.045	4.65E-14	-
Fibreboard 2	18	274	2068	0.047	1.37E-12	-
Foil	0.2	-	-	-	airtight	-
OSB	15	630	1880	0.06	8.20E-15	-
Mineral Wool	300	21.3	840	0.033	1.7E-09	6E-09[2]

[1] heat conductivity at measured 0°C

[2] large variation depending on specimen positiong (2 - 10 E-09 m²)

Table 2. Boundary condition in hotbox (HB) and coldbox (CB) during the consecutive measuring steps

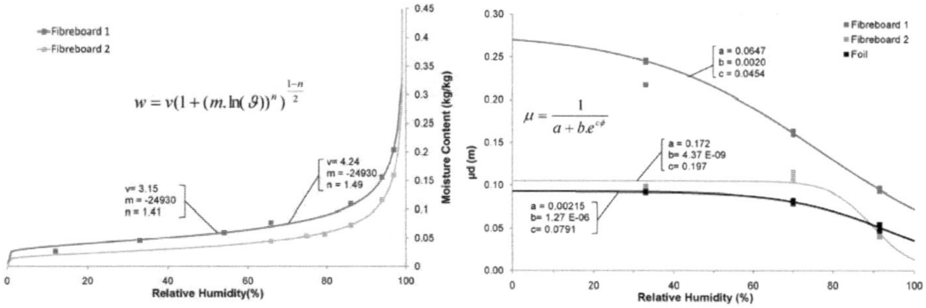

Figure 2. Sorption isotherm (left) and vapour diffusion resistance (right) of the different wind barriers as function of relative humidity.

TEST RESULTS

Thermal response

As a first step in analysing the results we focus on the thermal distribution in the test walls. Figure 3 shows the dimensionless temperature profiles of the four walls at the top and bottom row for measuring stage 1, 2 and 4. This corresponds with the situations where (1) the interior OSB sheathing is intact, (2) top and bottom gaps in the interior sheathing are introduced and (3) an overpressure of 10 Pa is realised in the warm chamber.

This figure clearly shows that during the first step (blue triangles) the temperature distribution bends upwards at the top (filled markers) and downwards at the bottom (open markers) which indicated the existence of natural convection within the walls. This was studied more in depth with additional numerical simulation in [13]. This study shows that the existance of very small vertical gaps between the mineral wool and the sheathing material combined with the (local) increase of the air permeability of the installed mineral wool has a great influence on the magnitude of natural convection inside the walls. During the second measuring step (red squares) this effect increases as a result of the introduction of the gaps in the interior sheathing. At this stage it was also noticed that the temperature profile for the least airtight wall (FIBREBOARD 2) bends also upwards at the bottom row. This means that the exterior sheathing is even so air permeable that as a result of the 1.6 Pa overpressure across this wall (introduced by the ventilators in the cold chamber) forced exfiltration already dominates the air flow in this section. When finally an overpressure of 10 Pa (green dots) is realised in the fourth step this effect becomes of course much more dominant in the 'FIBREBOARD 2' section. Also for FIBREBOARD 1 this effect is (to a minor degree) observed. For the most airtight wall (FOIL) this remains negligible.

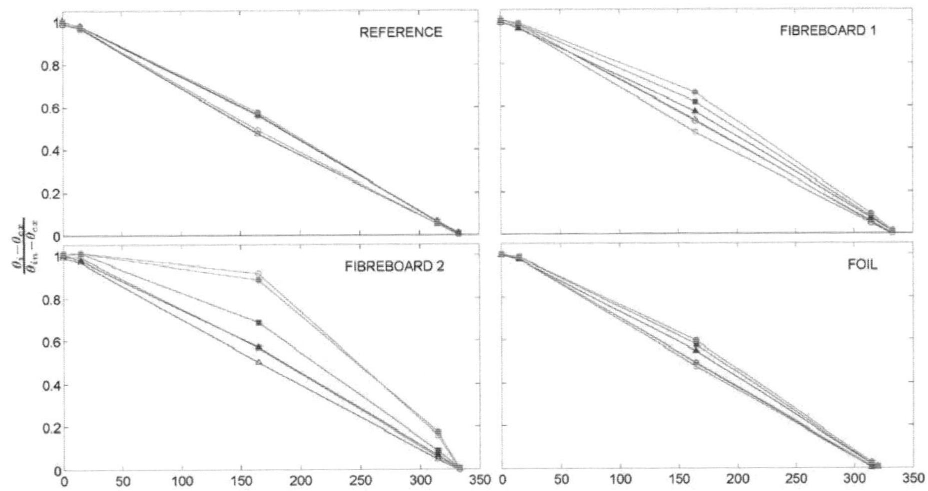

Figure 3. Dimensionless temperature profile of four test walls at top row (filled markers) and bottom row (open markers) during measuring step 1 (blue triangle), step 2 (red square) and step 4 (green dot).

Hygrothermal response

The same method and notation is used to present the dimensionless vapour pressure profiles across the test walls in Figure 4. During the first step all four walls show a similar vapour pressure profile: steep drop behind the vapour retarder (OSB) followed by a slight decrease towards the outer side. Only for FIBREBOARD 1 the vapour pressure at the top row is somewhat deviating from the expected values. At this row the vapour pressure is slightly higher than in the other walls. This cannot be the result of air exfiltration since this should also be noticed in the temperature distribution as well. A more plausible explanation might be found in a local decrease of the vapour resistance in the interior sheathing material or the sealed gap.

In the subsequent step, when the gaps in the interior sheathing are opened, the influence of natural convection on the moisture load becomes very pronounced. For all walls with interior gaps the vapour pressure at the top row increases while the vapour pressure at the bottom row remains the same. Only for FIBREBOARD 2 also the vapour pressure at the bottom row increases since forced exfiltration is then already dominant as a result of the high air permeance of the exterior sheathing as discussed in the previous section.

When an overpressure of 10 Pa is realised the vapour pressure profiles confirms the observation of the temperature profiles. For FIBREBOARD 1 a slight increase in vapour pressure is noticed as a result of forced exfiltration while this effect is much stronger in FIBREBOARD 2. For the wall with the exterior foil this influence is hardly noticed.

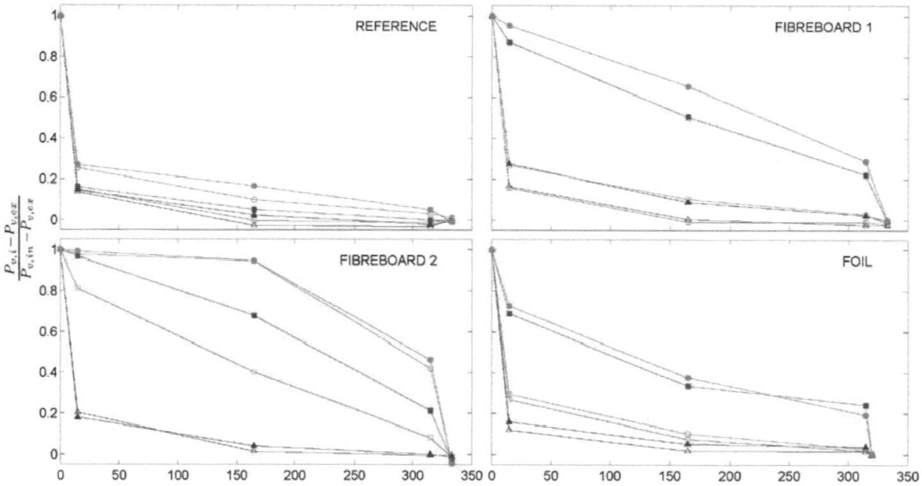

Figure 4. Dimensionless vapour pressure profile of four test walls at top row (filled markers) and bottom row (open markers) during measuring step 1 (blue triangle), step 2 (red square) and step 4 (green dot).

In addition to the vapour pressure profiles the evolution of the moisture content in the exterior sheathing material (Figure 5) gives valuable information about the continious increased moisture load introduced by natural convection.

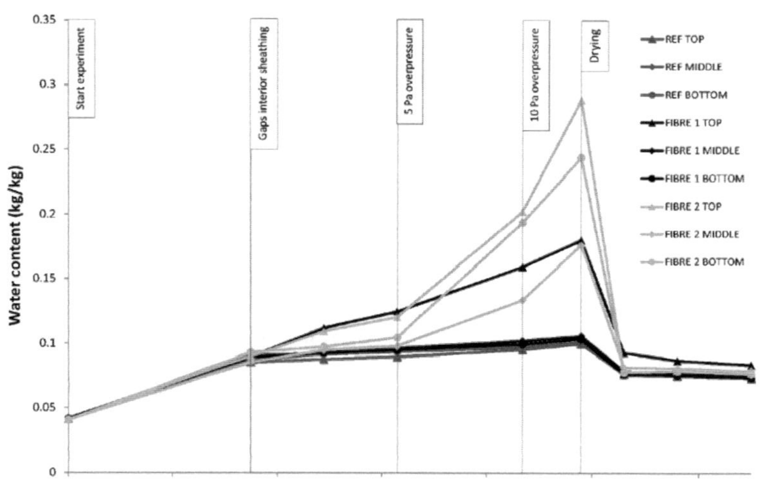

Figure 5. Moisture content evolution of the weight sample in the exterior sheathing. (top: triangle, middle: diamond, bottom: dot)

During the first stage all weight monster show the same moisture content evolution. However, from the moment the gaps are introduced a significant moisture increase is noticed at the top position. For FIBREBOARD 2 also the moisture content at the bottom and middle position slightly increases indicating the existence of forced convection. Creating in the next two steps an overpressure across the walls does not seems to influence the moisture content of

FIBREBOARD 1. For FIBREBOARD 2 on the other hand – of which the exterior sheathing is twenty time more air open – we can see a clear correlation between the magnitude of the overpressure and the moisture content of the weight samples at the three heights.
As a result of the high vapour permeance of the exterior sheathing an instant steep decrease of the moisture content is observed when drying condition are created in the final step.

DISCUSION

The current paper studies the hygrothermal impact of light weight highly insulated walls with an exterior air barrier. Three different potential exterior air barrier materials were tested. The results show that a sufficiently airtight material is a prerequisite in obtaining a safe building envelope. To this respect FIBREBOARD 2 (0.1 $m^3/m^2/h/Pa$) obviously fails, resulting in a forced exfiltration flow through the wall, and thus, increased heat losses and very high moisture contents of the exterior sheathing. For FIBREBOARD 2 (0.005 $m^3/m^2/h/Pa$) this effect was limited to a very minimum from which we can conclude that such levels of air permeance are sufficiently low to prevent harmful amounts of forced exfiltration.
On the other hand, the result show that even if forced convection is limited (FIBREBOARD 2 and FOIL), an increased moisture load is introduced into the structure by moving the air barrier to the exterior of the building envelope. For both tests sections the results show that water vapour driven by natural convection enters through the upper gap and deposits at the cold side of the insulation layer (Figure 6). The danger of this process is its continuity. Driven by the temperature difference across the wall, this convection loop provides a constant moisture flow towards the upper cold side of the structure.

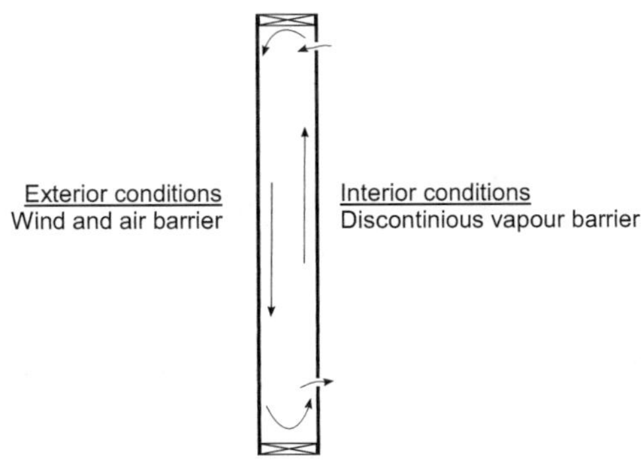

Figure 6. Air flow pattern as a result of natural convection inside a vertical wall with exterior air barrier.

CONCLUSION

The current paper presents the results of a laboratory experiment to study the hygrothermal behaviour of highly insulated walls with an exterior air barrier. The results show an increased moisture flow at the upper part of the walls driven by buoyancy forces. The magnitude of this flow is, apart from the position and size of the gaps, highly depending on the air permeability and the accuracy of the installation of the insulation layer. For the current study (very carefully installed) mineral wool is used which leads to a significant moisture increase. Further tests to investigate the importance of this effect for other insulation materials, such as cellulose or studying the influence of bad workmanship of the insulation layer would be an added value to this research.

ACKNOWLEDGEMENTS

Research funded by a Ph.D. grant (grant number 81153) of the Institute for the Promotion of Innovation through Science and Technology in Flanders (IWT-Vlaanderen).

REFERENCES

[1] Langmans, J., Klein, R., De Paepe, M., & Roels, S. (2010). Potential of wind barriers to assure airtightness of wood-frame low energy constructions. *Energy and Buildings*, *42*(12), 9.

[2] Jokisalo, J., Kurnitski, J., Korpi, M., Kalamees, T., & Vinha, J. (2009). Building leakage, infiltration, and energy performance analyses for Finnish detached houses. *Building and Environment*, *44*(2), 377-387.

[3] Kalamees, T. (2007). Air tightness and air leakages of new lightweight single-family detached houses in Estonia. *Building and Environment*, *42*(6), 2369-2377.

[4] Holøs, S. B., & Relander, T.-oskar. (2010). Airtightness Measurements of Wood Frame Low Energy Row Houses. *BEST conference* (pp. 1-11). Portland.

[5] Kalamees, T. (2007). Air tightness and air leakages of new lightweight single-family detached houses in Estonia. *Building and Environment*, *42*(6), 2369-2377.

[6] S. Uvsløkk, The importance of wind barriers for insulated timber frame constructions, Journal of Thermal Insulation and Building Envelopes 20 (1996) 40–62.

[7] Ojanen, T. 1993. Criteria for hygrothermal performranco for wind barrier structures,' proceedings of the 3rd symposium builidng physics in the Noridc Countries, B. Saxhof, ed., Thermal Insulation Laboratory, Technical University of Denmark, pp. 643-652

[8] Janssens, A., & Hens, H. (2007). Effects of wind on the transmission heat loss in duo-pitched insulated roofs: A field study. *Energy and Buildings*, *39*(9), 1047-1054.

[9] Myhre, L., & Aurlien, T. (2005). Measured air-tightness in low-energy houses. *Nordic Building Physics Symposium*. Reykjavik, Iceland.

[10] Langmans, J., Klein, R., & Roels, S. (2011). Hygrothermal response of highly insulated timber frame walls with an exterior air barrier system: laboratory investigation. *9th Nordic Symposium on Building Physics* (pp. 525-531). Tampere University of Technology.

State of the Art of Non-Residential Buildings Air-tightness and Impact on the Energy Consumption

Valerie Leprince*[1], Adeline Bailly [1], Rémi Carrié [2] and Myriam Olivier [1]

1 Centre d'Etudes Techniques de l'Equipement de Lyon.
46 rue Saint Théobald. F – 38081 L'Isle d'Abeau Cedex, France,
**Corresponding author*
valerie.leprince@developpement-durable.gouv.fr

2 INIVE,
Lozenberg 7
BE-1932 Sint-Stevens-Woluwe, Belgium

ABSTRACT

Starting January 1st, 2013 the French thermal regulation will impose a minimum requirement for residential buildings air-tightness. However, nothing is planed for non-residential building, for two reasons:

- There is no clear view on the level to be imposed on non-residential building air-tightness
- Air-tightness impact on energetic consumption is different in non -residential and in residential buildings.

Through the measurer's authorization process, it is possible to collect any measure done in France by certified measurer [1]. This paper first presents the analyse of those data and the non-residential buildings air tightness level according to their volume, kind of construction and use. It also compares these results with the level of residential buildings

Then the paper presents an estimation of the impact of air-tightness on energy consumption, according to various parameters (climatic zone, kind of ventilation, kind of building...). Calculations were realized with thermodynamic calculation tool included in the French national thermal regulation.

KEYWORDS

Air-tightness, Thermal regulation, energy performance, non-residential buildings

INTRODUCTION

Building airtight is a compulsory condition for low energy buildings. For residential building a minimum air-tightness is required for most low-energy labels such as BBC-Effinergie, Passiv'Haus, Minergie and shortly in the 2012 French thermal regulation (RT2012). Requirements are less easy to set for non-residential buildings.

Indeed, non-residential building air-tightness measurement may require very specific equipments. Moreover as there are much less measurements in non residential buildings, the state of the current building stock is little known. Finally, as there are various kinds of non-residential buildings, it is important to know the impact of air-tightness on energy consumption, according to the use and the location of the building before setting a requirement.

This article first presents the state of the art on non-residential buildings air-tightness from a trusted database. It will include the distribution by kind of building, year of construction, and kind of construction.

Secondly, this paper discusses the impact of air-tightness on non-residential buildings energy consumption according to their use, their localisation, the kind of heating and of ventilation. All calculations are done with the new EP-calculation tool included in the 2012 French thermal regulation.

The objective of this study is to estimate the feasibility and the opportunity to add a requirement on air-tightness for low-and very low-energy non-residential buildings.

STATE OF ART OF AIR-TIGHTNESS IN NON-RESIDENTIAL BUILDINGS

The analysed data

The 188 measurements analysed in this paper are extracted from the measurement databases of "licensed technicians" authorized to perform pressurization tests in low-energy (BBC-Effinergie certified) buildings. In fact, the authorization process described by Carrié et al (2010) [1] requires for each authorized technician to produce an annual report that includes results of all of his air leakage measurements. Therefore, the sample is heavily biased towards low-energy buildings: 29% of the tested non-residential buildings were involved in a BBC-Effinergie certification process, whereas the market share for this certification is only 3% of all new constructions. As a result, even if there is no air-tightness requirement for non-residential buildings, the distribution is certainly quite optimistic as BBC builder are mostly aware of air-tightness impact.

The 188 measurements include offices, schools, restaurant, etc. and present various constructive techniques, the repartition is given in Figure 1.

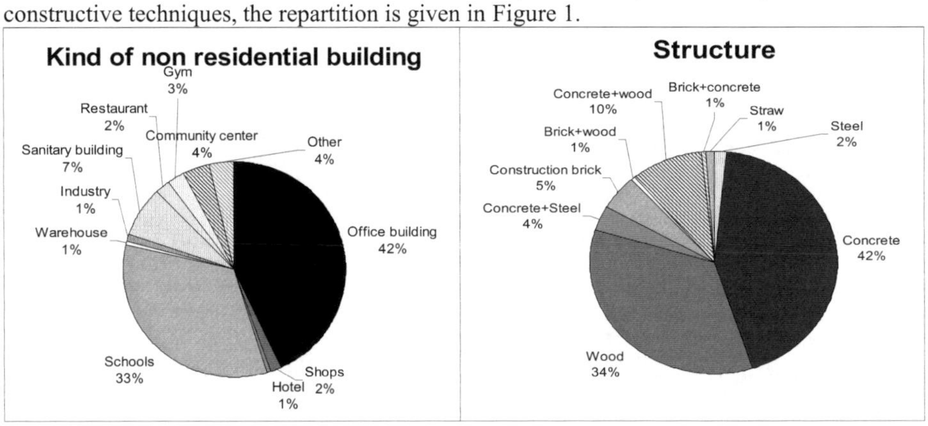

Figure 1: distribution of measured buildings

Most of the measured buildings are small and newly built, Figure 2 shows that more than 80% were built in the last ten years and 90% are less than 5000m³.

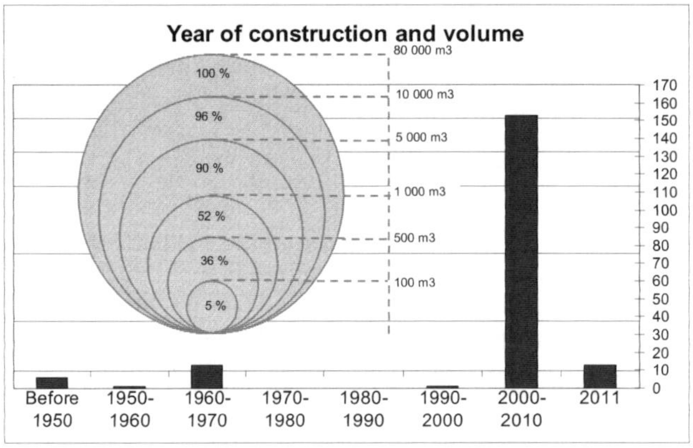

Figure 2: Volume and construction year of measured buildings

Those buildings are located all over France, their thermal insulation is either interior (43%), or exterior (20%) or else distributed (37%). They are equipped with various kind of ventilation system: ventilation with recovery system (67%), extraction only (28%) and natural ventilation (5%). Thus this sample is quite representative of every kind of construction in France.

Building air-tightness results

The average air-tightness in the 188 studied non-residential buildings is $Q_{4Pasurf}$= 2.28m^3/h.m², the median is $Q_{4Pasurf}$=1.28 m^3/h.m² and the standard deviation is 2.57m^3/h.m². Figure 3 and Figure 4 represent the distribution of measured air-tightness for each kind of building and for each constructive technology.

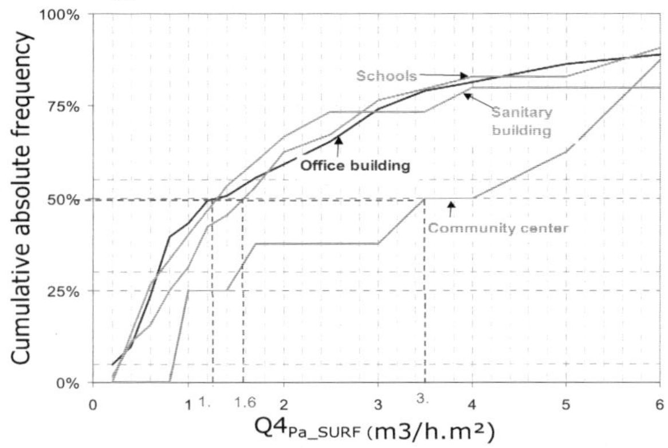

Figure 3: Distribution of measured air-tightness for each kind of building

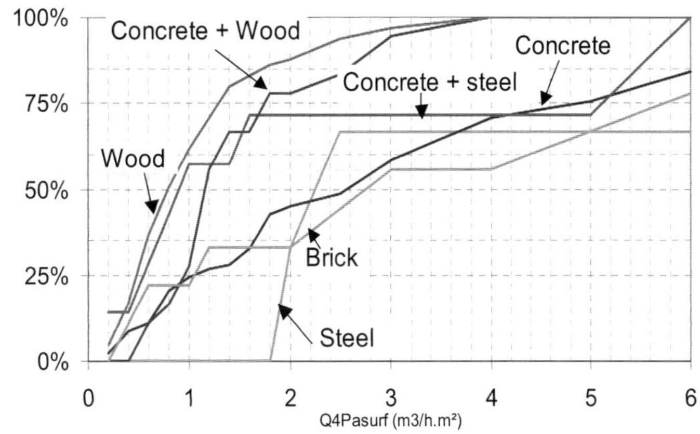

Figure 4: Distribution of measured air-tightness for each construction kind

Figure 3 shows that there is no real differences between each kind of buildings, excluding community center but we only have a small sample for this kind. However Figure 4 strongly highlights differences between each construction kind : wood and concrete + wood structure get much better results than others. Indeed the air-tightness average for wood structure is 1,00 m^3/h.m² whereas the average for all other structure is 3 m3/h.m². Nevertheless this difference is not irrecoverable as among the 188 measures the minimum value is obtain for a concrete structure and is 0.07 $m^3/h.m^2$.

Table 1 compare the previous values with those obtained in apartment buildings for which we have more than 400 measurements.

	Non residential building	Apartment building
Average	2.28 $m^3/h.m^2$	0.86 $m^3/h.m^2$
Standard deviation	2.57 $m^3/h.m^2$	0.80 $m^3/h.m^2$
Median	1.28 $m^3/h.m^2$	0.65 $m^3/h.m^2$

Table 1: Comparison between residential and non-residential building

Table 1 shows that apartment buildings get much better results than non-residential buildings, whereas both volume and construction kind are not much different. So it seems obvious that if an air-tightness level was compulsory in non-residential buildings (as in apartments building), results would easily be improved.

IMPACT OF AIR-TIGHNESS IN NON-RESIDENTIAL BUILDING

Procedures

The EP-calculation tool for the new French regulation, RT2012, is ready to make sensitive analysis So we developed an interface with an Excel spreadsheet to make automatic sensitive analysis for the following parameters:

- Building air-tightness
- Geographic localisation
- Duct air-tightness
- Ratio of duct outside the heating volume
- Recovery coefficient of the ventilation system.

To make the sensitive analysis tested buildings for the validation of the EP-calculation tool are used. For each building, we have at our disposal an xml file readable by the EP-calculation tool. The Excel tool automatically modify the xml file and launches the executable (dll) of the EP-calculation.

The sensitive analysis has been done on two non-residential buildings: a 1331 m^2 wood-heated primary school and a 613 m^2 gas-heated office building.

Parametric Values :

- two ventilations systems (extraction only and recovery system) and the three climatic zones (Oceanic (Nantes), Mediterranean (Marseille), Continental (Paris)) were tested.

- $Q_{4Pasurf}$=1,2 $m^3/h.m^2$ and 3 $m^3/h.m^2$. These 2 values represent the median of data and the largest default value for non-residential building in the French thermal regulation.

Consumption is evaluated in primary energy which mean that electric consumptions are multiplied by 2,58.

Results

Figure 5 shows that, depending of the ventilation system, improving air-tightness from $Q_{4Pasurf}$= 3 $m^3/h.m^2$ to 1,2 $m^3/h.m^2$ leads to a 13 to 37% decrease of energy consumption for those two low-energy buildings. It represents from 6 to 17 kWh/m².year.

If the sole heating consumption is estimated, air-tightness can be responsible for an over-consumption of almost 200% (Figure 5 - see Marseille, ventilation with recovery system). In fact, in this climatic zone, low-energy building offices have very low heating needs, as climate is mild and building get high internal and solar gains.

For both school and office building, air-tightness impact is strongest in oceanic climate, because it's the windiest French climate and it is less mild than Mediterranean's. Nevertheless, no cooling system has been modelled yet and this could change the impact.

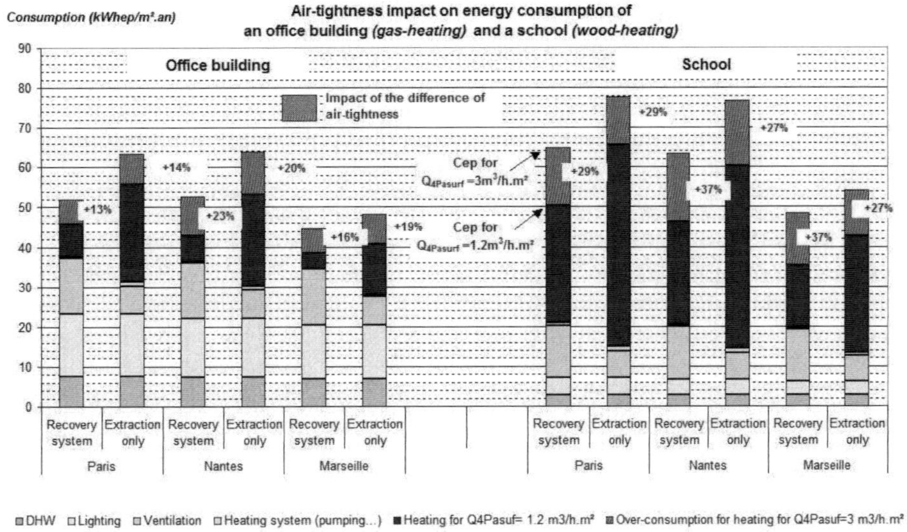

Figure 5: Air-tightness impact on energy consumption for two non-residential buildings

As expected, if heating consumption only is taken into account, the impact of air-tightness is much more important in buildings equipped with recovery systems ventilation than with extraction systems ventilation.

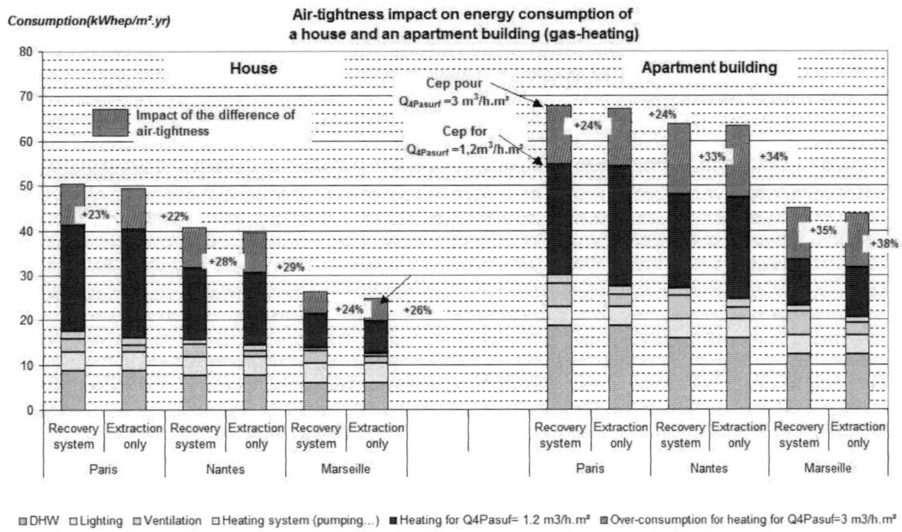

Figure 6: Air-tightness impact on energy consumption for two residential buildings

As far as the objective of this study is to estimate the opportunity of adding an air-tightness requirement in non-residential low energy buildings, its impact should be compared with residential buildings where this requirement does exist.
Figure 6 shows the same evaluation for an apartment and a house. It shows that the impact is almost the same in the apartment and in the school. The impact in the office building is smaller but as it can reach 23 % it still seems opportune to control the air-tightness.

CONCLUSION

The objective of this study was to estimate feasibility and opportunity to require air-tightness level for low- and very low-energy non-residential buildings.
On measurement feasibility, 95% of measures, performed on non-residential building, were made on less-than-6000m3 volume. Technically, this volume requires 3 classical blower-doors for an air-tightness of Q4Passurf=1,2m^3/h.m² when reaching 50Pa is required. Three blower-doors are easy to gather as measurers are an organised network, but more than three seems more difficult to gather, as there are only few measurement over 6000m3. Anyway, in such case, measure adjustments could be allowed (dividing the building, reaching only 25 Pa, etc.).
The second point was the level of requirement ; technically $Q_{4Pasurf}$= 1,2 m^3/h.m² seems a reasonable value for schools, office buildings, hotels, sanitary buildings, community centers and restaurants. Their building techniques are more or less equivalent to residential buildings and this value is close from the measures median we have. Nevertheless it's important to keep in mind that, according to our data, this value may be easier to reach with wooden-structure for example, than with steel-structure for example.
The third point was the requirement request opportunity : the sensitive analysis showed that such a requirement was as interesting for non-residential as for residential building, with a potential global consumption gain of more than 30% for some kind of buildings. Thanks to our automatic tool, this study will be easily extended to others kind of buildings, when their xml files (readable by the EP-calculation tool) will be available, including buildings with air-conditioning system.

ACKNOWLEDGEMENTS

The contribution of CETE de Lyon in the implementation of these quality frameworks is funded by the ministry of ecology, sustainable development, transport and housing (MEDDTL).
The sole responsibility for the content of this publication lies with the authors. It does not necessarily reflects the opinion of the Ministry.

REFERENCES

[1] Carrie, F.R., Charrier, S. and Leprince, V. 2010. *Implementation of measurement and quality frameworks in the French regulation for achieving airtight envelopes*, 31th AIVC Conference, 26-28 October 2010, Seoul Korea

[2]*Arrêté du 26 octobre 2010 relatif aux caractéristiques thermiques et aux exigences de performance énergétique des bâtiments nouveaux et des parties nouvelles de bâtiments*, JO 27 octobre 2010

[3]*Arrêté du 24 mai 2006 relatif aux caractéristiques thermiques des bâtiments nouveaux et des parties nouvelles de bâtiments*, JO 25 mai 2006

INVESTIGATIONS ON THE EFFECTS OF AIRTIGHT PERFORMANCE IMPROVEMENT AND ENERGY CONSUMPTION OF INSULATION RETROFIT IN DETACHED HOUSES

Hiroshi Yoshino[*1], Kenichi Hasegawa[2], Shinichi Matsumoto[2], Hayato Hosobuchi[2], Akiharu Uchida[1], Takuya Ino[1]

1 Tohoku University,
Department of Architecture and Building Science
6-6-11-1203, Aramaki-aza, Aoba, Aoba-ku,
Sendai, Miyagi, Japan

2 Akita Prefectural University,
Faculty of Systems Science and Technology
84-4, Ebinokuchi, Tsuchiya-aza,
Yurihonjyo, Akita, Japan

**Corresponding author:*
yoshino@sabine.pln.archi.tohoku.ac.jp

Note : the contact addresses may be re-arranged

ABSTRACT

Recently, insulation retrofits of existing houses have been thought to be one of the effective measures from the viewpoint of global warming prevention. However, the overall reduction effects of environmental loads by the insulation retrofits have not yet been clarified. This study intends to accumulate basic data concerning the insulation retrofits and to promote the energy saving of existing houses.

The environmental performances of 4 detached wooden houses in *Tohoku* region, Japan before and after the insulation retrofits were investigated. The indoor thermal environments, energy consumption, and performances of insulation and air tightness before and after the retrofit were analyzed. In addition, the effects of insulation retrofits were clarified.

The heat loss coefficient ("Q") and the equivalent air leakage area in proportion to the floor area ("C") were calculated in each house before and after the retrofit. After the retrofit, the values were changed to Q=1.2 W/m^2K & C=2.1 cm^2/m^2 in house_A and Q= 1.5W/m^2K & C=1.1 cm^2/m^2 in house_B.

During winter, in house_C, temperature differences between the living room and other rooms went up to 20 ℃ before the retrofit. After the retrofit, temperature differences were limited about 5 ℃, and the indoor vertical temperature difference was 2.5 ℃at a maximum.

In house_A, city gas was used twice and once a day before and after the retrofit respectively. The average values of city-gas consumption were 10 kW and 8 kW before and after the retrofit respectively.

Comparing annual energy consumption before and after the retrofit, in houses_A and B, energy consumption were decreased by 35% and 44% after the retrofit respectively. However, in house_D, energy consumption was increased by 42% because the insulation retrofit had partial effect.

KEYWORDS

Insulation retrofit, Air tightness, Energy consumption, Indoor thermal environment, Field measurement, Detached house

INTRODUCTION

The first commitment period of the Kyoto Protocol started in 2008. The government set the medium-term target for reducing greenhouse gas emissions in Japan; reducing emissions by 25% compared with 1990. Recently, insulation retrofits of existing houses have been thought to be one of the effective measures from the viewpoint of global warming prevention. In Japan, there are many existing houses without enough thermal insulation. However, the overall reduction effects of environmental loads by the insulation retrofits have not yet been clarified. Therefore, this study intends to accumulate basic data concerning the insulation retrofits and to promote the energy saving of existing houses based on the methods and effects of energy saving for retrofitted houses.

The authors investigated the environmental performances of 4 detached wooden houses in *Tohoku* region, Japan before and after the insulation retrofits [1]-[4]. In this paper, the indoor thermal environments, energy consumption, and performances of insulation and air tightness before and after the retrofit were analyzed. The effects of insulation retrofits were clarified.

OUTLINE OF INVESTIGATED HOUSES

The location of the investigated houses is shown in Figure 1. The description of each house is shown in Table 1.

In houses_A, B and C, the whole house was renovated while the occupants' living there. Heat insulators were added to the existing walls after the exterior materials were removed, and the interior materials remained except some changed parts. In house_A, a push-pull ventilation fan with total heat exchanger was set in each room during the retrofit. All rooms can be warmed up by existing heating panels, and the heat source is water heated by a co-generation system (gas engine type). Moreover, one month after the retrofit, the photovoltaic system was installed in house_A. In house_B, all rooms can be heated by new hydronic heating panels and existing heating equipments. In house_C, there is an existing air conditioner, and hydronic heating panels were set after the retrofit. In both houses_B and C, the heat source of hydronic heating panels is water heated during night time. Hot water is supplied by an air refrigerative heat pump boiler, and the pull ventilation system is used.

In house_D, heat insulators were partially added to only the walls and the base. The household equipments are the existing ones.

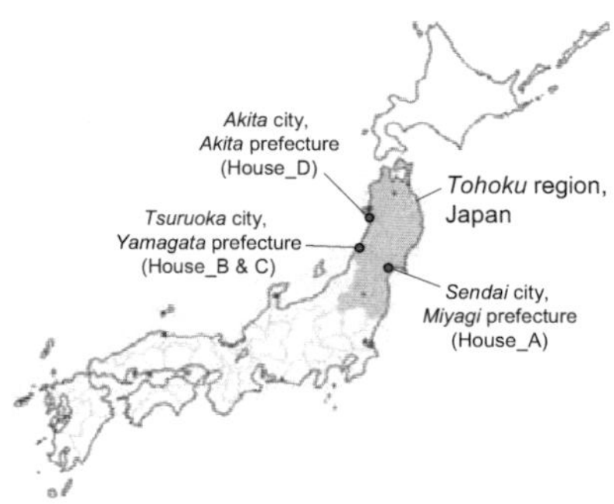

Figure 1. Location of investigated houses.

	House_A		House_B		House_C		House_D	
	Before	After	Before	After	Before	After	Before	After
Site	Sendai, Miyagi		Tsuruoka, Yamagata		Tsuruoka, Yamagata		Akita, Akita	
Household size	4	3	3		3	7	2	
Completion year	1984	2010	1991	2008	1980	2007	1995	2010
Floor area	124.2 m^2		141.2 m^2	164.0 m^2	229.4 m^2	244.3 m^2	134.3 m^2	
Heat loss coefficient	3.6 W/m^2K	1.2 W/m^2K	4.5 W/m^2K	1.5 W/m^2K	8.3 W/m^2K	1.7 W/m^2K	2.3 W/m^2K	2.1 W/m^2K
Equivarent leakeage area	3.6 cm^2/m^2	2.1 cm^2/m^2	8.2 cm^2/m^2	1.1 cm^2/m^2	--	1.2 cm^2/m^2	--	2.1 cm^2/m^2
Space heating*	Hydronic panel heater, AC		FF type stove heater, Hydronic heater, AC	Hydronic panel heater, Hydronic heater, AC	FF type stove heater, Fan heater	Hydronic panel heater, AC	FF type stove heater, Hydronic heater, AC	
Hot water supply**	Gas engine co-generation system		Combination boiler	Electric HP water heater	Combination boiler	Electric HP water heater	Combination boiler	
Ventilation type (in living space)	--	Push-pull (heat exchanger:62%)	Pull		--	Pull	Pull	
Energy source	City gas, electricity		Oil, electricity	Electricity	Oil, city gas, electricity	Electricity	Oil, city gas, electricity	

*FF type: Forced draught balanced Flue type, AC:Air Conditioner **HP:Heat Pump

Table 1. Description of investigated houses.

PERFORMANCES OF INSULATION AND AIR TIGHTNESS

The heat loss coefficient ("Q") and the equivalent air leakage area in proportion to the floor area ("C") were calculated in each house before and after the retrofit.

The Q values were calculated from the design documents of each house [5]. In house_A, the Q value before and after, respectively is 3.6 W/m^{2}K and 1.2 W/m^{2}K. With the retrofit, 100mm insulation boards (polystyrene foams or phenolic foams) were added to the existing walls. 400mm glass wool and 100mm polystyrene foams were added to the ceiling and floor respectively. As for the windows, Low-E triple-pane glasses and insulated plastic frames are used for the retrofit.

The Q values were changed from 4.5 W/m^{2}K to 1.5 W/m^{2}K in house_B, and from 8.3 W/m^{2}K to 1.7 W/m^{2}K in house_C. In the wall and ceiling, phenolic foams & high performance glass wool were added. Polystyrene foams were added to the existing base, and Low-E double-pane glasses and plastic frames are used after the retrofit.

In house_D, the Q value before and after, respectively is 2.3 W/m^{2}K and 2.1 W/m^{2}K. The insulation performance was not improved so much because added heat insulators were partial.

The C values were measured by the depressurization method using the airtight instrument (Figure 2). The C values were changed from 3.6 cm^2/m^2 to 1.2 cm^2/m^2 in house_A, and from 8.2 cm^2/m^2 to 1.1 cm^2/m^2 in house_B. In houses_C and D, the C values were not measured before the retrofit. After the retrofit, that is 1.2 cm^2/m^2 in house_C and 2.1 cm^2/m^2 in house_D.

Figure 2. Measuring air tightness.

AIRTIGHT CONSTRUCTION METHODS IN EACH HOUSE

Examples of airtight construction methods in each house are shown in Figure 3.

In house_A, insulation boards were added to the existing walls and floor from outside. In houses_B and C, glass wools were filled in the walls and insulation boards were added from outside. Then air tightness was ensured by the added insulation boards in each house. For example, airtight sheets were applied to the walls before insulation boards were added, and the connections of insulation boards were sealed with the airtight tape. The connections between the window frames and insulation boards were also sealed. In house_A, after that, damp-proof membranes were added from outside, and the vent layer was made.

Moreover, in houses_B and C, the gaps between insulation boards and the groundsill / roof rafters were filled with the foam insulation in the base / the attic. In house_A, the gaps between insulation boards and the base / pipes etc. were filled with urethane foam in the underfloor space. The foam insulation were also used places beyond the reach.

Figure 3. Examples of airtight construction methods.
(Left:house_A, Middle:house_B, Right:house_C)

PROFILES OF INDOOR THERMAL ENVIRONMENT IN WINTER

Temperature and humidity of each house in winter were measured. The changes of outdoor and living-room temperatures during 1 week in winter before and after the retrofit are shown in Figure 4. (In house_D, before the retrofit, the data for 5 days are used.) The data interval is 30 min. In house_B, the measurement was not carried out before the retrofit.

In houses_A and D, there were little differences of living-room temperatures before and after the retrofit. This is because thermal insulation performances before the retrofit were not so low, and the same heating equipments were used before and after the retrofit in houses_A and D.

In house_B, though not in comparison with temperatures before the retrofit, living-room temperature after the retrofit exhibited relatively less volatility than temperatures in other houses. In house_C, while living-room temperature fluctuated widely before the retrofit, the fluctuation range after the retrofit was the smallest in all houses.

Next, the measurement result in house_C is shown in detail. In house_C, temperature and humidity were measured in 6 places (Living-room, Bedroom, Guest-room, Toilet, Hallway, Outdoor), and the measurement interval was 10min. The profile of temperature in each room during 3 days in winter before and after the retrofit is shown in Figure 5.

Before the retrofit, temperature differences between living-room and other rooms were very large from morning (around 7 a.m.) till midnight, and went up to 20 ℃. Though living-room temperature varied around 22 ℃ in heating equipments operation, temperatures of other rooms fluctuated just like outdoor temperature. Moreover, after turning off heaters, living-room temperature went down soon.

On the other hand, after the retrofit, indoor temperatures varied from 14 ℃ to 21 ℃, and temperature difference between rooms was small.

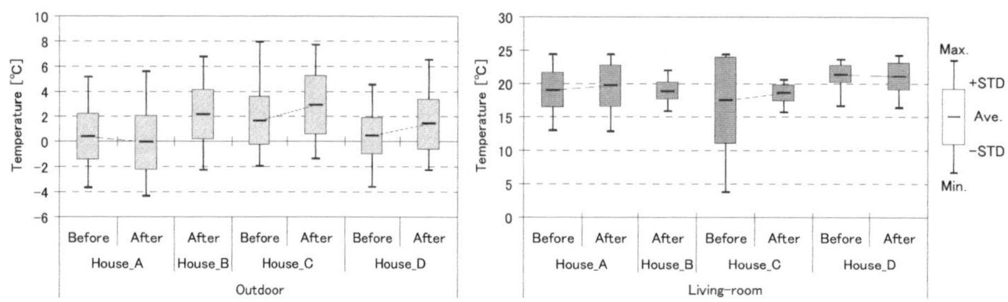

Figure 4. Changes of outdoor and living-room temperatures in winter.

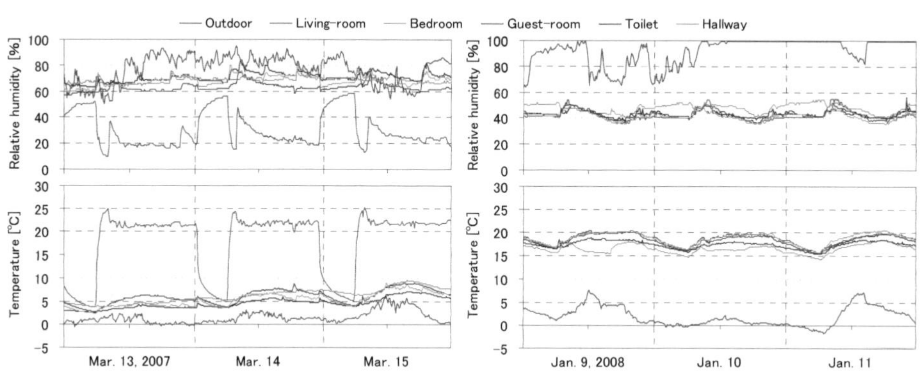

Figure 5. Profiles of temperatures in house_C (Left:before, Right:after).

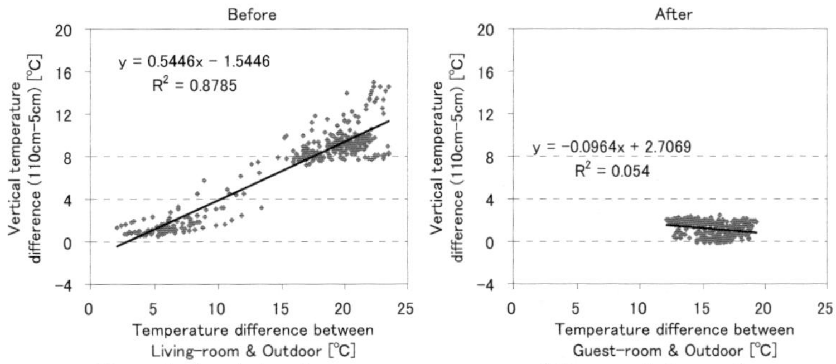

Figure 6. Correlations between two temperature differences in house_C.

The correlations between two temperature differences during 1 week in winter before and after the retrofit are shown in Figure 6. In this figure, the horizontal axis shows temperature difference between indoor and outdoor. The vertical axis shows indoor vertical temperature difference between 5cm and 110cm above the floor. The data interval is 30 min.

The vertical temperature difference was 15 ℃ at a maximum in house_C. Before the retrofit, the value of coefficient of determination (R^2) is 0.88, so outdoor temperature contributed to increment of vertical temperature difference. After the retrofit, the vertical temperature difference was 2.5 ℃ at a maximum. The upper limit of vertical temperature difference (0.1m-1.1m above the floor) is 3 ℃ in ISO 7730 [6]. Moreover, the value of coefficient of R^2 is very small, and outdoor temperature and vertical temperature difference have no correlationship.

PROFILES OF ENERGY CONSUMPTION IN WINTER

In house_A, energy consumption in winter was measured in detail. The profiles of energy consumption and temperatures during 3 days in February before and after the retrofit are shown in Figure 7 and Figure 8 respectively. The data interval was 15 min. 3 day profiles when daily-averaged outdoor temperatures were similar before and after the retrofit are compared.

Before the retrofit, outdoor temperature changed from 0 ℃ to 13 ℃, while living-room temperature varied from 16 ℃ to 24 ℃. Space heating was operated twice a day in the morning and evening. Living-room temperature rose by 5 ℃ during space heating. Bedroom temperature, which varied from 15 ℃ to 21 ℃, was a little lower than living-room temperature. The co-generation system ("CGS") generated electricity (max. 1 kW) when city gas and electricity were used so much.

In contrast, outdoor temperature widely varied from -4 ℃ to 17 ℃ after the retrofit. But living-room temperature changed from 18 ℃ to 25 ℃ and bedroom temperature changed from 17 ℃ to 21 ℃. Though space heating was not operated so much in the morning, indoor temperature change was smaller than that of before retrofit, and room temperature was kept more than 17 ℃. The photovaltaic system ("PV") generated electricity (max. 2.4 kW) during the day.

City gas was used twice a day before and once a day after the retrofit respectively. The average values of city-gas consumption were 10 kW and 8 kW before and after the retrofit respectively.

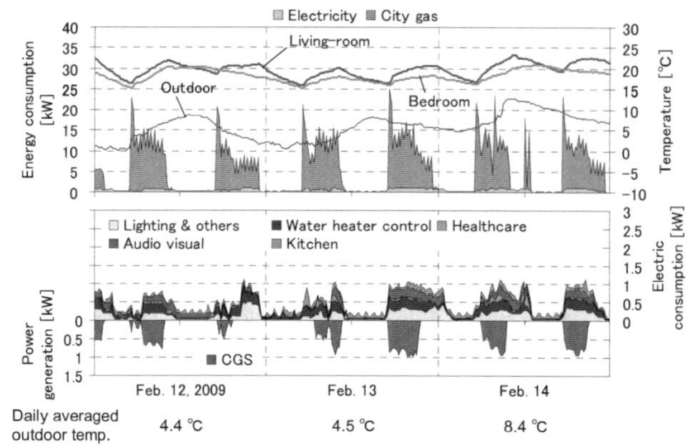

Figure 7. Profiles of energy consumption in house_A before the retrofit.

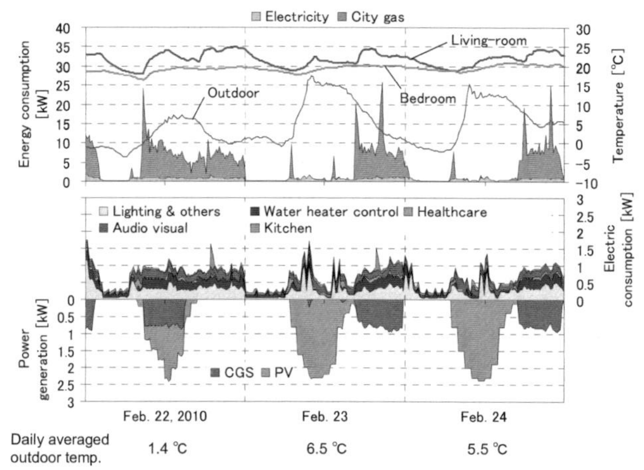

Figure 8. Profiles of energy consumption in house_A after the retrofit.

Based on the comparison of temperature change and city-gas consumption before and after the retrofit, it can be stated that insulation retrofit made the change of temperature smaller and decreased the city-gas consumption.

COMPARISON OF ANNUAL ENERGY CONSUMPTION BEFORE AND AFTER THE RETROFIT

Annual energy consumption in each house before and after the retrofit is shown in Figure 9. This was calculated by the receipts of energy bill (oil, city gas and electricity), and measurement results was considered in house_A. Energy coefficients of oil, city gas and electricity are shown in Table 2 [7].

In house_A, energy consumption decreased by 23% (24.8 GJ) after the retrofit. The power generation from CGS and PV system was 76% of the electric consumption ("Others" in Figure 9) after the retrofit. Considering the power generation as reduction of energy consumption, energy consumption decreased by 35% (35.7 GJ).

In house_B, the whole energy consumption decreased by 44% (48.4 GJ) after the retrofit. This is especially for space heating with energy consumption decreased by 33% (19.7 GJ).

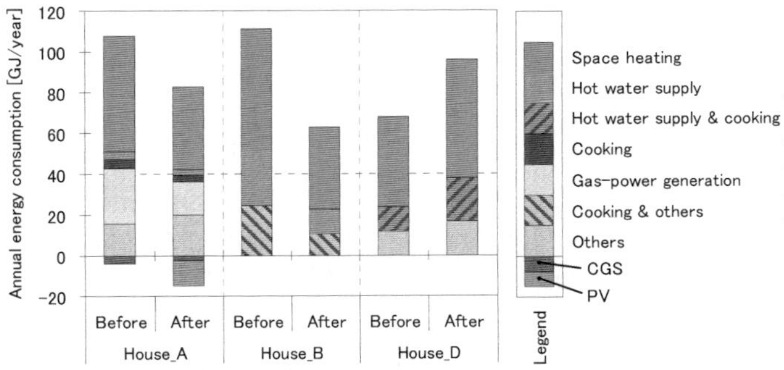

Figure 9. Annual energy consumption before and after the retrofit.

	Basic unit	Energy coefficient
Oil	L	37.26 MJ/L
City gas	m^3	46.05 MJ/m^3
Electricity	kWh	3.6 MJ/kWh

Table 2. Energy coefficients.

Hence, the performance of thermal insulation, energy efficiency of equipments and occupants' awareness of energy saving were improved after the retrofit.

In contrast, the whole energy consumption increased by 42% (28.5 GJ) after the retrofit in house_D. The space heating alone, energy consumption increased by 33% (14.5 GJ). After the retrofit, though reduction of energy consumption for space heating was expected, the insulation retrofit was ineffective for some parts of the house. Usually heat losses from the windows are large, so the athermalize of the windows may be more effective than that of the walls and the base.

CONCLUSION

This paper provided measurement results of 4 houses before and after the insulation retrofit in Tohoku region, Japan.

About profiles of indoor thermal environment and energy consumption in winter,

- In houses_A and D, there were little differences of living-room temperatures between before and after the retrofit. In houses_B and C, the fluctuation ranges were small after the retrofit.
- In house_C, temperature differences between rooms was small after the retrofit, and the vertical temperature difference was 2.5 ℃ at a maximum.
- In house_A, heating equipment operation hours and daily consumption of city gas decreased after the retrofit.

About annual energy consumption,

- In house_A, considering the power generation by CGS and PV as reduction of energy consumption, energy consumption decreased by 35% (35.7 GJ). In house_B, energy consumption decreased by 44% (48.4 GJ) after the retrofit.
- In house_D, energy consumption increased by 42% (28.5 GJ) after the retrofit. This is because the insulation retrofit was ineffective for some parts of the house.

ACKNOWLEDGEMENTS

SHINWASOUKEN Ltd., KIBA Corporation and the occupants in the investigated houses cooperated on the measurements. Mr. Yamato Ishiguchi, Ms. Risako Sano and Ms. Nanako Anbo, who were students in Akita Prefectural University, are acknowledged for help with the measurements and data analysis. A part of this study was carried out in cooperation with many people who participated in "The experiment project of an insulation retrofit aimed at making it carbon-neutral (No.1) (organized by Living & Environment Tohoku Forum)". The authors would like to express their appreciation for the cooperation of those who were involved in this study.

REFERENCES

[1] Akiharu UCHIDA, Hiroshi YOSHINO and Takuya INO. Aug. 2011. *Measurement of Energy Consumption in A Detached House Before/After Insulation Retrofit and PV Installation*, Summaries of Technical Papers of Annual Meeting Architectural Institute of Japan, D-2, 315-316. (in Japanese)

[2] Hiroshi Yoshino and Akiharu Uchida. Sep. 2011. *Energy Consumption In A Detached House Before/After Insulation Retrofit*, Proceedings of The 24th World Congress of Architecture, UIA2011 Tokyo, Japan, 180-184.

[3] Ken-ichi HASEGAWA, Shin-ichi MATSUMOTO, Kahori GENJO and Yamato ISHIGUCHI. Jun. 2008. *Evaluation of Energy Saving Performance in a Wooden Detached House Before and After Thermal Retrofit : Part2 Thermal Improvement in Indoor and Energy Saving Effect in a House Located in Tsuruoka City*, Proceedings of AIJ Tohoku Chapter Architectural Research Meeting, Vol.71, 17-20. (in Japanese)

[4] Ken-ichi HASEGAWA, Shin-ichi MATSUMOTO and Hayato HOSOBUCHI. Aug. 2011. *Case Study of Indoor Environmental Performance Improvement in Five Detached Houses Before and After Thermal Retrofit*, Summaries of Technical Papers of Annual Meeting Architectural Institute of Japan, D-2, 263-266. (in Japanese)

[5] Institute for Building Environment and Energy Conservation. 2002. Description of Energy Conservation Standards for Residential Buildings. (in Japanese)

[6] International Organisation for Standardization, Geneva. 1984. *Moderate thermal environments - Determination of the PMV and PPD indices and specification of the conditions for thermal comfort*, ISO 7730.

[7] 1999. Order for Enforcement of the Act on Promotion of Global Warming Countermeasures, Article 3. (in Japanese)

THE INFLUENCE OF AIR PERMEABILITY AND TYPE OF UNDERLAY ON THE HYGROTHERMAL PERFORMANCE OF AN INCLINED ROOF

Paul Steskens[1*], Filip Dobbels[1], Xavier Loncour[1], Gilles Flamant[1]

1 Belgian Building Research Institute
Avenue P. Holoffe 21
1342 Limelette,Belgium
**Corresponding author: paul.steskens@bbri.be*

ABSTRACT

The airtightness of inclined roofs is important in order to avoid hygrothermal problems and to guarantee the durability of the construction itself. Previous research in building physics showed that perfect airtightness of inclined roofs is difficult to achieve and maintain. In practice, air transport through the construction, i.e. in/exfiltration, cannot be avoided due to for example imperfections or bad workmanship. The heat and moisture conditions in the building component are strongly influenced by advection, i.e. air transport through the building component, resulting from air pressure differences between the indoor and outdoor environment.

Current models to predict heat, air and moisture (HAM) conditions in building components assume uniform boundary conditions, both for the temperature and relative humidity imposed at the internal and external surface of the building component. In such models, the heat and moisture fluxes due to advection are generally not considered. A more detailed description and prediction of the influence of the air transport through the building component on the HAM conditions in the building component would be desired.

In the work presented in this paper, the hygrothermal performance of an inclined roof is analyzed while taking into account the air transport through the construction. An airflow model is used to describe the advective transport through the building component. The airflow model is integrated into an existing HAM building component model. The analysis focuses on a case study which is based on a common roof design in Western Europe. Three configurations consisting of a vapour-retarding underlay foil (which was applied in the past), vapour-open underlay, and a fibre cement board underlay are investigated.

KEYWORDS
Moisture control, hygrothermal performance, vapour advection.

INTRODUCTION

Building energy consumption and sustainability are currently one of the most important issues worldwide. Improved energy performance of buildings cannot be achieved by additional insulation and more energy efficient building systems only. The airtightness of the building envelope has a major influence on a building's energy performance, thermal comfort, indoor air quality, and moisture damages [1]. Hence, sufficient airtightness is required to guarantee the adequate hygrothermal performance of a building and to avoid moisture problems.

In light-weight constructions, an airtight building envelope is usually realised by an interior air barrier system. The most important property of the air barrier is its continuity. Ensuring continuity in order to prevent (advective) airflow through the building component requires sealing of all the joints and intersections of this layer. In cold and moderate climates, such as in North-West European areas, the layer often combines the function of air barrier with the function of vapour retarder.

In current building practice, it is labour intensive and often difficult to construct a perfectly continuous air barrier, due to a large number of internal joints, intersections, and perforations, e.g. for electrical and plumbing services. Often air leakages due to unintended gaps and perforations cannot be avoided and moisture problems may develop due to these air leakages. Since the compliance with air leakage criteria in building practice is uncertain, it is important to include control measures in the envelope design to prevent severe moisture problems when the continuity of airtightness is not achieved. Such measures could reduce the effects of potential air leakages on the moisture performance of the building envelope.

For inclined roofs, several guidelines for condensation control have been developed, mainly influenced by national regulations. Guidelines for the selection of the air/vapour barrier, are based on the vapour diffusion resistance factor (μd_i) of the air/vapour barrier, and of the roof underlay (μd_e). A proper combination of the air/vapour barrier and the roof underlay is achieved as a function of the indoor environmental conditions. In Belgium, selection criteria have been published by Meert [2] and Janssens [3]. Similarly, German guidelines have been reported by the Fraunhofer Institute for Building Physics [4] [5], and in the German standard [6]. The guidelines are based on numerical studies where the influence of (advective) vapour flow on the interstitial condensation risk in an inclined roof is investigated while assuming steady-state boundary conditions. Moreover, a maximum allowable condensation due to air transport of respectively 200 g/m^2 roof area [3] and 250 g/m^2 roof area [5] [6] is assumed. In case the predicted condensate is exceeded, the component is expected to fail and the construction is not recommended.

While hygrothermal simulation models nowadays use transient boundary conditions for the indoor and outdoor climatic conditions, i.e. indoor and outdoor air temperature and relative humidity, solar and longwave radiation, and rain loads, advective airflows, i.e. the airflows through the building component, are often neglected. And, if the model takes into account in/exfiltration through a building component, the airflow is often assumed to be constant and steady-state. In reality, however, the airflow is strongly dependent of the wind induced pressure differences between inside and outside, and thus changing with time. Neglecting these transient influences, may lead to inaccuracies in the predicted hygrothermal conditions in the building component, resulting in an under- or overprediction of the hygrothermal conditions in the building component, and consequently damage and/or degradation of the construction may occur.

It is the objective of this paper to study the influence of transient (advective) airflow through a roof construction on the hygrothermal conditions in the component. A case study which is based on a common roof design in Western Europe is selected for analysis. Three configurations consisting of a vapour-retarding underlay foil, a vapour-open underlay, and a fibre cement board underlay are investigated. The hygrothermal conditions in the roof are simulated using the Delphin 5 software for transient heat, air and moisture transfer in building components [7] [8]. The hygrothermal performance of the different configurations is analyzed based on the predicted condensate that is generated during a year, the total moisture content in the construction evaluated over 4 years, and the maximum moisture content in the wood construction.

ANALYSIS AND METHODS

A common roof design in Western Europe is selected as a base case for analysis of the hygrothermal conditions. The roof contains the following layers (from outside to inside): concrete tiles on laths and battens, an underlay, 120mm of fibre glass insulation between the rafters (50x120mm), a vapour retarder, and painted gypsum board. The analyzed roof construction is presented in Figure 1. The roof is 1.8m wide and 4.55m long per pitch and oriented northeast with a slope of 30°. Additional information on the topology of the building can be found in [9]. The underlay (Figure 1) is installed directly on top of the insulation, continuously form eave to eave, with a sealed overlap. The design thermal properties of the roof are calculated based on the thermal conductivity of the applied building materials. Moreover, the design U-value of the roof corresponds to 0.27 W/(m^{2}K).

A more detailed description of the composition of the roof is presented in Table 1. Three configurations consisting of a vapour-retarding underlay foil, a vapour-open underlay, and a fibre cement board underlay are investigated. Regarding new buildings, the use of a vapour-open underlay foil as a protection for the insulation has become common practice in current residential construction. Moreover, considering the renovation of a roof construction, it may be possible that the underlay consists of a vapour-retarding underlay foil (which was used in the past). Alternatively, a capillary-active roof underlay, for example fibre cement board, could be applied. In this study, the influence of different roof underlays on the hygrothermal performance of the roof is investigated. Table 2 presents the material properties of the applied building materials. The air permeability of the fibre cement board is based on the measurements reported in Langmans et *al.* (2010) [10], while the air permeability of the other building materials are based on [11].

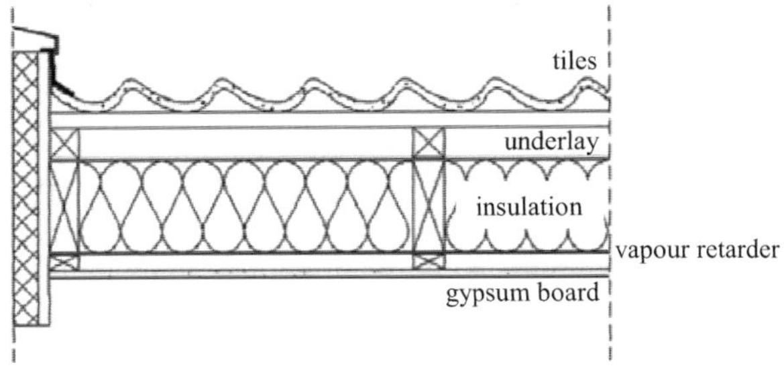

Figure 1: Roof construction with insulation between the rafters (and on top of the purlines)

Roof construction	**#1 - Vapour retarding underlay**	**#2 - Vapour-open underlay**	**#3 - Capillary-active underlay**
	Concrete tiles	Concrete tiles	Concrete tiles
Underlay	Vapour retarding foil	Vapour-open foil	Fibre cement board (3mm)
Insulation	Fibre glass (120mm)	Fibre glass (120mm)	Fibre glass (120mm)
Finishing layer	Painted gypsum board + vapour barrier (μd_i=2m)	Painted gypsum board + vapour barrier (μd_i=2m)	Painted gypsum board + vapour barrier (μd_i=2m)

Table 1: Analyzed roof constructions

	Thickness	Thermal conductivity	Vapour permeability	Air permeability (without joints)
	d [m]	λ [W/mK]	μ [-]	K_a [m s^{-1}Pa^{-1}]
Underlay foil:				
- vapour retarding	0.001	-	2000	$28.6 \cdot 10^{-7}$
- vapour-open	0.001	-	20	$28.6 \cdot 10^{-7}$
Fibre cement board	0.003	0.25	80	$6.4 \cdot 10^{-7}$
Fibre glass	0.12	0.035	1.5	
Vapour retarder	0.002	-	1000	$5.0 \cdot 10^{-7}$
Gypsum board	0.01	0.1	10	$6.9 \cdot 10^{-7}$

Table 2: Material properties

Hygrothermal modelling

A hygrothermal model of the presented roof construction has been developed using the hygrothermal software Delphin 5, which is an envelope model for the coupled simulation of heat, air, and moisture transport in a building component [7] [8]. First of all, the developed hygrothermal model has been verified using experimental data which has been obtained within the framework of a Belgian project on moisture problems in roof constructions [9]. Second, the model has been simulated during four years. External boundary conditions are applied using the Test Reference Year (TRY) for Belgian (Brussels) outdoor climatic conditions. Indoor environmental conditions are applied according to the Belgian classification for indoor climatic conditions. The indoor air temperature in the building is represented by Equation 1 and lies above 18°C during the entire year. The boundary conditions for the partial vapour pressure of the indoor air for respectively Belgian climate classes II (CC2) and III (CC3), are presented in Table 3 [2] [12].

Equation 1

$$T_i = \max(18; 8 + 0.8T_e)$$

where T_i and T_e are the indoor and outdoor air temperature [°C] respectively.

Climate class	**Type of building**		**Partial vapour pressure difference between the indoor and outdoor air [Pa]**
CC2	Buildings with moderate vapour production and more than sufficient ventilation	Large houses, schools, shops, etc.	$< 436 - 22T_e$
CC3	Buildings with significant vapour production and moderate ventilation	Small houses, restaurants, etc.	$< 713 - 22T_e$

Table 3: Belgian classification of indoor climate classes [2]

Airflow modelling

The air pressure difference across the construction is the driving force for the airflow through the construction. Air pressure differences are caused by buoyancy (stack effect), wind forces, pressure differences due to mechanical ventilation systems, open fireplaces, etc. The pressure difference across the construction is represented by Equation 2

Equation 2

$$\Delta p_a = C_p \rho_o \frac{U_h^2}{2} + \rho_o g H \frac{T_i - T_o}{T_i} + \Delta p_I$$

where Δp_a represents the total air pressure difference across a leak [Pa], C_p is the wind surface pressure coefficient [-], ρ_o the outdoor air density [kg/m^3], U_h the wind velocity [m/s], which is often taken at building height *h* [m] in the undisturbed upstream flow, *g* the gravitational acceleration [m/s^2], *H* the height above the reference plain [m], T_i and T_o represent the indoor and outdoor air temperature respectively [K], and Δp_I is the pressure that acts to balance inflows and outflows, including mechanical system flows [Pa].

In this study, only the pressure differences due to the stack effect and the wind forces are considered, since a mechanical ventilation system is not present in the building considered. The relative importance of wind and stack pressures in a building depends on the building height, internal resistance to vertical airflow, location and airflow resistance characteristics of envelope openings, local terrain, and the immediate shielding of the building. The wind pressure on the building envelope is usually expressed by pressure coefficients (C_p). Pressure coefficients on building facades are influenced by a wide range of parameters. As it is practically impossible to take into account the full complexity of C_p variation, building simulation models generally incorporate it in a simplified way. An intensive overview of wind pressure coefficient data in building energy simulation and air-flow network programs is reported by Cóstola (2009) [13].

Wind pressure coefficients for the building considered have been calculated using the web-based application C_p Generator, based on the geometry of the building, the building's surroundings, and sheltering [14]. The resulting C_p coefficients served as an input for the hygrothermal simulation model (Equation 2).

The airflow through a crack is approximated by the quadratic expression represented by Equation 3 [15]

Equation 3

$$\Delta p_a = \frac{12\mu z}{w d^3} Q + \frac{\rho C}{2 d^2 w^2} Q^2$$

where Δp_a is the air pressure difference across the crack [Pa], μ the dynamic viscosity [kg/(ms)], ρ the air density [kg/m^3], *Q* the air flow rate through the crack [m^3/s], and *z*, *w*, *d*, respectively the length, width, and height of the crack [m]. The parameter *C* is well approximated by $C = 1.5 + n_b$, where n_b is the number of right-angle bends in the crack [15].

In practice, detailed information on crack sizes and their distribution in buildings is limited. Leakage characteristics are generally expressed as an effective leakage area without specifying crack dimensions. Given the difficulty of this characterisation, the airflow through the roof construction is considered using a more general approach based on the overall air permeability of the roof construction (K_a). The resulting airflow through the roof is governed by Equation 4.

Equation 4

$$V_a = K_a(\Delta p_a)^n$$

where V_a is the airflow through the building component [$m^3/(m^2\ s)$], K_a is the airflow coefficient [m^3/m^2sPa], representing the air permeability of the building component, and n is the flow exponent, which in general has a value between 0.5 and 1.0, depending on the characteristics of the airflow.

Parameter analysis

Several values for the air permeability (K_a) of the roof construction have been investigated representing an airtight roof construction as well as a roof incorporating a relatively large number of leakages. Based on the observations obtained from literature [3] [16], typical values for the air permeability of the roof construction are applied for the different indoor environmental conditions (Table 4). The hygrothermal response of the roof is simulated using the presented values for the total air permeability of the roof, applied to the presented roof construction.

Table 4: Total air permeability of the roof construction

Cases	K_a [$\cdot 10^{-4}$ m^3/m^2sPa]
1	0.2
2	0.3
3	0.4
4	0.5
5	1.1
6	1.6
7	2.2

The resulting hygrothermal model incorporates the transient moisture sources due to the airflow through the roof construction calculated based on Equation 4. Depending on the specific air permeability of the component K_a and the environmental conditions around the building component, the hygrothermal conditions in the component are evaluated.

RESULTS

Figure 2 presents the predicted condensate per m^2 roof area during the second year of simulation, respectively between October 1 and June 1. Moreover, the figure presents the total moisture content in the roof construction over 4 years. In the figure, a comparison between a vapour retarding underlay, vapour-open underlay, and capillary-active underlay is shown. The roof constructions are susceptible to indoor climate class 2 and have an air permeability of $0.2 \cdot 10^{-4}$ $m^3/(m^2sPa)$. The figure shows that the roof construction with a vapour-retarding underlay is relatively sensitive to condensation, i.e. a maximum of approximately 700 g/m^2

condensate is developed, while the condensation occurring in the roof constructions with a vapour-open and capillary-active underlay is negligible. In addition, Figure 2 shows that the predicted total moisture contents in the vapour-retarding roof are relatively high compared to the moisture content observed for the other two roof constructions.

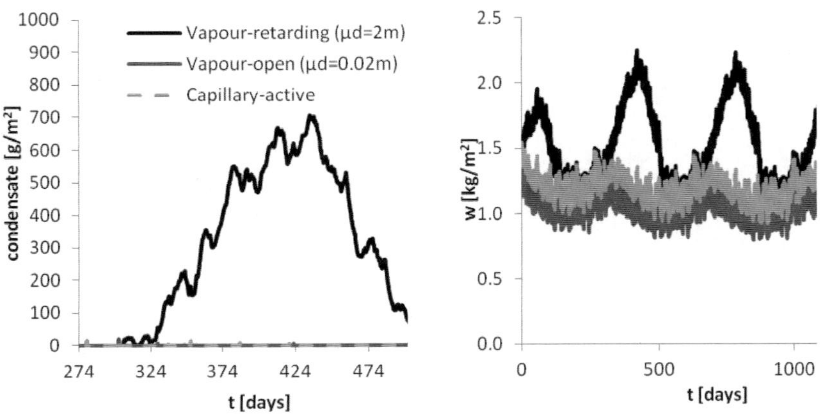

Figure 2: Predicted condensation [g/m^2] and moisture content [kg/m^2] in the roof construction for different construction configurations susceptible to indoor climate class 2 and an air permeability of $0.2 \cdot 10^{-4}$ m^3/(m^{2}sPa)

In Figure 3, the hygrothermal performance of a roof configuration with a vapour-open roof underlay and a capillary-active underlay are compared. The figure shows a roof susceptible to respectively indoor climate class 2, having an air permeability of $1.6 \cdot 10^{-4}$ m^3/(m^{2}sPa), and indoor climate class 3, having an air permeability of $0.5 \cdot 10^{-4}$ m^3/(m^{2}sPa). The figure shows that a roof with a capillary underlay is relatively sensitive to condensation induced by the airflow through the roof construction compared to the roof with the vapour-open underlay.

Figure 3 also shows a relatively large moisture content in the roof construction with a capillary underlay. It is obvious that the relatively high moisture content is caused by the ability of the capillary-active underlay to buffer moisture. Moreover, both roof designs are able to dry during summer and moisture accumulation over time does not occur.

The hygrothermal performance of the different configurations has been analyzed based on the condensate that is generated during winter after 3 years, the total moisture content after 1 year and after 4 years, and the maximum moisture content in the wood construction observed during the 4th year of simulation. The simulation results are summarized in respectively Table 5, Table 6, and Table 7.

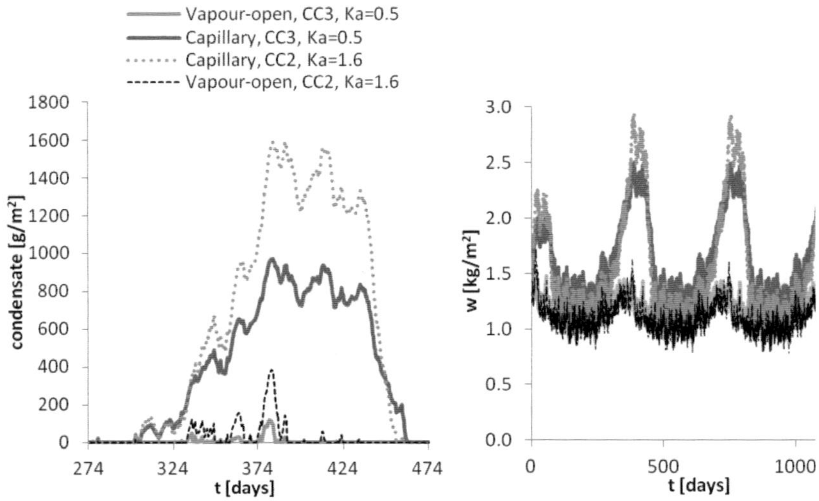

Figure 3: Simulation results for roof constructions with a vapour-open and a capillary-active underlay susceptible to respectively indoor climate class 2 and 3, and with an air permeability [$K_a \cdot 10^{-4}$] of $0.5 \cdot 10^{-4}$ $m^3/(m^2sPa)$ and $1.6 \cdot 10^{-4}$ $m^3/(m^2sPa)$

Roof construction	**#1 - Vapour retarding**		**#2 - Vapour-open**		**#3 - Capillary-active**	
	CC 2	**CC 3**	**CC 2**	**CC 3**	**CC 2**	**CC 3**
$\cdot$10-4 $m^3/(m^2sPa)$	C [kg/m^2]	C [kg/m^2]	C [kg/m^2]	C [kg/m^2]	C [kg/m^2]	C [kg/m^2]
0.2	0.70	4.03	0.02	0.03	0.03	0.04
0.3	5.66		0.03	0.03	0.04	0.16
0.4			0.03	0.05	0.07	0.40
0.5			0.03	0.12	0.15	0.97
1.1			0.16	1.20	0.92	3.24
1.6			0.39	3.62	1.59	9.58
2.2			0.79	7.38	2.77	13.83

Table 5: Predicted maximum condensate (C [kg/m^2]) in a winter season after 3 years

Roof construction	**#1 - Vapour retarding**				**#2 - Vapour-open**				**#3 - Capillary-active**			
	CC 2		**CC 3**		**CC 2**		**CC 3**		**CC 2**		**CC 3**	
$\cdot$10-4 $m^3/(m^2sPa)$	**W_{1y}**	**W_{4y}**	**W_{1y}**	**W_{4y}**	**W_{1y}**	**W_{4y}**	**W_{1y}**	**W_{4y}**	**W_{1y}**	**W_{4y}**	**W_{1y}**	**W_{4y}**
0.2	1.85	1.82	2.89	4.37	1.15	1.12	1.23	1.21	1.32	1.28	1.45	1.42
0.3	3.27	5.76			1.16	1.13	1.26	1.23	1.37	1.33	1.60	1.57
0.4					1.17	1.14	1.29	1.27	1.41	1.38	1.81	1.79
0.5					1.18	1.14	1.37	1.35	1.50	1.46	2.23	2.22
1.1					1.41	1.38	3.69	3.69	2.38	2.35	6.11	6.12
1.6					1.62	1.59	5.41	5.42	2.89	2.86	8.06	8.08
2.2					1.41	1.38	3.69	3.69	2.38	2.35	6.11	6.12

Table 6: Predicted moisture content [kg/m^2] in the roof construction after 1 year (w_{1y}) and after 4 years (w_{4y})

Roof construction	#1 - Vapour retarding		#2 - Vapour-open		#3 - Capillary-active	
	CC 2	CC 3	CC 2	CC 3	CC 2	CC 3
$\cdot$10-4 $m^3/(m^2sPa)$	u_{4y}	u_{4y}	u_{4y}	u_{4y}	u_{4y}	u_{4y}
0.2	17.0	21.1	17.0	19.0	16.1	16.6
0.3	22.3		17.5	19.1	16.3	16.9
0.4			18.1	19.1	16.5	17.1
0.5			18.3	19.1	16.7	17.3
1.1			19.1	19.2	16.9	17.5
1.6			19.5	19.2	17.4	17.9
2.2			19.7	19.2	17.6	18.1

Table 7: Predicted maximum moisture content [%-mass] in the wood construction during the 4th year of simulation

Regarding the vapour retarding roof underlay, Table 5 shows relatively large amounts of condensate for relatively small airflows through the construction. Moisture accumulation in case with a vapour retarding underlay is already observed at a relatively small air permeability of the roof construction. Therefore, the performance of the roof construction has not been analyzed for an air permeability of $0.4 \cdot 10^{-4}$ $m^3/(m^2sPa)$ and higher.

With respect to the capillary underlay, Table 5 shows that this roof configuration seems to be more sensitive to condensation compared to a vapour-open roof underlay. In addition, the simulation results presented in Table 6 show that relatively high total moisture contents are observed in the capillary-active roof, while the total moisture content in the vapour-open roof construction is lower. As has been mentioned previously, relatively high moisture contents are found in the roof with the capillary-active underlay and can be attributed to the moisture buffering ability of the roof underlay. Moreover, the predicted moisture content of the wood structure (Table 7) shows to be relatively low in the roof with the capillary-active underlay compared to the vapour-open roof. In summary, the simulation results show that the application of a capillary-active roof underlay results conditions which could be considered as more favourable hygrothermal conditions with respect to the moisture content of the wood.

DISCUSSION

Based on the simulation results, upper limits for the air permeability of the roof construction can be developed. Previously, guidelines for the air permeability of inclined roofs have been presented by Janssens (2005) [3]. These guidelines are presented in Table 8. It should be noticed that a roof with a capillary-active underlay has not been subject of Janssens' study.

Roof construction	#1 - Vapour retarding	#2 - Vapour-open
Indoor climate class	Janssens (2005) [3]	Janssens (2005) [3]
II	0.2	0.3
III	0.1	0.2

Table 8: Upper limits for the air permeability (K_a) [3]

Regarding the upper limits for the air permeability of a vapour retarding roof, the investigations show that the results of the present study and Janssens' study are comparable. In both studies, it is observed that a roof with a vapour-retarding underlay is sensitive to interstitial condensation, resulting in potential moisture accumulation and moisture problems.

A roof construction incorporating a vapour retarding roof underlay is not recommended. And, if a vapour retarding underlay is installed (for example in case of a building renovation without the possibility to replace the existing underlay), special attention should be paid to the air tightness of the roof construction.

With respect to the roof construction with a vapour-open underlay, Janssens [3] indicated relatively low upper limits for the critical air permeability, while the critical upper limits for the air permeability resulting from this study seem to be less strict. The results from the present study showed that moisture accumulation does not take place for air permeabilities larger than the limits reported by Janssens [3]. Additional research should be carried out in order to verify the research results and to make these generally applicable.

The critical upper limits of the air permeability for a roof with a capillary-active underlay are relatively high, compared to a vapour-open roof underlay. Due to the moisture buffering capacity of the fibre cement board, the moisture content of the wood construction is relatively small, which can be considered as more favourable hygrothermal conditions.

CONCLUSION

The influence of the airtightness of an inclined roof construction on the hygrothermal performance has been investigated. It was the objective of this paper to study the influence of transient (advective) airflow through a roof construction on the hygrothermal conditions in the component.

The main conclusion is that an inclined roof incorporating a vapour-open or capillary-active underlay is less sensitive to air leakages, for example due to poor workmanship, compared to a vapour retarding underlay. From this point of view, it is recommended to avoid the application of a vapour retarding underlay. When this kind of underlay is present, e.g. in a renovated building, special attention should be paid to the realisation of good air tightness. Moreover, it can be noted that this type of product has nearly disappeared from the Belgian construction market.

The study also showed that in the analyzed cases the application of a capillary-active underlay results in more favourable hygrothermal conditions and may increase the durability of the component. The capillary-active roof underlay has the potential to buffer moisture and to reduce the risk for condensation in the roof construction. Moreover, a capillary-active roof underlay makes the roof construction less susceptible to interstitial condensation and moisture accumulation and is from this point of view an indirect 'safety' against bad-workmanship.

As a practical guideline, it is recommended that inclined roofs should be as tight as possible. Potential moisture problems in inclined roofs are nearly only caused by air leakages at the level of the vapour retarder and are generally not influenced by the performance of the vapour retarder itself, as far as a vapour retarder is installed effectively. For new building constructions, the use of vapour-open underlay is recommended. This kind of product has also the advantage to be more tolerant to poor workmanship regarding the air tightness.

The research results presented in this paper are based on a case study of a building component using numerical simulation. The reader should notice that it may be questionable to what extent the numerical research results can be generalized. Other aspects as the type of insulation, the performance of the insulation, or the type of vapour-retarder have not been considered in this study. A more detailed research and investigation of the influence of these parameters combined with the influence of the (advective) airflow through inclined roof constructions is desired.

REFERENCES

1 Bankvall, C. 2007, *The effect and cost impact of poor airtightness - information for developers and clients*, Proceedings of Thermal Performance of the Exterior Envelopes of Whole Buildings X International Conference.

2 Meert, E. 2000, *Het platte dak : Opbouw, materialen, uitvoering, onderhoud.* Technische Voorlichting 215, Belgian Building Research Institute, Brussels.

3 Janssens, A., Dobbels, F. 2005, *Ontwikkeling en toepassing van luchtdichtheidscriteria voor hellende daken*, Bouwfysica, vol. 18, no. 2.

4 Kuenzel, H.M. 1996, *Kann bei voll gedämten, nach aussen diffusionsoffenen Steildachkonstruktionene auf eine Dampfsperre verzichtet werden?*, Bauphysik, vol. 18.

5 Kuenzel, H.M., Zirkelbach, D., Schafaczek, B. 2011, *Vapour control design of wooden structures including moisture sources due to air exfiltration*, 9th Nordic Symposium on Building Physics - NSB 2011.

6 DIN 4108 Teil 3 2001, *Wärmeschutz im Hochbau; Klimabendingter Feuchteschults; Anforderungen und Hinweise für Planung und Ausführung*, Deutsches Institut für Normung.

7 Nicolai, A., Grunewald, J. 2006, *Delphin 5: Coupled Heat, Air, Moisture and Salt Transport (User Manual and Program Reference)*, Dresden University of Technology.

8 Nicolai, A., 2007, *Modeling and numerical simulation of salt transport and phase transitions in unsaturated porous building materials*, PhD thesis, Dresden University of Technology.

9 WTCB, KUL, WenK, UGent. 2004, *Vochtproblemen in daken: impact van de actuele randvoorwaarden en bouwtechnieken in België (Eindrapport).*

10 Langmans, J., Roels, S. 2010, *Luchtdoorlatendheid Onderdaken*, Katholieke Universiteit Leuven, Leuven.

11 Houvenaghel, G., Horta, A., Hens, H. 2004, *The impact of airflow on the hygrothermal behavior of highly insulated pitched roof systems*, Proceedings, International Conference for Performance of Exterior Envelopes of Whole Buildings IXU.S. Department of Energy, Oak Ridge.

12 Janssens, A., 1999. *Internal condensation: causes, measures, design guidelines* (In Dutch). Syllabus van de Studiecyclus 'Vocht in de bouw', Deel 3. TI-KVIV.

13 Cóstola, D., Blocken, B., Hensen, J.L.M. 2009, *Overview of pressure coefficient data in building energy simulation and airflow network programs*, Building and Environment, vol. 44, no. 10, pp. 2027-2036.

14 Knoll, B., Phaff, J.C., de Gids, W.F. 1995, *Pressure simulation program*, Proceedings of the Conference on Implementing the Results of Ventilation ResearchAIVC, .

15 Baker, P.H., Sharples, S. Ward, I.C., 1987, *Air-flow through cracks*, Building and Environment 22, 293-304.

16 Maref, W., Elmahdy, A., Swinton, M., Tariku, F. 2009, *Assessment of energy rating of PU spray foam walls*, NRC-CNRC, Canada.

IMPACTS OF AIRTIGHTENING RETROFITS ON VENTILATION AND ENERGY IN A MANUFACTURED HOME

Andrew Persily, Steven Nabinger, and W. Stuart Dols

National Institute of Standards and Technology
100 Bureau Drive MS8633
Gaithersburg, MD 20899, USA

ABSTRACT

A retrofit study was conducted in an unoccupied manufactured house to investigate the impacts of airtightening on ventilation rates and energy consumption. This report describes the retrofits and the results of the pre- and post-retrofit assessment of building airtightness, ventilation, and energy use. Building envelope and air distribution systems airtightness were measured using fan pressurization. Air change rates were measured continuously using the tracer gas decay technique. Energy consumption associated with heating and cooling was monitored through measurement of gas consumption by the forced-air furnace for heating and electricity use by the air-conditioning system for cooling. The results of the study show that the retrofits reduced building envelope leakage by about 18 % and duct leakage by about 80 %. The reduction in the house infiltration rates depended on weather conditions and the manner in which the heating and cooling system was controlled, but in general these rates were reduced by about one third. The energy consumption of the house for heating and cooling was reduced by only about 10 %, which is relatively small but not unexpected given that infiltration only accounts for a portion of the heating and cooling load. An existing multizone airflow model of the building was modified to reflect the airtightening retrofits, and the predicted infiltration rates agreed well with the measured values over a range of weather and system operation conditions.

KEYWORDS
Airtightness; energy; manufactured homes; residential; retrofit; ventilation

INTRODUCTION
Single-family residential buildings have traditionally been ventilated via weather-driven infiltration through unintentional leakage sites in the building envelope. More recently, there has been a trend towards the use of mechanical ventilation to provide more predictable ventilation rates and air distribution that are less dependent on weather conditions. ASHRAE Standard 62.2, Ventilation and Acceptable Indoor Air Quality in Low-Rise Residential Buildings, requires mechanical ventilation in many U.S. climates [1]. However, only a small fraction of site-built, low-rise residential buildings employ mechanical ventilation. The situation is different in manufactured homes, for which the U.S. Department of Housing and Urban Development (HUD) Manufactured Home Construction and Safety Standards [2] contain requirements for mechanical ventilation for these dwellings. In the implementation of the HUD standards, and the implementation of mechanical ventilation in low-rise residential buildings in general, questions have arisen regarding the actual ventilation rates in homes built to the HUD and other standards, the approaches being used to provide mechanical

ventilation, and the energy and indoor air quality impacts of mechanical ventilation [3]. Specific questions also exist regarding how duct leakage, local exhaust fans and ventilation inlets affect ventilation rates, air movement patterns, and building pressures.

Based on 2005 statistics, manufactured homes constitute about 6 % of U.S. households and about 5 % of U.S. residential energy consumption [4]. These same data show that on average manufactured homes consume more energy per unit floor area on an annual basis, 850 MJ/m^2, than detached homes, which consume 450 MJ/m^2. Given the smaller size of manufactured homes, the average energy consumption per household is about 74 GJ/y compared with 114 GJ/y for detached homes). Low energy manufactured homes have been constructed, with annual energy consumption as low as 52 MJ/y [5]. Therefore, while manufactured homes constitute a small fraction of the national housing stock, they also provide an opportunity for significant energy savings through improved design and construction.

The manner in which residential buildings, including manufactured homes, are ventilated and the roles of weather-driven infiltration and mechanical ventilation are becoming increasingly important in the context of national efforts to reduce residential energy consumption in new and existing buildings. Working with HUD and various state energy offices, the U.S. Department of Energy is undertaking a major effort to retrofit large numbers of residences [6]. A number of other efforts are underway to increase the energy efficiency of new residential buildings. One key component of energy efficiency in both new and existing residences is the control of infiltration through tight envelope construction and the provision of adequate amounts of outdoor air to meet the health and comfort needs of the occupants. Efforts to improve building design and to implement effective retrofits are raising the same questions as noted previously about how to provide ventilation in residential buildings and how ventilation and energy performance are impacted by duct leakage, fan operation and other effects.

To obtain insight into the issues of residential infiltration and ventilation, a modeling study was performed on a manufactured home to investigate different ventilation scenarios [7]. The results of that study showed that assuming a single value of 0.25 h^{-1} for the weather-driven infiltration rate, as is done in the HUD standard, is inherently problematic given the strong dependence of infiltration on weather. The simulated infiltration rates varied by as much as 5 to 1 based on variations in weather conditions alone. Including the impacts of exhaust fan and forced-air fan operation more than doubled the range of variation in the air change rates. In addition, the predicted infiltration rates were lower than this assumed value under milder weather conditions. Therefore, assuming an infiltration rate of 0.25 h^{-1} in modern manufactured homes may be too high, but more importantly ignores variations due to weather and fan operation. The study also showed that employing an outdoor air intake duct on the forced-air return duct can be effective in raising air change rates and distributing ventilation air throughout the house. However, the overall impact on the building air change rate is a strong function of the operating schedule of the forced-air system, which in turn depends on the extent of system over-sizing and the specific strategies used to control the system operation such as manual switches and timers. While increased forced-air fan operation provides higher ventilation rates, there is an energy cost associated with the increased fan operation, particularly when the forced-air fan has a high wattage rating. Also, given the existence of significant duct leakage, this ventilation approach was associated with excessive air change rates (relative to the requirements in standards) particularly when weather-driven infiltration is high.

In order to investigate these residential ventilation issues, as well as a range of other indoor air quality issues, a manufactured house was installed on the NIST campus in 2002. The house and its airtightness, ventilation and energy performance, as installed, have been described previously [8]. Since that time, the building was subject to a series of retrofits to improve the airtightness of the building envelope and the air distribution system ductwork. This paper summarizes the results of these retrofit on the building airtightness, ventilation rates and energy consumption.

DESCRIPTION OF HOUSE AND VENTILATION SYSTEMS

The study was performed in a double-wide manufactured home installed on the NIST campus, shown in the photograph in Figure 1. This house was built to the HUD standard that applies nationwide to manufactured homes [2]. Additional details on the house can be found in Nabinger and Persily [8]. Figure 2 is a schematic floor plan of the test house, showing the three bedrooms, two baths, kitchen, and the family, dining and living area. The house has a floor area of 140 m^2, a volume of 340 m^3 and a cathedral ceiling over its full length that is 2.7 m high at the center and slopes down to 2.1 m at the front and back walls.

Figure 1. Photograph of the manufactured house.

The house's heating, ventilating and air-conditioning (HVAC) system consists of a 22 kW gas furnace, a 15 kW air conditioner, and a forced air re-circulation fan with a design airflow rate of 470 L/s. In addition there is a whole house, kitchen, and two bathroom exhaust fans in the house. The whole house and bathroom ventilation fans have an airflow capacity of 24 L/s, while the kitchen fan capacity is 47 L/s. Both of these airflow capacities are per manufacturer claims.

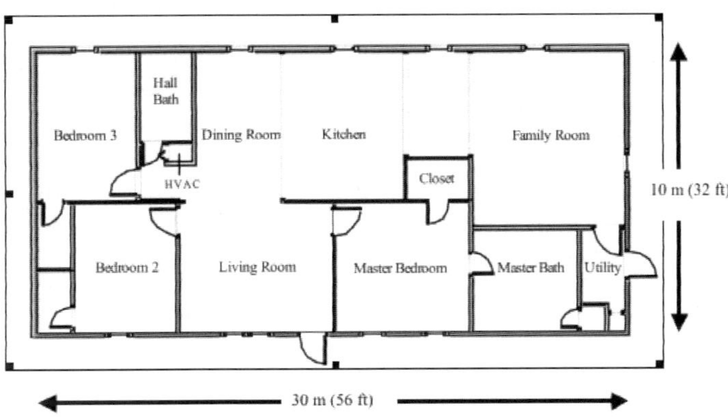

Figure 2. Schematic floor plan of manufactured house.

MEASUREMENT METHODS

This section summarizes the measurement techniques and instrumentation used in the test house, with additional details available in Nabinger and Persily [8].

Whole house air change rates

Whole house air change rates were measured using the tracer gas decay technique as described in ASTM test method E-741 [9]. These rates reflect the combination of the rate at which outdoor air enters the house from mechanical ventilation and due to infiltration through leaks in the house's envelope. The air change rate measurements in the house were made with an automated tracer gas monitor system employing sulfur hexafluoride (SF_6) as the tracer gas that injects tracer gas into the building and monitors the concentration at multiple locations.

Exterior envelope and duct leakage

Exterior envelope leakage of the house was measured using whole building pressurization testing per ASTM E779 using a blower door [10]. These whole building tests were performed with the air distribution system off but with all the vents in their normal operating positions (unsealed) and then again with all the supply and return air vents sealed with plastic (sealed). The latter test provides an indication of the envelope leakage independent of any leakage via the air distribution system, while the unsealed test provides an overall leakage for the envelope and air distribution system in combination. The measurements of airtightness using the blower door have an accuracy of roughly +/- 10 % of the measured value based on the calibrations performed. Pressurization tests were also used to determine the leakage from the air distribution system using ASTM E1554 [11], which also has a measurement uncertainty of about +/- 10%.

Energy consumption

Energy consumption for heating and cooling was monitored to determine any reductions due to the airtightness retrofits. The electrical energy consumption of the air conditioning system was measured using a 240 V power transducer, and the energy used by the forced-air fan was monitored with 120 V energy meters. The energy use by other items in the house, and therefore contributing to the interior heat gain, were monitored separately. The 240 V power transducers have an accuracy of +/- 0.5 % of full scale and a resolution of 0.1 W. The 120 V energy meters have an accuracy of +/- 0.2 % of full scale and a resolution of 3.6 W•h. The heating energy was measured using a calibrated gas flow meter that recorded the average value of natural gas flow rate into the furnace. The gas meter has a measurement range of 0 L/s to 0.94 L/s and an accuracy of 1 % of full scale and a resolution of 0.0002 L/s.

Environmental and System Parameters

Additional sensors were used to monitor air temperature, relative humidity, wind conditions and operating status of the forced-air system fan. Details on these measurements and the associated uncertainty are contained in Nabinger and Persily [8]. Temperatures of the indoor and outdoor air were measured using epoxy coated polymer thermistors, and the relative humidity was monitored with capacitive thin film polymer sensors. Wind speed and direction were measured at the top of a 10 m tower located approximately 5 m south of the southernmost wall of the house. Fan status switches were wired into the electrical circuits of the forced air fan and the four exhaust fans to detect and record whether each fan is on or off.

RETROFITS

The retrofits focused on increasing the airtightness of the building envelope and the air distribution system supply ductwork. The envelope retrofits included installation of a house wrap over the exterior walls and sealing of leaks in the belly and living space floor. The house wrap installation involved removing the siding from the house and installing the wrap from the top of the crawl space to the top of the walls, and then replacing the vinyl siding. The wrap was a flash spunbonded olefin, non-woven sheet material and was installed per the manufacturer's instructions. The second portion of the envelope airtightening effort involved sealing the leakage sites in the flooring of the house and in the insulated belly that encloses the ductwork. Sealing the flooring involved spraying a two-part foam, which expanded and hardened to seal the leaks. The leakage sites included the accessible portions of the marriage line between the front and back section of the house; holes in the floor made for the water drainage pipes, "P-traps" in the bathrooms; and gas and other utility lines.

Figure 3 shows several of these leakage sites, before and after sealing. The top two photographs show drain and water lines in the floor of the living area, viewed from below, showing these lines passing through large holes that constitute significant leakage sites. The existence of such leakage sites is not unusual in residential construction, though recent efforts to build tighter homes result in such airflow paths being sealed during construction. The photograph on the lower right shows both of these leakage sites sealed with spray foam. The photograph on the lower left shows a supply air register in the floor, viewed from above, after it was sealed with mastic. In this house, and many others with supply ductwork under the floor, significant leakage occurs where the vertical supply duct connects with the floor register.

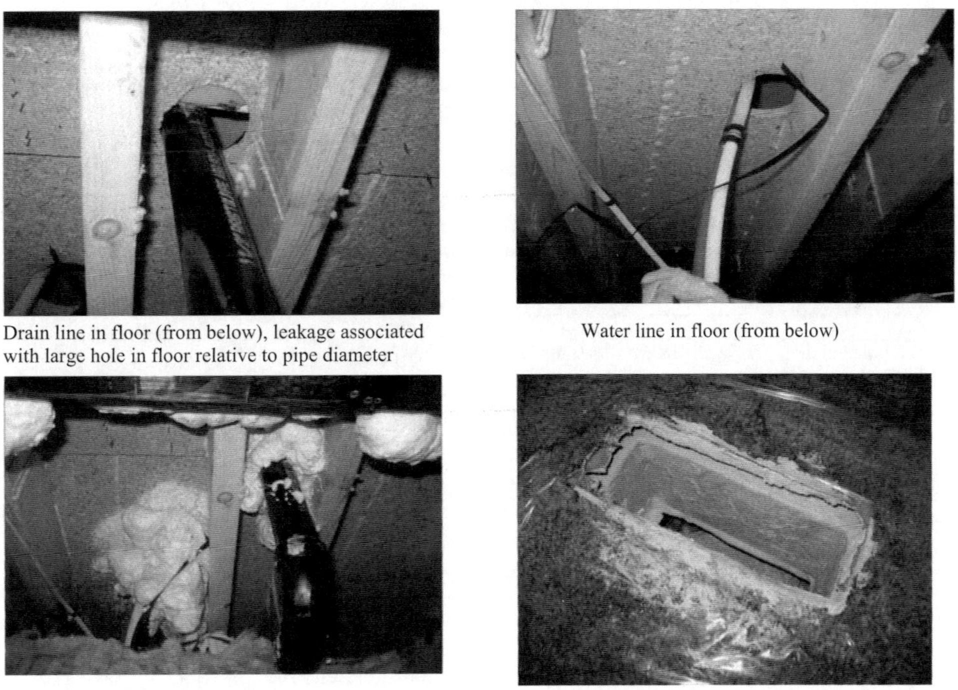

Drain line in floor (from below), leakage associated with large hole in floor relative to pipe diameter

Water line in floor (from below)

Drain and water lines after sealing

Floor register (from above) after sealing with mastic

Figure 3. Schematic floor plan of manufactured house.

Additional air sealing was performed in the air distribution system in the belly space, including large leaks in the four ends of the two main supply ducts in the front and rear halves of the house and the large connection to the underside of the HVAC system. Foam was also used to seal leaks at the ends of the two crossover ducts joining the two main HVAC ducts at each end of the house. Such duct leakage is not uncommon in manufactured homes and can have a major, negative effect on the overall system efficiency and thereby the energy consumption for heating and cooling [12].

As noted elsewhere in this paper, it is much easier and cost-effective to achieve a tight envelope during construction than as part of a retrofit effort. The retrofits reported on here were thorough, even to the extent of removing the siding to install a house wrap, but some leakages sites were inaccessible. No leaks in the ceiling of the house could be accessed or sealed, and there were probably additional leaks in the floor and the belly that could not be repaired.

RESULTS

This section presents the results of the measurements of the impact of the retrofits on the building and air distribution system tightness, the ventilation system airflow rates, the whole building air change rates and the energy consumption for heating and cooling.

Airtightness

As noted above, pressurization test methods were used to measure the exterior envelope leakage, the leakage between the living space and the crawl space, and the leakage of the air distribution system ductwork. The results of these measurements are shown in Table 1. In terms of the air change rate at 50 Pa, the whole building pressurization leakage was reduced by 24 % relative to the pre-retrofit results with the system unsealed and by about 11 % relative to pre-retrofit leakage with the system sealed. As noted before, the unsealed tests were conducted with the air distribution system off but with all vents in their normally open positions, while the sealed tests were conducted with all supply and return vents sealed. The leakage reduction due to the retrofit in terms of the effective leakage area (ELA) at 4 Pa was similar on a percentage basis. Relative to the average of the unsealed and sealed pre-retrofit values, the house leakage was reduced by 18 % by the airtightening retrofits. The duct leakage reduction is quite significant, with a reduction of about 82 % relative to the pre-retrofit value.

	Pre-retrofit		**Post-retrofit**	
Whole building pressurization	Airflow at 50 Pa, h^{-1}	ELA at 4 Pa, cm^2	Airflow at 50 Pa, h^{-1}	ELA at 4 Pa, cm^2
Forced-air system unsealed	11.8	728	9.0*	555*
Forced-air system sealed	10.1	636		
	ELA at 25 Pa, cm^2			
Duct leakage		320		58

* Post-retrofit values with the system sealed and unsealed are not significantly different relative to measurement uncertainty.

Table 1. Pre- and post-retrofit airtightness results.

Air change rates

The air change rates in the house decreased after the airtightening retrofits as expected. Figure 4 shows the pre- and post-retrofit air change rates measured with the forced-air system off (Condition 0) as a function of the indoor-outdoor temperature difference under low wind speed conditions (less than 2 m/s). While the data exhibit a fair bit of scatter, the post-retrofit rates are roughly 20 % lower than the pre-retrofit values, which is close to the reduction in the whole building leakage measured by pressurization testing as seen in Table 1. Figure 5 shows the pre- and post-retrofit air change rates with the forced-air system off as a function of wind speed (u) under low indoor-outdoor temperature differences (ΔT), i.e., absolute values less than 10 °C. There are only a relatively small number of post-retrofit points, but the reduction in air change rates is again roughly 20 % relative to the pre-retrofit values.

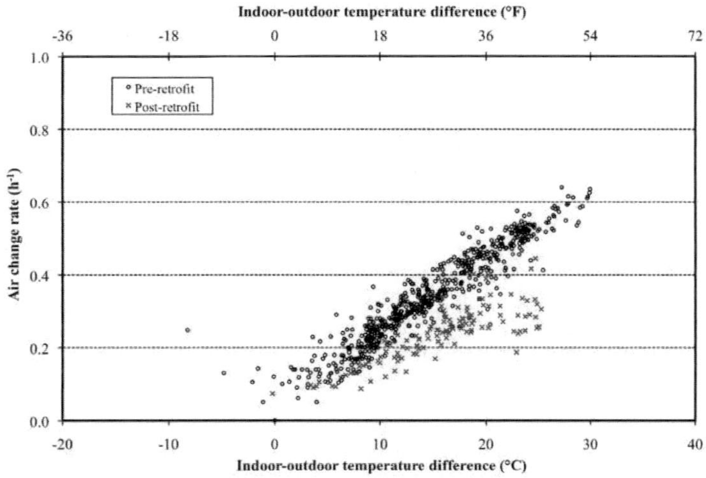

Figure 4. Pre- and Post-Retrofit measured air change rates as a function of temperature difference (low wind speed): forced-air fan off (Condition 0)

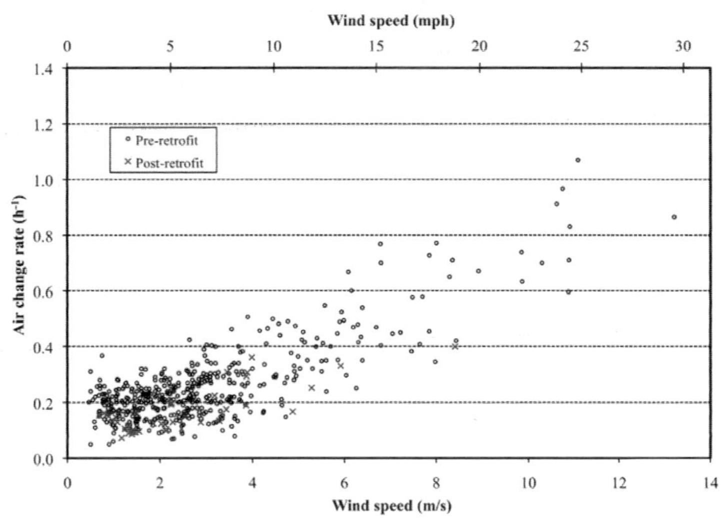

Figure 5. Pre- and Post-Retrofit measured air change rates as a function of wind speed (low ΔT): forced-air fan off (Condition 0)

Figures 6 and 7 show the pre- and post-retrofit air change rates with the forced-air fan on and the outdoor air intake closed (Condition 1a), plotted against temperature difference and wind speed respectively. As seen in Figure 6, the air change rate reduction with the system on is much larger than the reduction with the system off due to the impacts of the reduced duct leakage, particularly at low temperature differences. The pre-retrofit data exhibit an unusual dependence on temperature difference as discussed previously [8] because of the duct leakage pressurizing the volume under the living space. With the improved airtightness of the ductwork and the belly volume, the dependence of air change rate on temperature difference is more consistent with the pattern seen in other buildings. Figure 7 shows a significant reduction in the post-retrofit air change rates as a function of wind speed, larger than that seen for the fan-off data in Figure 5. The reduction is more pronounced with the fan on because these data correspond to low temperature-differences, where the reduced duct leakage has a large impact on the post-retrofit rates. The pre- and post-retrofit air change rates for Condition 1b (forced-air fan on and outdoor air intake open) are similar to the results seen for Condition 1a and plots of those data are not presented in this paper.

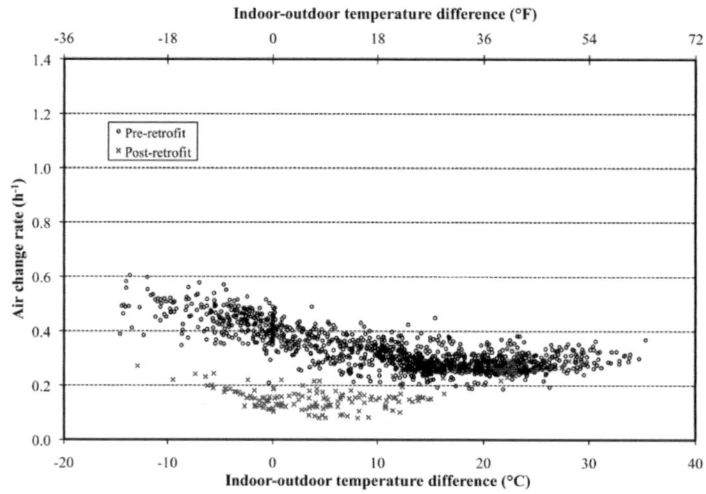

Figure 6. Pre- and Post-Retrofit measured air change rates as a function of temperature difference (low wind speed): forced-air fan on, outdoor air intake sealed (Condition 1a)

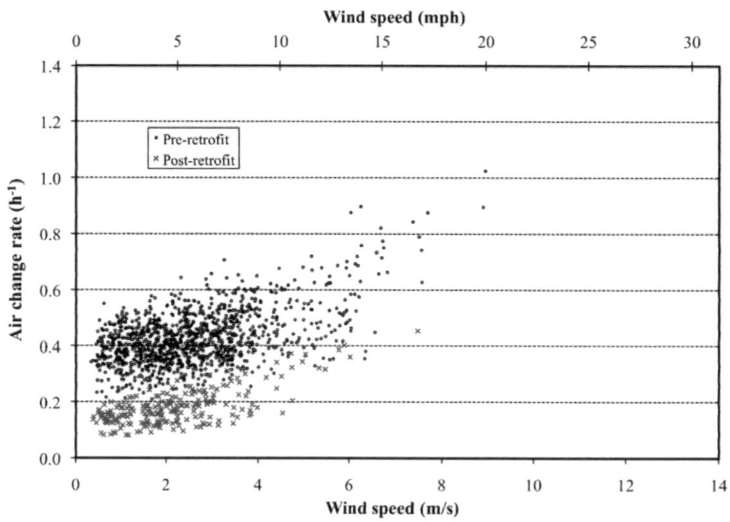

Figure 7. Pre- and Post-Retrofit measured air change rates as a function of wind speed (low ΔT): forced-air fan on, outdoor air intake sealed (Condition 1a)

Table 2 summarizes the air change rate reductions for the various conditions of fan operation and for specific ranges of outdoor weather. Each mean air change rate in the table is calculated for the test condition and the noted weather condition. The last column shows the percentage reduction in the post-retrofit air change rate relative to the pre-retrofit value for that particular case. Most of the mean air change rates decrease by about 25 % to 35 %, with some exceptions. There is very little reduction for high ΔT and low wind speed for conditions 1a and 1b as discussed earlier. The reductions for low ΔT with the fan on tend to be larger than 30 %, as large as 50 % in one case. As noted earlier, these reductions are impacted by the decrease in duct leakage more than the other cases.

	Mean air change rate (h^{-1})		
	Pre-retrofit	**Post-retrofit**	**% reduction**
Condition 0 (fan off)			
ΔT = 10 °C to 20 °C, u < 2 m/s	0.34	0.24	29
u = 4 m/s to 6 m/s, |ΔT| < 10 °C	0.36	0.25*	31
Condition 1a (fan on, intake sealed)			
ΔT = 0 °C to 10 °C, u < 2 m/s	0.35	0.24	31
ΔT = 20 °C to 30 °C, u < 2 m/s	0.27	0.26	4
u = 4 m/s to 6 m/s, |ΔT| < 10 °C	0.50	0.31	38
Condition 1b (fan on, intake open)			
ΔT = 0 °C to 10 °C, u < 2 m/s	0.39	0.19	51
ΔT = 20 °C to 30 °C, u < 2 m/s	0.34	0.34	0
u = 4 m/s to 6 m/s, |ΔT| < 10 °C	0.52	0.31	41
Condition 2a (fan controlled by thermostat, intake sealed)			
ΔT = 10 °C to 20 °C, u < 2 m/s	0.30	0.23	23
u = 4 m/s to 6 m/s, |ΔT| < 10 °C	0.39	0.29	26
Condition 2b (fan controlled by thermostat, intake open)			
ΔT = 10 °C to 20 °C, u < 2 m/s	0.34	0.28	18
u = 4 m/s to 6 m/s, |ΔT| < 10 °C	0.39	0.33	15

* Only 3 post-retrofit air changes rates in this range of weather conditions.

Table 2. Summary of pre- and post-retrofit air change rates.

Energy Consumption

After the retrofits were completed, the energy used to heat and cool the building was measured for several months in order to compare it with the levels of energy use before the retrofits. Figures 8 and 9 show the pre and post gas (heating) and electrical (cooling) energy use respectively. Each point in these plots corresponds to 24 h of heating or cooling energy consumption, versus the average indoor-outdoor temperature difference during that same 24-h period. The gas energy consumption values are based on the rate of gas consumption times the heating system's claimed value of efficiency of 85 % and therefore correspond to the energy to heat the house. The electrical energy consumption values are based on the system's rated COP of 2.93 to yield the cooling energy required. The plots show, but do not distinguish between, data collected when the forced air fan is always on (condition 1), and when the fan operation is controlled by the thermostat (condition 2).

The data in both Figures 8 and 9 exhibit a lot of scatter, making it difficult to see the difference in the pre and post energy consumption. In order to estimate the difference, the average heating energy use before and after the retrofit was determined for a temperature difference range of 25 °C to 30 °C. As seen in Figure 8, this temperature range appears to be better behaved than other ranges and has sufficient points to determine reliable averages. In this temperature range, the daily average heating energy consumption pre-retrofit is 137.5 MJ and the average post-retrofit is 125.6 MJ, corresponding to a reduction of 8.6 %. In the case of cooling, Figure 9, the average energy use was calculated for temperature differences between

-6 °C and -1 °C. Again, the data in this range exhibit less scatter than at the higher temperature differences. In this temperature range, the daily average cooling energy consumption pre-retrofit is 371.9 MJ and the average post-retrofit is 328.7 MJ, corresponding to a reduction of 11.6 %. Based on the standard deviations of the mean energy consumption values for heating and cooling, the uncertainty in the energy reduction for heating and cooling are quite large because of the large uncertainty in the difference between the pre- and post-retrofit energy consumption. In both cases the uncertainties are estimated to be larger than the reductions themselves.

Analysis of the pre-retrofit heating and cooling yielded slopes of about 22 MJ/°C and 10 MJ/°C respectively [8]. These two values should have been the same based on conduction heat losses (gains) alone, but the factor of two difference is not unreasonable given the use of assumed values for the system efficiencies, the impact of solar and other internal gains, the lack of consideration of latent loads, and the variation of infiltration with weather. A simple heat loss calculation for the building, assuming an air change rate of 0.5 h^{-1}, yields a heat loss rate through the envelope of 15 MJ/°C, which is roughly halfway between the two measured, pre-retrofit values. The same heat loss calculation was used to examine the energy impacts of the infiltration rate reductions. Considering the air change rates plotted in Figures 4 through 7, the heat loss was calculated for a pre-retrofit air change rate of 0.4 h^{-1} and a post-retrofit rate of 0.2 h^{-1}. The corresponding reduction in heat loss for the building is 13.5 %, which is reasonably close to the estimated energy reductions above of 8.6 % and 11.6 %.

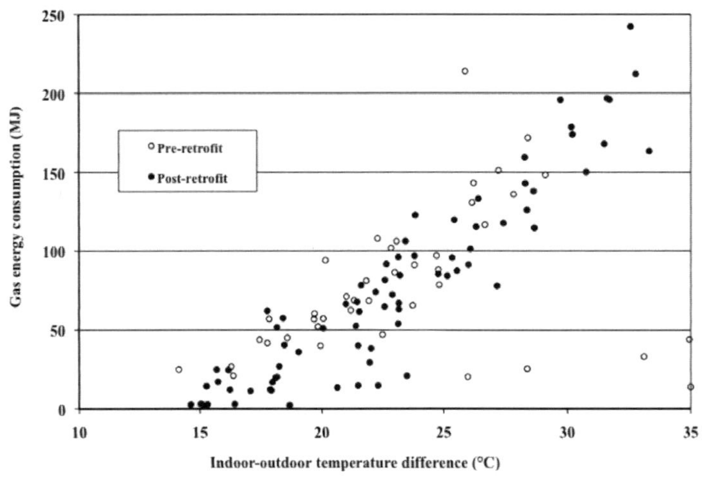

Figure 8. Pre-post gas heating energy versus indoor-outdoor température difference

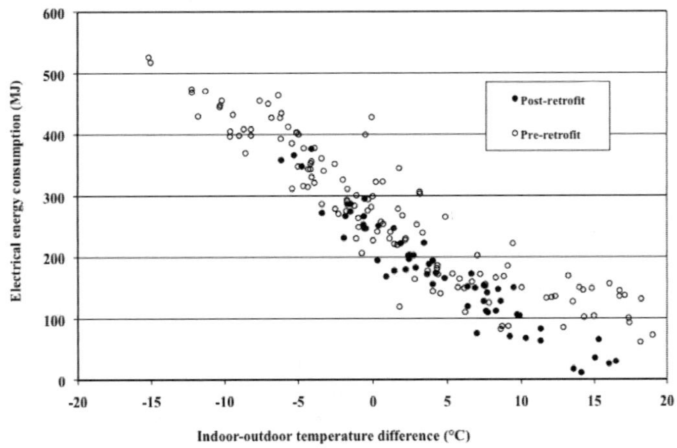

Figure 9. Pre-post electric cooling energy versus indoor-outdoor température difference

AIR CHANGE RATE PREDICTIONS

In the previous report on the pre-retrofit assessment of the house [8], the multizone airflow model CONTAM [13] was used to predict the air change rates under various conditions of fan operation and weather. Those predictions were in fairly good agreement with the air change rates measured with the tracer gas system in the house. The model was used again after the retrofits to predict the air change rates.

Table 3 shows the input values used in the CONTAM model of the house for both pre- and post-retrofit conditions. Only those airflow paths that were impacted by the retrofit efforts are seen to be different post-retrofit. The leakage values for the exterior envelope of the living space that are above floor level only decrease by about 10 %, but the leakage to the belly, the belly to the crawl space, and the duct leakage decrease by a much larger fraction.

Figure 10 shows the measured and predicted post-retrofit air change rates under condition 0 (forced-air fan off) plotted against indoor-outdoor temperature difference. The values in this plot correspond to wind speeds less than 2 m/s. The predictions match the measurements very well, though the measured values exhibit some scatter due in part to variations in wind speed and direction. Figure 11 shows the measured and predicted rates plotted against wind speed, again for condition 0, corresponding to indoor-outdoor temperature differences between -10 °C and +10 °C. Again, the predictions match the measurements fairly well.

Figures 12 and 13 show the measured and predicted air change rates as a function of temperature difference and wind speed respectively for condition 1a (forced-air fan on, outdoor air intake closed). The predictions match the general trend seen in the measurements but tend to under-predict, particularly for near-zero and negative temperature differences. The predicted air change rates in Figure 13 are somewhat higher than the measured values at low wind speeds. The plots of measured and predicted air change rates for condition 1b (the same as 1a except the outdoor air intake is open) show very similar trends to those seen in Figures 12 and 13 and are not shown in this paper.

		ELA at 4 Pa	
	Airflow path	Pre-retrofit	Post-retrofit
Living space envelope	Exterior wall	0.14 cm^2/m^2	0.13 cm^2/m^2
	Ceiling wall interface	0.81 cm^2/m	0.73 cm^2/m
	Floor wall interface	1.24cm^2/m	1.12 cm^2/m
	Window #1	5.00 cm^2	*
	Window #2	1.94 cm^2	*
	Corner interface	0.81 cm^2/m	0.73 cm^2/m
	Exterior doors	18.7 cm^2	*
	Living space to belly	3.65 cm^2/m^2	1.43 cm^2/m^2
Interior airflow paths	Interior walls	2 cm^2/m^2	*
	Bedroom doorframe	410 cm^2	*
	Open interior doors	2 m x 0.9 m	*
	Bathroom doorframe	330 cm^2	*
	Interior doorframe	250 cm^2	*
	Closet doorframe	4.6 cm^2	*
Attic	Attic floor	2 cm^2/m^2	*
	Roof vents	0.135 m^2/each	*
	Eave vents	106 cm^2/m	296 cm^2/m **
Crawl space and belly	Exterior walls of crawl space	25 cm^2/m^2	*
	Rear crawl space vents	323 cm^2	*
	Front crawl space vents	465 cm^2	*
	Crawl space access door	206 cm^2	*
	Crawl space to "belly"	258 cm^2	181 cm^2
	Duct leak into belly	320 cm^2	58 cm^2

* Same as pre-retrofit

** The eave vent values were remeasured and the values were corrected in the post-retrofit model.

Table 3. Model inputs for pre- and post-retrofit conditions.

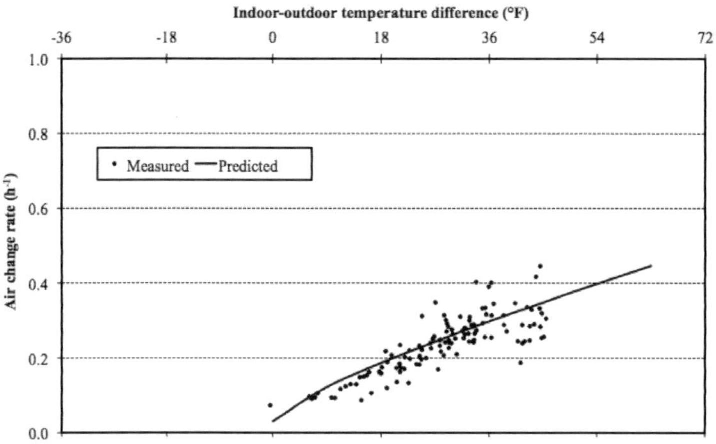

Figure 10. Measured and predicted post-retrofit air change rates as a function of temperature difference (low wind speed): Forced-air fan off (Condition 0).

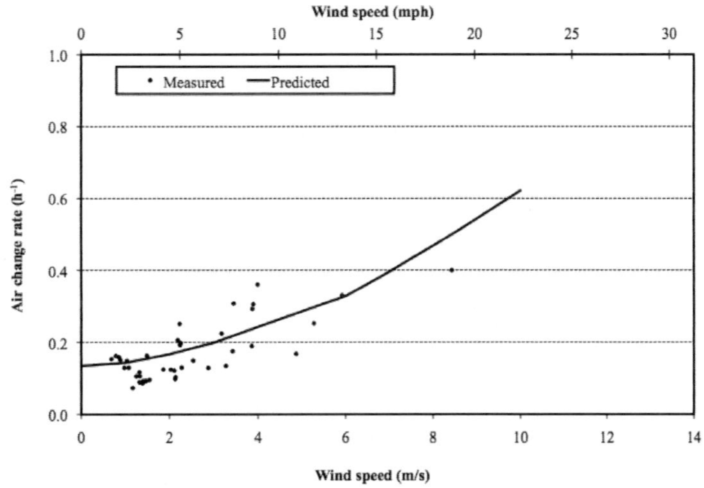

Figure 11. Measured and predicted post-retrofit air change rates as a function of wind speed (low temperature difference): Forced-air fan off (Condition 0).

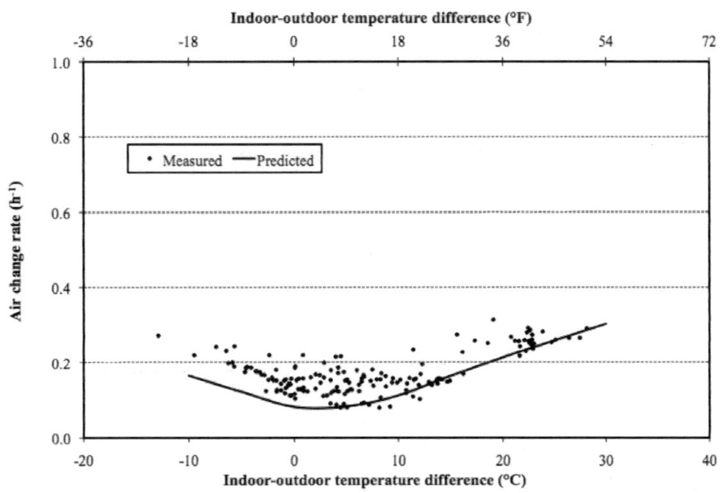

Figure 12. Measured and predicted post-retrofit air change rates as a function of temperature difference (low wind speed): Forced-air fan on, outdoor air intake sealed (Condition 1a).

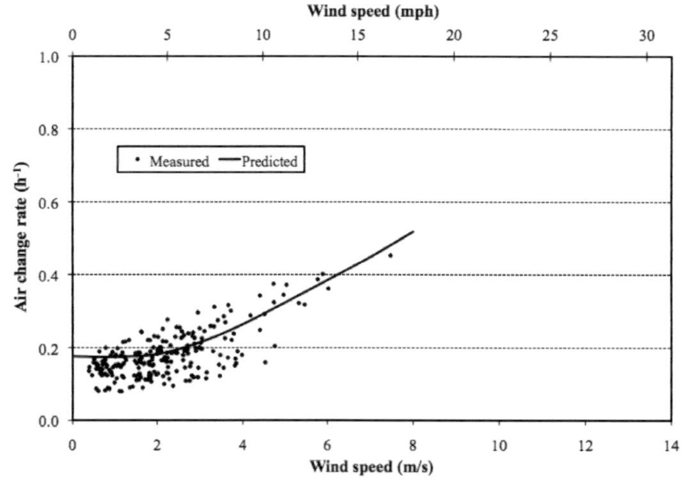

Figure 13. Measured and predicted post-retrofit air change rates as a function of wind speed (low temperature difference): Forced-air fan on, outdoor air intake sealed (Condition 1a).

Overall, as was the case with the pre-retrofit data, the air change rates predicted with CONTAM are in fairly good agreement with the measurements. Even with different conditions of fan operation and weather, the good agreement is sustained. Table 5 summaries the agreement of the predicted and measured air change rates for the different cases of system operation. For each case, a linear regression of the predicted air change rate against the measured rate was performed for temperature and wind dominated conditions. The values of R-squared and the standard error of the regression are presented for each of the five cases. The values of R-squared range from 0.54 to 0.78 with one exception and the standard error of regression values are all about 0.04. The agreement tends to be better for the temperature dominated conditions, which is not surprising given that it is generally more challenging to predict wind-driven air change rates.

	Temperature difference		**Wind speed**	
Case	**R-squared**	**Std error of regression**	**R-squared**	**Std error of regression**
0 – Fan off	0.685	0.04	0.572	0.05
1a – Fan on, intake sealed	0.615	0.03	0.542	0.04
1b - Fan on, intake open	0.696	0.03	0.308	0.05
2a – T-stat control, intake sealed	0.784	0.04	0.597	0.05
2b - T-stat control, intake open	0.755	0.05	0.752	0.04

Table 4. Summary Statistics of Predicted versus Measured Air Change Rates.

DISCUSSION

This study was conducted to evaluate the impacts of airtightening retrofits on building airtightness, ventilation rates and energy consumption in an existing manufactured home. In this study, a manufactured home constructed in 2002 was subjected to a series of airtightening retrofits including installing house wrap over the exterior walls, sealing a number of leakage sites in the living space floor, tightening the insulated belly layer, and sealing leaks in the air distribution system. These retrofits reduced the whole house leakage, as determined by a fan pressurization test, by about 18 % and the duct leakage by about 80 %. Whole house infiltration rates were reduced by about one-third, with the specific reduction dependent on weather conditions and how the forced-air system was operating. The energy consumption rate for heating and cooling was reduced by about 10 %.

While the retrofits did improve the airtightness of the house and reduce the energy consumption, the effectiveness of the effort was limited by the challenges of airtightening an existing building. In general, it is easier to construct a tight building envelope than to achieve one through retrofits [14]. Manufactured homes in particular have the potential for high levels of airtightness performance given the quality control that can be achieved in the factory. Similarly, quality design and construction of these homes has been shown to yield high levels of energy performance [15].

It is important to note that the post-retrofit infiltration rates were often below the target ventilation rate of 0.35 h^{-1} in the HUD manufactured housing standard [2]. As seen in Table 3 under conditions 1b and 2b, the mean air change rates are all below 0.35 h^{-1} with the outdoor air intake open. Theses conditions of under-ventilation occur in part because the airflow through the intake is less than half of the HUD requirement. Also, in the case of condition 2b when the forced-air fan is controlled by the thermostat, the intake only operates when there is a demand for heating or cooling and is otherwise not bringing any outdoor air into the building.

While the mechanical ventilation system is not providing the rates required by the HUD standard, the reduced duct leakage combined with the tighter envelope does provide much better control of the envelope infiltration rates. Figures 8 through 11 show the reduced infiltration rates with the intake sealed, which highlights the potential to provide much better ventilation control through a mechanical approach than was possible before the retrofits. However, a mechanical ventilation system must still be provided to meet the overall ventilation requirements of house, either using an adequately sized intake, a whole house exhaust fan or some other approach. Ideally, the mechanical ventilation will be controlled independent of the need for heating or cooling.

This study also demonstrated the ability of multizone airflow modeling, in this case using the CONTAM model, to predict whole building air change rates with good accuracy. Before the retrofits, the air change rate exhibited an unusual dependence on temperature due to the duct leakage pressurizing the belly space, but this behavior was predicted by the model. After the retrofits, the new leakage values were entered into the model and the predicted air change rates matched the measurements quite well. These results show the potential value of building airflow modeling for analyzing the ventilation and infiltration performance of residences, which can be extended to simulating indoor contaminant levels as a means of understanding the impacts of building design, construction, and operation on indoor air quality.

ACKNOWLEGEMENTS

The authors wish to express their appreciation to Michael Lubliner of Washington State University and the DOE Building America Program for his assistance in this study.

REFERENCES

[1] ASHRAE, 2010. ANSI/ASHRAE Standard 62.1-2010 Ventilation for Acceptable Indoor Air Quality, American Society of Heating, Refrigerating and Air-Conditioning Engineers, Inc.

[2] HUD, 1994. Part 3280, Manufactured Home Construction and Safety Standards, in, U.S. Department of Housing and Urban Development.

[3] M. Lubliner, D.T. Stevens, B. Davis, 1997. Mechanical Ventilation in HUD-Code Manufactured Housing in Pacific Northwest, *ASHRAE Transactions*,103(1):693-705.

[4] DOE, 2005. Residential Energy Consumption Survey (RECS), Energy Information Administration, U.S. Department of Energy.

[5] M. Lubliner, A. Hadley, A. Gordon, 2004. Manufactured Home Performance Case Study: A Preliminary Comparison of Zero Energy and Energy Star, ASHRAE Building Thermal Envelope Conference, Clearwater Beach, Florida.

[6] DOE, 2010. Weatherization & Intergovernmental Program.

[7] A.K. Persily, S.R. Martin, 2000. A Modeling Study of Ventilation in Manufactured Homes. Nnational Institute of Standards and Technology, NISTIR 6455.

[8] S. Nabinger, A. Persily, 2008. Airtightness, Ventilation and Energy Consumption in a Manufactured House: Pre-Retrofit Results, National Institute of Standards and Technology, NISTIR 7478.

[9] ASTM, 2006. E741 Standard Test Method for Determining Air Change in a Single Zone by Means of a Tracer Gas Dilution, American Society for Testing and Materials.

[10] ASTM, 2008. E779 Standard Test Method for Determining Air Leakage Rate by Fan Pressurization, American Society for Testing and Materials.

[11] ASTM, 2003. E1554 Standard Test Methods for Determining External Air Leakage of Air Distribution Systems by Fan Pressurization, American Society for Testing and Materials.

[12] P.W. Francisco, L. Palmiter, 2007. Thermal Characterization and Duct Losses of Belly Spaces in Manufactured Homes, *ASHRAE Transactions*, 113(2): 81-89.

[13] Walton, G. N. and W. S. Dols, 2005. CONTAMW 2.4 User Guide and Program Documentation. National Institute of Standards and Technology, NISTIR 7251.

[14] S.D. Hale, B. Davis, R.B. Peeks, 2007. Effect of Mastic on Duct Tightness in Energy-Efficient Manufactured Homes, *ASHRAE Transactions*, 113(2): 77-80.

[15] R. Lucas, P. Fairey, R. Garcia, M. Lubliner, 2007. National Energy Savings Potential in HUD-Code Housing from Thermal Envelope and HVAC Equipment Improvements, *ASHRAE Transactions*, 113(2): 62-69.

Thursday 13 October 2011

15.50 – 17.00 Closing session

- Summing up of airtightness track (Willem de Gids, Expert, VentGuide, Netherlands)
- Summing up of ventilation track (Martin Liddament, Expert, Veetech Ltd, UK)
- Perspectives for AIVC and TightVent projects (Rémi Carrié, Senior Consultant, INIVE, France)

- Can we meet the ventilation required in international standards in an energy efficient way? (Bjarne Olesen, Professor, Technical University of Denmark

17.00 End of the conference

CAN WE MEET THE VENTILATION REQUIRED IN INTERNATIONAL STANDARDS IN AN ENERGY EFFICIENT WAY?

Bjarne W. Olesen

International Centre for Indoor Environment and Energy,
Department of Civil Engineering, Technical University of Denmark
Nils Koppels Alle, building 402
DK-2800 Kgs. Lyngby, Denmark

ABSTRACT

Today an acceptable indoor air quality is mainly defined by specifying the required level of ventilation in air changes per hour or the outside air supply rate. This would be equivalent to defining the requirements for thermal comfort by specifying the level of heating or cooling in Watts. The increasing societal need for energy efficiency will often result in very tight buildings. This means that the amount of outside air supplied by infiltration is not enough to provide the required ventilation. In some standards the required ventilation is based on adapted people (occupants) while other standards refer to un-adapted persons, who have just entered a room. Which approach is correct? Or should it depend on the type of space or occupancy? Furthermore, the level of ventilation will depend on the criteria for acceptability, like health, comfort (perceived air quality) or occupant performance. The required outside air supply rate will be the same or higher than the required ventilation rate depending on the ventilation effectiveness. Existing standards do not or only in a limited way acknowledge the use of air cleaning as substitute for outside air. Furthermore the concept of demand controlled ventilation is in many cases not taken into account.

The present paper provides an overview and discusses the criteria used for specifying required ventilation rates, and suggest ways of meeting the criteria in a more energy efficient way by means of improved ventilation effectiveness, use of air cleaning and by means of demand controlled ventilation

KEYWORDS

Ventilation, energy, indoor environment, air purification, ventilation effectiveness,

INTRODUCTION

All over the world there is today a lot of focus on energy conservation in buildings, possible use of renewable energy sources and search for new energy sources. It is very important to realize that the reason for building buildings and install heating, cooling and ventilation systems is to provide a healthy and comfortable indoor environment that does not impair the productivity of the occupants. Therefore the research trends have been to find energy efficient measures and technologies that will provide an acceptable indoor environment.

In many cases it is possible to improve the indoor environment and at the same time reduce energy consumption. An example is the increase in better insulation of windows and external walls that result in a more uniform indoor thermal environment with less risk of radiant asymmetry, risk of draught and vertical air temperature differences. Unfortunately the tighter buildings have often resulted in too low ventilation rates in many residential buildings. Recent studies [1] have found that it is quite acceptable to provide a drifting temperature during the day, which can result in reduction of energy consumption and peak loads.

Finally many research projects are studying new technologies like personalized HVAC systems, use of air cleaning technologies to substitute for fresh outside air, use of low emitting building materials, and improved control systems. These technologies will in most cases provide an acceptable indoor environment with reduced energy consumption.

REQUIRED VENTILATION RATES IN INTERNATIONAL STANDARDS

The requirements to an acceptable indoor air quality is in most cases specifyed as a required level of ventilation, assuming that the outside air is fresh clean). The required rates depends on the type of building and space.

Non-residential buildings

Originally, most standards and guidelines for required ventilation rates were given as the required outdoor air supply rate per person. Laboratory and field studies have subsequently shown[4] that people and their activity (smoking, activity level), building and furnishing (floor covering, paint, furniture, cleaning, electronic equipment, etc.) and ventilation systems (filters, humidifiers, ducts etc.) may also contribute to indoor pollution. Even the outside air may be a source of poor indoor air quality.

Both people and building-related sources of pollution are taken into account in newer standards for the required ventilation rates in buildings, which include ASHRAE 62.1 [2], and EN15251 [3]. In all of the standards more than one procedure is specified. They all include a prescriptive method, where the minimum ventilation rates can be found in a table listing values for different types of space, as well as an analytical procedure for calculating the required ventilation rate. By means of the analytical procedure, the required minimum ventilation rates can be calculated on the basis of type of pollutant, emission rates and acceptable concentration. All of the proposed standards deal with the health issue and as well as the comfort issue.

Prescriptive procedure

For the prescriptive method, a minimum ventilation rate per person and a minimum ventilation rate per square metre floor area are required. The two ventilation rates are then added. The person-related ventilation rate should take care of pollution emitted from the person (odour and other bioeffluents) and the ventilation rate based on the person's activity and the floor area should cover emissions from the building, furnishing, HVAC system, etc.

The design outdoor airflow required in the breathing zone of the occupied space or spaces in a zone, i.e., the breathing zone outdoor airflow (V_{bz}), is determined in accordance with the equation:

$$V_{bz} = R_p P_z + R_a A_z \qquad (1)$$

Where:

A_z = Zone floor area: the net occupied floor area of the zone m^2,

P_z = Zone population: the greatest number of people expected to occupy the zone during typical usage.

R_p = Outdoor airflow rate required per person: these values are based on adapted occupants in EN15251 and un-adapted in ASHRAE-62.1.

R_a = Outdoor airflow rate required per unit area.

In the standards different methods for calculating the recommended ventilation rate are included. As a minimum it must be ventilated to dilute the bioeffluents from the occupants (people component, R_p, see table 1). These rates are in EN15251 specified for three categories of indoor air quality, based on the prediction that a certain percentage of visitors will find the air quality unacceptable (see Table 1). The design levels are thus adequate for people who walk into a space. It is debatable if this should always be the case. People adapt very quickly to the odour (bioeffluents) in a space while there is less adaption to emissions from building materials and tobacco smoke (odour and irritants, [5]). To provide an acceptable perceived air quality for occupants (who have adapted to the air quality for at least 15 min.) it is estimated that one third of the ventilation rate is sufficient i.e. for category II 2, 5 instead of 7 l/s per person. The ASHRAE Standard 62.1 for ventilation and indoor air quality defines ventilation levels for adapted persons (occupants).

In addition, the minimum recommended ventilation is increased with a building-related ventilation rate, in order to take into account the emissions from the building and its systems (see Table 1). There is, however no general agreement on whether the contribution from the building should be added in full. Several studies indicate this is the best approximation, but it may not be valid for all types of pollutants. Here it is the contribution to the odour and irritation (perceived air quality) which must be taken into account. So it can be argued they all influence one organ (the nose) and so should be added. When the health risk is considered a simple addition can only be made for the same chemical component. Therefore in some countries it is recommended that the minimum ventilation rate should be specified as a value between the minimum level for people (bioeffluents) and the higher rate that takes account of both occupant and building emissions.

Category	Expected Percentage Dissatisfied	People component l/s person	Building component		
			Very low polluting l/s m²	Low polluting l/s m²	Not Low polluting l/s m²
I	15	10	0,5	1,0	2,0
II	20	7	0,35	0,7	1,4
III	30	4	0,2	0,4	0,8

Table 1. Basic required ventilation rates for diluting emissions (bio effluents) from people and three types of buildings for different categories of indoor environmental quality

Criteria for the ventilation rate may also be expressed as total rates per m^2 floor area (l/s per m^2) or per occupant (l/s per occupant). By expressing it as a people part and as a building part it becomes easier to calculate required ventilation rates for non-typical levels of occupancy. One issue is the rate for the building component for different types of buildings. For new buildings (Wargocki-2004 [6]), where attention has been paid to the selection of building materials, no smoking etc., lower levels corresponding to 0,14 l/s per m^2 can be obtained. A very-low polluting building is therefore included in Table 1 and 2. There is, however, a need to establish procedures and test methods for materials on the basis of which the type of building pollution level can be determined. In the annex to the standard some suggested criteria have been included.

Table 2 shows the required ventilation rates from standard EN15251 compared to ASHRAE 62.1. There are quite big differences between the European recommendations and those listed by ASHRAE. One major reason is that ASHRAE requirements are minimum code requirements, where the basis for design is adapted people, while the European recommendations are for un-adapted people (visitors).

Ventilation effectiveness

The ventilation rates specified in the standards (Table 2) are the required rates at breathing level in the occupied zone. The required ventilation rate at the room supply diffusers are calculated as:

$$\text{Total ventilation rate V} = V_{bz} / \varepsilon_v \quad (2)$$

Where :

V_{bz} = *breathing zone ventilation*

$$\varepsilon_v = \frac{C_e - C_s}{C_i - C_s} \qquad (3)$$

Where:

ε_v = Ventilation effectiveness

C_e = Pollutant concentration in extract air

C_s = Pollutant concentration in supply air

C_i = Pollutant concentration at breathing level

The ventilation effectiveness depends on the air distribution efficiency and the type and position of the pollution source(s), so this value is not only a system characteristic. In most cases it is assumed that the pollutant emission is uniform, so the ventilation effectiveness is the same as the air distribution effectiveness. For a fully-mixed ventilation system the value is 1 and the ventilation rates in Table 2 can be used for the design of the supply grills. The ventilation effectiveness or air distribution efficiency is a function of the position and type of supply and return grills, and depends on the difference between supply and room temperature and on the total amount of airflow through the supply grill. The air distribution effectiveness can be calculated numerically or measured experimentally. Typical examples of ventilation effectiveness/air distribution effectiveness are shown in Figure 1 [7].

Mixing ventilation		Mixing ventilation		Displacement ventilation		Personalized ventilation	
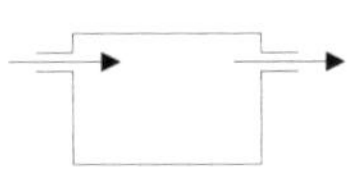		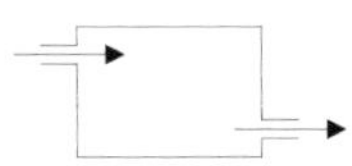		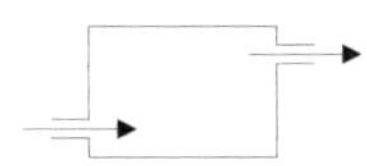			
T supply - T inhal °C	Vent. effect. -	T supply - T inhal °C	Vent. effect. -	T supply - T inhal °C	Vent. effect. -	T supply - T room °C	Vent. effect. -
< 0	0,9 - 1,0	< -5	0,9	<0	1,2 - 1,4	-6	1,2 - 2,2
0 - 2	0,9	-5 - 0	0,9 - 1,0	0-2	0,7 - 0,9	-3	1,3 - 2,3
2 - 5	0,8	> 0	1	>2	0,2 - 0,7	0	1,6 - 3,5
> 5	0,4 - 0,7						

Figure 1. Typical examples of ventilation/air distribution effectiveness [8]

The air distribution effectiveness takes into account the air distribution in a space, but does not take into account how effectively the outside air is transported through the ducts to the space. If the system has any air leakage, the amount of ventilation air must be increased. This is not dealt with in en15251, but is mentioned in ASHRAE 62.1.

Analytical procedure

All of the listed standards specify an analytical procedure, either in the standard text or in an informative appendix. In this procedure the required ventilation rate is calculated on a comfort basis (perceived odour and/or irritation) as well as on a health basis. The highest calculated value, which in most cases will be the comfort value, is then used as the required minimum ventilation rate. The basis for the calculation is in all standards based on a mass balance calculation.

The required ventilation rate is calculated as:

$$Q = \frac{G}{(C_i - C_o) \cdot E_v} \quad \mathrm{l\,s^{-1}} \qquad (4)$$

where G = Total emission rate mg s^{-1}
C_i = Concentration limit mg l^{-1}
C_o = Concentration in outside air mg l^{-1}
E_v = Ventilation effectiveness

Type of building space	Occupancy person/m^2	Category CEN	Minimum ventilation rate, i.e. for occupants only l/s person		Additional ventilation for building (add only one) l/s·m^2				Total l/s·m^2	
			ASH-RAE R_p	CEN	CEN Very low-pollut.	CEN Low-pollut.	CEN *Not* low-pollut.	ASH-RAE R_a	CEN Low Pol.	ASH-RAE
Single office (cellular office)	0,1	I	2,5	10	10	1,0	2,0	0,3	2	0,55
		II		7	7	0,7	1,4		1,4	
		III		4	4	0,4	0,8		0,8	
Land-scaped office	0,07	I	2,5	10	10	1,0	2,0	0,3	1,7	0,48
		II		7	7	0,7	1,4		1,2	
		III		4	4	0,4	0,8		0,7	
Confe-rence room	0,5	I	2,5	10	10	1,0	2,0	0,3	6	1,55
		II		7	7	0,7	1,4		4,2	
		III		4	4	0,4	0,8		2,4	
Audito-rium	1,5	I	3,8	10	10	1,0	2,0	0,3	16	6
		II		7	7	0,7	1,4		11,2	
		III		4	4	0,4	0,8		6,4	
Class-room	0,5	I	3,8	10	10	1,0	2,0	0,3	6	2,2
		II		7	7	0,7	1,4		4,2	
		III		4	4	0,4	0,8		2,4	
Kinder-garten	0,5	I	5,0	12	12	1,0	2,0	0,9	7	3,4
		II		8,4	8,4	0,7	1,4		4,9	
		III		4,8	4,8	0,4	0,8		2,8	
Depart-ment store	0,15	I	3,8	14,7	14,7	2,0	3,0	0,6	4,1	1,17
		II		10	10	1,4	2,1		2,9	
		III		6	6	0,8	1,2		1,7	

Table 2 Smoking free spaces in commercial buildings according to ASHRAE 62.1[3] and EN15251 [4]

In all of the standards, however, knowledge concerning emission rates (G) and concentration limits (C_i) that safeguard health, is very limited. Within the next few years, knowledge will increase and data will be available from ongoing research projects and from testing by manufacturers of building materials and furnishing.

Residential buildings

In residential buildings the requirements are often expressed as required air change per hour, ach and additional required exhaust in kitchen, bathrooms and toilets. (Table2).

Category	Air change rate [1)]		Living room and bedrooms, mainly outdoor air flow		Exhaust air flow, l/s		
	l/s,m² (1)	ach	l/s, pers[2)] (2)	l/s/m² (3)	Kitchen (4a)	Bathrooms (4b)	Toilets (4)
I	0,49	0,7	10	1,4	28	20	14
II	0,42	0,6	7	1,0	20	15	10
III	0,35	0,5	4	0,6	14	10	7

[1)] The air change rates expressed in l/sm² and ach correspond to each other when the ceiling height is 2,5 m
[2)]The number of occupants in a residence can be estimated from the number of bedrooms. The assumptions made at national level have to be used when existing, they may vary for energy and for IAQ calculations.
Table 2 Ventilation rates for residential buildings with mechanical ventilation. Continuous operation of ventilation during occupied hours. Complete mixing. EN15251:

In ASHRAE 62.2 [5] the requirements are specifyed as a basic mechanical ventilation of:

$$0.05l/s*FloorArea+3.5l/s*number\ of\ persons \qquad (4)$$

where it is assume that the number of people=1+number of bedrooms
It is also assumed that you get 10 L/s per 100 m² of floor area of extra ventilation from infiltration. For a US sized house this leads to a total of about 0.35 ACH half of which is coming from infiltration and half from the mechanical system. This corresponds to category III in EN15251.

TECHNOLOGIES AND METHODS TO IMPROVE INDOOR AIR QUALITY

Removal of Sources
The most efficient way of reducing energy consumption for ventilation and increase the indoor environmental quality is reducing the sources of emissions in buildings. There is an urgent need for better certification and labeling of the materials used in buildings and the ventilation standards must include methods that favor the manufacturers of "good" (low polluting) materials. A start has been made by defining three types of buildings in EN15251, but the method for evaluating to which type an existing or projected building should belong is not good enough.
In many countries a material labelling system and certification of emissions from material have been introduced ; but without any direct link to the influence on required ventilation in a space.
Increased ventilation effectiveness
In most cases it is assumed that the pollutant emission is uniform, so the ventilation effectiveness is the same as the air distribution effectiveness. The air distribution efficiency is a function of the position and type of supply and return grills, and depends on the difference

between supply and room temperature and on the total amount of airflow through the supply grill [7]. The air distribution effectiveness can be calculated numerically or measured experimentally.
To use ventilation effectiveness can be one way of taking into account personalized ventilation; but we need also to look at issues like acceptance of higher air velocities and less strict requirements to the general environment if the person is equipped with a personal system.

Demand controlled ventilation
A well known method to reduce energy use for ventilation is only to ventilate when occupants are present. As also the number of occupnats are the source strength that varies the most in a building the level of ventilation should preferably corresponds to the number of occupants. This is basiccally what demand controlled ventilation (DCV) is about.
In residential buildings DCV can be controlled by time schedule (occupation of the house), humidity or CO_2 sensors. As some of the major sources in residential buildings may not be people a CO_2 sensor is not always the best choise. There has been for many years an interest in developping sensors, which reponses to many polutants, like a human nose ; but until now it has been very difficult to find a good sensor. As there is a significant background level of pollutant in a residential building the onset of ventilation must be before the occupants arrive, which makes it even more difficult. Some countries require a ventilation rate of 0.5 ach in every room all the time. It would be more energy efficient if this requirement was only valid when the room is occupied. This would then also open up for the possibility to vary the ventilation rate from room to room i.e.ventilate in kitchen living room during day and evening and bedrooms during the night.
In commercial buildings DCV could first of all be based on a schedule for working hours, secondly it could be based on a person detector and finally what is becoming more common on a CO_2 sensor. In commercial buildings a CO_2 sensor is normally a good choise as the dominant source is people.

Air cleaning
Gaseous air cleaning is not taken into account at all in EN15251, while ASHRAE 62.1 by using the analytical procedure can allow some credits for air cleaning. There is an increased interest in the development of air purification equipment. This may be an acceptable way of reducing the amount of outside air, saving energy and still having an acceptable indoor air quality. However, better test methods for air cleaners are required, because at present the test is usually based on chemical measurements and the resulting effect on odour or perceived air quality is not taken into account [8,9]. It is also very important to specify which kind of "pollutants" should be used when testing. Some air purification systems may work well on VOC's (emission from materials) but have zero or even a negative effect if the source is people (bio effluents).

Adapted or Non-adapted
Who should we ventilate for? For people just entering the room (un-adapted) or for people already occupying a room (adapted)? Here the philosophy adopted by ASHRAE 62.1 and EN15251 differs. But should it really be one or the other? In a conference room, auditorium or lecture room most people enter at the same time. It then takes some time before the odour level has reached an unacceptable level and meanwhile people adapt. In this case it may be appropriate to require a ventilation rate based on adapted persons. There may be other spaces where you would design for un-adapted people, e.g. in a first class restaurant, offices, and department stores. It seems logical that more differentiated criteria could be used.

Occupant behaviour

A factor that can influence the energy consumption by ventilation significantly is the occupant behaviour. How often do they open windows? Why do they open windows, bad air quality or feeling too warm? Do they block the supply grills because of draught? Or do they shut off the fans because of noise? This factor is often the reason why predicted energy consumption does not match with measured energy consumption. Here the control of the ventilation, feed-back to the occupants and a guide for operation of the ventilation can in most cases significantly improve the energy efficiency by ventilation and at the same time guarantee an acceptable indoor air quality.

DISCUSSIONS

Even if we today have standards and guidelines for estimating the required minimum ventilation rate, they are fare form being complete. The goal is of course to be able to calculate the required ventilation rate as straightforwardly as in cooling load calculations. We need to know the requirements for acceptable indoor air quality based on health, comfort and performance and we need to know the emission rates from all the sources. Unfortunately, this is not as easy as in cooling load calculations, where room and outside temperature (^oC), energy emission (watts), heat storage, solar radiation (watts) are all evaluated with similar units and all affect the same parameter of the human body (heat balance). For indoor air quality, we have thousand of substances that are emitted from people, furnishing, systems, from outside etc., each of which may affect one or more organs of the body.

We have good knowledge about the required ventilation for the "people" component, while the "building" component is not very well documented. There is an urgent need for better certification and labeling of the materials used in buildings and we must also develop ventilation standards that favour the manufacturers of "good" (low polluting) materials

The rates given in the Tables are based on full mixing and in practice the ventilation effectiveness is very seldom taken into account. One complication is that some systems may have a different ventilation effectiveness summer and winter. If the supply temperature is lower than room temperature the ventilation effectiveness is normally 1 or higher, but if the ventilation system is used for heating in winter the ventilation effectiveness could be as low as 0.5, and the ventilation rates should really be doubled. More information and a greater emphasis on this factor is required.

Better test methods for gaseous air cleaners are required, because at present the test is usually based on chemical measurements and the resulting effect on odour or perceived air quality is not taken into account

One serious problem is how to ventilate if a building is located in an area with poor outside air quality or if there is a time of the day (e.g. rush hour) when the outside air quality is unacceptable. In some cases it might even be better to reduce ventilation under these circumstances, and the use of air cleaning technologies can be a better solution.

CONCLUSION

Recent research has resulted in improved personalized ventilation and conditioning systems, which can improve the indoor environment at a work place in an energy efficient way.

The results of research on improved air distribution in spaces demand control ventilation and the use of air cleaning will be future technologies that reduce energy consumption and may improve comfort.

Occupant behaviour is a major factor for the energy performance of a building.

REFERENCES

[1] Kolarik, J,Toftum, Jørn; Olesen, Bjarne W.; Shitzer, Avraham (2008)"Human subjects' perception of indoor environment and their office work performance during exposures to moderate operative temperature ramps" Presented at: Indoor Air. Copenhagen, 2008

[2] ASHRAE 62.1 (2010), Ventilation for acceptable indoor air quality, Atlanta, GA, American Society of Heating, Refrigerating and Air-Conditioning Engineers.

[3] EN 15251 (2007) Indoor environmental input parameters for design and assessment of energy performance of buildings- addressing indoor air quality, thermal environment, lighting and acoustics.

[4] ASHRAE 62.2-2010, Ventilation and Acceptable Indoor Air Quality in Low-Rise Residential Buildings

[5] Gunnarsen, L., Fanger, P.O. (1992) "Adaptation to indoor air pollution", *Environment International*, 18:43-54.

[6] Wargocki, P. (2004) "Sensory pollution sources in buildings", Indoor Air, Vol.14 (Suppl 7), 82-91

[7] CR 1752, 1996: "Ventilation for Buildings: Design Criteria for the Indoor Environment"

[8] ASHRAE 145.2 "Method of Testing Gaseous Contaminant Air Cleaning Devices for Removal Efficiency" 2nd draft.

[9] prEN ISO DIS 10121-2: Test methods for assessing the performance of gas-phase air cleaning media and devices for general ventilation - Part 2: Gas phase air cleaning devices (GPACD) (ISO/DIS 10121-2:2011)

AUTHOR INDEX

AUTHOR INDEX